Hyperbolic Partial Differential Equations and Geometric Optics

Hyperbolic Partial Differential Equations and Geometric Optics

Jeffrey Rauch

Graduate Studies in Mathematics

Volume 133

American Mathematical Society
Providence, Rhode Island

2010 *Mathematics Subject Classification.* Primary 35A18, 35A21, 35A27, 35A30, 35Q31, 35Q60, 78A05, 78A60, 78M35, 93B07.

For additional information and updates on this book, visit
www.ams.org/bookpages/gsm-133

Library of Congress Cataloging-in-Publication Data

Rauch, Jeffrey.
Hyperbolic partial differential equations and geometric optics / Jeffrey Rauch.
p. cm. — (Graduate studies in mathematics ; v. 133)
Includes bibliographical references and index.
ISBN 978-0-8218-7291-8 (alk. paper)
1. Singularities (Mathematics). 2. Microlocal analysis. 3. Geometrical optics—Mathematics. 4. Differential equations, Hyperbolic. I. Title.

QC20.7.S54R38 2012
535′.32—dc23

2011046666

∞ The paper used in this book is acid-free and falls within the guidelines
established to ensure permanence and durability.
Visit the AMS home page at `http://www.ams.org/`

10 9 8 7 6 5 4 3 2 1   17 16 15 14 13 12

To Geraldine.

Contents

Preface

P.1. How this book came to be, and its peculiarities

This book presents an introduction to hyperbolic partial differential equations. A major subtheme is linear and nonlinear geometric optics. The two central results of linear microlocal analysis are derived from geometric optics. The treatment of nonlinear geometric optics gives an introduction to methods developed within the last twenty years, including a rethinking of the linear case.

Much of the material has grown out of courses that I have taught. The crucial step was a series of lectures on nonlinear geometric optics at the Institute for Advanced Study/Park City Mathematics Institute in July 1995. The Park City notes were prepared with the assistance of Markus Keel and appear in [Rauch, 1998]. They presented a straight line path to some theorems in nonlinear geometric optics. Graduate courses at the University of Michigan in 1993 and 2008 were important. Much of the material was refined in invited minicourses:

- École Normale Supérieure de Cachan, 1997;
- Nordic Conference on Conservation Laws at the Mittag-Leffler Institute and KTH in Stockholm, December 1997 (Chapters 9–11);
- Centro di Ricerca Matematica Ennio De Giorgi, Pisa, February 2004;
- Université de Provence, Marseille, March 2004 (§3.4, 5.4, Appendix 2.I);
- Università di Pisa, February–May 2005, March–April 2006 (Chapter 3, §6.7, 6.8), March–April 2007 (Chapters 9–11);
- Université de Paris Nord, February 2006–2010 (§1.4–1.7).

The auditors included many at the beginnings of their careers, and I would like to thank in particular R. Carles, E. Dumas, J. Bronski, J. Colliander, M. Keel, L. Miller, K. McLaughlin, R. McLaughlin, H. Zag, G. Crippa, A. Figalli, and N. Visciglia for many interesting questions and comments.

The book is aimed at the level of graduate students who have studied one hard course in partial differential equations. Following the lead of the book of Guillemin and Pollack (1974), there are exercises scattered throughout the text. The reader is encouraged to read with paper and pencil in hand, filling in and verifying as they go. There is a big difference between passive reading and active acquisition. In a classroom setting, correcting students' exercises offers the opportunity to teach the writing of mathematics.

To shorten the treatment and to avoid repetition with a solid partial differential equations course, basic material such as the fundamental solution of the wave equation in low dimensions is not presented. Naturally, I like the treatment of that material in my book *Partial Differential Equations* [Rauch, 1991].

The choice of subject matter is guided by several principles. By restricting to symmetric hyperbolic systems, the basic energy estimates come from integration by parts. The majority of examples from applications fall under this umbrella.

The treatment of constant coefficient problems does not follow the usual path of describing classes of operators for which the Cauchy problem is weakly well posed. Such results are described in Appendix 2.I along with the Kreiss matrix theorem. Rather, the Fourier transform is used to analyse the dispersive properties of constant coefficient symmetric hyperbolic equations including Brenner's theorem and Strichartz estimates.

Pseudodifferential operators are neither presented nor used. This is not because they are in any sense vile, but to get to the core without too many pauses to develop machinery. There are several good sources on pseudodifferential operators and the reader is encouraged to consult them to get alternate viewpoints on some of the material. In a sense, the expansions of geometric optics are a natural replacement for that machinery. Lax's parametrix and Hörmander's microlocal propagation of singularities theorem require the analysis of oscillatory integrals as in the theory of Fourier integral operators. The results require only the method of nonstationary phase and are included.

The topic of caustics and caustic crossing is not treated. The sharp linear results use more microlocal machinery and the nonlinear analogues are topics of current research. The same is true for supercritical nonlinear geometric optics which is not discussed. The subjects of dispersive and diffractive nonlinear geometric optics in contrast have reached a mature state. Readers of this book should be in a position to readily attack the papers describing that material.

The methods of geometric optics are presented as a way to understand the qualitative behavior of partial differential equations. Many examples proper to the theory of partial differential equations are discussed in the text, notably the basic results of microlocal analysis. In addition two long examples, stabilization of waves in §5.6 and dense oscillations for inviscid compressible fluid flow in Chapter 11 are presented. There are many important examples in science and technology. Readers are encouraged to study some of them by consulting the literature. In the scientific literature there will not be theorems. The results of this book turn many seemingly ad hoc approximate methods into rigorous asymptotic analyses.

Only a few of the many important hyperbolic systems arising in applications are discussed. I recommend the books [Courant, 1962], [Benzoni-Gavage and Serre, 2007], and [Métivier, 2009]. The asymptotic expansions of geometric optics explain the physical theory, also called geometric optics, describing the rectilinear propagation, reflection, and refraction of light rays. A brief discussion of the latter ideas is presented in the introductory chapter that groups together elementary examples that could be, but are usually not, part of a partial differential equations course. The WKB expansions of geometric optics also play a crucial role in understanding the connection of classical and quantum mechanics. That example, though not hyperbolic, is presented in §5.2.2.

The theory of hyperbolic mixed initial boundary value problems, a subject with many interesting applications and many difficult challenges, is not discussed. Nor is the geometric optics approach to shocks.

I have omitted several areas where there are already good sources; for example, the books [Smoller, 1983], [Serre, 1999], [Serre, 2000], [Dafermos, 2010], [Majda, 1984], [Bressan, 2000] on conservation laws, and the books [Hörmander, 1997] and [Taylor, 1991] on the use of pseudodifferential techniques in nonlinear problems. Other books on hyperbolic partial differential equations include [Hadamard, 1953], [Leray, 1953], [Mizohata, 1965], and [Benzoni-Gavage and Serre, 2007]. Lax's 1963 Stanford notes occupy a special place for me. I took a course from him in the late 1960s that corresponded to the enlarged version [Lax, 2006]. When I approached him to ask if he'd be my thesis director he asked what interested me. I indicated two subjects from the course, mixed initial boundary value problems and the section on waves and rays. The first became the topic of my thesis, and the second is the subject of this book and at the core of much of my research. I owe a great intellectual debt to the lecture notes, and to all that Peter Lax has taught me through the years.

The book introduces a large and rich subject and I hope that readers are sufficiently attracted to probe further.

P.2. A bird's eye view of hyperbolic equations

The central theme of this book is hyperbolic partial differential equations. These equations are important for a variety of reasons that we sketch here and that recur in many different guises throughout the book.

The first encounter with hyperbolicity is usually in considering scalar real linear second order partial differential operators in two variables with coefficients that may depend on x,

$$a\,u_{x_1x_1} \;+\; b\,u_{x_1x_2} \;+\; c\,u_{x_2x_2} \;+\; \text{lower order terms}\,.$$

Associate the quadratic form $\xi \mapsto a\,\xi_1^2 + b\,\xi_1\xi_2 + c\,\xi_2^2$. The differential operator is *elliptic* when the form is positive or negative definite. The differential operator is *strictly hyperbolic* when the form is indefinite and nondegenerate.[1]

In the elliptic case one has strong local regularity theorems and solvability of the Dirichlet problem on small discs. In the hyperbolic cases, the initial value problem is locally well set with data given at noncharacteristic curves and there is finite speed of propagation. Singularities or oscillations in Cauchy data propagate along characteristic curves.

The defining properties of hyperbolic problems include well posed Cauchy problems, finite speed of propagation, and the existence of wave like structures with infinitely varied form. To see the latter, consider in $\mathbb{R}^2_{t,x}$ initial data on $t = 0$ with the form of a short wavelength wave packet, $a(x)\,e^{ix/\epsilon}$, localized near a point p. The solution will launch wave packets along each of two characteristic curves. The envelopes are computed from those of the initial data, as in §5.2, and can take any form. One can send essentially arbitrary amplitude modulated signals.

The infinite variety of wave forms makes hyperbolic equations the preferred mode for communicating information, for example in hearing, sight, television, and radio. The model equations for the first are the linearized compressible inviscid fluid dynamics, a.k.a. acoustics. For the latter three it is Maxwell's equations. The telecommunication examples have the property that there is propagation with very small losses over large distances. The examples of wave packets and long distances show the importance of short wavelength and large time asymptotic analyses.

[1] The form is nondegenerate when its defining symmetric matrix is invertible.

Well posed Cauchy problems with finite speed lead to hyperbolic equations.[2] Since the fundamental laws of physics must respect the principles of relativity, finite speed is required. This together with causality requires hyperbolicity. Thus there are many equations from physics. Those which are most fundamental tend to have close relationships with Lorentzian geometry. D'Alembert's wave equation and the Maxwell equations are two examples. Problems with origins in general relativity are of increasing interest in the mathematical community, and it is the hope of hyperbolicians that the wealth of geometric applications of elliptic equations in Riemannian geometry will one day be paralleled by Lorentzian cousins of hyperbolic type.

A source of countless mathematical and technological problems of hyperbolic type are the equations of inviscid compressible fluid dynamics. Linearization of those equations yields linear acoustics. It is common that viscous forces are important only near boundaries, and therefore for many phenomena inviscid theories suffice. Inviscid models are often easier to compute numerically. This is easily understood as a small viscous term $\epsilon^2\partial^2/\partial x^2$ introduces a length scale $\sim\epsilon$, and accurate numerics require a discretization small enough to resolve this scale, say $\sim\epsilon/10$. In dimensions $1+d$ discretization of a unit volume for times of order 1 on such a scale requires $10^4\epsilon^{-4}$ mesh points. For ϵ only modestly small, this drives computations beyond the practical. Faced with this, one can employ meshes which are only locally fine or try to construct numerical schemes which resolve features on longer scales without resolving the short scale structures. Alternatively, one can use asymptotic methods like those in this book to describe the boundary layers where the viscosity cannot be neglected (see for example [Grenier and Guès, 1998] or [Gérard-Varet, 2003]). All of these are active areas of research.

One of the key features of inviscid fluid dynamics is that smooth large solutions often break down in finite time. The continuation of such solutions as nonsmooth solutions containing shock waves satisfying suitable conditions (often called entropy conditions) is an important subarea of hyperbolic theory which is not treated in this book. The interested reader is referred to the conservation law references cited earlier. An interesting counterpoint is that for suitably dispersive equations in high dimensions, small smooth data yield global smooth (hence shock free) solutions (see §6.7).

The subject of geometric optics is a major theme of this book. The subject begins with the earliest understanding of the propagation of light. Observation of sunbeams streaming through a partial break in clouds or a

[2]See [Lax, 2006] for a proof in the constant coefficient linear case. The necessity of hyperbolicity in the variable coefficient case dates to [Lax, Duke J., 1957] for real analytic coefficients. The smooth coefficient case is due to Mizohata and is discussed in his book.

flashlight beam in a dusty room gives the impression that light travels in straight lines. At mirrors the lines reflect with the usual law of equal angles of incidence and reflection. Passing from air to water the lines are bent. These phenomena are described by the three fundamental principles of a physical theory called *geometric optics*. They are rectilinear propagation and the laws of reflection and refraction.

All three phenomena are explained by Fermat's principle of least time. The rays are locally paths of least time. Refraction at an interface is explained by positing that light travels at different speeds in the two media. This description is purely geometrical involving only broken rays and times of transit. The appearance of a minimum principle had important philosophical impact, since it was consistent with a world view holding that nature acts in a best possible way. Fermat's principle was enunciated twenty years before Römer demonstrated the finiteness of the speed of light based on observations of the moons of Jupiter.

Today light is understood as an electromagnetic phenomenon. It is described by the time evolution of electromagnetic fields, which are solutions of a system of partial differential equations. When quantum effects are important, this theory must be quantized. A mathematically solid foundation for the quantization of the electromagnetic field in $1 + 3$ dimensional space time has not yet been found.

The reason that a field theory involving partial differential equations can be replaced by a geometric theory involving rays is that visible light has very short wavelength compared to the size of human sensory organs and common physical objects. Thus, much observational data involving light occurs in an asymptotic regime of very short wavelength. The short wavelength asymptotic study of systems of partial differential equations often involves significant simplifications. In particular there are often good descriptions involving rays. We will use the phrase *geometric optics* to be synonymous with short wavelength asymptotic analysis of solutions of systems of partial differential equations.

In optical phenomena, not only is the wavelength short but the wave trains are long. The study of structures which have short wavelength and are in addition very short, say a short pulse, also yields a geometric theory. Long wave trains have a longer time to allow nonlinear interactions which makes nonlinear effects more important. Long propagation distances also increase the importance of nonlinear effects. An extreme example is the propagation of light across the ocean in optical fibers. The nonlinear effects are very weak, but over 5000 kilometers, the cumulative effects can be large. To control signal degradation in such fibers, the signal is treated about every 30 kilometers. Still, there is free propagation for 30 kilometers which needs to be understood. This poses serious analytic, computational, and engineering challenges.

A second way to bring nonlinear effects to the fore is to increase the amplitude of disturbances. It was only with the advent of the laser that sufficiently intense optical fields were produced so that nonlinear effects are routinely observed. The conclusion is that for nonlinearity to be important, either the fields or the propagation distances must be large. For the latter, dissipative losses must be small.

The ray description as a simplification of the Maxwell equations is analogous to the fact that classical mechanics gives a good approximation to solutions of the Schrödinger equation of quantum mechanics. The associated ideas are called the *quasiclassical approximation*. The methods developed for hyperbolic equations also work for this important problem in quantum mechanics. A brief treatment is presented in §5.2.2. The role of rays in optics is played by the paths of classical mechanics. There is an important difference in the two cases. The Schrödinger equation has a small parameter, Planck's constant. The quasiclassical approximation is an approximation valid for small Planck's constant. The mathematical theory involves the limit as this constant tends to zero. The Maxwell equations apparently have a small parameter too, the inverse of the speed of light. One might guess that rays occur in a theory where this speed tends to infinity. This is not the case. For the Maxwell equations in a vacuum the small parameter that appears is the wavelength which is introduced via the initial data. It is not in the equation. The equations describing the dispersion of light when it interacts with matter do have a small parameter, the inverse of the resonant frequencies of the material, and the analysis involves data tuned to this frequency just as the quasiclassical limit involves data tuned to Planck's constant. Dispersion is one of my favorite topics. Interested readers are referred to the articles [Donnat and Rauch, 1997] (both) and [Rauch, 2007].

Short wavelength phenomena cannot simply be studied by numerical simulations. If one were to discretize a cubic meter of space with mesh size 10^{-5} cm so as to have five mesh points per wavelength, there would be 10^{21} data points in each time slice. Since this is nearly as large as the number of atoms per cubic centimeter, there is no chance for the memory of a computer to be sufficient to store enough data, let alone make calculations. Such brute force approaches are doomed to fail. A more intelligent approach would be to use radical local mesh refinement so that the fine mesh was used only when needed. Still, this falls far outside the bounds of present computing power. Asymptotic analysis offers an alternative approach that is not only powerful but is mathematically elegant. In the scientific literature it is also embraced because the resulting equations sometimes have exact solutions and scientists are well versed in understanding phenomena from small families of exact solutions.

Short wavelength asymptotics can be used to great advantage in many disparate domains. They explain and extend the basic rules of linear geometric optics. They explain the dispersion and diffraction of linear electromagnetic waves. There are nonlinear optical effects, generation of harmonics, rotation of the axis of elliptical polarization, and self-focusing, which are also well described.

Geometric optics has many applications within the subject of partial differential equations. They play a key role in the problem of solvability of linear equations via results on propagation of singularities as presented in §5.5. They are used in deriving necessary conditions, for example for hypoellipticity and hyperbolicity. They are used by Ralston to prove necessity in the conjecture of Lax and Phillips on local decay. Via propagation of singularities they also play a central role in the proof of sufficiency. Propagation of singularities plays a central role in problems of observability and controlability (see §5.6). The microlocal elliptic regularity theorem and the propagation of singularities for symmetric hyperbolic operators of constant multiplicity is treated in this book. These are the two basic results of linear microlocal analysis. These notes are not a systematic introduction to that subject, but they present an important part *en passant.*

Chapters 9 and 10 are devoted to the phenomenon of *resonance* whereby waves with distinct phases interact nonlinearly. They are preparatory for Chapter 11. That chapter constructs a family of solutions of the compressible $2d$ Euler equations exhibiting three incoming wave packets interacting to generate an infinite number of oscillatory wave packets whose velocities are dense in the unit circle.

Because of the central role played by rays and characteristic hypersurfaces, the analysis of conormal waves is closely related to geometric optics. The reader is referred to the treatment of progressing waves in [Lax, 2006] and to [Beals, 1989] for this material.

Acknowledgments. I have been studying hyperbolic partial differential equations for more that forty years. During that period, I have had the pleasure and privilege to work for extended periods with (in order of appearance) M. Taylor, M. Reed, C. Bardos, G. Métivier, G. Lebeau, J.-L. Joly, and L. Halpern. I thank them all for the things that they have taught me and the good times spent together. My work in geometric optics is mostly joint with J.-L. Joly and G. Métivier. This collaboration is the motivation and central theme of the book. I gratefully acknowledge my indebtedness to them.

My research was partially supported by a sequence of National Science Foundation grants; NSF-DMS-9203413, 9803296, 0104096, and 0405899. The early stages of the research on geometric optics were supported by the Office of Naval Research under grant OD-G-N0014-92-J-1245. My visits to Paris have been supported by the CNRS, the Fondation Mathématiques de Paris, and the Institut Henri Poincaré. My visits to Pisa have been supported by the the Università di Pisa, INDAM, GNAMPA, and the Centro di Ricerca Matematica Ennio De Giorgi. Invitations as visiting professor at many departments (and notably l'Université de Paris Nord, l'École Normale Supérieure, and l'École Polytéchnique for multiple invitations) have been crucially important in my research. I sincerely thank all these organizations and also the individuals responsible.

Chapter 1

Simple Examples of Propagation

This chapter presents examples of wave propagation governed by hyperbolic equations. The ideas of propagation of singularities, group velocity, and short wavelength asymptotics are introduced in simple situations. The method of characteristics for problems in dimension $d = 1$ is presented as well as the method of nonstationary phase. The latter is a fundamental tool for estimating oscillatory integrals. The examples are elementary. They could each be part of an introductory course in partial differential equations, but often are not. This material can be skipped. If needed later, the reader may return to this chapter.[1] In sections 1.3, 1.5, and 1.6 we derive in simple situations the three basic laws of physical geometric optics.

Wave like solutions of partial differential equations have spatially localized structures whose evolution in time can be followed. The most common are solutions with propagating singularities and solutions that are modulated *wave trains* also called *wave packets*. They have the form

$$a(t,x)\, e^{i\phi(t,x)/\epsilon}$$

with smooth profile a, real valued smooth phase ϕ with $d\phi \neq 0$ on $\operatorname{supp} a$.

The parameter ϵ is a wavelength and is small compared to the scale on which a and ϕ vary. The classic example is light with a wavelength on the order of 5×10^{-5} centimeter. Singularities are often restricted to varieties of lower codimension, hence of width equal to zero, which is infinitely small compared to the scales of their other variations. Real world waves modeled

[1] Some ideas are used which are not formally presented until later, for example the Sobolev spaces $H^s(\mathbb{R}^d)$ and Gronwall's lemma.

by such solutions have their singular behavior spread over very small lengths, not exactly zero.

The path of a localized structure in space-time is curvelike, and such curves are often called *rays*. When phenomena are described by partial differential equations, linking the above ideas with the equation means finding solutions whose salient features are localized and in simple cases are described by transport equations along rays. For wave packets such results appear in an asymptotic analysis as $\epsilon \to 0$.

In this chapter some introductory examples are presented that illustrate propagation of singularities, propagation of energy, group velocity, and short wavelength asymptotics. That energy and singularities may behave very differently is a consequence of the dichotomy that, up to an error as small as one likes in energy, the data can be replaced by data with compactly supported Fourier transform. In contrast, up to an error as smooth as one likes, the data can be replaced by data with Fourier transform vanishing on $|\xi| \le R$ with R as large as one likes. Propagation of singularities is about short wavelengths while propagation of energy is about wavelengths bounded away from zero. When most of the energy is carried in short wavelengths, for example the wave packets above, the two tend to propagate in the same way.

1.1. The method of characteristics

When the space dimension is equal to one, the method of characteristics reduces many questions concerning solutions of hyperbolic partial differential equations to the integration of ordinary differential equations. The central idea is the following. When $c(t,x)$ is a smooth real valued function, introduce the ordinary differential equation

$$\frac{dx}{dt} = c(t,x)\,. \tag{1.1.1}$$

Solutions $x(t)$ satisfy

$$\frac{dx(t)}{dt} = c(t,x(t)).$$

For a smooth function u,

$$\frac{d}{dt}u(t,x(t)) = \big(\partial_t u + c\,\partial_x u\big)\big|_{(t,x(t))}\,.$$

Therefore, solutions of the homogeneous linear equation

$$\partial_t u + c(t,x)\,\partial_x u = 0$$

are exactly the functions u that are constant on the integral curves $(t,x(t))$. These curves are called *characteristic curves* or simply *characteristics*.

Example 1.1.1. If $c \in \mathbb{R}$ is constant, then $u \in C^\infty([0,T] \times \mathbb{R})$ satisfies

$$\partial_t u + c\,\partial_x u \;=\; 0 \tag{1.1.2}$$

if and only if there is an $f \in C^\infty(\mathbb{R})$ so that $u = f(x - ct)$. The function f is uniquely determined.

Proof. For constant c the characteristics along which u is constant are the lines $(t, x + ct)$. Therefore, $u(t,x) = u(0, x - ct)$ proving the result with $f(x) := u(0,x)$. $\square$

This shows that the Cauchy problem consisting of (1.1.2) together with the initial condition $u|_{t=0} = f$ is uniquely solvable with solution $f(x - ct)$. The solutions are waves translating rigidly with velocity equal to c.

Exercise 1.1.1. *Find an explicit solution formula for the solution of the Cauchy problem*

$$\partial_t u + c\,\partial_x u + z(t,x)u \;=\; 0, \qquad u|_{t=0} = g,$$

where $z \in C^\infty$.

Example 1.1.2 (D'Alembert's formula)**.** If $c \in \mathbb{R} \setminus 0$, then $u \in C^\infty([0,T] \times \mathbb{R})$ satisfies

$$u_{tt} \;-c^2\; u_{xx} \;=\; 0 \tag{1.1.3}$$

if and only if there are smooth $f, g \in C^\infty(\mathbb{R})$ so that

$$u \;=\; f(x - ct) \;+\; g(x + ct)\,. \tag{1.1.4}$$

The set of all pairs $\tilde f, \tilde g$ so that this is so is of the form $\tilde f = f + b\,,\ \tilde g = g - b$ with $b \in \mathbb{C}$.

Proof. Factor

$$\partial_t^2 - c^2\partial_x^2 \;=\; (\partial_t - c\partial_x)\,(\partial_t + c\partial_x) \;=\; (\partial_t + c\partial_x)\,(\partial_t - c\partial_x)\,.$$

Conclude that

$$u_+ \;:=\; \partial_t u - c\,\partial_x u \qquad \text{and} \qquad u_- \;:=\; \partial_t u + c\,\partial_x u$$

satisfy

$$(\partial_t \;\pm\; c\,\partial_x)u_\pm \;=\; 0\,. \tag{1.1.5}$$

Example 1.1.1 implies that there are smooth $F_\pm$ so that

$$u_\pm \;=\; F_\pm(x \mp ct)\,. \tag{1.1.6}$$

In order for (1.1.4) and (1.1.6) to hold, one must have

$$F_+ = (\partial_t - c\partial_x)u \;=\; (1 + c^2)f', \quad F_- = (\partial_t + c\partial_x)u \;=\; (1 + c^2)g'. \tag{1.1.7}$$

Thus if $G_\pm$ are primitives of $F_\pm$ that vanish at the origin, then one must have

$$f = \frac{G_+}{(1+c^2)} + C_+, \qquad g = \frac{G_+}{(1+c^2)} + C_-, \qquad C_+ + C_- = u(0,0)\,.$$

Reversing the process shows that if G_+, f, g are defined as above, then $\tilde{u} := f(x-ct) + g(x+ct)$ yields a solution of D'Alembert's equation with

$$(\partial_t \mp c\,\partial_x)\tilde{u} = F_\pm \qquad \text{so} \qquad (\partial_t \mp c\,\partial_x)(u - \tilde{u}) = 0\,.$$

Adding and subtracting this pair of equations shows that

$$\nabla_{t,x}(u - \tilde{u}) = 0.$$

Since $u(0,0) = \tilde{u}(0,0)$, it follows by connectedness of $[0,T] \times \mathbb{R}$ that $u = \tilde{u}$, and the proof is complete. □

For speeds $c(t,x)$ that are not bounded, it is possible that characteristics escape to infinity with interesting consequences.

Example 1.1.3 (Nonuniqueness in the Cauchy problem). Consider $c(t,x) := x^2$. The characteristic through $(0, x_0)$ is the solution of

$$x' = x^2, \qquad x(0) = x_0\,.$$

Then,

$$1 = \frac{x'}{x^2} = \frac{d}{dt}\left(\frac{-1}{x}\right).$$

Integrating from $t = 0$ yields

$$\frac{-1}{x(t)} - \frac{-1}{x_0} = t \qquad \text{and, therefore,} \qquad x(t) = \frac{x_0}{1 - x_0\,t}\,.$$

Through each point t, x with $t \geq 0$ there is a unique characteristic tracing backward to $t = 0$. Therefore, given initial data $u(0,x) = g(x)$, the solution $u(t,x)$ is uniquely determined in $t \geq 0$ by requiring u to be constant on characteristics.

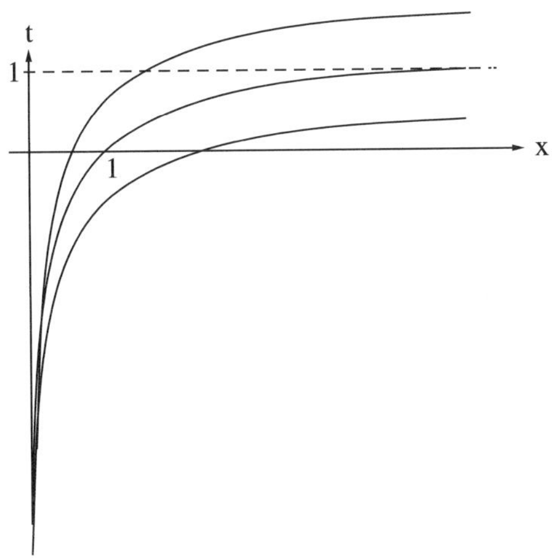

FIGURE 1.1.1. Characteristics diverge in finite time

As indicated in Figure 1.1.1, the characteristics through $(0, \pm 1)$ diverge to $\pm\infty$ at time $t = 1$. Thus all the backward characteristics starting in $t \geq 1$ meet $\{t = 0\}$ in the interval $]-1, 1[$. The data for $|x| \geq 1$ does not influence the solution in $t \geq 1$. There has been a loss of information. Another manifestation of this is that the initial values do not uniquely determine a solution in $t < 0$.

The characteristics starting at $t = 0$ meet $\{t = -1\}$ in the interval $]-1, 1[$. Outside that interval, the values of a solution are not determined, not even influenced by the initial data. There are many solutions in $t < 0$ which have the given Cauchy data. They are constant on characteristics which diverge to infinity, but their values on these characteristics is otherwise arbitrary.

To avoid this phenomenon we make the following assumption that not only prevents characteristics from diverging, but avoids some technical difficulties that occur for unbounded c with characteristics that do not diverge.

Hypothesis 1.1.1. *Suppose that for all $T > 0$*

$$\partial^\alpha_{t,x} c \in L^\infty([0,T] \times \mathbb{R}) .$$

The coefficient $d(t,x)$ satisfies analogous bounds.

For arbitrary $f \in C^\infty(\mathbb{R}^2)$ and $g \in C^\infty(\mathbb{R})$, there is a unique solution of the Cauchy problem

$$\Big(\partial_t + c(t,x)\,\partial_x + d(t,x)\Big)u = f, \qquad u(0,x) = g.$$

Its values along the characteristic $(t, x(t))$ are determined by integrating the nonhomogeneous linear ordinary differential equation

$$\frac{d}{dt}\, u(t, x(t)) + d(t, x(t))\, u(t, x(t)) \;=\; f(t, x(t))\,. \tag{1.1.8}$$

There are finite regularity results too. If f, g are k times differentiable with $k \geq 1$, then so is u. Though the equation is first order, u is in general not smoother than f. This is in contrast to the elliptic case.

The method of characteristics also applies to systems of hyperbolic equations. Consider vector valued unknowns $u(t, x) \in \mathbb{C}^N$. The simplest generalization is diagonal real systems

$$u_t \;+\; \mathrm{diag}(c_1(t, x), \ldots, c_N(t, x))\, u \;=\; 0.$$

Here u_j is constant on characteristics with speed $c_j(t, x)$. This idea extends to some systems

$$L \;:=\; \partial_t \;+\; A(t, x)\, \partial_x \;+\; B(t, x),$$

where A and B are smooth matrix valued functions so that

$$\forall T,\ \forall \alpha, \quad \partial^\alpha_{t,x}\{A, B\} \;\in\; L^\infty([-T, T] \times \mathbb{R})\,.$$

The method of characteristics applies when the following hypothesis is satisfied. It says that the matrix A has real eigenvalues and is smoothly diagonalizable. The real spectrum as well as the diagonalizability are related to a part of the general theory of constant coefficient hyperbolic systems sketched in the Appendix 2.I to Chapter 2.

Hypothesis 1.1.2. *There is a smooth matrix valued function, $M(t, x)$, so that*

$$\forall T,\ \forall \alpha, \quad \partial^\alpha_{t,x} M \ \text{ and } \ \partial^\alpha_{t,x}(M^{-1}) \ \text{ belong to } \ L^\infty([0, T] \times \mathbb{R})$$

and

$$M^{-1}\, A\, M \;=\; \text{diagonal and real.} \tag{1.1.9}$$

Examples 1.1.4. 1. The hypothesis is satisfied if for each t, x the matrix A has N distinct real eigenvalues $c_1(t, x) < c_2(t, x) < \cdots < c_N(t, x)$. Such systems are called *strictly hyperbolic*. To guarantee that the estimates on M, M^{-1} are uniform as $|x| \to \infty$, it suffices to make the additional assumption that

$$\inf_{(t,x) \in [0,T] \times \mathbb{R}} \ \min_{2 \leq j \leq N} \ c_j(t, x) \;-\; c_{j-1}(t, x) \;>\; 0\,.$$

2. More generally the hypothesis is satisfied if for each (t, x), A has uniformly distinct real eigenvalues and is diagonalizable. It follows that the multiplicity of the eigenvalues is independent of t, x.

3. If A_1 and A_2 satisfy Hypothesis 1.1.2. then so does

$$\begin{pmatrix} A_1 & 0 \\ 0 & A_2 \end{pmatrix} \qquad \text{with} \qquad M := \begin{pmatrix} M_1 & 0 \\ 0 & M_2 \end{pmatrix}.$$

In this way one constructs examples with variable multiplicity.

Exercise 1.1.2. *Prove assertions* **1** *and* **2**.

Since $A = MDM^{-1}$ with $D = \operatorname{diag}(c_1, \ldots, c_N)$,

$$L = \partial_t + MDM^{-1}\partial_x + B .$$

Define v by $u = Mv$ so

$$M^{-1}Lu = M^{-1}\Big(\partial_t + MDM^{-1}\partial_x + B\Big)M\,v .$$

When the derivatives on the right fall on v, the product $M^{-1}M = I$ simplifies. This shows that

$$M^{-1}Lu = \Big(\partial_t + D\,\partial_x + \widetilde{B}\Big)u := \widetilde{L}\,v,$$

where

$$\widetilde{B} := M^{-1}\,B\,M + M^{-1}\,M_t + M^{-1}\,A\,M_x .$$

This change of variable converts the equation $Lu = f$ to $\widetilde{L}v = M^{-1}f$ where $\widetilde{L}$ has the same form as L but with leading part that is a set of directional derivatives.

Theorem 1.1.1. *Suppose that Hypothesis* 1.1.2 *is satisfied,* $k \geq 1$, $f \in C^k([0,T]\times\mathbb{R})$, $g \in C^k(\mathbb{R})$, *and for all* α, β *with* $|\alpha| \leq k$ *and* $|\beta| \leq k$,

$$\partial_{t.x}^{\alpha} f \in L^\infty([0,T]\times\mathbb{R}) \qquad \textit{and} \qquad \partial_x^\beta g \in L^\infty([0,T]\times\mathbb{R}) .$$

Then there is a unique solution $u \in C^k([0,T]\times\mathbb{R})$ *to the initial value problem* $Lu = f$, $u|_{t=0} = g$ *so that all partial derivatives of* u *of order* $\leq k$ *are in* $L^\infty([0,T]\times\mathbb{R})$.

The crux is the following estimate called *Haar's inequality.* For a vector valued function $w(x) = (w_1(x), \ldots, w_N(x))$ on $\mathbb{R}$, the L^∞ norm is taken to be

$$\|w\|_{L^\infty(\mathbb{R})} := \max_{1\leq j\leq N} \|w_j(x)\|_{L^\infty(\mathbb{R})} .$$

Haar's Lemma 1.1.2. *Suppose that Hypothesis* 1.1.2 *is satisfied.*

i. *There is a constant* $C = C(T, L)$ *so that if* u *and* Lu *are bounded continuous functions on* $[0,T]\times\mathbb{R}$, *then for* $t \in [0,T]$

$$\|u(t)\|_{L^\infty(\mathbb{R})} \leq C\left(\|u(0)\|_{L^\infty(\mathbb{R})} + \int_0^t \|Lu(\sigma)\|_{L^\infty(\mathbb{R})}\,d\sigma\right).$$

ii. *More generally, there is a constant $C(k,T,L)$ so that if for all $|\alpha|\le k$, $\partial^\alpha_{t,x}u$ and $\partial^\alpha_{t,x}Lu$ are bounded continuous functions on $[0,T]\times\mathbb{R}$, then*

$$m_k(u,t) \;:=\; \sum_{|\alpha|\le k} \|\partial^\alpha_{t,x}u(t)\|_{L^\infty(\mathbb{R})}$$

satisfies for $t\in[0,T]$,

$$m_k(u,t) \;\le\; C\Big(m_k(u,0) \;+\; \int_0^t m_k(Lu,\sigma)\,d\sigma\Big)\,.$$

Proof of Lemma 1.1.2. The change of variable shows that it suffices to consider the case of a real diagonal matrix $A=\operatorname{diag}(c_1(t,x),\dots,c_N(t,x))$.

i. For $\underline{t}\in[0,T]$ and $\epsilon>0$ choose j and $\underline{x}$ so that

$$\|u(\underline{t})\|_{L^\infty(\mathbb{R})} \;\le\; \|u_j(\underline{t},\underline{x})\|_{L^\infty(\mathbb{R})} \;+\; \epsilon\,.$$

Choose $(t,x(t))$ the integral curve of $x'=c_j(t,x)$ passing through $\underline{t},\underline{x}$. Then

$$u_j(\underline{t},\underline{x}) \;=\; u_j(0,x(0)) \;+\; \int_0^{\underline{t}} (\partial_t + c_j(t,x)\partial_x)u_j(\sigma,x(\sigma))\;d\sigma\,.$$

Therefore

$$\|u(\underline{t})\|_{L^\infty(\mathbb{R})} \;\le\; \|u(0)\|_{L^\infty(\mathbb{R})} \;+\; \int_0^{\underline{t}} \|(Lu \;-\; B\,u)(\sigma)\|_{L^\infty(\mathbb{R})}\;d\sigma \;+\; \epsilon.$$

Since this is true for all ϵ, one has

$$\|u(\underline{t})\|_{L^\infty(\mathbb{R})} \;\le\; \|u(0)\|_{L^\infty(\mathbb{R})} \;+\; \int_0^{\underline{t}} \|(Lu)(\sigma)\|_{L^\infty(\mathbb{R})} \;+\; C\,\|u(\sigma)\|_{L^\infty(\mathbb{R})}\;d\sigma\,,$$

and **i** follows using Gronwall's Lemma 2.1.3.

ii. Apply the inequality of **i** to $\partial^\alpha_{t,x}u$ with $|\alpha|\le k$.

$$\|\partial^\alpha u(t)\|_{L^\infty(\mathbb{R})} \;\le\; C\Big(\|\partial^\alpha u(0)\|_{L^\infty(\mathbb{R})} \;+\; \int_0^t \|L\partial^\alpha u(\sigma)\|_{L^\infty(\mathbb{R})}\Big)\;d\sigma.$$

Compute

$$L\,\partial^\alpha u \;=\; \partial^\alpha Lu \;+\; [L,\partial^\alpha]\,u\,.$$

The commutator is a differential operator of order k with bounded coefficients, so

$$\|[L,\partial^\alpha]\,u(\sigma)\|_{L^\infty(\mathbb{R})} \;\le\; C\,m_k(u,\sigma)\,.$$

Therefore,

$$\|\partial^\alpha u(t)\|_{L^\infty(\mathbb{R})} \le C\Big(\|\partial^\alpha u(0)\|_{L^\infty(\mathbb{R})} + \int_0^t \|\partial^\alpha Lu(\sigma)\|_{L^\infty(\mathbb{R})} \;+\; m_k(u,\sigma)\;d\sigma\Big).$$

Sum over $|\alpha|\le k$ to find

$$m_k(u,t) \le C\Big(m_k(u,0) + \int_0^t m_k(u,\sigma) \;+\; m_k(Lu,\sigma)\,d\sigma\Big)\,.$$

Gronwall's lemma implies **ii.** □

Proof of Theorem 1.1.1. The change of variable shows that it suffices to consider the case of $A = \operatorname{diag}(c_1, \dots, c_N)$.

The solution u is constructed as a limit of approximate solutions u^n. The solution u^0 is defined as the solution of the initial value problem

$$\partial_t u^0 \ + \ A\,\partial_x u^0 \ = \ f, \qquad u^0|_{t=0} = g\,.$$

The solution is explicit by the method of characteristics. It is C^k with bounded derivatives on $[0,T] \times \mathbb{R}$, so

$$\exists C_1, \ \ \forall t \in [0,T], \ \ m_k(u^0, t) \ \le \ C_1\,. \tag{1.1.10}$$

For $n > 0$ the solution u^n is again explicit by the method of characteristics in terms of u^{n-1},

$$\partial_t u^n \ + \ A\,\partial_x u^n \ + \ B\,u^{n-1} \ = \ f\,, \qquad u^{n-1}|_{t=0} = g\,. \tag{1.1.11}$$

Using (1.1.10) and Haar's inequality yields,

$$\exists C_2, \ \ \forall t \in [0,T], \ \ m_k(u^1, t) \ \le \ C_2\,. \tag{1.1.12}$$

For $n \ge 2$ estimate $u^n - u^{n-1}$ by applying Haar's inequality to

$$\tilde{L}\,(u^n - u^{n-1}) \ + \ B\,(u^{n-1} - u^{n-2}) \ = \ 0, \qquad (u^n - u^{n-1})\big|_{t=0} \ = \ 0\,,$$

to find

$$m_k(u^n - u^{n-1}, t) \ \le \ C \int_0^t m_k(u^{n-1} - u^{n-2}, \sigma)\, d\sigma\,. \tag{1.1.13}$$

For $n = 2$, this together with (1.1.10) and (1.1.12) yields

$$m_k(u^2 - u^1, t) \ \le \ (C_1 + C_2) C t\,.$$

Injecting this in (1.1.13) yields

$$m_k(u^3 - u^2, t) \ \le \ (C_1 + C_2) C^2 t^2 / 2\,.$$

Continuing yields

$$m_k(u^n - u^{n-1}, t) \ \le \ (C_1 + C_2) C^{n-1} t^{n-1} / (n-1)!\,. \tag{1.1.14}$$

The summability of the right-hand side implies (Weierstrass's M-test) that u^n and all of its partials of order $\le k$ converge uniformly on $[0,T] \times \mathbb{R}$. The limit u is C^k with bounded partials. Passing to the limit in (1.1.11) shows that u solves the initial value problem.

To prove uniqueness, suppose that u and v are solutions. Haar's inequality applied to $u - v$ implies that $u - v = 0$. □

The proof also yields finite speed of propagation of signals. Define $\lambda_{\min}(t,x)$ and $\lambda_{\max}(t,x)$ to the smallest and largest eigenvalues of $A(t,x)$. Then the functions λ are uniformly Lipschitzean on $[0,T]\times\mathbb{R}$. To prove this, observe that equation (1.1.9) shows that the diagonal elements $c_j(t,x)$ of the right-hand side are the eigenvalues of A. Their expression by the left-hand side shows that their partial derivatives of first order (in fact any order) are bounded on $[0,T]\times\mathbb{R}$. Therefore the c_j are uniformly Lipschitzean on $[0,T]\times\mathbb{R}$. It follows that $\lambda_{\min} := \min_j \{c_j\}$ is uniformly Lipschitzean.

The characteristics have speeds bounded below by $\lambda_{\min}$ and above by $\lambda_{\max}$. The next result shows that these are lower and upper bounds, respectively, for the speeds of propagation of signals.

Corollary 1.1.3. *Suppose that $-\infty < x_l < x_r < \infty$ and γ_l (resp. γ_r) is the integral curve of $\partial_t + \lambda_{\min}(t,x)\partial_x$ (resp. $\partial_t + \lambda_{\max}(t,x)\partial_x$) passing through x_l (resp. x_r). Denote by Q the four sided region in $0\le t\le T$ bounded on the left by γ_l and the right by γ_r. If g is supported in $[x_l,x_r]$ and $f\big|_{0\le t\le T}$ is supported in Q, then for $0\le t\le T$ the solution u is supported in Q.*

Proof. The explicit formulas of the method of characteristics show that the approximate solutions u^n are supported in Q. Passing to the limit proves the result. □

Consider next the case of $f=0$ and $g\in C^1(\mathbb{R})$ whose restrictions to $]-\infty,\underline{x}[$ and $]\underline{x},\infty[$ are each smooth with uniformly bounded derivatives of every order. Such a function is called *piecewise smooth.*

The simplest case is that of an operator $\partial_t + c(t,x)\partial_x$. Denote by γ the characteristic through $\underline{x}$. The values of u to the left of γ are determined by g to the left of $\underline{x}$. Choose a $\tilde g\in C^\infty(\mathbb{R})$ which agrees with g to the left and has bounded derivatives of all orders. The solution $\tilde u$ then agrees with u to the left of γ and $\tilde u$ has bounded partials of all orders for $0\le t\le T$. An analogous argument works for the right-hand side, and one sees that u is piecewise C^∞ in the decomposition of $[0,T]\times\mathbb{R}$ into two pieces by γ.

Suppose now that A satisfies Hypothesis 1.1.2, and for all $(t,x)\in[0,T]\times\mathbb{R}$ has N distinct real eigenvalues ordered so that $c_j<c_{j+1}$. Denote by γ_j the corresponding characteristics through $\underline{x}$. Define open wedges,

$$W_1 := \big\{(t,x)\ :\ 0<t<T,\quad -\infty<x<\gamma_1(t)\big\},$$

$$W_{N+1} := \big\{(t,x)\ :\ 0<t<T,\quad \gamma_N(t)<x<\infty\big\},$$

and for $2<j<N$,

$$W_j := \big\{(t,x)\ :\ 0<t<T,\quad \gamma_{j-1}(t)<x<\gamma_j(t)\big\}.$$

They decompose $[0,T]\times\mathbb{R}$ into pie shaped wedges with vertex at $(0,\underline{x})$ and numbering from left to right.

Definition 1.1.4. *For $\mathbb{Z} \ni k \geq 1$, the set PC^k consists of functions which are piecewise C^k as the set of bounded continuous functions u on $[0,T] \times \mathbb{R}$ so that for $\alpha \leq k$ and $1 \leq j \leq N$, the restriction $u|_{W_j}$ belongs to $C^k(W_j)$ and for all $\alpha \leq k, \partial^\alpha(u|_{W_j})$ extends to a bounded continuous function on the closure $\overline{W}_j$. It is a Banach space with the norm*

$$\|u\|_{L^\infty([0,T]\times\mathbb{R})} + \sum_{|\alpha|\leq k} \sum_{1\leq j\leq N+1} \|\partial^\alpha_{t,x}(u|_{W_j})\|_{L^\infty(W_j)} .$$

The next result asserts that for piecewise smooth data with singularity at $\underline{x}$, the solution is piecewise smooth with its singularities restricted to the characteristics through $\underline{x}$.

Theorem 1.1.5. *Suppose in addition to Hypothesis 1.1.2, that A has N distinct real eigenvalues for all (t,x). If $f \in PC^k$ and $g \in L^\infty(\mathbb{R})$ have bounded continuous derivatives up to order k on each side of $\underline{x}$, then the solution u belongs to PC^k.*

Sketch of Proof. The construction of u yielded an $L^\infty([0,T]\times\mathbb{R})$ estimate. In addition we need estimates for the derivatives of order $\leq k$ on the wedge W_j. Introduce

$$\mu_k(u,\sigma) := \|u(\sigma)\|_{L^\infty(\mathbb{R})} + \sum_{2\leq|\alpha|\leq k} \sum_{1\leq j\leq N+1} \|\partial^\alpha_{t,x}(u|_{W_j})(\sigma)\|_{L^\infty(W_j\cap\{t=\sigma\})} .$$

To estimate $u^n - u^{n-1}$ use the following lemma. □

Lemma 1.1.6. *Assume the hypotheses of the theorem and that $c_j(t,x)$ is one of the eigenvalues of $A(t,x)$. Then, there is a constant $C(j,T,L)$ so that if $f \in PC^k$ and*

$$(\partial_t + c_j(t,x)\,\partial_x)w = f, \qquad w\big|_{t=0} = 0,$$

then $w \in PC^k$ and

$$\mu_k(w,t) \leq C\Big(\mu_k(w,0) + \int_0^t \mu_k(f,\sigma)\, d\sigma\Big).$$

Exercise 1.1.3. *Prove the lemma. Then finish the proof of the theorem.*

Exercise 1.1.4. *Suppose that u is as in the theorem, $f = 0$, and that for some $\epsilon > 0$ and j, the derivatives of u of order $\leq k$ are continuous across $\gamma_j \cap \{0 \leq t < \epsilon\}$. Prove that they are continuous across $\gamma_j \cap \{0 \leq t \leq T\}$.*
Hints. Show that the set of times $\underline{t}$ for which the solution is C^k on $\gamma_j \cap \{0 \leq t \leq \underline{t}\}$ is both open and closed. Use finite speed.

Denote by $\Phi_j(t,x)$ the flow of the ordinary differential equation $x' = c_j(t,x)$. That is $x(t) = \Phi_j(t,\underline{x})$ is the solution with $x(0) = \underline{x}$. The solution operator for the pure transport equation $(\partial_t + c_j\partial_x)u = 0$ with initial value g is then

$$u(t) \;=\; g(\Phi_j(-t,x))\,.$$

The values at time t are the rearrangements by the diffeomorphism $\Phi(-t,\cdot)$ of the initial function. Because of the uniform boundedness of the derivatives of c_j on slabs $[0,T]\times\mathbb{R}$, one has

$$\partial^\alpha_{t,x}\Phi \;\in\; L^\infty([0,T]\times\mathbb{R})\,.$$

The derivative $\partial_x\Phi$ measures the expansion or contraction by the flow. It is the length of the image of an infinitesimal interval divided by the original length. In particular Φ can at most expand lengths by a bounded quantity. The inverse of $\Phi(t,\cdot)$ is the flow by the ordinary differential equation from time t to time 0, so the inverse also cannot expand by much. This is equivalent to a lower bound,

$$\big(\partial_x\Phi\big)^{-1} \;\in\; L^\infty([0,T]\times\mathbb{R})\,.$$

The diffeomorphism $\Phi(t,\cdot)$ can neither increase nor decrease length by much. Therefore the maps $u(0)\mapsto u(t)$ are uniformly bounded maps from $L^p(\mathbb{R})$ to itself for all $p\in[1,\infty]$. The case $p=\infty$ is equivalent to the Haar inequalities. There are analogous estimates

$$\|u(t)\|_{L^p(\mathbb{R})} \;\le\; C\left(\|u(t)\|_{L^p(\mathbb{R})} \;+\; \int_0^t \|Lu(\sigma)\|_{L^p(\mathbb{R})}\,d\sigma\right),$$

with constant independent of p. This in turn leads to an existence theory like that just recounted but $m_k(u,t)$ is replaced by $\sum_{|\alpha|\le k}\|\partial^\alpha_{t,x}u(t)\|_{L^p(\mathbb{R})}$. For these one dimensional hyperbolic Cauchy problems, there is a wide class of spaces for which the evolution is well posed. The case of $p=1$ is particularly important for the theory of shock waves in $d=1$. Brenner's Theorem 3.3.5 shows that only the case $p=2$ remains valid for typical hyperbolic equations in dimension $d>1$.

1.2. Examples of propagation of singularities using progressing waves

D'Alembert's solution (see Example 1.1.2) of the one-dimensional wave equation,

$$u_{tt} - u_{xx} = 0\,, \tag{1.2.1}$$

is the sum of progressing waves

$$f(x-t) + g(x+t)\,. \tag{1.2.2}$$

The rays are the integral curves of

$$\partial_t \pm \partial_x \,. \tag{1.2.3}$$

Structures are rigidly transported at speeds ± 1.

There is an energy law. If u is a smooth solution whose support intersects each time slab $a \le t \le b$ in a compact set, one has

$$\begin{aligned} \frac{d}{dt}\int_{\mathbb{R}} u_t^2 + u_x^2 \, dx &= \int \partial_t (u_t^2 + u_x^2)\, dx \\ &= \int 2u_t(u_{tt} - u_{xx}) + \partial_x(2u_t\, u_x)\, dx \;=\; 0\,, \end{aligned}$$

since the first summand vanishes and the second is the x derivative of a function vanishing outside a compact set.

The fundamental solution that solves (1.2.3) together with the initial values

$$u(0,x) = 0\,, \qquad u_t(0,x) = \delta(x)\,, \tag{1.2.4}$$

is given by the explicit formula

$$u(t,x) \;=\; \frac{\operatorname{sgn} t}{2}\, \chi_{[-t,t]} \;=\; \frac{1}{2}\Big(h(x-t) - h(x+t)\Big)\,, \tag{1.2.5}$$

where h denotes Heaviside's function, the characteristic function of $]0,\infty[$.

Exercise 1.2.1. i. *Derive (1.2.5) by solving the initial value problem using the Fourier transform in x.* **Hint.** You will likely decompose an expression regular at $\xi = 0$ into two that are not. Use a principal value to justify this step.

ii. *The proof of D'Alembert's formula (1.2.2) shows that every distribution solution of (1.2.1) is given by (1.2.2) for f, g distributions on $\mathbb{R}$. Derive (1.2.5) by finding the f, g that yield the solution of (1.2.4).* **Hint.** You will need to find the solutions of $du/dx = \delta(x)$.

The singularities of the solution (1.2.5) lie on the characteristic curves through $(0,0)$. This is a consequence of Theorem 1.1.4. In fact, define v as the solution of

$$v_{tt} - v_{xx} = 0, \qquad v(0,x) = 0, \qquad v_t(0,x) = x_+^2/2, \qquad x_+ := \max\{x, 0\}\,.$$

Introduce

$$V \;:=\; (v_1, v_2, v), \qquad v_1 := \partial_t v - \partial_x v, \quad v_2 := \partial_t v + \partial_x v$$

to find

$$\partial_t V \;+\; \begin{pmatrix} 1 & 0 & 0 \\ 0 & 1 & 0 \\ 0 & 0 & -1 \end{pmatrix} \partial_x V \;+\; \begin{pmatrix} 0 & 0 & 1 \\ 0 & 0 & 0 \\ 0 & 0 & 0 \end{pmatrix} V \;=\; 0\,.$$

The Cauchy data, $V(0,x)$ are continuous, piecewise smooth, and singular only at $x = 0$. Theorem 1.1.4 shows that V is piecewise smooth with singularities only on the characteristics through $(0,0)$. In addition $u = \partial_x^3 v$ (in the sense of distributions) since they both satisfy the same initial value problem. Thus v and $u = \partial_x^3 v$ have singular support only on the characteristics through $(0,0)$.

Interesting things happen if one adds a lower order term. For example, consider the Klein–Gordon equation

$$u_{tt} - u_{xx} + u = 0\,. \tag{1.2.6}$$

In sharp contrast with equation (1.2.2), there are hardly any undistorted progressing wave solutions.

Exercise 1.2.2. *Find all solutions of (1.2.6) of the form $f(x - ct)$ and all solutions of the form $e^{i(\tau t - x\xi)}$.* **Discussion.** The solutions $e^{i(\tau t - x\xi)}$ with $\xi \in \mathbb{R}$ are particularly important since the general solution is a Fourier superposition of these special *plane waves.* The equation $\tau = \tau(\xi)$ defining such solutions is called the *dispersion relation* of (3.1.6).

There is an energy conservation law. Denote by $\mathcal{S}(\mathbb{R}^d)$ the Schwartz space of rapidly decreasing smooth functions. That is, functions such that for all α, β,

$$\sup_{x\in\mathbb{R}^d} \left|x^\beta \partial_x^\alpha \psi(x)\right| < \infty\,.$$

Exercise 1.2.3. *Prove that if $u \in C^\infty(\mathbb{R} : \mathcal{S}(\mathbb{R}))$ is a real valued solution of the Klein–Gordon equation, then*

$$\int u_t^2 + u_x^2 + u^2 \; dx$$

is independent of t. This quantity is called the **energy**. **Hint.** Justify carefully differentiation under the integral sign and integration by parts. If you find weaker hypotheses which suffice, that is good.

The fundamental solution of the Klein–Gordon equation with initial data (1.2.4), is not as simple as the fundamental solutions of the wave equation. Theorem 1.1.4 implies that the singular support lies on $\{x = \pm t\}$. The proof is as for the wave equation except that the zeroth order term in the equation for V is replaced by

$$\begin{pmatrix} 1 & 0 & -1 \\ 0 & 1 & -1 \\ 0 & 0 & -1 \end{pmatrix} V\,.$$

The singularities are computed by the method of progressing waves. For $n \in \mathbb{N}$, introduce

$$(1.2.7) \qquad h_n(x) := \begin{cases} x^n/n! & \text{for} \quad x \geq 0, \\ 0 & \text{for} \quad x \leq 0. \end{cases}$$

Then

$$(1.2.8) \qquad \frac{d}{dx} h_{n+1} = h_n \,, \qquad \text{for} \quad n \geq 0 \,.$$

Exercise 1.2.4. *Show that there are uniquely determined functions $a_n(t)$ satisfying*

$$a_0(0) = 1/2 \quad \textit{and} \quad a_n(0) = 0 \quad \textit{for} \quad n \geq 1 \,,$$

and so that for all $N \geq 2$,

$$(1.2.9) \qquad \left(\partial_t^2 - \partial_x^2 + 1\right) \sum_{n=0}^{N} a_n(t)\, h_n(x-t) \in C^{N-2}(\mathbb{R}^2) \,.$$

In this case, we say that the series

$$\sum_{n=0}^{\infty} a_n(t)\, h_n(t-x)$$

is a formal solution of $(\partial_t^2 - \partial_x^2 + 1)u \in C^\infty$. **Hints.** Pay special attention to the most singular term(s). In particular show that, $\partial_t a_0 = 0$.

Exercise 1.2.5. *Suppose that u is the fundamental solution of the Klein–Gordon equation and $M \geq 0$. Find a distribution w_M such that $u - w_M \in C^M(\mathbb{R}^2)$. Show that the fundamental solution of the wave equation and that of the Klein–Gordon equation differ by a Lipschitz continuous function. Show that the singular supports of the two fundamental solutions are equal.* **Hint.** Add (1.2.9) to its spatial reflection and choose initial values for the two solutions to match the initial data.

Exercise 1.2.6. *Study the fundamental solution for the dissipative wave equation*

$$(1.2.10) \qquad u_{tt} - u_{xx} + 2u_t = 0 \,.$$

Use Theorem 1.1.4 to show that the singular support is contained in the characteristics through $(0,0)$. Show that it is not a continuous perturbation of the fundamental solution of the wave equation. **Hint.** Find solutions of $(\partial_t^2 - \partial_x^2 + 2\partial_t)u \in C^\infty$ of the form $\sum_n b_n(t)\, h_n(t-x)$ as in Exercises 1.2.4 and 1.2.5. Use two such solutions as in Exercise 1.2.5.

The method of *progressing wave expansions* from these examples is discussed in more generality in chapter 6 of Courant and Hilbert Vol. 2, and in [Lax, 2006]. The higher dimensional analogue of these solutions are singular along codimension one characteristic hypersurfaces in space-time. The singularities propagate satisfying transport equations along rays generating the hypersurface. The general class goes under the name *conormal solutions.* M. Beals' book [Beals, 1989] is a good reference. They describe propagating wavefronts. Luneberg's book [Luneberg, 1944] recounts his discovery that the propagation laws for fronts of singularities coincide with the physical laws of geometric optics.

1.3. Group velocity and the method of nonstationary phase

The Klein–Gordon equation has constant coefficients, and so it can be solved explicitly using the Fourier transform. The computation of the singularities of the fundamental solution of the Klein–Gordon equation in Exercise 1.2.5 suggests that the main part of solutions travel with speed equal to 1. One might expect that the energy in a disk growing linearly in time at a speed < 1 would be small for $t \gg 1$. For compactly supported data, such a disk would contain no singularities for large time. Thus it is not unreasonable to guess that for each $\sigma < 1$ and $R > 0$,

$$\limsup_{t\to\infty} \int_{|x|<R+\sigma t} u_t^2 + u_x^2 + u^2\, dx \;=\; 0. \tag{1.3.1}$$

Either the method of characteristics or the energy method shows that speeds are no larger than one. The idea about the Klein–Gordon energy expressed in (1.3.1) is dead wrong. The main part of the energy travels strictly slower than speed 1, even though singularities travel with speed exactly equal to 1.

The solution of the Cauchy problem for the Klein–Gordon equation in dimension d,

$$u_{tt} \;-\; \Delta u \;+\; u \;=\; 0\,, \qquad (t,x) \;\in\; \mathbb{R}^{1+d}\,,$$

is given by

$$u \;=\; \sum_{\pm} (2\pi)^{-d/2} \int a_\pm(\xi)\; e^{i(\pm\langle\xi\rangle t + x\xi)}\; d\xi\,, \qquad \langle\xi\rangle := \big(1+|\xi|^2\big)^{1/2},$$

$$\hat u(0,\xi) \;=\; a_+(\xi) + a_-(\xi)\,, \qquad \hat u_t(0,\xi) \;=\; i\,\langle\xi\rangle\,\big(a_+(\xi) - a_-(\xi)\big)\,.$$

The conserved energy is equal to

$$\frac12 \int u_t^2 \;+\; |\nabla_x u|^2 \;+\; u^2\; dx \;=\; \int \langle\xi\rangle^2\,\big(\,|a_+(\xi)|^2 \;+\; |a_-(\xi)|^2\,\big)\; d\xi\,.$$

Exercise 1.3.1. *Verify these formulas. Verify conservation of energy by an integration by parts argument as in Exercise* 1.2.3. **Hint.** Follow the computation that starts §1.4.

Consider the behavior for large times. The phases $\phi_\pm(t,x,\xi) := \pm\langle\xi\rangle t + x\xi$ have gradients

$$\nabla_\xi \phi_\pm(t,x,\xi) \ := \ \nabla_\xi\Big(\pm\langle\xi\rangle t + x\xi\Big) \ = \ \frac{\pm t\xi}{\langle\xi\rangle} + x \ = \ t\Big(\frac{\pm\xi}{\langle\xi\rangle} + \frac{x}{t}\Big).$$

At space-time points (t,x) with $t \gg 1$ and

$$\frac{\pm\xi}{\langle\xi\rangle} + \frac{x}{t} \neq 0,$$

the phase oscillates rapidly and the contribution to the integral is expected to be small. The contribution to the $a_\pm$ integral from $\xi \sim \underline{\xi}$ is felt predominantly at points where $x/t \sim \mp\underline{\xi}/\langle\underline{\xi}\rangle$. Setting $\tau_\pm(\xi) := \pm\langle\xi\rangle$, one has

$$\frac{\mp\underline{\xi}}{\langle\underline{\xi}\rangle} \ = \ -\nabla_\xi \tau_\pm(\underline{\xi}).$$

This agrees with the formula for the group velocity (re)introduced on purely geometric grounds in §2.4.

For $t \to \infty$ the contributions of the plane waves $a_\pm(\xi)e^{i(\tau_\pm(\xi)t + x\xi)}$ with $\xi \sim \underline{\xi}$ are expected to be felt at points with $x/t \sim -\nabla_\xi \tau_\pm(\underline{\xi})$. A precise version is proved using the method of nonstationary phase.

Proposition 1.3.1. *Suppose that $a_\pm(\xi) \in \mathcal{S}(\mathbb{R}^d)$, and define*

$$\mathbf{V} \ := \ \bigcup_\pm \ \big\{\mathbf{v} \ : \ \mathbf{v} = -\nabla_\xi \tau_\pm(\xi) \ \ \textit{for some } \xi \in \operatorname{supp} a_\pm\big\}$$

to be the closed set of group velocities that appear in the plane wave decomposition of u. For $\mu > 0$, let $\mathbf{K}_\mu \subset \mathbb{R}^d$ denote the set of points at distance $\geq \mu$ from $\mathbf{V}$. Denote by Γ_μ the cone

$$\Gamma_\mu \ := \ \Big\{(t,x) \ : \ t > 0 \ \ \textit{and} \ \ x/t \in \mathbf{K}_\mu\Big\}.$$

Then for all $N > 0$ and α,

$$(1 + t + |x|)^N \ \partial^\alpha_{t,x} u(t,x) \in L^\infty(\Gamma_\mu).$$

Proof. The solution u is smooth with values in $\mathcal{S}$ so one need only consider $\{t \geq 1\}$. We estimate the u_+ summand. The u_- summand is treated similarly. The subscript $+$ is suppressed in u_+, ϕ_+, a_+ and τ_+.

Introduce the first order differential operator

$$(1.3.2) \quad \ell(t,x,\partial) \ := \ \frac{1}{i|\nabla_\xi \phi|^2} \sum_j \frac{\partial\phi}{\partial\xi_j}\frac{\partial}{\partial\xi_j}, \quad \text{so} \quad \ell(t,x,\partial_\xi)\, e^{i\phi} \ = \ e^{i\phi}.$$

The operator is only defined where $\nabla_\xi \phi \neq 0$. The coefficients are smooth functions on a neighborhood of Γ_μ, and are homogeneous of degree minus one in (t,x) and satisfy

$$\frac{1}{|\nabla_\xi \phi|^2}\left|\frac{\partial \phi}{\partial \xi_j}\right| \le C(t+|x|)^{-1} \quad \text{for} \quad (t,x,\xi)\in \Gamma_\mu \times \operatorname{supp} a\,.$$

The identity $\ell\, e^{i\phi} = e^{i\phi}$ implies

$$\int a(\xi)\ e^{i\phi}\ d\xi = \int a(\xi)\ \ell^N e^{i\phi}\ d\xi\,.$$

Denote by $\ell^\dagger$ the transpose of ℓ and integrate by parts to find

$$\int a(\xi)\ e^{i\phi}\ d\xi = \int \left[(\ell^\dagger)^N a(\xi)\right]\ e^{i\phi}\ d\xi\,.$$

The operator

$$(\ell^\dagger)^N = \sum_{|\alpha|\le N} c_\alpha(t,x,\xi)\,\partial_\xi^\alpha$$

with coefficients c_α smooth on a neighborhood of Γ_μ and homogeneous of degree $-N$ in t,x. Therefore,

$$|c_\alpha(t,x)| \le C(\alpha)(1+t+|x|)^{-N} \quad \text{for} \quad (t,x,\xi)\in\Gamma_\mu\times \operatorname{supp} a\,.$$

It follows that

$$\left|\int a(\xi)\ e^{i\phi}\ d\xi\right| \le C\,(1+t+|x|)^{-N}\,.$$

Since the t,x derivatives of this integral are again integrals of the same form, this suffices to prove the proposition. □

Example 1.3.1. Introduce for $0<\mu\ll 1$, $\tilde{\mathbf{V}}_\mu := \mathbb{R}^d \setminus \mathbf{K}_\mu$ an open set slightly larger than $\mathbf{V}$. For $t\to\infty$ virtually all the energy of a solution is contained in the cone $\{(t,x)\ :\ x/t\in\tilde{\mathbf{V}}\}$. This is particularly interesting when $a_\pm$ are supported in a small neighborhood of a fixed $\underline{\xi}$. For large times virtually all the energy is localized in a small conic neighborhood of the pair of lines $x=-t\,\nabla_\xi\tau_\pm(\underline{\xi})$ that travel with the group velocities associated to $\underline{\xi}$.

The integration by parts method introduced in this proof is very important. The next estimate for nonstationary oscillatory integrals is a straightforward application. The fact that the estimate is uniform in the phases is useful.

Lemma of Nonstationary Phase 1.3.2. *Suppose that Ω is a bounded open subset of $\mathbb{R}^d$, $m\in\mathbb{N}$, and $C_1>1$. Then there is a constant $C_2>0$ so that for all $f\in C_0^m(\Omega)$ and $\phi\in C^m(\Omega\,;\,\mathbb{R})$ satisfying*

$$\forall |\alpha|\le m,\quad \|\partial^\alpha\phi\|_{L^\infty}\le C_1, \quad \textit{and} \quad \forall x\in\Omega,\quad C_1^{-1}\le|\nabla_x\phi|\le C_1,$$

one has the estimate

$$\left| \int e^{i\phi/\epsilon} \, f(x) \, dx \right| \;\le\; C_2 \, \epsilon^m \sum_{|\alpha| \le m} \|\partial^\alpha f\|_{L^1} \,.$$

Exercise 1.3.2. *Prove the lemma.* **Hint.** Use (1.3.2).

Example 1.3.2. Applied to the phases $\phi = x\xi := \sum_j x_j \xi_j$ with ξ belonging to a compact subset of $\mathbb{R}^d \setminus 0$, the lemma implies the rapid decay of the Fourier transform of smooth compactly supported functions. Conversely, the lemma can be reduced to the special case of the Fourier transform. Since the gradient of ϕ in the lemma does not vanish, for each $\underline{x} \in \operatorname{supp} f$, there is a neighborhood and a nonlinear change of coordinates so that in the new coordinates ϕ is equal to x_1. Using a partition of unity, one can suppose that f is the sum of a finite number of functions each supported in one of the neighborhoods. For each such function, a change of coordinates yields an integral of the form

$$\int e^{ix_1/\epsilon} \, g(x) \, dx \;=\; c\,\hat{g}(1/\epsilon, 0, \ldots, 0)\,,$$

which is rapidly decaying since it is the transform of an element of $C_0^\infty(\mathbb{R}^d)$. Care must be taken to obtain estimates uniform in the family of phases of the lemma.

Exercise 1.3.3. *Suppose that* $f \in H^1(\mathbb{R})$ *and* $g \in L^2(\mathbb{R})$ *and that* u *is the unique solution of the Klein–Gordon equation with initial data*

$$u(0,x) = f(x)\,, \qquad u_t(0,x) = g(x)\,. \tag{1.3.3}$$

Prove that for any $\epsilon > 0$ *and* $R > 0$, *there is a* $\delta > 0$ *such that*

$$\limsup_{t \to \infty} \int_{|x| > (1-\delta)t - R} u_t^2 + u_x^2 + u^2 \, dx \;<\; \epsilon\,. \tag{1.3.4}$$

Hint. Replace $\hat{f}, \hat{g}$ by compactly supported smooth functions making an error at most $\epsilon/2$ in energy. Then use Lemma 1.3.2 noting that the group velocities are uniformly smaller than those for ξ belonging to the supports of $a_\pm$. **Discussion.** Note that as $\xi \to \infty$, the group velocities approach ± 1. High frequencies will propagate at speeds nearly equal to one. In particular they travel at the same speed. High frequency signals stay together better than low frequency signals. Since singularities of solutions are made of only the high frequencies (modifying the Fourier transform of the data on a compact set modifies the solution by a smooth term), one expects singularities to propagate at speeds ± 1. That is proved for the fundamental solution in Exercise 1.2.5. Once this is known for the fundamental solution, it follows for all. The simple proof is an exercise in my book [Rauch, 1991, pp. 164–165].

The analysis of Exercise 1.3.3 does not apply to the fundamental solution since the latter does not have finite energy. However it belongs to $C^j(\mathbf{R} : H^{s-j}(\mathbb{R}))$ for all $s < 1/2$ and $j \in \mathbb{N}$. The next result provides a good replacement for (1.3.4).

Exercise 1.3.4. *Suppose that u is the fundamental solution of the Klein–Gordon equation (1.1.6) and that $s < 1/2$. If $0 \le \chi \in C^\infty(\mathbb{R})$ is a plateau cutoff supported on the positive half line, that is*

$$\chi(x) = 0 \quad \textit{for} \quad x \le 0 \qquad \textit{and} \quad \chi(x) = 1 \quad \textit{for} \quad x \ge 1\,,$$

then for all $\epsilon > 0$ and $R > 0$ there is a $\delta > 0$ so that

$$\limsup_{t \to \infty} \| \, \chi(R + |x| - (1-\delta)t)\, u(t,x) \, \|_{H^s(\mathbb{R}_x)} < \epsilon\,. \tag{1.3.5}$$

Hints. Prove that

$$\| \, \chi \, u(t) \, \|_{H^s(\mathbb{R})} \le C\Big(\, \|u(0)\|_{H^s(\mathbb{R})} + \|u_t(0)\|_{H^{s-1}(\mathbb{R})} \Big)$$

with C independent of t and the initial data. Conclude that it suffices to prove (1.3.4) with initial data $u(0), u_t(0)$ dense in $H^s \times H^{s-1}$. Take the dense set to be data with Fourier transform in $C_0^\infty(\mathbb{R})$.

These examples illustrate the important observation that the propagation of singularities in solutions and the propagation of the majority of the energy may be governed by different rules. For the Klein–Gordon equation, both answers can be determined from considerations of group velocities.

1.4. Fourier synthesis and rectilinear propagation

For equations with constant coefficients, solutions of the initial value problem are expressed as Fourier integrals. Injecting short wavelength initial data and performing an asymptotic analysis yields the approximations of geometric optics. This is how such approximations were first justified in the nineteenth century. It is also the motivating example for the more general theory. The short wavelength approximations explain the *rectilinear propagation of waves* in homogeneous media. This is the first of the three basic physical laws of geometric optics. It explains, among other things, the formation of shadows. The short wavelength solutions are also the building blocks in the analysis of the laws of reflection and refraction presented in §1.6 and §1.7.

Consider the initial value problem

$$0 \;=\; \Box u \;:=\; \frac{\partial^2 u}{\partial t^2} - \sum_{j=1}^{d} \frac{\partial^2 u}{\partial x_j^2}\,, \qquad u(0,x) = f, \quad u_t(0,x) = g. \tag{1.4.1}$$

Fourier transformation with respect to the x variables yields

$$\partial_t^2 \hat{u}(t,\xi) \;+\; |\xi|^2\,\hat{u}(t,\xi) \;=\; 0\,, \qquad \hat{u}(0,\xi) = \hat{f}\,, \quad \partial_t \hat{u}(0,\xi) = \hat{g}\,.$$

Solve the ordinary differential equations in t to find

$$\hat{u}(t,\xi) \;=\; \hat{f}(\xi)\,\cos t|\xi| + \hat{g}(\xi)\,\frac{\sin t|\xi|}{|\xi|}\,.$$

For $t \neq 0$ the function $\xi \mapsto \sin(t|\xi|)/|\xi|$ is a real analytic function on $\mathbb{R}^d$ with derivatives uniformly bounded on bounded time intervals. Write

$$\cos t|\xi| \;=\; \frac{e^{it|\xi|} + e^{-t|\xi|}}{2}, \qquad \sin t|\xi| \;=\; \frac{e^{it|\xi|} - e^{-t|\xi|}}{2i},$$

to find

(1.4.2) $$\hat{u}(t,\xi)\, e^{ix\xi} \;=\; a_+(\xi)\, e^{i(x\xi - t|\xi|)} \;-\; a_-(\xi)\, e^{i(x\xi + t|\xi|)}\,,$$

with

(1.4.3) $$a_+ \;:=\; \frac{1}{2}\Big(\hat{f} \;+\; \frac{\hat{g}}{i|\xi|}\Big), \qquad a_- \;:=\; \frac{1}{2}\Big(\hat{f} \;-\; \frac{\hat{g}}{i|\xi|}\Big).$$

For each ξ, the right-hand side of (1.4.2) is a linear combination of the plane wave solutions of the wave equation $e^{i(x\xi + t\tau(\xi))}$ with dispersion relation $\tau = \mp|\xi|$ and amplitude $a_\pm(\xi)$. The group velocities associated to $a_\pm$ are

$$\mathbf{v} \;=\; -\nabla_\xi \tau \;=\; -\nabla_\xi(\mp|\xi|) \;=\; \pm\frac{\xi}{|\xi|}\,.$$

The solution is the sum of two terms,

$$u_\pm(t,x) \;:=\; \frac{1}{(2\pi)^{d/2}} \int a_\pm(\xi)\; e^{i(x\xi \mp t|\xi|)}\, d\xi\,.$$

The conserved energy for the spring equation satisfied by $\hat{u}(t,\xi)$ shows that

$$\frac{1}{2}\Big(|\hat{u}_t(t,\xi)|^2 \;+\; |\xi|^2|\hat{u}(t,\xi)|^2\Big)$$
$$=\; |\xi|^2\big(|a_+(\xi)|^2 \;+\; |a_-(\xi)|^2\big) \;=\; \text{independent of } t\,.$$

Integrate $d\xi$ and use $\mathcal{F}\big(\partial u/\partial x_j\big) \;=\; i\xi_j \hat{u}$, and Parseval's theorem to show that the quantity

$$\int |\xi|^2\big(|a_+(\xi)|^2 \;+\; |a_-(\xi)|^2\big)\; d\xi \;=\; \frac{1}{2}\int |u_t(t,x)|^2 \;+\; |\nabla_x u(t,x)|^2\; dx$$

is independent of time for the solutions of the wave equation.

The formula for $a_\pm$ are potentially singular at $\xi = 0$. The energy for the wave equation is expressed in terms of the pair of functions $|\xi| a_\pm(\xi)$. They are given by nonsingular expressions in terms of $|\xi|\hat{f}$ and $\hat{g}$.

There are conservations of all orders. Multiplying the spring energy by $\langle\xi\rangle^{2s}$ and integrating $d\xi$ shows that each of the following quantities is independent of time:

$$\frac{1}{2}\,\|\nabla_{t,x}u(t)\|^2_{H^s(\mathbb{R}^d)} \;=\; \int \langle\xi\rangle^{2s}\;|\xi|^2\;\big(|a_+(\xi)|^2\;+\;|a_-(\xi)|^2\big)\;d\xi\,.$$

Consider initial data a wave packet with wavelength of order ϵ and phase equal to x_1/ϵ,

$$(1.4.4)\quad u^\epsilon(0,x)\;=\;\gamma(x)\,e^{ix_1/\epsilon}\,,\qquad u^\epsilon_t(0,x)\;=\;0\,,\qquad \gamma\;\in\;\bigcap_s H^s(\mathbb{R}^d)\,.$$

The choice $u_t = 0$ postpones dealing with the factor $1/|\xi|$ in (1.4.3). When ϵ is small, the initial value is an envelope or profile γ multiplied by a rapidly oscillating exponential.

Applying (1.4.3) with $g = 0$ and

$$\hat f(\xi)\;=\;\hat u(0,\xi)\;=\;\mathcal{F}\big(\gamma(x)\,e^{ix_1/\epsilon}\big)\;=\;\hat\gamma(\xi-\mathbf{e}_1/\epsilon)$$

yields $u = u_+ + u_-$ with

$$u^\epsilon_\pm(t,x)\;:=\;\frac{1}{2}\,\frac{1}{(2\pi)^{d/2}}\int\hat\gamma(\xi-\mathbf{e}_1/\epsilon)\;e^{i(x\xi\mp t|\xi|)}\,d\xi\,.$$

Analyze u^ϵ_+. The other term is analogous. For ease of reading, the subscript plus is omitted. Introduce

$$\zeta\;:=\;\xi\;-\;\mathbf{e}_1/\epsilon,\qquad \text{so}\qquad \xi\;=\;\frac{\mathbf{e}_1\;+\;\epsilon\zeta}{\epsilon}\,,\qquad \text{and}$$

$$(1.4.5)\qquad u^\epsilon(t,x)\;=\;\frac{1}{2}\,\frac{1}{(2\pi)^{d/2}}\int\hat\gamma(\zeta)\;e^{ix(\mathbf{e}_1+\epsilon\zeta)/\epsilon}\;e^{-it|\mathbf{e}_1+\epsilon\zeta|/\epsilon}\;d\zeta\,.$$

The approximation of geometric optics comes from injecting the first order Taylor approximation,

$$\big|\mathbf{e}_1+\epsilon\zeta\big|\;\approx\;1+\epsilon\zeta_1\,,$$

yielding

$$u^\epsilon_{\text{approx}}\;:=\;\frac{1}{2}\,\frac{1}{(2\pi)^{d/2}}\int\hat\gamma(\zeta)\;e^{ix(\mathbf{e}_1+\epsilon\zeta)/\epsilon}\;e^{-it(1+\epsilon\zeta_1)/\epsilon}\;d\zeta\,.$$

The rapidly oscillating terms $e^{i(x_1-t)/\epsilon}$ do not depend on ζ, so

$$(1.4.6)\quad u_{\text{approx}} = e^{i(x_1-t)/\epsilon}\;A(t,x),\qquad A(t,x) := \frac{1}{2}\,\frac{1}{(2\pi)^{d/2}}\int\hat\gamma(\zeta)\,e^{i(x\zeta-t\zeta_1)}\,d\zeta.$$

Write $x\zeta - t\zeta_1 = (x-t\mathbf{e}_1)\zeta$ to find

$$A(t,x)\;=\;\frac{1}{2}\,\frac{1}{(2\pi)^{d/2}}\int\hat\gamma(\zeta)\,e^{i(x-t\mathbf{e}_1)\zeta}\,d\zeta\;=\;\frac{1}{2}\,\gamma(x-t\mathbf{e}_1)\,.$$

The approximation is a wave packet with envelope A and wavelength ϵ. The wave packet translates rigidly with velocity equal to $\mathbf{e}_1$. The waveform γ is arbitrary. The approximate solution resembles the columnated light from a flashlight. If the support of γ is small, the approximate solution resembles a light ray.

The amplitude A satisfies the transport equation

$$\frac{\partial A}{\partial t} + \frac{\partial A}{\partial x_1} = 0,$$

so it is constant on the *rays* $x = \underline{x} + t\mathbf{e}_1$. The construction of a family of short wavelength approximate solutions of D'Alembert's wave equations requires only the solution of a simple transport equation.

The dispersion relation of the family of plane waves,

$$e^{i(x\xi + \tau t)} = e^{i(x\xi - |\xi| t)},$$

is $\tau = -|\xi|$. The velocity of transport, $\mathbf{v} = (1, 0, \ldots, 0)$, is the group velocity $\mathbf{v} = -\nabla_\xi \tau(\underline{\xi}) = \underline{\xi}/|\underline{\xi}|$ at $\underline{\xi} = (1, 0, \ldots, 0)$. For the opposite choice of sign, the dispersion relation is $\tau = |\xi|$, the group velocity is $-\mathbf{e}_1$, and the rays are the lines $x = \underline{x} - t\mathbf{e}_1$.

Had we taken data with oscillatory factor $e^{ix\xi/\epsilon}$, then the approximate solution would have been the sum of two wave packets with group velocities $\pm\xi/|\xi|$,

$$\frac{1}{2}\Big(e^{i(x\xi - t|\xi|)/\epsilon}\, \gamma\Big(x - t\frac{\xi}{|\xi|}\Big) + e^{i(x\xi + t|\xi|)/\epsilon}\, \gamma\Big(x + t\frac{\xi}{|\xi|}\Big)\Big).$$

The approximate solution (1.4.6) is a function $H(x - t\mathbf{e}_1)$ with $H(x) = e^{ix_1/\epsilon} h(x)$. When h has compact support, or more generally tends to zero as $|x| \to \infty$, the approximate solution is localized and has velocity equal to $\mathbf{e}_1$. The next result shows that when $d > 1$, no exact solution can have this form. In particular the distribution $\delta(x - \mathbf{e}_1 t)$ that is the most intuitive notion of a light ray is *not* a solution of the wave equation or Maxwell's equation.

Proposition 1.4.1. *If $d > 1$, $s \in \mathbb{R}$, $K \in H^s(\mathbb{R}^d)$ and $u = K(x - \mathbf{e}_1 t)$ satisfies $\Box u = 0$, then $K = 0$.*

Exercise 1.4.1. *Prove Proposition 1.4.1.* **Hint.** Prove the following lemma.

Lemma. *If $k \le d$, $s \in \mathbb{R}$, and $w \in H^s(\mathbb{R}^d)$ satisfies $0 = \sum_k^d \partial^2 w/\partial^2 x_j$, then $w = 0$.*

Next, analyze the error in (1.4.6). Reintroduce the subscripts. Then extract the rapidly oscillating factor in (1.4.5) to find exact amplitudes,

$$
\begin{aligned}
u^\epsilon_\pm(t,x) &= e^{i(x_1 \mp t)/\epsilon}\, A^\pm_{\text{exact}}(\epsilon,t,x)\,, \\
A^\pm_{\text{exact}}(\epsilon,t,x) &:= \frac{1}{(2\pi)^{d/2}2} \int \hat\gamma(\zeta)\, e^{ix.\zeta}\, e^{\mp it(|\mathbf{e}_1+\epsilon\zeta|-1)/\epsilon}\, d\zeta\,.
\end{aligned}
\tag{1.4.7}
$$

Proposition 1.4.2. *The exact and approximate solutions of $\square u^\epsilon = 0$ with Cauchy data (1.4.4) are given by*

$$
u^\epsilon = \sum_\pm e^{i(x_1\mp t)/\epsilon}\, A^\pm_{\text{exact}}(\epsilon,t,x)\,, \qquad u^\epsilon_{\text{approx}} = \sum_\pm e^{i(x_1\mp t)/\epsilon}\, \frac{\gamma(x\mp \mathbf{e}_1 t)}{2}\,,
$$

as in (1.4.7) and (1.4.6). The error is $O(\epsilon)$ on bounded time intervals. Precisely, there is a constant $C>0$ so that for all s,ϵ,t,

$$
\Big\| A^\pm_{\text{exact}}(\epsilon,t,x) \;-\; \frac{\gamma(x\mp\mathbf{e}_1 t)}{2}\Big\|_{H^s(\mathbb{R}^N)} \;\le\; C\,\epsilon\,|t|\,\big\|\gamma\big\|_{H^{s+2}(\mathbb{R}^d)}\,.
$$

Proof. It suffices to estimate the error with the plus sign. The definitions yield

$$
A^+_{\text{exact}}(\epsilon,t,x) - \gamma(x-\mathbf{e}_1t)/2 \;=\; C\int \hat\gamma(\zeta)\, e^{ix.\zeta}\,\big(e^{-it(|\mathbf{e}_1+\epsilon\zeta|-1)/\epsilon)} - e^{-it\zeta_1}\big)\, d\zeta\,.
$$

The definition of the $H^s(\mathbb{R}^d)$ norm yields

$$
\begin{aligned}
\Big\|A^+_{\text{exact}}(\epsilon,t,x) \;-\; \gamma(x-\mathbf{e}_1t)/2\Big\|_{H^s(\mathbb{R}^N)} \\
= \Big\|\langle\zeta\rangle^s\, \hat\gamma(\zeta)\,\big(e^{-it(|\mathbf{e}_1+\epsilon\zeta|-1)/\epsilon} - e^{-it\zeta_1}\big)\Big\|_{L^2(\mathbb{R}^N)}\,.
\end{aligned}
\tag{1.4.8}
$$

Taylor expansion yields for $|\beta|\le 1/2$,

$$
|\,\mathbf{e}_1+\beta\,| \;=\; 1+\beta_1+r(\beta)\,, \qquad |r(\beta)|\le C\,|\beta|^2\,.
$$

Increasing C if needed, the same inequality is true for $|\beta|\ge 1/2$ as well.

Applied to $\beta=\epsilon\zeta$ this yields

$$
\Big| t\big(|\mathbf{e}_1+\epsilon\zeta| - 1\big)/\epsilon - t\,\zeta_1\, x_1\Big| \;\le\; C\,\epsilon\,|t|\,|\zeta|^2
$$

so

$$
\Big| e^{-it(|\mathbf{e}_1+\epsilon\zeta|-1)/\epsilon} \;-\; e^{-it\zeta_1}\Big| \;\le\; C\,\epsilon\,|t|\,|\zeta|^2\,.
$$

Therefore,

$$
\Big\|\langle\zeta\rangle^s\,\hat\gamma(\zeta)\,\big(e^{-it(|\mathbf{e}_1+\epsilon\zeta|-1)/\epsilon} - e^{-it\zeta_1}\big)\Big\|_{L^2(\mathbb{R}^d)} \;\le\; C\,\epsilon\,|t|\,\Big\|\langle\zeta\rangle^s|\zeta|^2\,\hat\gamma(\zeta)\Big\|_{L^2}\,.
\tag{1.4.9}
$$

Combining (1.4.8) and (1.4.9) yields the estimate of the proposition. □

The approximation retains some accuracy so long as $t = o(1/\epsilon)$.

The approximation has the following geometric interpretation. One has a superposition of plane waves $e^{i(x\xi - t|\xi|)}$ with $\xi = (1/\epsilon, 0, \ldots, 0) + O(1)$. Replacing ξ by $(1/\epsilon, 0, \ldots, 0)$ and $|\xi|$ by $1/\epsilon$ in the plane waves yields the approximation (1.4.6).

The wave vectors, ξ, make an angle $O(\epsilon)$ with $\mathbf{e}_1$. The corresponding rays have velocities which differ by $O(\epsilon)$ so the rays remain close for times small compared with $1/\epsilon$. For longer times the fact that the group velocities are not parallel is important. The wave begins to spread out. Parallel group velocities are a reasonable approximation for times $t = o(1/\epsilon)$.

The example reveals several scales of time. For times $t \ll \epsilon$, u and its gradient are well approximated by their initial values. For times $\epsilon \ll t \ll 1$ $u \approx e^{i(x-t)/\epsilon} a(0,x)$. The solution begins to oscillate in time. For $t = O(1)$ the approximation $u \approx A(t,x)\, e^{i(x-t)/\epsilon}$ is appropriate. For times $t = O(1/\epsilon)$ the approximation ceases to be accurate. Refined approximations valid on this longer time scale are called *diffractive geometric optics*. The reader is referred to [Donnat, Joly, Métiver, and Rauch, 1995–1996] for an introduction in the spirit of Chapters 7–8.

It is typical of the approximations of geometric optics that

$$\Box\big(u_{\text{approx}} - u_{\text{exact}}\big) \;=\; \Box u_{\text{approx}} = O(1)$$

is not small. The error $u_{\text{approx}} - u_{\text{exact}} = O(\epsilon)$ is smaller by a factor of ϵ. The residual $\Box u_{\text{approx}}$ oscillates on the scale ϵ, and after applying $\Box^{-1}$ it is smaller by a factor ϵ.

The analysis just performed can be carried out without fundamental change for initial oscillations with nonlinear phase. An excellent description, including the phase shift on crossing a focal point, can be found in [Hörmander 1983, §12.2].

Next the approximation is pushed to higher accuracy with the result that the residuals can be reduced to $O(\epsilon^N)$ for any N. Taylor expansion to higher order yields

$$|\mathbf{e}_1 + \eta| \;=\; 1 + \eta_1 + \sum_{|\alpha|\ge 2} c_\alpha \eta^\alpha\,, \qquad |\eta| < 1, \tag{1.4.10}$$

so

$$\big(|\mathbf{e}_1 + \epsilon\zeta| - 1\big)/\epsilon \;\sim\; \zeta_1 \;+\; \sum_{|\alpha|\ge 2} \epsilon^{|\alpha|-1}\, c_\alpha\, \zeta^\alpha,$$

$$e^{it(|\mathbf{e}_1+\epsilon\zeta|-1)/\epsilon} \;\sim\; e^{it\zeta_1}\, e^{\sum_{|\alpha|\ge 2} it\epsilon^{|\alpha|-1}\, c_\alpha\zeta^\alpha} \;\sim\; e^{it\zeta_1}\Big(1 + \sum_{j\ge 1} \epsilon^j\, h_j(t,\zeta)\Big)\,.$$

Here, $h_j(t,\zeta)$ is a polynomial in t,ζ. Injecting in the formula for $A_{\text{exact}}(\epsilon,t,x)$ yields an expansion

$$(1.4.11)\qquad \begin{aligned} A_{\text{exact}}(\epsilon,t,x) &\sim A_0(t,x)+\epsilon\,A_1(t,x)+\epsilon^2 A_2(t,x)+\cdots\,,\\ A_0(t,x)&=\gamma(x-\mathbf{e}_1t)/2, \end{aligned}$$

$$(1.4.12)\qquad \begin{aligned} A_j &= \frac{1}{(2\pi)^{-d/2}\,2}\int \hat\gamma(\zeta)\;e^{i(x\zeta-t\zeta_1)}\;h_j(t,\zeta)\;d\zeta\\ &= \frac{1}{2}\Big(h_j(t,\partial/i)\gamma\Big)(x-\mathbf{e}_1t)\,. \end{aligned}$$

The series is asymptotic as $\epsilon\to 0$ in the sense of Taylor series. For any s,N, truncating the series after N terms yields an approximate amplitude which differs from A_{exact} by $O(\epsilon^{N+1})$ in H^s *uniformly on compact time intervals.*

Exercise 1.4.2. *Compute the precise form of the first corrector* a_1.

Formula (1.4.11) implies that if the Cauchy data are supported in a set $\mathcal{O}$, then the amplitudes A_j are all supported in the tube of rays

$$(1.4.13)\qquad \mathcal{T}\;:=\;\Big\{(t,x)\;:\;x=\underline{x}+t\mathbf{e}_1,\quad \underline{x}\in\mathcal{O}\Big\}\,.$$

Warning 1. Though the A_j are supported in this tube, it is not true, when $d\ge 2$, that $A^\epsilon_{\text{exact}}$ is supported in the tube. If it were, then u^ϵ would be supported in the tube. When $d\ge 2$, the function $u=0$ is the only solution of D'Alembert's equation with support in a tube of rays with compact cross section (see Exercise 5.2.9).

Warning 2. By a closer inspection of (1.4.11) or by the analysis after Exercise 5.2.9, one can show that $\|A_j(t,\cdot)\|_{L^\infty}\sim (j!)^{-1}\sum_{|\beta|\le 2j}\|\partial^\beta\gamma\|_{L^\infty}$. So, for a typical analytic γ, the series $\sum\epsilon^j A_j$ have terms of size $\epsilon^j C^j(2j)!/j!$ so they diverge no matter how small is ϵ. For a nonanalytic γ, for example $\gamma\in C_0^\infty$, matters are worse still. *The series* $\sum\epsilon^j A_j$ *is a divergent series that gives an accurate asymptotic expansion as* $\epsilon\to 0$. It is a nonconvergent Taylor expansion of $A_{\text{exact}}(\epsilon,t,x)$.

To analyze the oscillatory initial value problem with $u(0)=0$, $u_t(0)=\beta(x)\,e^{ix_1/\epsilon}$ requires one more idea to handle the contributions from $\xi\approx 0$ in the expression

$$u(t,x)\;=\;(2\pi)^{-d/2}\int\frac{\sin t|\xi|}{|\xi|}\;\hat\beta\Big(\xi-\frac{\mathbf{e}_1}{\epsilon}\Big)\;e^{ix\xi}\;d\xi\,.$$

Choose $\chi\in C_0^\infty(\mathbb{R}^d_\xi)$ with $\chi=1$ on a neighborhood of $\xi=0$. The cutoff integrand is equal to

$$\chi(\xi)\,\frac{\sin t|\xi|}{|\xi|}\;\frac{1}{\langle\xi-\mathbf{e}_1/\epsilon\rangle^s}\;k_s(\xi-\mathbf{e}_1/\epsilon)\,e^{ix\xi}\,,\quad k_s(\xi)\;:=\;\langle\xi\rangle^s\,\hat\beta(\xi)\;\in\;L^2(\mathbb{R}^d_\xi)\,.$$

The $\sin t|\xi|/|\xi|$ factor is $\le |t|$. For ϵ small, the distance of $\mathbf{e}_1/\epsilon$ to the support of χ is $\ge C/\epsilon$. Therefore,

$$\Big\| \chi(\xi)\, \frac{\sin t|\xi|}{|\xi|}\, \frac{1}{\langle \xi - \mathbf{e}_1/\epsilon\rangle^s} \Big\|_{L^\infty(\mathbb{R}^d_\xi)} \le C_s\, |t|\, \epsilon^s\,, \qquad 0 < \epsilon \le 1\,.$$

It follows that

$$\Big\| \chi(\xi)\, \frac{\sin t|\xi|}{|\xi|}\, \frac{1}{\langle \xi - \mathbf{e}_1/\epsilon\rangle^s}\, k_s(\xi - \mathbf{e}_1/\epsilon) \Big\|_{L^2(\mathbb{R}^d)} \le C_s\, |t|\, \epsilon^s\, \big\| k_s \big\|_{L^2(\mathbb{R}^d)}\,,$$

with s arbitrarily large. The small frequency contribution is negligible in the limit $\epsilon \to 0$. It is removed with a cutoff as above, and then the analysis away from $\xi = 0$ proceeds by decomposition into plane wave as in the case with $u_t(0) = 0$. It yields left and right moving waves with the same phases as before.

Exercise 1.4.3. *Solve the Cauchy problem for the anisotropic wave equation,* $u_{tt} = u_{xx} + 4u_{yy}$ *with initial data given by*

$$u^\epsilon(0,x) = \gamma(x)\, e^{ix\xi/\epsilon}\,, \qquad u^\epsilon_t(0,x) = 0\,, \qquad \gamma \in \bigcap_s H^s(\mathbb{R}^d)\,.$$

Find the leading term in the approximate solution to u_+. *In particular, find the velocity of propagation as a function of* ξ. **Discussion.** The velocity is equal to the group velocity from §1.3.

1.5. A cautionary example in geometric optics

A typical science text discussion of a mathematics problem involves simplifying the underlying equations. The usual criterion applied is to ignore terms which are small compared to other terms in the equation. It is striking that in many of the problems treated under the rubric of geometric optics, such an approach can lead to completely inaccurate results. It is an example of an area where more careful mathematical consideration is not only useful but necessary.

Consider the initial value problems

$$\partial_t u^\epsilon + \partial_x u^\epsilon + u^\epsilon = 0\,, \qquad u^\epsilon\big|_{t=0} = a(x)\cos(x/\epsilon)$$

in the limit $\epsilon \to 0$. The function a is assumed to be smooth and to vanish rapidly as $|x| \to \infty$, so the initial value has the form of a wave packet. The initial value problem is uniquely solvable, and the solution depends continuously on the data. The exact solution of the general problem

$$\partial_t u + \partial_x u + u = 0\,, \qquad u\big|_{t=0} = f(x)\,,$$

is $u(t,x) = e^{-t} f(x-t)$, so the exact solution u^ϵ is

$$u^\epsilon(t,x) = e^{-t}\, a(x-t)\, \cos((x-t)/\epsilon)\,.$$

In the limit as $\epsilon \to 0$, one finds that both $\partial_t u^\epsilon$ and $\partial_x u^\epsilon$ are $O(1/\epsilon)$ while $u^\epsilon = O(1)$ is negligibly small in comparison. Dropping this small term leads to the simplified equation for an approximation v^ϵ,

$$\partial_t v^\epsilon + \partial_x v^\epsilon = 0, \qquad v^\epsilon\big|_{t=0} = a(x)\cos(x/\epsilon).$$

The solution is

$$v^\epsilon(t,x) = a(x-t)\cos\big((x-t)/\epsilon\big),$$

which misses the exponential decay. It is *not* a good approximation except for $t \ll 1$. The two large terms compensate so that the small term is not negligible compared to their sum.

1.6. The law of reflection

Consider the wave equation $\Box u = 0$ in the half-space $\mathbb{R}^d_- := \{x_1 \le 0\}$. At $\{x_1 = 0\}$ a boundary condition is required. The condition encodes the physics of the interaction with the boundary.

Since the differential equation is of second order, one might guess that two boundary conditions are needed as for the Cauchy problem. An analogy with the Dirichlet problem for the Laplace equation suggests that one condition is required.

A more revealing analysis concerns the case of dimension $d = 1$. D'Alembert's formula shows that at all points of space-time, the solution consists of the sum of two waves, one moving toward the boundary and the other toward the interior. The waves approaching the boundary will propagate to the edge of the domain. At the boundary one does not know what values to give to the waves which move into the domain. The boundary condition must give the value of the incoming wave in terms of the outgoing wave. That is one boundary condition.

Factoring

$$\partial_t^2 - \partial_x^2 = (\partial_t - \partial_x)(\partial_t + \partial_x) = (\partial_t + \partial_x)(\partial_t - \partial_x)$$

shows that $(\partial_t - \partial_x)(u_t + u_x) = 0$, so $u_t + u_x$ is transported to the left. Similarly, $u_t - u_x$ moves to the right. Thus from the initial conditions, $u_t - u_x$ is determined everywhere in $x \le 0$, including the boundary $x = 0$. The boundary condition at $\{x = 0\}$ must determine $u_t + u_x$. The conclusion is that half of the information needed to find all the first derivatives is already available and one needs only one boundary condition.

Consider the Dirichlet condition,

$$u(t,x)\big|_{x_1=0} = 0. \tag{1.6.1}$$

Differentiating (1.6.1) with respect to t shows that $u_t(t,0) = 0$, so at $t = 0$ $(u_t + u_x) = -(u_t - u_x)$. The incoming wave at the boundary has amplitude equal to -1 times the amplitude of the outgoing wave.

We next analyze the mixed initial boundary value problem for a function $u(t,x)$ defined in $x_1 \le 0$,

$$\text{(1.6.2)} \qquad \Box u = 0, \qquad u\big|_{x_1=0} = 0, \qquad u(0,x) = f\,, \qquad u_t(0,x) = g\,.$$

If the data are supported in a compact subset of $\mathbb{R}^d_-$, then for small time the support of the solution does not meet the boundary. When waves hit the boundary, they are reflected. We analyze this reflection process.

Uniqueness of solutions and finite speed of propagation for (1.6.2) are both consequences of a local energy identity. A function is a solution if and only if the real and imaginary parts are solutions. Thus it suffices to treat the real case for which

$$u_t\,\Box u \;=\; \partial_t e - \sum_{j\ge 1} \partial_j(u_t \partial_j u)\,, \qquad e := \frac{u_t^2 + |\nabla_x u|^2}{2}\,.$$

Denote by Γ a backward light cone

$$\Gamma \;:=\; \Big\{(t,x)\;:\;|x - \underline{x}|^2 < \underline{t} - t\Big\}$$

and by $\tilde\Gamma$ the part in $\{x_1 < 0\}$,

$$\tilde\Gamma \;:=\; \Gamma \,\cap\, \big\{x_1 < 0\big\}\,.$$

For any $0 \le s < \underline{t}$, the section at time s is denoted

$$\tilde\Gamma(s) \;:=\; \tilde\Gamma \,\cap\, \big\{t = s\big\}\,.$$

Both uniqueness and finite speed are consequences of the following estimate.

Proposition 1.6.1. *If u is a smooth solution of (1.6.2), then for $0 < t < \underline{t}$,*

$$\phi(t) \;:=\; \int_{\tilde\Gamma(t)} e(t,x)\;dx$$

is a nonincreasing function of t.

Proof. Translating the time, if necessary, it suffices to show that for $s > 0$, $\phi(s) \le \phi(0)$.

In the identity

$$0 \;=\; \int_{\tilde\Gamma\cap\{0\le t\le s\}} u_t\,\Box u\;dt\,dx\,,$$

integrate by parts to find integrals over four distinct parts of the boundary. The tops and bottoms contribute $\phi(s)$ and $-\phi(0)$, respectively. The intersection of $\tilde{\Gamma}(s)$ with $x_1 = 0$ yields

$$\int_{\tilde{\Gamma}(s)\cap\{x_1=0\}} u_t\, \partial_1 u\; dt\, dx_2 \cdots dx_d\,.$$

The Dirichlet condition implies that $u_t = 0$ on this boundary, so the integral vanishes.

The contribution of the sides $|x - \underline{x}| = \underline{t} - t$ yield an integral of

$$n_0\, e \;+\; \sum_{j=1}^{d} n_j\, u_t\, \partial_j u\,,$$

where $(n_0, n_1, n_2, \dots, n_d)$ is the outward unit normal. Since the cone has sides of slope one,

$$n_0 \;=\; \Big(\sum_{j=1}^{d} n_j^2\Big)^{1/2} \;=\; \frac{1}{\sqrt{2}}\,.$$

The Cauchy–Schwarz inequality yields

$$\Big|\sum_{j=1}^{d} n_j\, u_t\, \partial_j u\Big| \;\le\; \frac{1}{\sqrt{2}}\, |u_t||\nabla_x u| \;\le\; \frac{1}{\sqrt{2}}\, e\,.$$

Thus the integrand from the contributions of sides is nonnegative, so the integral over the sides is nonnegative.

Combining yields

$$0 \;=\; \int_{\tilde{\Gamma}\cap\{0\le t\le s\}} u_t\, \Box u\; dt\, dx \;\ge\; \phi(s) - \phi(0)\,,$$

completing the proof. □

1.6.1. The method of images. Introduce the notations

$$x = (x_1, x'), \quad x' := (x_2, \dots, x_d), \qquad \xi = (\xi_1, \xi'), \quad \xi' := (\xi_2, \dots, \xi_d).$$

Definitions 1.6.2. *A function f on $\mathbb{R}^{1+d}$ is* **even** *(resp.,* **odd***) in x_1 when*

$$f(t, x_1, x') = f(t, -x_1, x'), \qquad \textit{respectively,} \quad f(t, -x_1, x') = -f(t, x_1, x').$$

Define the **reflection operator** *R by*

$$(Rf)(t, x_1, x') \;:=\; f(t, -x_1, x')\,.$$

The even (resp., odd) parts of a function f are defined by

$$\frac{f + Rf}{2}, \qquad \textit{respectively,} \qquad \frac{f - Rf}{2}.$$

Proposition 1.6.3. i. *If $u \in C^\infty(\mathbb{R}^{1+d})$ satisfies $\Box u = 0$ and is odd in x_1, then the restriction of u to $\{x_1 \le 0\}$ is a smooth solution of $\Box u = 0$ satisfying the Dirichlet boundary condition (1.6.1).*

ii. *Conversely, if $u \in C^\infty(\{x_1 \le 0\})$ is a smooth solution of $\Box u = 0$ satisfying (1.6.1), define $\tilde u$ to be the odd extension of u to $\mathbb{R}^{1+d}$. Then $\tilde u$ is a smooth odd solution of $\Box \tilde u = 0$.*

Proof. i. Setting $x_1 = 0$ in the identity $u(t, x_1, x') = -u(t, -x_1, x')$ shows that (1.6.1) is satisfied.

ii. First prove by induction on n that

$$\forall n \ge 0, \qquad \frac{\partial^{2n} u}{\partial^{2n} x_1}\bigg|_{x_1=0} = 0 . \tag{1.6.3}$$

The case $n = 0$ is (1.6.1).

Since the derivatives ∂_t and ∂_j for $j > 1$ are parallel to the boundary along which $u = 0$, it follows that u_{tt} and $\partial_j^2 u$ with $j > 1$ vanish at $x_1 = 0$. The equation $\Box u = 0$ implies

$$\frac{\partial^2 u}{\partial x_1^2} = \frac{\partial^2 u}{\partial t^2} - \sum_{j=2}^{d} \frac{\partial^2 u}{\partial x_j^2}.$$

The right-hand side vanishes on $\{x_1 = 0\}$ proving the case $n = 1$.

If the case $n \ge 1$ is known, consider $v := \partial_1^{2n} u$. It satisfies the wave equation in $x_1 \le 0$ and, by the inductive hypothesis, satisfies the Dirichlet boundary condition at $x_1 = 0$. The case $n = 1$ applied to v proves the case $n + 1$. This completes the proof of (1.6.3).

It is not hard to prove using Taylor's theorem that (1.6.3) is a necessary and sufficient for the odd extension $\tilde u$ to belong to $C^\infty(\mathbb{R}^{1+d})$. The equation $\Box \tilde u = 0$ for $x_1 \ge 0$ follows from the equation in $x_1 \le 0$ since $\Box \tilde u$ is odd. $\Box$

Example 1.6.1. Suppose that $d = 1$ and that $f \in C_0^\infty(]-\infty, 0[)$ so that for $0 \le t$ small $u = f(x - t)$ is a solution of the wave equation supported to the left of and approaching the boundary $x_1 = 0$. To describe the continuation as a solution satisfying the Dirichlet condition, use the method of images as follows. The solution in $\{x < 0\}$ is the restriction to $x < 0$ of an odd solution of the wave equation. For $t < 0$ the odd extension is equal to the given function in $x < 0$ and to minus its reflection in $\{x > 0\}$,

$$\tilde u = f(x - t) - f(-x - t) .$$

The formula on the right is the unique odd solution of the wave equation that is equal to u in $\{t < 0\} \cap \{x < 0\}$. The solution u is the restriction of $\tilde u$ to $x < 0$.

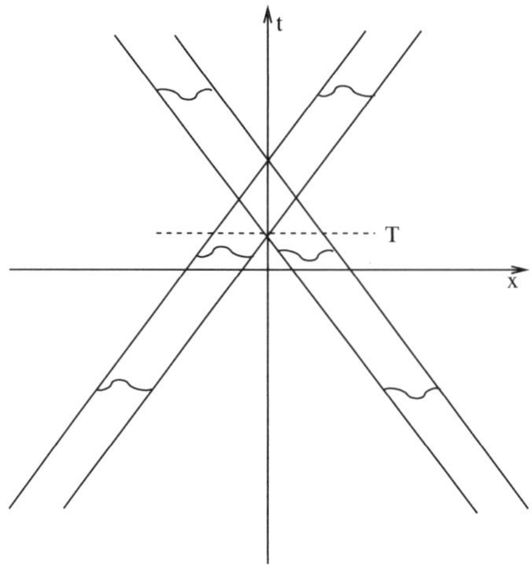

FIGURE 1.6.1. Reflection in dimension $d = 1$

An example is sketched in the Figure 1.6.1. In $\mathbb{R}^{1+1}$ one has an odd solution of the wave equation. There is a rightward moving wave with positive profile and a leftward moving wave with negative profile equal to -1 times the reflection of the first.

Viewed from $x < 0$, there is a wave with positive profile which arrives at the boundary at time T. At that time a leftward moving wave seems to emerge from the boundary. It is the reflection of the wave arriving at the boundary. If the wave arrives at the boundary with amplitude a on an incoming ray, the reflected wave on the reflected ray has amplitude $-a$. The *coefficient of reflection* is equal to -1 (see Figure 1.6.1).

Example 1.6.2. Suppose that $d = 3$ and in $t < 0$ one has an outgoing spherically symmetric wave centered at a point $\underline{x}$ with $\underline{x}_1 < 0$.[2] Until it reaches the boundary, the boundary condition does not play a role. The reflection is computed by extending the incoming wave to an odd solution consisting of the given solution and its negative in mirror image. The moment when the original wave reaches the boundary from the left, its image arrives from the right.

[2]The smooth rotationally symmetric solutions u of $\square_{1+3} u = 0$ centered at the origin are given by (see [Rauch, 1991])

$$u(t,x) = \frac{f(t+|x|) - f(t-|x|)}{|x|}, \quad \text{when} \quad x \neq 0, \qquad u(t,0) = 2f'(t),$$

where $f \in C^\infty(\mathbb{R})$ is arbitrary. Equivalently, $ru(t,r)$ is an odd solution of $\square_{1+1}\big(ru(t,r)\big) = 0$.

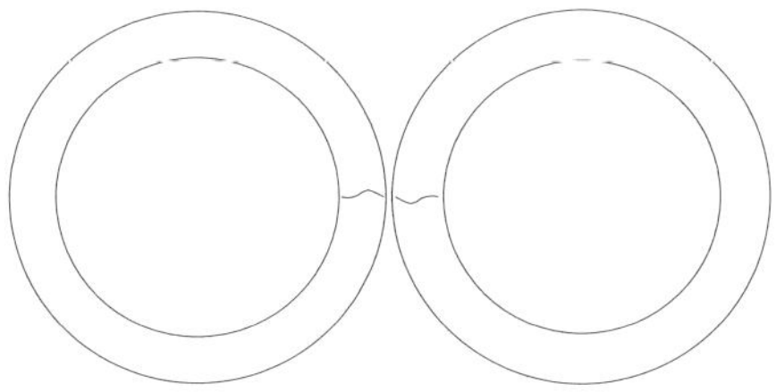

FIGURE 1.6.2. Spherical wave arrives at the boundary

In Figure 1.6.2 the wave on the left has positive profile; and that on the right, a negative profile.

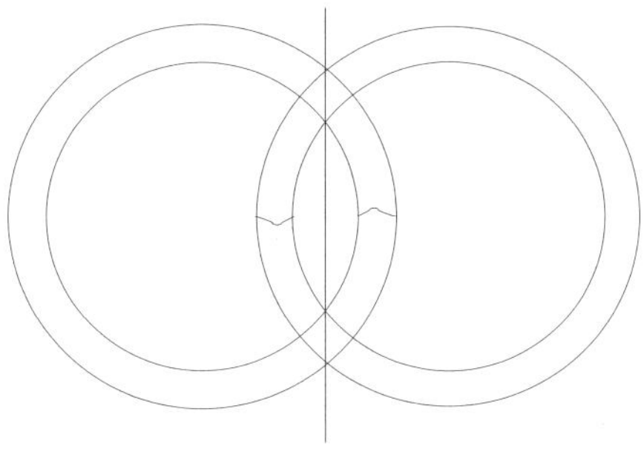

FIGURE 1.6.3. Spherical wave with reflection

In Figure 1.6.3. the middle line represents the boundary. Viewed from $x_1 < 0$, the wave on the left disappears into the boundary and a reflected spherical wave emerges with profile flipped. The profiles of outgoing spherical waves in three-space preserve their thickness and shape. They decrease in amplitude as time increases.

1.6.2. The plane wave derivation. In many texts you will find a derivation that goes as follows. Begin with the plane wave solution

$$e^{i(x\xi - t|\xi|)}, \qquad \xi \in \mathbb{R}^d.$$

The solution is everywhere of modulus one, so it cannot satisfy the Dirichlet boundary condition.

Seek a solution of the initial boundary value problem which is a sum of two plane waves,

$$e^{i(x\xi - t|\xi|)} + A\, e^{i(x\eta + t\sigma)}, \qquad A \in \mathbb{C}.$$

To satisfy the wave equation, one must have $\sigma^2 = |\eta|^2$. In order that the plane waves sum to zero at $x_1 = 0$, it is necessary and sufficient that $\eta' = \xi'$, $\sigma = -|\xi|$, and $A = -1$. Since $\sigma^2 = |\eta|^2$, it follows that $|\eta| = |\xi|$, so

$$\eta = (\pm\xi_1, \xi_2, \ldots, \xi_d).$$

The sign + yields the solution $u = 0$. Denote

$$\tilde{x} := (-x_1, x_2, \ldots, x_d), \qquad \tilde{\xi} := (-\xi_1, \xi_2, \ldots, \xi_d).$$

The sign $-$ yields the interesting solution

$$e^{i(x\xi - t|\xi|)} - e^{i(x\tilde{\xi} - t|\tilde{\xi}|)},$$

which is twice the odd part of $e^{i(x\xi - t|\xi|)}$.

The textbook interpretation of the solution is that $e^{i(x\xi - t|\xi|)}$ is a plane wave approaching the boundary $x_1 = 0$, and $e^{i(x\tilde{\xi} - t|\tilde{\xi}|)}$ moves away from the boundary. The first is an incident wave and the second is a reflected wave. The factor $A = -1$ is the reflection coefficient. The direction of motions are computed using the group velocity computed from the dispersion relation.

Both waves are of infinite extent and of modulus one everywhere in space-time. They have finite energy density but infinite energy. They both meet the boundary at all times. It is questionable to think of either one as incoming or reflected. The next subsection shows that there are localized waves which are clearly incoming and reflected waves with the property that when they interact with the boundary, the local behavior resembles the sum of plane waves just constructed.

To analyze reflections for more general mixed initial boundary value problems, wave forms more general than plane waves need to be included. All solutions of the form $e^{i(x\xi + t\tau)}$ with ξ', τ real and $\operatorname{Im} \xi_1 \le 0$ must be considered. When $\operatorname{Im} \xi_1 < 0$, the associated waves are localized near the boundary. The Rayleigh waves in elasticity are a classic example. They carry the devastating energy of earthquakes. Waves of this sort are needed to analyze total reflection described at the end of §1.7. The reader is referred to [Benzoni-Gavage and Serre, 2007], [Chazarain and Piriou, 1982], [Taylor, 1981], [Hörmander, 1983, v.II], and [Sakamoto, 1982] for more information.

1.6.3. Reflected high frequency wave packets. Consider functions that for small time are equal to high frequency solutions from §1.3,

$$u^\epsilon = e^{i(x\xi - t|\xi|)/\epsilon}\, a(\epsilon, t, x), \quad a(\epsilon, t, x) \sim a_0(t, x) + \epsilon\, a_1(t, x) + \cdots, \tag{1.6.4}$$

with

$$\xi = (\xi_1, \xi_2, \ldots, \xi_d), \qquad \xi_1 > 0.$$

Then $a_0(t,x) = h(x - t\xi/|\xi|)$ is constant on the rays $\underline{x} + t\xi/|\xi|$. If the Cauchy data are supported in a set $\mathcal{O} \Subset \{x_1 < 0\}$, then the amplitudes a_j are supported in the tube of rays

$$\mathcal{T} := \Big\{(t, x) \ : \ x = \underline{x} + t\xi/|\xi|, \quad \underline{x} \in \mathcal{O}\Big\}. \tag{1.6.5}$$

Finite speed shows that the wave as well as the geometric optics approximation stays strictly to the left of the boundary for small $t > 0$.

The method of images computes the reflection. Define v^ϵ to be the reversed mirror image solution,

$$v^\epsilon(t, x_1, x_2, \ldots, x_d) \ := \ -u^\epsilon(t, -x_1, x_2, \ldots, x_d) .$$

The solution of the Dirichlet problem is then equal to the restriction of $u^\epsilon + v^\epsilon$ to $\{x_1 \le 0\}$.

Then

$$v^\epsilon = -\, e^{i(\tilde{x}\xi - |\xi| t)/\epsilon}\, h(\tilde{x} - t\xi/|\xi|) + \text{h.o.t.} = e^{i(x\tilde{\xi} - |\tilde{\xi}| t)/\epsilon}\, (-Rh)(x - t\tilde{\xi}/|\tilde{\xi}|) + \text{h.o.t} .$$

To leading order, $u^\epsilon + v^\epsilon$ is equal to

$$e^{i(x\xi - t|\xi|)/\epsilon}\, h(x - t\xi/|\xi|) \ - \ e^{i(x\tilde{\xi} - |\tilde{\xi}| t)/\epsilon}\, (Rh)(x - t\tilde{\xi}/|\tilde{\xi}|) . \tag{1.6.6}$$

The wave represented by u^ϵ has a leading term which moves with velocity $\xi/|\xi|$. The wave corresponding to v^ϵ has a leading term with velocity $\tilde{\xi}/|\tilde{\xi}|$ that comes from $\xi/|\xi|$ by reversing the first component. At the boundary $x_1 = 0$, the tangential components of $\xi/|\xi|$ and $\tilde{\xi}/|\tilde{\xi}|$ are equal and their normal components are opposite. The directions are related by the standard law that the angle of incidence equals the angle of reflection. The amplitude of the reflected wave v^ϵ on the reflected ray is equal to -1 times the amplitude of the incoming wave u^ϵ on the incoming wave. This is summarized by the statement that the reflection coefficient is equal to -1.

Suppose that $\underline{t}, \underline{x}$ is a point on the boundary and $\mathcal{O}$ is a neighborhood of size large compared to the wavelength ϵ and small compared to the scale on which h varies. Then, on $\mathcal{O}$, the solution is approximately equal to

$$e^{i(x\xi - t|\xi|)/\epsilon}\, h(\underline{x} - \underline{t}\xi/|\xi|) \ - \ e^{i(x\tilde{\xi} - t|\tilde{\xi}|)/\epsilon}\, \tilde{h}(\underline{x} - \underline{t}\tilde{\xi}/|\tilde{\xi}|) .$$

This recovers the reflected plane waves of §1.6.2. An observer on such an intermediate scale sees the structure of the plane waves. Thus, even though the plane waves are completely nonlocal, the asymptotic solutions of geometric optics shows that they predict the local behavior at points of reflection.

The method of images also solves the Neumann boundary value problem in a half-space using an *even* mirror reflection in $x_1 = 0$. It shows that for the Neumann condition, the reflection coefficient is equal to 1.

Proposition 1.6.4. i. *If $u \in C^\infty(\mathbb{R}^{1+d})$ is an even solution of $\Box u = 0$, then its restriction to $\{x_1 \le 0\}$ is a smooth solution of $\Box u = 0$ satisfying the Neumann boundary condition*

$$\partial_1 u|_{x_1 = 0} = 0. \tag{1.6.7}$$

ii. *Conversely, if $u \in C^\infty(\{x_1 \le 0\})$ is a smooth solution of $\Box u = 0$ satisfying (1.6.8), then the even extension of u to $\mathbb{R}^{1+d}$ is a smooth even solution of $\Box u = 0$.*

The analogue of (1.6.3) in this case is

$$\forall n \geq 0, \qquad \frac{\partial^{2n+1} u}{\partial x_1^{2n+1}}\bigg|_{x_1=0} = 0\,. \tag{1.6.8}$$

Exercise 1.6.1. *Prove Proposition 1.6.4.*

Exercise 1.6.2. *Prove uniqueness of solutions by the energy method.* **Hint.** Use the local energy identity.

Exercise 1.6.3. *Verify the assertion concerning the reflection coefficient by following the examples above. That is, consider the case of dimension $d = 1$, the case of spherical waves with $d = 3$, and the behavior in the future of a solution which near $t = 0$ is a high frequency asymptotic solution approaching the boundary.*

1.7. Snell's law of refraction

Refraction is the bending of waves as they pass through media whose propagation speeds vary from point to point. The simplest situation is when media with different speeds occupy half-spaces, for example $x_1 < 0$ and $x_1 > 0$. The classical physical situations are when light passes from air to water or from air to glass. Snell observed that for fixed materials, the ratio of the sines of the angles of incidence and refraction $\sin\theta_i/\sin\theta_r$ is independent of the incidence angle. Fermat observed that this would hold if the speed of light were different in the two media and the light chose a path of least time. In that case, the quotient of sines is equal to the ratio of the speeds, c_i/c_r. In this section we derive this behavior for a model problem quite close to the natural Maxwell equations.

The simplified model with the same geometry is

$$u_{tt} - \Delta\, u = 0 \text{ in } x_1 < 0\,, \quad u_{tt} - c^2\,\Delta\, u = 0 \text{ in } x_1 > 0\,, \quad 0 < c < 1\,. \tag{1.7.1}$$

In $x_1 < 0$, the speed is equal to 1 which is greater than the speed c in $x > 0$.[3]

A transmission condition is required at $x_1 = 0$ to encode the interaction of waves with the interface. In the one-dimensional case, there are waves that approach the boundary from both sides. The waves that move from the boundary into the interior must be determined from the waves that arrive from the interior. There are two arriving waves and two departing waves. One needs two boundary conditions.

[3]To see that c is the speed of the latter equation, one can (in order of increasing sophistication) factor the one-dimensional operator $\partial_t^2 - c^2\partial_x^2 = (\partial_t + c\partial_x)(\partial_t - c\partial_x)$, or use the formula for group velocity with dispersion relation $\tau^2 = c^2|\xi|^2$, or prove finite speed using the differential law of conservation of energy or Fritz John's Global Hölmgren Theorem.

We analyze the transmission condition that imposes continuity of u and $\partial_1 u$ across $\{x_1 = 0\}$. Seek solutions of (1.7.1) satisfying the transmission condition

$$u(t,0^-,x') \;=\; u(t,0^+,x')\,, \qquad \partial_1 u(t,0^-,x') \;=\; \partial_1 u(t,0^+,x')\,. \tag{1.7.2}$$

Denote by square brackets the jump

$$\big[u\big](t,x') \;:=\; u(t,0^+,x') \;-\; u(t,0^-,x')\,.$$

The transmission condition is then

$$\big[u\big] \;=\; 0\,, \qquad \big[\partial_1 u\big] \;=\; 0\,.$$

For solutions that are smooth on both sides of the boundary $\{x_1 = 0\}$, the transmission condition (1.7.2) can be differentiated in t or $x_2, \dots, x_d$ to find

$$\Big[\partial^\beta_{t,x'} u\Big] \;=\; 0\,, \qquad \Big[\partial^\beta_{t,x'}\partial_1 u\Big] \;=\; 0\,. \tag{1.7.3}$$

The partial differential equations then imply that in $x_1 < 0$ and $x_1 > 0$, respectively, one has

$$\frac{\partial^2 u}{\partial x_1^2} \;=\; \frac{\partial^2 u}{\partial t^2} \;-\; \sum_{j=2}^{d} \frac{\partial^2 u}{\partial x_j^2}\,, \qquad \frac{\partial^2 u}{\partial x_1^2} \;=\; \frac{1}{c^2}\frac{\partial^2 u}{\partial t^2} \;-\; \sum_{j=2}^{d} \frac{\partial^2 u}{\partial x_j^2}.$$

Therefore at the boundary

$$\left[\frac{\partial^2 u}{\partial x_1^2}\right] \;=\; \left(1 - \frac{1}{c^2}\right)\frac{\partial^2 u}{\partial t^2}\,,$$

the second derivative $\partial_1^2 u$ is expected to be discontinuous at $\{x_1 = 0\}$.

The physical conditions for Maxwell's equations at an air-water or air-glass interface can be analyzed in the same way. In that case, the dielectric constant is discontinuous at the interface.

Define

$$\gamma(x) \;:=\; \begin{cases} 1 & \text{when } x_1 > 0, \\ c^{-2} & \text{when } x_1 < 0\,, \end{cases} \qquad e(t,x) \;:=\; \frac{\gamma\, u_t^2 + |\nabla_x u|^2}{2}\,.$$

From (1.7.1) it follows that solutions suitably small at infinity satisfy

$$\partial_t \int_{x_1<0} e\; dx \;=\; \int u_t(t,0^-,x')\;\partial_1 u(t,0^+,x')\; dx'\,,$$

$$\partial_t \int_{x_1>0} e\; dx \;=\; -\int u_t(t,0^+,x')\;\partial_1 u(t,0^+,x')\; dx'\,.$$

The transmission condition guarantees that the terms on the right compensate exactly so

$$\partial_t \int_{\mathbb{R}^3} e\; dx \;=\; 0\,.$$

This suffices to prove uniqueness of solutions. A localized argument as in §1.6.1, shows that signals travel at most at speed one.

Exercise 1.7.1. *State and prove this finite speed result.*

A function $u(t,x)$ is called *piecewise smooth* if its restriction to $x_1 < 0$ (resp., $x_1 > 0$) has a C^∞ extension to $x_1 \le 0$ (resp., $x_1 \ge 0$). The Cauchy data of piecewise smooth solutions must be piecewise smooth (with the analogous definition for functions of x only). They must, in addition, satisfy conditions analogous to (1.6.3).

Proposition 1.7.1. *If u is a piecewise smooth solution u of the transmission problem, then the partial derivatives satisfy the sequence of compatibility conditions, for all $j \ge 0$,*

$$\begin{aligned}\Delta^j\{u,u_t\}(t,0^-,x_2,x_3) &= (c^2\Delta)^j\{u,u_t\}(t,0^+,x_2,x_3)\,,\\ \Delta^j\partial_1\{u,,u_t\}(t,0^-,x_2,x_3) &= (c^2\Delta)^j\partial_1\{u,u_t\}(t,0^+,x_2,x_3)\,.\end{aligned}$$

ii. *Conversely, if the piecewise smooth f, g satisfy for all $j \ge 0$,*

$$\begin{aligned}\Delta^j\{f,g\}(0^-,x_2,x_3) &= (c^2\Delta)^j\{f,g\}(0^+,x_2,x_3)\,,\\ \Delta^j\partial_1\{f,g\}(0^-,x_2,x_3) &= (c^2\Delta)^j\partial_1\{f,g\}(0^+,x_2,x_3)\,,\end{aligned}$$

then there is a piecewise smooth solution with these Cauchy data.

Proof. i. Differentiating (1.7.2) with respect to t yields

$$\big[\partial_t^j u\big] = 0\,, \qquad \big[\partial_t^j\partial_1 u\big] = 0\,. \tag{1.7.4}$$

Compute for $k \ge 1$,

$$\partial_t^{2k}u = \begin{cases}\Delta^k u & \text{when } x_1<0,\\ (c^2\Delta)^k u & \text{when } x_1>0,\end{cases} \qquad \partial_t^{2k}u_t = \begin{cases}\Delta^k u_t & \text{when } x_1<0,\\ (c^2\Delta)^k u_t & \text{when } x_1>0.\end{cases}$$

The transmission conditions (1.7.4) prove **i.**

The proof of **ii** is technical, interesting, and omitted. One can construct solutions using finite differences almost as in §2.2. The shortest existence proof to state uses the spectral theorem for selfadjoint operators.[4] The general regularity theory for such transmission problems can be obtained by

[4]For those with sufficient background, the Hilbert space is $\mathcal{H} := L^2(\mathbb{R}^d\,;\,\gamma\,dx)$.

$$D(\mathcal{A}) := \Big\{w \in H^2(\mathbb{R}^d_+)\cap H^2(\mathbb{R}^d_-)\ :\ [w] = [\partial_1 w] = 0\Big\}\,,$$

$$\mathcal{A}w := \Delta w \quad \text{in} \quad x_1<0, \qquad \mathcal{A}w := c^2\Delta \quad \text{in} \quad x_1>0\,.$$

Then,

$$(\mathcal{A}u, v)_{\mathcal{H}} = (u, \mathcal{A}v)_{\mathcal{H}} = -\int \nabla u\cdot\nabla v\,dx\,,$$

so $-\mathcal{A} \ge 0$. The Elliptic Regularity Theorem implies that $\mathcal{A}$ is selfadjoint. The regularity theorem is proved, for example, by the methods in [Rauch, 1991, Chapter 10]. The solution of the initial folding them to a boundary value problem and using the results of [Rauch and Massey, 1974] and [Sakamoto, 1982]. □

Next consider the mathematical problem whose solution explains Snell's law. The idea is to send a wave in $x_1 < 0$ toward the boundary and ask how it behaves in the future. Suppose

$$\xi \in \mathbb{R}^d, \qquad |\xi| = 1, \qquad \xi_1 > 0,$$

and consider a short wavelength asymptotic solution in $\{x_1 < 0\}$ as in §1.6.3,

$$\text{(1.7.5)} \quad I^\epsilon \sim e^{i(x\xi - t)/\epsilon}\, a(\epsilon, t, x), \qquad a(\epsilon, t, x) \sim a_0(t, x) + \epsilon\, a_1(t, x) + \cdots,$$

where for $t < 0$ the support of the a_j is contained in a tube of rays with compact cross section and moving with speed ξ. Take a to vanish outside the tube. Since the incoming waves are smooth and initially vanish identically on a neighborhood of the interface $\{x_1 = 0\}$, the compatibilities are satisfied and there is a family of piecewise smooth solutions u^ϵ defined on $\mathbb{R}^{1+d}$. We construct an infinitely accurate description of the family of solutions u^ϵ.

Seek an asymptotic solution that at $\{t = 0\}$ is equal to this incoming wave. A first idea is to find a transmitted wave which continues the incoming wave into $\{x_1\} > 0$.

Seek the transmitted wave in $x_1 > 0$ in the form

$$T^\epsilon \sim e^{i(x\eta + t\tau)/\epsilon}\, d(\epsilon, t, x), \qquad d(\epsilon, t, x) \sim d_0(t, x) + \epsilon\, d_1(t, x) + \cdots.$$

On the interface, the incoming wave oscillates with phase $(x'\xi' - t)/\epsilon$ and the proposed transmitted wave oscillates with phase $(x'\eta' + t\tau)/\epsilon$. In order that there be any chance at all of satisfying the transmission conditions one must take

$$\eta' = \xi', \qquad \tau = -1,$$

so that the two expressions oscillate together. In order that the transmitted wave be an approximate solution on the right with positive velocity, one must have

$$\tau^2 = c^2 |\eta|^2, \qquad \eta_1 > 0.$$

The equation $\tau^2 = c^2|\eta|^2$ implies

$$\eta_1^2 = \frac{\tau^2}{c^2} - |\eta'|^2 = \frac{1}{c^2} - |\xi'|^2, \qquad \text{so} \qquad \eta_1 = \Big(\frac{1}{c^2} - |\xi'|^2\Big)^{1/2} > \xi_1 > 0.$$

Thus,

$$\text{(1.7.6)} \qquad T^\epsilon \sim e^{i(x\eta - t)/\epsilon}\, d(\epsilon, t, x), \qquad \eta = \big((c^{-2} - |\xi'|^2)^{1/2}, \, \xi' \big).$$

---

value problem is

$$u = \cos t\sqrt{-\mathcal{A}}\, f + \frac{\sin t\sqrt{-\mathcal{A}}}{\sqrt{-\mathcal{A}}}\, g.$$

For piecewise H^∞ data, the sequence of compatibilities is equivalent to the data belonging to $\bigcap_j D(\mathcal{A}^j)$.

From section 1.6.3, the leading amplitude d_0 must be constant on the rays $t \mapsto (t\,,\, \underline{x} + c\,t\,\eta/|\eta|)$. To determine d_0, it suffices to know the values $d_0(t, 0^+, x')$ at the interface. One could choose d_0 to guarantee the continuity of u or of $\partial_1 u$, but not both. One cannot construct a good approximate solution consisting of just an incident and transmitted wave.

Add to the recipe a reflected wave. Seek a reflected wave in $x_1 \geq 0$ in the form

$$R^\epsilon \;\sim\; e^{i(x\zeta + t\sigma)/\epsilon}\, b(\epsilon, t, x)\,, \qquad b(\epsilon, t, x) \;\sim\; b_0(t, x) + \epsilon\, b_1(t, x) + \cdots .$$

In order that the reflected wave oscillate with the same phase as the incident wave in the interface $x_1 = 0$, one must have $\zeta' = \xi'$ and $\sigma = -1$. To satisfy the wave equation in $x_1 < 0$ requires $\sigma^2 = |\zeta|^2$. Together these imply $\zeta_1^2 = \xi_1^2$. To have propagation away from the boundary requires $\zeta_1 = -\xi_1$ so $\zeta = \tilde{\xi}$. Therefore,

$$\text{(1.7.7)}\quad R^\epsilon \;\sim\; e^{i(x\tilde{\xi} - t)/\epsilon}\, b(\epsilon, t, x)\,, \qquad b(\epsilon, t, x) \;\sim\; b_0(t, x) + \epsilon\, b_1(t, x) + \cdots .$$

Summarizing, seek

$$v^\epsilon \;=\; \begin{cases} I^\epsilon + R^\epsilon & \text{in} \quad x_1 < 0, \\ \quad T^\epsilon & \text{in} \quad x_1 > 0. \end{cases}$$

The continuity required at $x_1 = 0$ forces

$$\text{(1.7.8)}\quad e^{i(x'\xi' - t)/\epsilon}\, \big(a(\epsilon, t, 0, x') + b(\epsilon, t, 0, x')\big) \;=\; e^{i(x'\xi' - t)/\epsilon}\, d(\epsilon, t, 0, x')\,.$$

The continuity of u and $\partial_1 u$ hold if and only if at $x_1 = 0$ one has

$$\text{(1.7.9)}\quad a \,+\, b \;=\; d\,, \qquad \text{and} \qquad \frac{i\xi_1}{\epsilon} a + \partial_1 a \;-\; \frac{i\xi_1}{\epsilon} b + \partial_1 b = \frac{i\eta_1}{\epsilon} d + \partial_1 d\,.$$

The first of these relations yields

$$\text{(1.7.10)}\qquad \Big(a_j \,+\, b_j \,-\, d_j\Big)_{x_1 = 0} = 0, \qquad j = 0, 1, 2, \ldots .$$

The second relation in (1.7.9) is expanded in powers of ϵ. The coefficients of ϵ^j must match for all all $j \geq -1$. The leading order is ϵ^{-1} and yields

$$\text{(1.7.11)}\qquad \big(a_0 - b_0 - (\eta_1/\xi_1) d_0\big)_{x_1 = 0} = 0\,.$$

Since a_0 is known, the $j = 0$ equation from (1.7.10) together with (1.7.11) is a system of two linear equations for the two unknown b_0, d_0,

$$\begin{pmatrix} -1 & 1 \\ 1 & \eta_1/\xi_1 \end{pmatrix} \begin{pmatrix} b_0 \\ d_0 \end{pmatrix} \;=\; \begin{pmatrix} a_0 \\ a_0 \end{pmatrix}.$$

Since the matrix is invertible, this determines the values of b_0 and d_0 at $x_1 = 0$.

The amplitude b_0 (resp., d_0) is constant on rays with velocity $\tilde{\xi}$ (resp., $c\eta/|\eta|$). Thus the leading amplitudes are determined throughout the half-spaces on which they are defined.

Once these leading terms are known, the ϵ^0 term from the second equation in (1.7.9) shows that on $x_1 = 0$,

$$a_1 - b_1 - d_1 = \text{known}\,.$$

Since a_1 is known, this together with the case $j = 2$ from (1.7.10) suffices to determine b_1, d_1 on $x_1 = 0$. Each satisfies a transport equation along rays that are the analogue of (1.4.12). Thus from the initial values just computed on $x_1 = 0$, they are determined everywhere. The higher order correctors are determined analogously.

Once the b_j, d_j are determined, one can choose b, c as functions of ϵ with the known Taylor expansions at $x = 0$. They can be chosen to have supports in the appropriate tubes of rays and to satisfy the transmission conditions (1.7.9) exactly.

The function u^ϵ is then an infinitely accurate approximate solution in the sense that it satisfies the transmission and initial conditions exactly while the residuals

$$v^\epsilon_{tt} - \Delta\, v^\epsilon := r^\epsilon \quad \text{in} \quad x_1 < 0\,, \qquad v^\epsilon_{tt} - c^2\,\Delta\, v^\epsilon := \rho^\epsilon$$

satisfy for all N, s, T there is a C so that

$$\big\|r^\epsilon\big\|_{H^s([-T,T]\times\{x_1<0\})} + \big\|\rho^\epsilon\big\|_{H^s([-T,T]\times\{x_1>0\})} \le C\,\epsilon^N\,.$$

From the analysis of the transmission problem, it follows that with new constants,

$$\big\|u^\epsilon - v^\epsilon\big\|_{H^s([-T,T]\times\{x_1>0\})} \le C\,\epsilon^N\,.$$

The proposed problem of describing the family of solutions u^ϵ is solved.

The angles of incidence and refraction, θ_i and θ_r, are computed from the directions of propagation of the incident and transmitted waves. From Figure 1.7.1 one finds

$$\sin\theta_i = \frac{|\xi'|}{|\xi|}, \qquad \text{and} \qquad \sin\theta_r = \frac{|\eta'|}{|\eta|} = \frac{|\xi'|}{|\xi|/c}\,.$$

Therefore,

$$\frac{\sin\theta_i}{\sin\theta_r} = \frac{1}{c}\,,$$

is independent of θ_i. The high frequency asymptotic solutions explain Snell's law. This is the last of the three basic laws of geometric optics. The derivation of Snell's law only uses the phases of the incoming and transmitted waves. The phases are determined by the requirement that the restriction of the phases to $x_1 = 0$ are equal. They do not depend on the precise transmission condition that we chose. It is for this reason that the conclusion is the same for the correct transmission problem for Maxwell's equations.

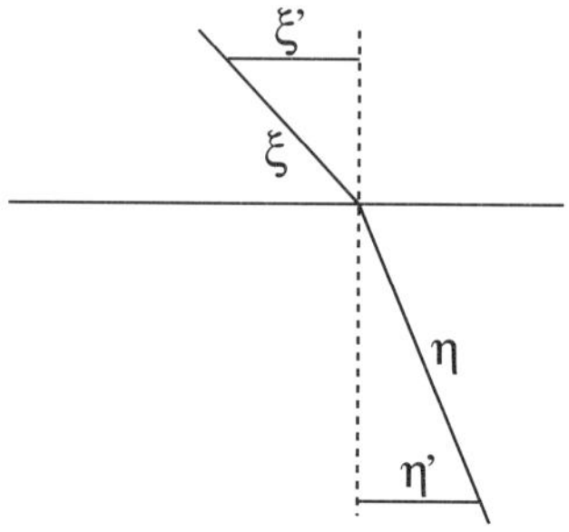

FIGURE 1.7.1.

On a neighborhood $(\underline{t}, \underline{x}) \in \{x_1 = 0\}$ that is small compared to the scale on which a, b, c vary and large compared to ϵ, the solution resembles three interacting plane waves. In science texts one usually computes for which such triples the transmission condition is satisfied in order to find Snell's law. The asymptotic solutions of geometric optics show how to overcome the criticism that the plane waves have modulus independent of (t, x) so cannot reasonably be viewed as either incoming or outgoing.

For a more complete discussion of reflection and refraction, see [Taylor, 1981] and [Benzoni-Gavage and Serre, 2007]. In particular these treat the phenomenon of *total reflection* that can be anticipated as follows. From Snell's law one sees that $\sin \theta_r < 1/c$ and approaches $1/c$ as θ_i approaches $\pi/2$. The refracted rays lie in the cone $\theta_r < \arcsin(1/c)$. Reversing time shows that light rays from below approaching the surface at angles smaller than this critical angle traverse the surface tracing backward the old incident rays. For angles larger than $\arcsin(1/c)$ there is no possible continuation as a ray above the surface. One can show by constructing infinitely accurate approximate solutions that there is total reflection. Below the surface there is a reflected ray with the usual law of reflection. The role of a third wave is played by a boundary layer of thickness $\sim \epsilon$ above which the solution is $O(\epsilon^\infty)$.

Chapter 2

The Linear Cauchy Problem

Hyperbolic initial value problems with constant coefficients are efficiently analyzed using the Fourier transform. Such equations describe problems where the medium is identical at all points. When the physical properties of the medium vary from point to point, the corresponding models have variable coefficients. The initial value problem in such situations is usually not explicitly solvable. Constant coefficient systems also arise as the linearization at constant solutions of translation invariant nonlinear equations.

The key idea of stability or well-posedness, emphasized by Hadamard, is that for a model to be reasonable, small changes in the data, for example the initial data, can only result in small changes in predictions. For a linear equation, this is equivalent to showing that small data yields small solutions. In the linear case, normed vector spaces are often appropriate settings to describe this continuity. In the linear case, continuity is equivalent to showing that the norm of solutions is bounded by at most a constant times a norm of the data.

For hyperbolic problems, for example Maxwell's equations or D'Alembert's wave equation, L^2 norms and associated L^2 Sobolev spaces yield better estimates (for example without loss of derivatives) than L^p or Hölder spaces. Initial data in Sobolev spaces H^s yield solutions with values in H^s, while the analogous statement for C^α or $W^{s,p}$, $p \neq 2$ is false in space dimension greater than one.

The analysis of constant coefficient hyperbolic systems is summarized in Appendix 2.I.

2.1. Energy estimates for symmetric hyperbolic systems

2.1.1. The constant coefficient case. Three classic examples of hyperbolic equations are D'Alembert's equation of vibrating strings, Maxwell's equations of electrodynamics, and Euler's equations of inviscid compressible fluid flow. The first two have constant coefficients. Linearizing the third at a constant state also yields a constant coefficient system.

Maxwell's equations describe electric and magnetic fields $E(t,x)$, $B(t,x)$ that are vector fields defined on $\mathbb{R}^{1+3}$. There are two dynamic equations

$$(2.1.1)\quad E_t = c\,\mathrm{curl}\,B - 4\pi\mathbf{j}\,,\qquad B_t = -c\,\mathrm{curl}\,E\,,\qquad c = 3\times 10^{10}\,\mathrm{cm/sec}.$$

The vector field $\mathbf{j}(t,x)$ is the current density measuring the flow of charge. It is a source term that is assumed to be given. These equations determine E, B from their initial data once $\mathbf{j}$ is known. Not all initial data and sources are physically relevant. The physical solutions are a subset of the dynamics defined by the additional Maxwell equations

$$(2.1.2)\qquad \mathrm{div}\,E = 4\pi\rho \quad\text{and}\quad \mathrm{div}\,B = 0\,,$$

where $\rho(t,x)$ is the charge density.

Taking the divergence of the first equation in (2.1.1) and the time derivative of the first equation in (2.1.2) shows that the *continuity equation*,

$$(2.1.3)\qquad \partial_t\rho \;=\; -\mathrm{div}\,\mathbf{j}\,,$$

follows from the Maxwell system. This equation expresses the *conservation of charge* and must be satisfied by the given source terms ρ and $\mathbf{j}$.

Taking the divergence of (2.1.1) and using (2.1.3) yields

$$\partial_t\,\mathrm{div}\,B \;=\; \partial_t\,\mathrm{div}\,(E - 4\pi\rho) \;=\; 0\,.$$

Thus, when the continuity equation is satisfied, the constraint equations (2.1.2) hold for all time as soon as they are satisfied at time $t = 0$. In summary the sources must satisfy (2.1.3) and the solutions of interest are those that satisfy (2.1.2).[1]

[1] This argument is historically very important. Experiments with charges and currents yielded Ampère's law and Faraday's law of induction,

$$0 \;=\; c\,\mathrm{curl}\,B - 4\pi\mathbf{j}\,,\qquad B_t \;=\; -c\,\mathrm{curl}\,E\,,$$

together with (2.1.2) and (2.1.3). Taking the divergence of the first and using (2.1.2) yields $\rho_t = 0$ showing that the equations are inconsistent in the case of nonstatic ρ. The analysis of the problem of a steadily charging condensor yields conflicting predictions (see section II.18.2 of [Feynman et al., 1963]). Modifying the first equation by inserting an unknown term F on the left of the first equation, one finds that the equations are coherent exactly when $\mathrm{div}\,F = \mathrm{div}\,E_t$. This together with the symmetry of the equations in the pair E, B led Maxwell to propose the equations with E_t on the left of the first equation. This resolves the condensor paradox and other test experiments. It led to Maxwell's prediction of electromagnetic waves verified experimentally by Hertz.

The system (2.1.1) is a symmetric hyperbolic system in the following sense. Introduce the $\mathbb{R}^6$ valued unknown $u := \big(E\,,\,B\big)$. Equation (2.1.1) has the form

$$\frac{\partial u}{\partial t} + \sum_{j=1}^{3} A_j\,\frac{\partial u}{\partial x_j} = f\,, \qquad f := \big(\mathbf{j}\,,\,0\big)\,, \tag{2.1.4}$$

with constant 6×6 matrices A_j.

Exercise 2.1.1. *Compute the matrices A_j. In particular, verify that they are real and symmetric.*

Definition. *A constant coefficient operator*

$$\frac{\partial u}{\partial t} \;+\; \sum_{j=1}^{d} A_j\,\frac{\partial}{\partial x_j}$$

on $\mathbb{R}^d$ is **symmetric hyperbolic** *when the coefficient matrices are hermitian symmetric matrices.*

The importance of symmetry is that it leads to simple L^2 and more generally H^s estimates.

Symmetric systems with constant coefficients are efficiently analyzed using the Fourier transform. The transform and its inverse are given by

$$u(x) \;=\; (2\pi)^{-d/2}\int_{\mathbb{R}^d} e^{ix\xi}\,\hat u(\xi)\;d\xi\,,$$

where

$$\hat u(\xi) \;:=\; (2\pi)^{-d/2}\int_{\mathbb{R}^d} e^{-ix\xi}\,u(\xi)\;d\xi\,.$$

The Fourier transform is also denoted by $\mathcal{F}$.

Consider the case $f=0$ and take the Fourier transform in x to find

$$\partial_t\hat u(t,\xi) \;+\; \sum i\,A_j\,\xi_j\,\hat u(t,\xi) \;=\; 0\,.$$

For ξ fixed, integrate the ordinary differential equation in time to find

$$\hat u(t,\xi) \;=\; e^{-it\sum A_j\xi_j}\,\hat u(0,\xi)\,.$$

The symmetry implies that $\exp(-it\sum A_j\xi_j)$ is a unitary matrix-valued function of t,ξ. Thus for all t,ξ

$$\|\hat u(t,\xi)\|^2 \;=\; \|\hat u(0,\xi)\|^2\,.$$

Example 2.1.1. For Maxwell's equations this asserts that for all ξ,

$$|\hat E(t,\xi)|^2 + |\hat B(t,\xi)|^2 \;=\; \text{independent of time}\,.$$

This expresses the conservation of energy at every frequency.

Integrating $d\xi$ and using the Plancherel theorem implies that the L^2 norm is conserved; that is for all t,

$$\|u(t)\|_{L^2(\mathbb{R}^d)} = \|u(0)\|_{L^2(\mathbb{R}^d)}.$$

For Maxwell's equations this asserts that

$$\int_{\mathbb{R}^d} |E|^2 + |B|^2 \, dx = \text{independent of time}.$$

This is the physical law of *conservation of energy.*

More generally, the Sobolev $H^s(\mathbb{R}^d)$ norms defined for $s \in \mathbb{R}$ by

$$\|v\|^2_{H^s(\mathbb{R}^d)} := \int_{\mathbb{R}^d} (1+|\xi|^2)^s \, |\hat{v}(\xi)|^2 \, d\xi$$

are conserved. When $s \neq 0$, these norms have no natural physical interpretation. They are important in mathematical analysis. These estimates yield the following result.

Proposition 2.1.1. *Suppose that* $L = \partial_t + \sum A_j \partial_j$ *is a constant coefficient symmetric hyperbolic operator. For* $g \in \bigcap_s H^s(\mathbb{R}^d) := H^\infty(\mathbb{R}^d)$ *there is one and only one solution*

$$u \in \bigcap_k C^k((\mathbb{R}\, ;\, H^k(\mathbb{R}^d)) := C^\infty(\mathbb{R}\, ;\, H^\infty(\mathbb{R}^d))$$

of the initial value problem,

$$Lu = 0, \qquad u\big|_{t=0} = g.$$

The solution is given by $\hat{u}(t) = e^{-it\sum A_j \xi_j}\, \hat{g}$. *For all* t, s, $\|u(t)\|_{H^s(\mathbb{R}^d)} = \|u(0)\|_{H^s(\mathbb{R}^d)}$.

A proof of the conservation of the $H^s(\mathbb{R}^d)$ norm without using the Fourier transform is based on local conservation laws. Denote with brackets $\langle\, ,\, \rangle$ the scalar product in $\mathbb{C}^N$, and by ∂ either ∂_t or ∂_j for some j. When A is hermitian symmetric and $u \in C^1$,

$$\begin{aligned} \partial\langle Au\,,\,u\rangle &= \langle A\partial u\,,\,u\rangle + \langle Au\,,\,\partial u\rangle \\ &= \langle A\partial u\,,\,u\rangle + \langle u\,,\,A\partial u\rangle = 2\mathrm{Re}\,\langle A\partial u\,,\,u\rangle, \end{aligned}$$

where the symmetry is used at the second step. Summing shows that if L is the differential operator on the left in (2.1.4), then

$$\partial_t\langle u\,,\,u\rangle + \sum_j \partial_j \langle A_j u\,,\,u\rangle = 2\mathrm{Re}\,\langle Lu\,,\,u\rangle.$$

This is the infinitesimal or differential version of the L^2 conservation law. When $Lu = 0$ and u decays sufficiently rapidly as $x \to \infty$, integrating this identity over $[0,t] \times \mathbb{R}^d$ proves the conservation of the L^2 norm.

Similarly, if $Lu = 0$, then also $Lw = 0$, where $w := \partial_x^\alpha u$. The L^2 conservation shows that $\|\partial_x^\alpha w(t)\|_{L^2(\mathbb{R}^d)}$ is independent of time. Summing over $|\alpha| \le s$ yields an H^s conservation law when s is an integer. For the noninteger case, the L^2 conservation for $w := (1-\Delta_x)^{s/2}u$ gives the desired identity.

2.1.2. The variable coefficient case. Analogous results are valid for variable coefficient operators (for example, Maxwell's equations in a non-homogeneous dielectric) satisfying a symmetry hypothesis. The introduction of this class of operators and the observation that it is ubiquitous in mathematical physics is due to K. O. Friedrichs (1954).

Definition. *In $\mathbb{R}^{1+d}$ introduce coordinates $y=y_0, y_1, \ldots, y_d := t, x_1, \ldots, x_d$. A partial differential operator*

$$L(y,\partial) = \sum_{\mu=0}^{d} A_\mu(y)\,\frac{\partial}{\partial y_\mu} + B(y) \tag{2.1.5}$$

is called **symmetric hyperbolic** *if and only if the following hold:*

i. *The coefficient matrices A_μ, and B have uniformly bounded derivatives,*

$$\forall \alpha, \qquad \sup_{y\in\mathbb{R}^{1+d}} \left\| \partial_y^\alpha \Big(A_\mu(y), B(y) \Big) \right\| < \infty; \tag{2.1.6}$$

ii. *The A_μ are hermitian symmetric valued; and*

iii. *A_0 is strictly positive in the sense that there is a $c > 0$ so that for all y,*

$$A_0(y) \ge c\,I\,. \tag{2.1.7}$$

The L^2 estimate has a generalization to such problems. *For a first version suppose that $A_0 = I$*, and define by

$$G(t) := \sum_j A_j(t,x)\,\partial_j + B(t,x)\,.$$

Denote the $L^2(\mathbb{R}^d)$ scalar product by

$$(f,h) := \int_{\mathbb{R}^d} \langle f(x), h(x)\rangle\, dx = \int_{\mathbb{R}^d} f(x)\cdot\overline{h(x)}\, dx\,.$$

The adjoint differential operator is defined by

$$G^*(t) := -\sum_j A_j^*(t,x)\,\partial_j + B(t,x)^* - \sum_j (\partial_j A_j^*)\,,$$

so for ϕ and ψ belonging to $C_0^\infty(\mathbb{R}^d)$,

$$(G(t)\,\phi\,,\,\psi) = (\phi\,,\,G^*(t)\,\psi)\,.$$

Then

$$G(t) + G(t)^* \;=\; B(y) + B^*(y) - \sum_{j=1}^{3} (\partial_j A_j(y))$$

is multiplication by a uniformly bounded matrix. As an operator on $L^2(\mathbb{R}^d)$ one has

$$\| G(t) + G(t)^* \| \;\le\; 2C\,. \tag{2.1.8}$$

The operator G is nearly antiselfadjoint. If

$$u \in C(\mathbb{R}\,;\, H^1(\mathbb{R}^d)), \qquad u_t \in C(\mathbb{R}\,;\, L^2(\mathbb{R}^d)), \qquad \text{and} \qquad f \;:=\; Lu\,,$$

then

$$u'(t) + G(t)\, u(t) \;=\; f(t). \tag{2.1.9}$$

By hypothesis $\|u(t)\|^2 \in C^1(\mathbb{R})$ and

$$\frac{d}{dt}\, \|u(t)\|^2 \;=\; (u, u') + (u', u)\,.$$

Using (2.1.9) yields

$$\frac{d}{dt}\, \|u(t)\|^2 \;=\; (u, -Gu) + (-Gu, u) + 2\mathrm{Re}\,(u, f)\,. \tag{2.1.10}$$

The Cauchy–Schwarz inequality shows that

$$2\mathrm{Re}\,(u, f) \le 2\, \|u(t)\|\, \|f(t)\|\,, \tag{2.1.11}$$

and the near antisymmetry implies that

$$(u, -Gu) + (-Gu, u) \;=\; -\big(u, (G + G^*)u\big) \le 2C\|u\|^2\,. \tag{2.1.12}$$

The left-hand side of (2.1.10) is $\frac{d}{dt}\, \|u(t)\|^2 \;=\; 2\, \|u(t)\|\, \frac{d}{dt}\, \|u(t)\|$, so,

$$2\, \|u(t)\|\, \frac{d}{dt}\, \|u(t)\| \;\le\; 2C\, \|u(t)\|^2 \;+\; 2\, \|u(t)\|\, \frac{d}{dt}\, \|f(t)\|\,.$$

Where $u(t) \ne 0$, dividing (2.1.10) by $\|u(t)\|$ yields

$$\frac{d\|u(t)\|}{dt} \le C\|u(t)\| + \|f(t)\|\,, \quad \text{equivalently} \quad \frac{d\big(e^{-Ct}\, \|u(t)\|\big)}{dt} \le e^{-Ct}\, \|f(t)\|\,. \tag{2.1.13}$$

If $u(s) \ne 0$ on $[0, t]$, then integrating the second inequality in (2.1.13) yields

$$\|u(t)\| \le e^{Ct}\, \|u(0)\| + \int_0^t e^{C(t-\sigma)}\, \|f(\sigma)\|\, d\sigma\,. \tag{2.1.14}$$

If $u(t)=0$, then (2.1.14) holds. In the remaining case, there is an $s\in[0,t]$ so that $u(s)\neq 0$ on $]s,t]$ and $u(s)=0$. Integrating (2.1.13) from $s+\epsilon$ to t and letting $\epsilon\to 0$ proves that

$$\|u(t)\| \le \int_s^t e^{C(t-\sigma)}\,\|f(\sigma)\|\,d\sigma\,,$$

implying that (2.1.14) holds in this case too. So, (2.1.14) holds for all $t>0$.

We next describe three methods for extending this argument to the case $A_0\neq I$. The first is by reduction to that case. Let $v:=A_0^{1/2}u$. The equation

$$\tilde L\, v \;:=\; A_0^{-1/2}\,L\,A_0^{-1/2}v \;=\; A_0^{-1/2}f \;:=\; \tilde f$$

is equivalent to the original equation $Lu=f$. The operator $\tilde L$ is symmetric hyperbolic since its coefficient matrices are $A_0^{-1/2}\,A_\mu\,A_0^{-1/2}$. Setting $\mu=0$ shows that the coefficient of ∂_t is equal to I. In this way the general case is reduced to the case $A_0=I$.

Next revisit the proof using integration by parts. Equation (2.1.10) shows that the estimate (2.1.14) is proved by taking the real part of the scalar product $\big(u(t)\,,\,(Lu)(t)\big)$. This argument generalizes to the case of an equation of the form

$$A_0(t)\,\frac{du}{dt}\;+\;G\,u=f\,, \tag{2.1.15}$$

where A_0 is strictly positive with $\|dA_0/dt\|\le C'$. The starting point is either

$$\frac{d}{dt}\,\big(u(t),A_0(t)\,u(t)\big)\,, \tag{2.1.16}$$

or equivalently

$$0=\mathrm{Re}\,\Big(u\,,\,A_0(t)\frac{du}{dt}+G\,u-f\Big)\,.$$

One finds that

$$\|u(t)\|\le C\,e^{Ct}\,\Big(\,\|u(0)\|+\int_0^t e^{-C(t-\sigma)}\,\|f(\sigma)\|\,d\sigma\Big)\,. \tag{2.1.17}$$

Exercise 2.1.2. *Carry out the last two derivations of the estimate (2.1.17).*

Proposition 2.1.2. *For every $s\in\mathbb{R}$ there is a constant $C(s,L)$ so that for all $t\ge 0$, for all $u\in C^1(\mathbb{R}\,;\,H^s(\mathbb{R}^d))\cap C(\mathbb{R}\,;\,H^{s+1}(\mathbb{R}^d))$,*
(2.1.18)

$$\|u(t)\|_{H^s(\mathbb{R}^d)}\le C\,e^{Ct}\|u(0)\|_{H^s(\mathbb{R}^d)}+\int_0^t C\,e^{C(t-\sigma)}\|(Lu)(\sigma)\|_{H^s(\mathbb{R}^d)}\,d\sigma\,.$$

Proof. The procedure of the paragraph after (2.1.14), reduces to the case $A_0=I$. The main step is to prove (2.1.18) for integer $s\ge 0$ when $A_0=I$. The case $s=0$ is (2.1.17).

For any $\alpha \in \mathbb{N}^d$ with $|\alpha| \le s$, the basic L^2 estimate (2.1.14) implies that

$$\|\partial_x^\alpha u(t)\| \le C\,e^{Ct}\,\|\partial_x^\alpha u(0)\| + \int_0^t C\,e^{C(t-\sigma)}\,\|L\,\partial_x^\alpha u\,(\sigma)\|\,d\sigma\,. \tag{2.1.19}$$

Using the product rule for differentiation and the fact that $A_0 = I$, one finds that

$$L\,\partial_x^\alpha\,u = \partial_x^\alpha\,L\,u + \sum_{|\beta|\le s} C_{\alpha,\beta}(y)\,\partial_x^\beta\,u$$

with smooth bounded matrix valued functions $C_{\alpha,\beta}$.

Definition. *For an integer $\sigma \ge 0$, $\mathrm{Op}(\sigma, \partial_x)$ denotes the family of partial differential operators in x of degree σ whose coefficients have derivatives bounded on $\mathbb{R}^{1+d}$. $\mathrm{Op}(\sigma, \partial_{t,x})$ is defined similarly.*

Then $[L, \partial^\alpha] = [G, \partial_x^\alpha] \in \mathrm{Op}(|\alpha|, \partial_x)$, more generally,

$$\big[\mathrm{Op}(m, \partial_x)\,,\,\partial_x^\alpha\big] \in \mathrm{Op}(m + |\alpha| - 1, \partial_x).$$

Define

$$\psi(t) \;:=\; \sum_{|\alpha|\le s} \big\|\partial_x^\alpha u(t)\big\|_{L^2(\mathbb{R}^d)}\,.$$

Summing (2.1.19) over all $|\alpha| \le s$ yields

$$\psi(t) \;\le\; C\,e^{Ct}\,\psi(0) + \int_0^t C\,e^{C(t-\sigma)}\,\psi(\sigma)\,d\sigma + \int_0^t C\,e^{C(t-\sigma)}\,\|f(\sigma)\|_{H^s(\mathbb{R}^d)}\,d\sigma\,.$$

If follows from Gronwall's Lemma below that one has the same estimate with a larger constant C' and without the middle term on the right-hand side.

Gronwall's Lemma 2.1.3. *If $0 \le g, \psi \in L^\infty_{\mathrm{loc}}(\overline{\mathbb{R}}_+)$, $0 \le h \in L^1_{\mathrm{loc}}(\overline{\mathbb{R}}_+)$, and*

$$\psi(t) \;\le\; g(t) \;+\; \int_0^t h(\sigma)\,\psi(\sigma)\,d\sigma\,, \qquad \text{a.e. } t > 0\,, \tag{2.1.20}$$

then, with $H(t) := \int_0^t h(\sigma)\,d\sigma$,

$$\psi(t) \;\le\; g(t) + e^{H(t)} \int_0^t e^{-H(\sigma)}\,h(\sigma)\,g(\sigma)\,d\sigma\,, \qquad \text{a.e. } t > 0\,.$$

Proof of Gronwall's Lemma. Denote by γ the absolutely continuous function

$$\gamma(t) \;:=\; \int_0^t h(\sigma)\,\psi(\sigma)\,d\sigma\,.$$

Then

$$\gamma'(t) \;=\; h(t)\psi(t) \;\le\; h(t)g(t) + h(t)\gamma(t)\,,$$

where the integral inequality is used in the last step.

Therefore

$$\big(e^{-H(t)}\gamma(t)\big)' \;=\; e^{-H(t)}\big(\gamma' - h\,\gamma\big) \;\le\; e^{-H(t)}\, h(t)\, g(t)\,.$$

Since $\gamma(0) = 0$, integrating this inequality from $t = 0$ to t yields

$$e^{-H(t)}\,\gamma(t) \;\le\; \int_0^t e^{-H(\sigma)}\, h(\sigma)\, g(\sigma)\, d\sigma\,.$$

This together with (2.1.21), which asserts that $\psi \le g + \gamma$, completes the proof. □

Exercise 2.1.3. *Show how Gronwall's Lemma suffices to erase the middle term in the estimate at the cost of increasing the constant C.*

Following [Lax, 1955], we prove (2.1.9) for $0 > s \in \mathbb{Z}$. With integer $s < 0$, introduce $\sigma := |s| > 0$ and $u := (1-\Delta)^{\sigma} v$, so

$$v \;=\; (1-\Delta)^{-\sigma} u, \qquad \|u(t)\|_s \;=\; \|u(t)\|_{-\sigma} \;=\; \|(1-\Delta)^{-\sigma} u\|_{\sigma} \;=\; \|v(t)\|_{\sigma}\,.$$

Use the estimate for the positive integer value σ on v. Toward that end, we need to estimate $\|Lv(t)\|_{\sigma}$.

Since $(1-\Delta)^{\sigma}$ is a scalar differential operator of degree 2σ,

$$\Big[L\,,\,(1-\Delta)^{\sigma}\Big] \;\in\; \mathrm{Op}(2\sigma, \partial_x)\,.$$

In particular, $\big[L\,,\,(1-\Delta)^{\sigma}\big]$ is for each t a continuous map $H^{\sigma} \to H^{-\sigma}$ with bound independent of t.

Compute

$$Lu = L\big((1-\Delta)^{\sigma} v\big) = (1-\Delta)^{\sigma} Lv + [L, (1-\Delta)^{\sigma}]v = (1-\Delta)^{\sigma} Lv + \mathrm{Op}(2\sigma, \partial_x) v\,.$$

Therefore

$$Lv \;=\; (1-\Delta)^{-\sigma}\Big(Lu \;+\; \mathrm{Op}(2\sigma, \partial_x) v\Big)\,,$$

so, with $\|\;\|_s$ denoting the $H^s(\mathbb{R}^d)$ norm,

$$\|Lv(t)\|_{\sigma} \le \|Lu(t)\|_{-\sigma} + C\|v(t)\|_{\sigma}\,.$$

Insert this in the inequality

$$\|v(t)\|_{\sigma} \;\le\; C\Big(\|v(0)\|_{\sigma} + \int_0^t \|Lv(r)\|_{\sigma}\, dr\Big)$$

to find

$$\|u(t)\|_{-\sigma} \;\le\; C\bigg(\|u(0)\|_{-\sigma} + \int_0^t \|u(r)\|_{-\sigma} + \|Lu(r)\|_{-\sigma}\, dr\bigg)\,.$$

Gronwall's Lemma yields the desired estimate (2.1.19) for the case $s = -\sigma$.

This completes the proof of (2.1.9) for integer values of s. The estimate for s not equal to an integer follows by interpolation. □

The a priori estimates show that solutions of $Lu = 0$ grow at most exponentially in time. The simple example $u_t = u$ shows that such growth can occur. The derivation of estimates for the derivatives shows that derivatives grow at most exponentially but perhaps at a faster rate. The next example shows that derivatives may grow more rapidly.

Examples 2.1.2. 1. Let $a(x) := -\arctan(x)$, and $L := \partial_t + a(x)\partial_x$. Solutions of $Lu = 0$ are constant on the characteristic curves which converge exponentially rapidly to $x = 0$ as $t \to +\infty$. The $L^\infty(\mathbb{R})$ norm of solutions compactly supported in x is independent of time, and the $L^p(\mathbb{R})$ norm of solutions tends to zero for all $1 \le p < \infty$. For such p,

$$\int |u(t,x)|^p \, dx \sim e^{-t} \qquad \text{as} \qquad t \to +\infty .$$

For $u(0,\cdot) \in C_0^\infty(\mathbb{R}) \setminus 0$ and t large, the solution is compressed into an interval of width $\sim e^{-t}$ so the derivatives are $\sim e^t$, and one finds

$$\int |\partial_x u(t,x))|^p \, dx \sim e^{-t}\, e^{pt} \qquad \text{as} \qquad t \to +\infty .$$

For $1 \le p < \infty$ the rate of growth of the derivatives is different than that of the solutions.

2. For $L := \partial_t + a(x)\partial_x + a'(x)/2$, $G(x,\partial_x) = a\partial_x + a'/2$ is antiselfadjoint and the time evolution conserves the $L^2(\mathbb{R})$ norm. The compression occurs as in the preceding example but the amplitudes grow so that the L^2 norm of the solution between any pair of characteristic curves is conserved. The amplitudes of the derivatives grow even faster by a factor e^t. For this modification, the L^2 norm is constant and the L^2 norm of first derivatives grow as e^t. The L^2 norm of derivatives of order s grow as e^{st}.

2.2. Existence theorems for symmetric hyperbolic systems

As is the case with many good estimates, the corresponding existence theorem lingers not far behind.

Friedrichs' Theorem 2.2.1. *If $g \in H^s(\mathbb{R}^d)$ and $f \in L^1_{\rm loc}(\mathbb{R}; H^s(\mathbb{R}^d))$ for some $s \in \mathbb{R}$, then there is one and only one solution $u \in C(\mathbb{R}\,;\, H^s(\mathbb{R}^d))$ to the initial value problem*

$$Lu = f, \qquad u|_{t=0} = g. \tag{2.2.1}$$

In addition, there is a constant $C = C(L,s)$ independent of f, g so that for all $t > 0$,

$$\|u(t)\|_{H^s(\mathbb{R}^d)} \le C\, e^{Ct} \|u(0)\|_{H^s(\mathbb{R}^d)} + \int_0^t C\, e^{C(t-\sigma)} \|f(\sigma)\|_{H^s(\mathbb{R}^d)} \, d\sigma , \tag{2.2.2}$$

with a similar estimate for $t < 0$.

Theorem 2.2.2 gives additional regularity in time assuming that f is smoother in t.

The solution u is constructed as the limit of approximate solutions u^h. The u^h are solutions of a differential-difference equation obtained by replacing x derivatives by centered difference quotients. This replaces the generator of the dynamics by a bounded linear operator so the existence of the approximate solution follows from existence for ordinary differential equations. The important step is to prove uniform bounds for the u^h as $h \to 0$.

As a warm up consider the simple initial value problem

$$\partial_t u + \partial_x u = 0\,, \qquad u(0,x) = g(x)\,, \qquad x \in \mathbb{R}^1\,.$$

Define the centered difference operator by

$$\delta^h \phi(x) = \frac{\phi(x+h) - \phi(x-h)}{2h}\,.$$

Approximate solutions are defined as solutions of

$$\partial_t u^h + \delta^h u^h = 0\,, \qquad u^h(0,x) = g(x)\,.$$

Note that as operators on H^s, the norms of the generators diverge to infinity as $h \to 0$. This corresponds to the fact that the difference operators δ^h converge to the unbounded operator ∂_x. The next exercises are strongly recommended for those without experience with finite difference approximations.

Exercise 2.2.1. *Show that for any $s \in \mathbb{R}$ and $g \in H^s(\mathbb{R})$, this recipe determines a sequence of approximate solutions which, as $h \to 0$, converge in $C(]-\infty,\infty[\,;\, H^s(\mathbb{R}))$ to the exact solution.* **Hint** (Von Neumann's method). Use the Fourier transform in x and compare the formulas for $\mathcal{F}u^h(t)$ and $\mathcal{F}u(t)$.

One does not have similar good behavior for the finite difference approximations

$$\partial_t u^h + i\,\delta^h u^h = 0\,, \qquad u^h(0,x) = g(x)\,,$$

to the nonhyperbolic initial value problem

$$\partial_t u + i\,\partial_x u = 0\,, \qquad u(0,x) = g(x)\,.$$

For this initial value problem and generic g there is nonexistence (see [Rauch, 1991, Chap. 3]).

Exercise 2.2.2. *For the approximations to the nonhyperbolic initial value problem above, prove that there is a $c > 0$ so that*

$$\liminf_{h \to 0}\ \|u^h(t)\|^2_{L^2(\mathbb{R})}\ \geq\ c \int_{\mathbb{R}} e^{2t\xi}\, |\hat g(\xi)|^2\, d\xi\,.$$

In particular, if the right-hand side is infinite, $u^h(t)$ does not converge in $L^2(\mathbb{R})$ as h tends to zero. The right-hand side is infinite for generic $g \in C_0^\infty(\mathbb{R})$. In the same way, prove that for generic smooth g, $u^h(t)$ is unbounded in $H^s(\mathbb{R})$ for all $t \neq 0$ and $s < 0$.

Proof of Friedrich's Theorem. It suffices to consider the case $A_0 = I$.

Step 1. Existence for $g \in \bigcap_s H^s(\mathbb{R}^n)$ and $f \in \bigcap_s C^s(\mathbb{R}\,;\, H^s(\mathbb{R}^d))$. For such f, g, define approximate solutions u^h as solutions of

$$L^h u^h = f\,, \qquad u^h(0) = g\,,$$

where for $h > 0$, L^h comes from L upon replacing the unbounded antiselfadjoint operators ∂_j with $j \geq 1$ by the bounded antiselfadjoint finite difference operators δ_j^h defined by

$$\delta_j^h \phi \;:=\; \frac{\phi(x_1, \dots, x_j + h, \dots, x_d) - \phi(x_1, \dots, x_j - h, \dots x_d)}{2h}\,.$$

Then $L^h = \partial_t + G^h(t)$ and for every s, $t \mapsto G^h(t)$ is a smooth function with values in $\mathcal{L}(H^s(\mathbb{R}^d))$. The norm of G^h is $O(1/h)$. It follows that the u^h are uniquely determined as solutions of ordinary differential equations in time and satisfy the crude estimate $\|u^h(t)\|_{H^s} \leq C(s)\, e^{C(s)/h} \|u(0)\|_{H^s(\mathbb{R}^d)}$.

This would be true for any system of partial differential operators L even those with ill-posed Cauchy problems. For the symmetric hyperbolic systems, the operators L^h satisfy estimates like (2.1.19) with constants independent of h. That is, for each $s \in \mathbb{R}$ there is a constant $C(s, L)$, so that for all $0 < h < 1$ and all $u \in C^1(\mathbb{R}\,;\, H^s(\mathbb{R}^d))$, and all $t > 0$,

$$(2.2.3) \quad \|u(t)\|_{H^s(\mathbb{R}^d)} \leq e^{Ct} \|u(0)\|_{H^s(\mathbb{R}^d)} + \int_0^t e^{C(t-\sigma)} \|(L^h u)(\sigma)\|_{H^s(\mathbb{R}^d)}\, d\sigma\,.$$

The proof of (2.2.3) for $s = 0$ mimics the proof for L. The key ingredient is almost antiselfadjointness expressed by the bound

$$\|(A_j \delta_j^h \;+\; (A_j \delta_j^h)^*) w\|_{L^2(\mathbb{R}^d)} \;\leq\; \big(\sup_{\mathbb{R}^{1+d}} \|\nabla_x A_j\|\big)\; \|w\|_{L^2(\mathbb{R}^d)}.$$

The right-hand side is independent of h. To prove this bound, use the fact that δ_j^j is antiselfadjoint to find

$$\big(A_j \delta_j^h\big)^* \;=\; (\delta_j^h)^* A_j^* \;=\; -\delta_j^h\, A_j \;=\; -A_j \delta_j^h - [A_j, \delta_j^h]\,.$$

Denoting by $\mathbf{e}_j$ the unit vector in the j direction and suppressing the j's for ease of reading, $[A_j, \delta_j^h]\, w$ is equal to

$$A(y)\, \frac{w(y + h\mathbf{e}) - w(y - h\mathbf{e})}{2h} \;-\; \frac{A(y + h\mathbf{e})\, w(y + h\mathbf{e}) - A(y - h\mathbf{e}) w(y - h\mathbf{e})}{2h}.$$

Regrouping yields

$$[A_j, \delta_j^h]\, w = \frac{1}{2}\left[\frac{A(y) - A(y+h\mathbf{e})}{h}\right] w(y+h\mathbf{e})$$
$$+\frac{1}{2}\left[\frac{A(y-h\mathbf{e}) - A(y)}{h}\right] w(y-h\mathbf{e})\,.$$

The bound follows since the sup norm of each of the factors in square brackets is equal to $\|\partial_j A\|_{L^\infty}$. Writing $L^h = \partial_t + G^h$, (2.2.3) for $s = 0$ follows from $\|G^h + (G^h)^*\| \le 2C$ with C independent of h.

The proof of (2.2.3) for $s \ge 0$ integer, uses the $s = 0$ result to estimate $\partial_x^\alpha u^h$ for $|\alpha| \le s$. Use the equation,

$$L^h \partial_x^\alpha u^h \;=\; \partial_x^\alpha L^h u^h \;+\; [L^h, \partial_x^\alpha] u^h\,.$$

For $|\alpha| = 1$, direct computation of the commutator yields

$$[A_j\delta_j^h, \partial_x] \;=\; (\partial_x A_j)\delta_j^h\,, \qquad [B, \partial_x] \;=\; (\partial_x B)\,.$$

Therefore

$$[L^h, \partial_x] \;=\; \mathrm{Op}(0) \;+\; \mathrm{Op}(0)\delta_x^h,$$

where this means a finite sum of terms of the type described. The general case is

$$[L^h, \partial_x^\alpha] \;=\; \mathrm{Op}(|\alpha| - 1, \partial_x) \;+\; \mathrm{Op}(|\alpha| - 1, \partial_x)\delta_x^h\,. \tag{2.2.4}$$

Exercises 2.2.3. **i.** *Prove (2.2.4) by induction on s.*

ii. *Prove (2.2.3) for integer $s \ge 0$ using (2.2.4).*

iii. *Prove (2.2.3) for a negative integer s by Lax's method as in the proof of the estimate for L.*

Discussion. The case of a noninteger s follows by interpolation.

Apply (2.2.3) to find that for all α and T the family

$$\{\partial_x^\alpha u^h\} \quad \text{is bounded in} \quad C([-T,T]\,;\, L^2(\mathbb{R}^d)) \subset L^\infty([-T,T]\,:\, L^2(\mathbb{R}^d))\,.$$

The differential equation shows, by induction on the order of the time derivative in β using the differentiability of f with respect to time, that for all T and β

$$\{\partial_{t,x}^\beta u^h\} \quad \text{is bounded in} \quad L^\infty([-T,T]\,:\, L^2(\mathbb{R}^d))\,.$$

Since H^s is a Hilbert space, $L^\infty([-T,T] \,:\, L^2(\mathbb{R}^d))$ is the dual of $L^1([-T,T] \,:\, L^2(\mathbb{R}^d))$. Use the weak star compactness of bounded sets in duals. The Cantor diagonal process allows one to extract a subsequence $u^{h(k)}$ with $h(k) \to 0$ as $k \to \infty$ so that for all T and β, as $k \to \infty$,

$$\partial_{t,x}^\beta u^{h(k)} \;\to\; \partial_{t,x}^\beta u \qquad \text{weak star in} \quad L^\infty([-T,T]\,:\, L^2(\mathbb{R}^d))\,.$$

In particular, $u^{h(k)}$ converges in $H^s([-T,T]\times\mathbb{R}^d)$ so $u^{k(h)}|_{t=0}$ converges to $u|_{t=0}$ in $H^{s-1/2}(\mathbb{R}^d)$. So $u(0)=g$.

Passing to the limit in the equation satisfied by $u^{h(k)}$ shows that u satisfies $Lu=f$. This completes the proof of Step 1.

Step 2. Existence in the general case.

Choose

$$g_n \in C_0^\infty(\mathbb{R}^d)\,, \qquad g_n \to g \ \text{ in } \ H^s(\mathbb{R}^d)\,,$$
$$f_n \in C_0^\infty(\mathbb{R}^{1+d})\,, \qquad f_n \to f \ \text{ in } \ L^1_{\mathrm{loc}}(\mathbb{R}\,;\,H^s(\mathbb{R}^d)\,.$$

Let u^n be the solution from the first step with data f_n, g_n.

Estimate (2.2.3) implies that for all T and as $n,m\to\infty$,

$$u^n - u^m \ \to 0 \ \text{ in } \ C\big([-T,T]\,;\,H^s(\mathbb{R}^d)\big). \tag{2.2.5}$$

By completeness there is a $u\in C(\mathbb{R}\,;\,H^s(\mathbb{R}^d))$ which is the limit uniformly on compact time intervals of the u^n. It follows that $Lu=f$ and $u|_{t=0}=g$.

Step 3. Uniqueness.

Suppose that $u\in C(\mathbb{R}\,;\,H^s(\mathbb{R}^d))$ satisfies $Lu=0$ and $u|_{t=0}=0$. The differential equation implies that $u_t\in C(\mathbb{R}\,;\,H^{s-1}(\mathbb{R}^d))$. So $u\in C^1(\mathbb{R}\,;\,H^{s-1}(\mathbb{R}^d))$. This is sufficient to apply the case $s-1$ of Proposition 2.1.1. □

Using difference approximations to prove existence goes back at least to the work of Cauchy (1840) and most notably Peano (1886) on ordinary differential equations. In the context of partial differential equations, note the seminal paper of Courant, Friedrichs, and Lewy (1928). The method has the advantages of being constructive and widely applicable. In Appendix 2.II, we present an alternate functional analytic method for passing from the a priori estimates to existence. It is elegant but limited to linear problems.

The next result shows that when the source term f is differentiable in time, then so is u.

Corollary 2.2.2. *If $m\geq 1$ and in addition to the hypotheses of Theorem 2.2.1, $\partial_t^k f\in C(\mathbb{R}\,;\,H^{s-k-1}(\mathbb{R}^d))$ for $k=0,1,\dots,m-1$, then*

$$u\in C^k\big(\mathbb{R}\,;\,H^{s-k}(\mathbb{R}^d)\big) \qquad \text{for} \quad k=0,1,\dots,m\,.$$

Proof. It suffices to consider $A_0=I$. For $m=1$, write

$$u_t \ = \ -\sum A_j\,\partial_j u - Bu + f.$$

The hypothesis together with Theorem 2.2.1 shows that the right-hand side is continuous with values in H^{s-1}. For $m=2$,

$$u_{tt} \ = \ \partial_t\Big(-\sum A_j\,\partial_j u - Bu + f\Big).$$

From the case $m=1$, the first terms are continuous with values in H^{s-2}. The hypothesis on f treats the last term. An induction completes the proof. □

We next prove a formula that expresses solutions of the inhomogeneous equation $Lu = f$ in terms of solutions of the Cauchy problem for $Lu = 0$. The formula is motivated as follows. Let $h(t)$ denote Heaviside's function. For f supported in $t \geq 0$ seek u also supported in the future with $Lu = f$. The source term f is the sum on σ of the singular sources $f\delta(t-\sigma)$ with $\sigma \geq 0$. The solution of $Lv = f\delta(t-\sigma)$ with a response supported in $t \geq \sigma$ is equal to $v = h(t-\sigma)w$, where w is the solution to the Cauchy problem

$$Lw \;=\; 0, \qquad w(\sigma) \;=\; A_0^{-1} f(\sigma)\,.$$

Summing the solutions v yields the formula of the next proposition.

Define the operator $S(t,\sigma)$ from $H^{-\infty}(\mathbb{R}^d) := \bigcup_s H^s(\mathbb{R}^d)$ to itself by $S(t,\sigma)g := u(t)$, where u is the solution of

$$Lu \;=\; 0\,, \qquad u(\sigma) \;=\; g\,.$$

The operator $S(t,\sigma)$ is the operator that marches from time σ to time t. Corollary 2.2.2 implies that for $g \in H^s(\mathbb{R}^d)$, $S(t,\sigma)g \in C^k(\mathbb{R}\,;\, H^{s-k}(\mathbb{R}^d))$ and, by definition, $L(t,x,\partial_{t,x})S = 0$. For any R, $\{S(t,\sigma) \,:\, |t,\sigma| \leq R\}$ is bounded in $\mathrm{Hom}(H^s, H^s)$.

Duhamel's Proposition 2.2.3. *If $f \in L^1_{\mathrm{loc}}([0,\infty[\;;\, H^s(\mathbb{R}^d))$, then the solution of the initial value problem*

$$Lu \;=\; f \quad \text{on} \quad [0,\infty[\,\times\,\mathbb{R}^d, \qquad u(0) \;=\; 0, \tag{2.2.6}$$

is given by

$$u(t) \;=\; \int_0^t S(t,\sigma)\, A_0^{-1}\, f(\sigma)\, d\sigma\,. \tag{2.2.7}$$

Proof. It suffices to prove (2.2.7) for $f \in C([0,\infty[\,;\, H^s(\mathbb{R}^d))$ since the general result then follows by approximation. For such f define u by (2.2.7). On compact sets of t,σ one has

$$\|S(t,\sigma)\|_{H^s \to H^s} \;+\; \|S_t(t,\sigma)\|_{H^s \to H^{s-1}} \;\leq\; C\,. \tag{2.2.8}$$

It follows that $u \in C([0,\infty[\,;\, H^s(\mathbb{R}^d))$, and differentiating under the integral sign,

$$\begin{aligned} u_t &= S(t,t)(A_0^{-1} f(t)) \;+\; \int_0^t S_t(t,\sigma)\, f(\sigma)\, d\sigma \\ &= A_0^{-1} f(t) \;+\; \int_0^t S_t(t,\sigma)\, f(\sigma)\, d\sigma \end{aligned}$$

and

$$\partial_j u \;=\; \int_0^t \partial_j S(t,\sigma)\, f(\sigma)\, d\sigma\,.$$

It follows from $LS = 0$ that $Lu = f$. Since $u(0) = 0$, the result follows from uniqueness of solutions to (2.2.6). □

2.3. Finite speed of propagation

The proof of Theorem 2.2.1 works as well for the operator $L + i\Delta_x$ as for L since the additional operator is exactly antisymmetric and commutes with ∂_y. However, the resulting evolution equations do not have finite speed of propagation. An important aspect of the L^2 estimates that form the basis of the results in §2.2 is that the local form of the energy law implies finite speed. That is not the case for the energy law for $L + i\Delta$.

One of the goals of the theory of partial differential equations is to be able to derive accurate qualitative properties of solutions from properties of the symbol. The precise speed estimate proved in §2.5 is a striking example.

2.3.1. The method of characteristics. Corollary 2.1.3 already addressed finite speed using the method of characteristics. Here we use this method in the simplest case of one dimensional homogeneous constant coefficient systems,

$$\partial_t u + A\partial_x u \;=\; 0\,, \qquad A = A^*\,.$$

The change of variable $w = Wv$ with unitary W yields $\partial_t Wv + \partial_x AWv = 0$. Multiplying by W^* yields

$$\partial_t w \;+\; W^* A W\,\partial_x w \;=\; 0\,.$$

Choose W that diagonalises A to find

$$\partial_t w + D\,\partial_x w \;=\; 0\,, \qquad D = \operatorname{diag}\big(\lambda_1, \lambda_2, \ldots, \lambda_d\big)\,, \qquad \lambda_1 \le \lambda_2 \le \cdots \le \lambda_N\,.$$

The exact solution is given by undistorted traveling waves,

$$w_j(t,x) \;=\; g_j(x - \lambda_j\, t)\,.$$

If $g = 0$ on an interval $I :=]a, b[$, then for $t \ge 0$, u vanishes on the triangle

$$\Big\{(t,x)\;:\; t \ge 0\,, \;\; \text{and} \;\; a + \lambda_N\, t \;<\; x \;<\; b - \lambda_1\, t\Big\}\,.$$

A degenerate case is when $\lambda_1 = \lambda_N$, in which case the triangle becomes a strip bounded by parallel lines. The triangle is called a *domain of determinacy* of the interval because the initial values of u on I determine the values of u on the triangle. Considering traveling waves with speeds λ_N and λ_1 shows that this result is sharp. Figure 2.3.1 sketches the case $\lambda_1 < 0 < \lambda_N$.

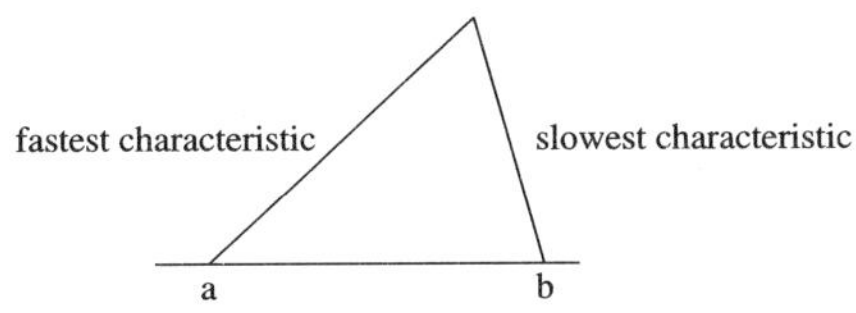

FIGURE 2.3.1. Domain of determinacy of I

The left-hand boundary moves at the largest velocity and the right-hand boundary at the smallest.

Viewed another way, the values of f on an interval $J = [A, B]$ influence the future values of the solution u only on the set

$$\Big\{(t,x)\ :\ t \geq 0\,,\quad \text{and}\ \ A + \lambda_1 t\ \leq\ x\ \leq\ B + \lambda_N t\Big\}.$$

It is called a *domain of influence* of J and is sketched in Figure 2.3.2 for the case $\lambda_1 < 0 < \lambda_N$.

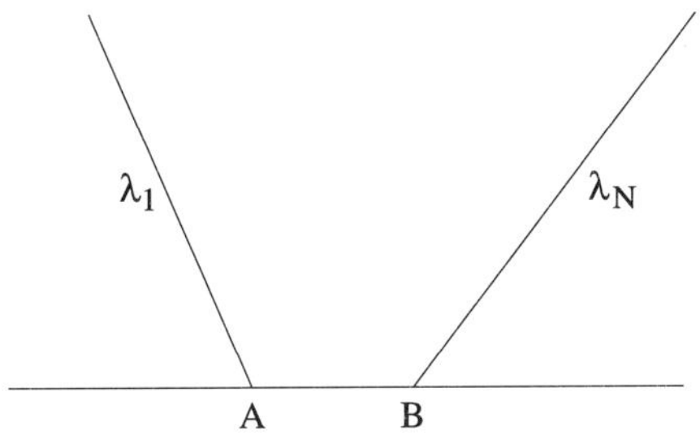

FIGURE 2.3.2. Domain of influence of J

2.3.2. Speed estimates uniform in space. The estimates in this section yield bounds on propagation speeds that are uniform in space-time. In subsequent sections precise bounds taking into account the variation of speed with y will be given. The results in this section are sharp for operators with constant coefficient principal part. That case is the backbone of the precise results of §2.5.

Usually one thinks of speed in terms of distance traveled divided by time. However, for a general symmetric hyperbolic operator, there is no natural metric to measure distance. Introducing an artificial metric, for example the Euclidean metric, can lead to imprecision in the results. The problem we analyze in this section depends only on affine geometry. We find the smallest space-time half-space that contains the support of solutions whose Cauchy data have support in a given half-space in $\{t = 0\}$. The case of the initial half-space $\{x_1 \leq 0\}$ yields the general result.

Define

$$\Lambda\ :=\ \inf\Big\{\ell\ :\ \forall y\,,\ \ A_0(y)^{-1/2}\,A_1(y)\,A_0(y)^{-1/2}\ \leq\ \ell\, I\Big\}. \tag{2.3.1}$$

Λ is the smallest upper bound for the eigenvalues of the family of matrices $A_0(y)^{-1/2}\, A_1(y)\, A_0(y)^{-1/2}$. Section 2.3.3 gives an intrinsic description in terms of the characteristic variety.

Lemma 2.3.1. *Suppose that L is a symmetric hyperbolic operator, $s \in \mathbb{R}$, Λ is defined by (2.3.1),*

$$u \in C([0,\infty[\,;\, H^s(\mathbb{R}^d)) \quad \textit{satisfies} \quad Lu = 0\,,$$

and

$$\operatorname{supp} u(0,x) \;\subset\; \{x_1 \;\leq\; 0\}\,.$$

Then for $0 \leq t$,

$$\operatorname{supp} u(t,x) \;\subset\; \{x_1 \;\leq\; \Lambda\, t\}\,.$$

Example 2.3.1. If $A_0^{-1/2} A_1 A_0^{-1/2}$ is independent of y, one obtains the same estimate for propagation in the x_1 that is proved for $\partial_t + A_0^{-1/2} A_1 A_0^{-1/2} \partial_x$ in the preceding section.

Proof. Choose $f_n \in C_0^\infty(\mathbb{R}^d)$ supported in $\{x_1 \leq 0\}$ with $f_n \to u(x,0)$ in $H^s(\mathbb{R}^d)$. Denote by u_n the solution of $L\,u_n = 0$ with $u_n(0,x) = f_n(x)$. For all $T > 0$, $u_n \to u$ in $C\big([0,T] \,:\, H^s(\mathbb{R}^d)\big)$. Therefore, it suffices to show that u_n is supported in $\{x_1 \leq \Lambda\, t\}$. Thus, it suffices to consider the case when the initial data of u belongs to $C_0^\infty(\mathbb{R}^d)$.

Use a local version of the basic energy law. When $A_\mu = A_\mu^*$, and $\langle\ ,\ \rangle$ is the scalar product in $\mathbb{C}^N$,

$$\big\langle A_\mu \partial_\mu u\,,\, u\big\rangle \;+\; \big\langle u\,,\, A_\mu \partial_\mu u\big\rangle \;=\; \partial_\mu \big\langle A_\mu u\,,\, u\big\rangle \;-\; \big\langle (\partial_\mu A_\mu) u\,,\, u\big\rangle\,.$$

Denote $L_1(y,\partial) := \sum_\mu A_\mu \partial_\mu$. Summing on μ yields

$$2\mathrm{Re}\,\big\langle L_1(y,\partial) u\,,\, u\big\rangle \;=\; \sum_\mu \partial_\mu \big\langle A_\mu u\,,\, u\big\rangle \;+\; \Big\langle \Big(\sum_\mu \partial_\mu A_\mu\Big) u\,,\, u\Big\rangle\,.$$

Adding the lower order terms yields the energy balance law

$$\begin{aligned}(2.3.2)\quad \partial_t \big\langle A_0(t,x)\, u(t,x), u(t,x)\big\rangle + \sum_{j=1}^{d} \partial_j \big\langle A_j(t,x)\, u(t,x), u(t,x)\big\rangle \\ = \big\langle Z(t,x)\, u(t,x), u(t,x)\big\rangle + 2\,\mathrm{Re}\,\big\langle (Lu)(t,x), u(t,x)\big\rangle\,,\end{aligned}$$

where Z is the smooth matrix valued function

$$(2.3.3) \qquad Z(y) \;:=\; -B(y) - B^*(y) + \sum_{\mu=0}^{d} \frac{\partial A_\mu(y)}{\partial y_\mu}\,.$$

Integrate this identity over the region where we want to prove that $u = 0$,

$$(2.3.4) \qquad \Omega(\underline{t}) \;:=\; \Big\{(t, x_1, x_2, \ldots, x_d) \;:\; 0 \leq t \leq \underline{t} \;\; \text{and} \;\; x_1 \geq t\Lambda\Big\}\,.$$

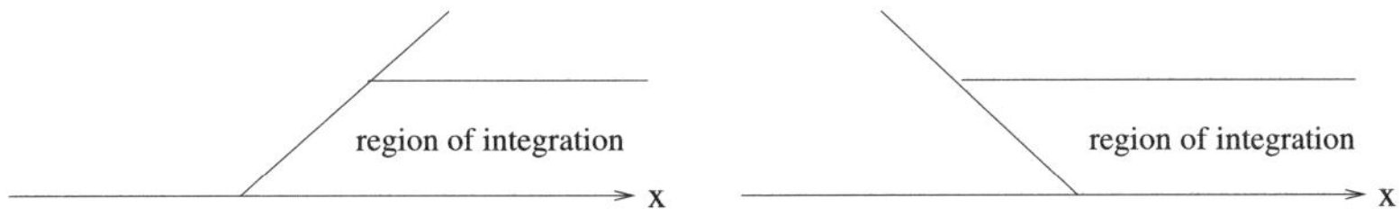

FIGURE 2.3.3.

The case $\Lambda > 0$ (resp., $\Lambda < 0$) is sketched on the left (resp., right) in Figure 2.3.3.

Define

$$\Phi(t) \ := \ \Big(\int_{x_1 \geq \Lambda t} \big\langle A_0(t,x)\, u(t,x)\,,\, u(t,x) \big\rangle \, dx \Big)^{1/2},$$

so $\Phi(t)$ is equivalent to the L^2 norm of $u(t)$ on $\{x_1 \geq \Lambda\, t\}$. Integrate (2.3.2) over $\Omega(\underline{t})$. Denote the lateral boundary of $\Omega(\underline{t})$ by

$$\mathcal{B}(\underline{t}) \ := \ \Big\{ (t,x) \ : \ 0 \leq t \leq \underline{t}, \quad \text{and} \quad x_1 = \Lambda\, t \Big\}.$$

Integrate by parts[2] to find

$$\Phi(\underline{t})^2 \ = \ \Phi(0)^2 \ - \ \int_{\Omega(\underline{t})} \langle Z\, u\,,\, u \rangle \; dx\, dt \ - \ \int_{\mathcal{B}(\underline{t})} \big\langle L_1(y,\nu)\, u\,,\, u \big\rangle \; d\sigma\,,$$

where ν is a the unit outward normal and $d\sigma$ is the surface area element. The definition of the principal symbol $L_1(y,\eta)$ is recalled in §2.3.3.

The function $\phi(t,x) := \Lambda\, t - x_1$ is negative in the interior of the region and positive in the interior of the complement. The outward pointing normals, ν, to the lateral boundary are positive multiples of

$$d\phi \ = \ \Big(\frac{\partial \phi}{\partial t}\,,\, \frac{\partial \phi}{\partial x_1}\,,\, 0, \ldots, 0 \Big) \ = \ \Big(\Lambda\,,\, -1\,,\, 0\,,\, \ldots\,, 0 \Big).$$

The boundary matrix $L_1(y,\nu) = \sum_\mu A_\mu(y)\nu_\mu$ is a positive multiple of

$$\Lambda A_0 - A_1 \ = \ A_0^{1/2} \Big(\Lambda I - A_0^{-1/2}\, A_1\, A_0^{-1/2} \Big) A_0^{1/2} \ \geq \ 0$$

from the definition of Λ. Therefore, the integral over $\mathcal{B}(\underline{t})$ is nonnegative.

---

[2]This integration by parts involves smooth functions u all of whose derivatives are square integrable. This is sufficient. For example, we show that

$$\int\!\!\int_{\Omega(\underline{t})} \partial_1\, g \; dx\, dt \ = \ - \int_{\mathcal{B}(\underline{t})} g \; d\sigma\,, \qquad \text{with} \qquad g \ = \ \langle A_1 u\,,\, u \rangle.$$

Denote by B the Banach space of functions g so that $\{g, \partial_1 g\} \subset L^1(\Omega(\underline{t}))$. $C^\infty_{(0)}(\Omega(\underline{t}) \subset B$ is dense. For the trace at $\mathcal{B}$ for elements of the dense set,

$$\int_{\mathcal{B}} g \; dt\, dx_2 \, \cdots \, dx_d = \int_0^t \int_{\mathbb{R}^{d-1}} \Big| \int_{\Lambda t}^{\infty} \frac{\partial g}{\partial x_1} \, dx_1 \Big| dx' \; dt \leq \|\partial_1 g\|_{L^1(\Omega(\underline{t}))}\,.$$

Therefore, the trace $g \mapsto g|_{\mathcal{B}}$ extends as a continuous map $B \to L^1(\mathcal{B})$. And the linear map

$$g \mapsto \int\!\!\int_{\Omega} \partial_1 g \, dx\, dt + \int_{\mathcal{B}} g \, d\sigma$$

is continuous and vanishes on a dense subset.

By hypothesis, $\Phi(0) = 0$. The volume integral satisfies

$$\Big| \int_{\Omega(\underline{t})} \langle Z\, u\,,\, u\rangle \; dx\, dt \Big| \;\le\; C \int_0^{\underline{t}} \Phi(t)^2 \; dt\,.$$

Combining yields

$$\Phi(\underline{t})^2 \;\le\; C \int_0^{\underline{t}} \Phi(\sigma)^2 \; d\sigma\,.$$

Gronwall's Lemma implies that $\Phi \equiv 0$ and the proof is complete. □

Remark. The same proof fails for the operator $L + i\Delta_x$ because there is no choice of Λ that guarantees that the boundary terms from the $i\Delta$ are nonnegative.

Theorem 2.3.2. *Suppose that L is a symmetric hyperbolic operator, $\xi \in \mathbb{R}^d \setminus 0$, and*
(2.3.5)

$$\Lambda(\xi) \;:=\; \inf\Big\{\ell \;:\; \forall y\,,\;\; A_0(y)^{-1/2}\Big(\sum_j A_j(y)\xi_j\Big)\, A_0(y)^{-1/2} \;\le\; \ell\, I\Big\}\,.$$

If $s \in \mathbb{R}$ and $u \in C([0,\infty[\,;\, H^s(\mathbb{R}^d))$ satisfies $Lu = 0$ and

$$\mathrm{supp}\; u(0,x) \;\subset\; \big\{x\xi \;\le\; 0\big\}\,,$$

then for $0 \le t$,

$$\mathrm{supp}\; u(t,x) \;\subset\; \big\{x\xi \;\le\; \Lambda(\xi)\, t\big\}\,.$$

Proof. Choose linear spatial coordinates

$$\tilde{x} \;:=\; M\,x\,, \qquad \tilde{x}_k \;=\; \sum M_{kj}x_j\,,$$

so that $\sum \xi_j dx_j = d\tilde{x}_1$. Since

$$d\tilde{x}_1 \;=\; \sum \frac{\partial \tilde{x}_1}{\partial x_k}\, dx_k \;=\; \sum M_{1k}\, dx_k,$$

this is equivalent to $M_{1k} = \xi_k$.

The partial derivatives with respect to x and $\tilde{x}$ satisfy

$$\frac{\partial}{\partial x_j} \;=\; \sum_k \frac{\partial \tilde{x}_k}{\partial x_j}\,\frac{\partial}{\partial \tilde{x}_k} \;=\; \sum_k M_{kj}\,\frac{\partial}{\partial \tilde{x}_k}\,.$$

Therefore, the expression of the differential operator in terms $\partial/\partial\tilde{x}$ is

$$\sum A_j\,\frac{\partial}{\partial x_j} \;=\; \sum_{j,k} A_j\, M_{kj}\,\frac{\partial}{\partial \tilde{x}_k}\,.$$

The coefficient of $\partial/\partial\tilde{x}_1$ is $\sum_j A_j\, M_{1j}\frac{\partial}{\partial_{x_k}} = \sum_j A_j\xi_j$. Since the M_{kj} are real, this is symmetric as soon as the A_j are. Therefore this system remains hyperbolic. The result follows upon applying Lemma 2.3.1 in the $\tilde{x}$ coordinates. □

Sketch of alternate proof. Modify the proof of Lemma 2.3.1 as follows. Use the energy method in the region

$$\Big\{0 \le t \le \underline{t} \ \text{ and } \ x\xi \ge t\Lambda(\xi)\Big\}.$$

The lateral boundary has equation

$$\phi(t,x) = 0\,, \qquad \phi(t,x) \ := \ t\Lambda(\xi) \ - \ x\xi\,.$$

The lateral boundary matrix $L_1(y,\nu)$ is a nonnegative multiple of

$$\begin{aligned} L_1(y, d\phi) = L_1(y, \Lambda, -\xi) &= \Lambda\, A_0 - \sum_j A_j \xi_j \\ &= A_0^{1/2}\Big(\Lambda I - A_0^{-1/2}\big(\sum_j A_j\xi_j\big)A_0^{-1/2}\Big)A_0^{1/2}\,. \end{aligned}$$

This is nonnegative by definition of Λ. □

Example 2.3.2. The fundamental solution u is the solution of $Lu = 0$ with $u|_{t=0} = \delta(x)$. Then for $t \ge 0$,

$$\text{supp}\ u \ \subset \ \Big\{(t,x) \ : \ \forall\,\xi\,, \ \ x\xi \ \le \ t\Lambda(\xi)\Big\}\,. \tag{2.3.6}$$

The set on the right is a convex cone in $t \ge 0$ whose section at $t = 1$ is compact. In the $d = 1$ case, $\text{supp}\ u \subset \big\{(t,x) \ : \ t\lambda_1 \le x \le t\lambda_N\big\}$.

Exercise 2.3.1. *Prove the estimate measuring speed using euclidean distance. Define*

$$c_{\max} \ := \ \max_{|\xi|=1} \ \Lambda(\xi) \qquad (\textit{n.b. the euclidean } |\xi|).$$

i. *If $Lu = 0$ and $u(0)$ is supported in $\{|x| \le r\}$, then $u(t)$ is supported in $\{|x| \le r + c_{\max}|t|\}$.*

ii. *For the solution u in Theorem 2.2.1, let*

$$K \ := \ \text{supp}\ u(0) \ \cup \ \big(\text{supp}\ f \cap \{t \ge 0\}\big).$$

Prove that in $\{t \ge 0\}$,

$$\text{supp}\ u \ \subset \ \Big\{(t,x) \ : \ \exists(\underline{t},\underline{x}) \in K, \ \ |x - \underline{x}| \le c_{\max}(t - \underline{t})\Big\}\,.$$

Hint. Use Duhamel's formula.

There are distributional right-hand sides that are not covered by the sources $f \in L^1_{\text{loc}}(\mathbb{R}\,;\,H^s(\mathbb{R}^d))$. An interesting example is $f = \delta(t,x)$. The next theorem covers general distribution sources. It includes sources with no decay as $x \to \infty$. For those, finite speed is used.

Theorem 2.3.3. *For any distribution $f \in \mathcal{D}'(\mathbb{R}^{1+d})$ supported in $\{t \ge 0\}$, there is one and only one distribution $u \in \mathcal{D}'(\mathbb{R}^{1+d})$ supported in $\{t \ge 0\}$ so that $Lu = f$.*

Proof. Since f has support in $\{t \geq 0\}$, the linear functional

$$C_0^\infty(\mathbb{R}^{1+d}) \ni v \quad \mapsto \quad \langle f, v \rangle$$

extends uniquely to a sequentially continuous functional on

$$\Big\{ v \in C^\infty(\mathbb{R}^{1+d}) \ : \ \text{supp}\, v \ \cap\ \{t \geq -1\} \text{ is compact} \Big\}. \tag{2.3.7}$$

Here sequential convergence $v_j \to v$ means that there is compact set K independent of j with $\text{supp}\, v_j \ \cap\ \{t \geq -1\} \ \subset\ K$, and v_j and each of its partial derivatives converge uniformly on compacts. Since $C_0^\infty(\mathbb{R}^{1+d})$ is sequentially dense in (2.3.7), there can be at most one extension.

An extension is constructed by choosing $\chi \in C^\infty(\mathbb{R})$ with $\chi = 1$ for $t \geq -1/3$ and $\chi = 0$ for $t \leq -2/3$. Since f is supported in $\{t \geq 0\}$, $\langle f, v \rangle = \langle f, \chi v \rangle$ for all $v \in C_0^\infty$. The right-hand side defines the desired extension. Similarly, the formula $\langle u, \chi\, v \rangle$ extends the functional $\langle u, v \rangle$ to the set of v in (2.3.7).

It follows that if u is a solution, then the identity

$$\big\langle u, L^\dagger v \big\rangle \ = \ \big\langle f, v \big\rangle \tag{2.3.8}$$

extends to v belonging to (2.3.7).

For a test function $\psi \in C_0^\infty(\mathbb{R}^{1+d})$, let v be the solution of

$$L^\dagger v \ = \ \psi\,, \qquad v = 0 \quad \text{for} \quad t \gg 1\,.$$

Exercise 2.3.1.ii shows that v belongs to (2.3.7), so formula (2.3.8) implies that

$$\big\langle u, \psi \big\rangle \ = \ \big\langle f, v \big\rangle\,. \tag{2.3.9}$$

This determines the value of any solution u on ψ proving uniqueness of solutions.

Exercise 2.3.2. *Show that the recipe given by formula (2.3.9) defines a solution u, proving existence.* □

2.3.3. Time-like and propagation cones. The results of the preceding section have a geometric interpretation in terms of the affine geometry of the characteristic variety of L.

Definition. *The* **principal symbol** $L_1(y, \eta)$ *of* $L(y, \partial)$ *is the function defined by*

$$L_1(y, \eta) \ := \ \sum_\mu A_\mu(y)\, \eta_\mu \ = \ A_0(y)\, \tau \ + \ \sum_j A_j(y)\, \xi_j\,.$$

The principal symbol arises by dropping the zero order term and replacing ∂_μ by η_μ. It is an $N \times N$ matrix valued function of (y, η). An alternative definition replaces ∂_μ by $i\, \eta_\mu$, and so differs by a factor i from the above.

The choice we take is natural if one expresses differential operators using the partial derivatives ∂_μ rather than $\frac{1}{i}\,\partial_\mu$. The advantage of the latter is that it is the Fourier multiplier by η_μ.

The principal symbol is invariantly defined on the cotangent bundle of $\mathbb{R}^{1+d}$. In fact, if ϕ is a smooth real valued function with $d\phi(y) = \eta$ and $v \in \mathbb{R}^N$, then as $\sigma \to \infty$,

$$\partial_\mu e^{i\sigma\phi} = i\sigma\,\frac{\partial\phi}{\partial y_\mu}\,e^{i\sigma\phi} + O(1) \quad \text{so}$$

$$L_1(y, d\phi(y))\,v \;=\; \lim_{\sigma\to\infty}\;\frac{1}{i\sigma}\,e^{-i\sigma\phi}\,L(y,\partial)\big(e^{i\sigma\phi}\,v\big)\,.$$

The right-hand side of the second identity is independent of coordinates and $(y, d\phi(y))$ is a well-defined element of the cotangent bundle.

Definitions. *The* **characteristic polynomial** *of* $L(y,\partial)$ *is the polynomial in* η, $p(y,\eta)$, *defined by*

$$p(y,\eta) \;:=\; \det L_1(y,\eta)\,.$$

The **characteristic variety** *of* L, *denoted* $\operatorname{Char} L$, *is the set of pairs* $(y,\eta) \in \mathbb{R}^{1+d} \times \mathbb{R}^{1+d} \setminus 0$ *such that* $p(y,\eta) = 0$. *Points in the complement of* $\operatorname{Char} L$ *are called* **noncharacteristic**.

The characteristic variety is a well-defined subset of the cotangent bundle. It is homogeneous in the sense that

$$r \in \mathbb{R}\setminus 0 \quad \text{and} \quad (y,\eta) \in \operatorname{Char} L \Longrightarrow (y\,,\, r\,\eta) \in \operatorname{Char} L\,.$$

Definition. *For a symmetric hyperbolic* $L(y,\partial)$, *the cone of* **forward time-like codirections at** y *is*

$$\mathcal{T}(y) \;:=\; \Big\{(\tau,\xi) \;:\; L_1(y,\tau,\xi) \;>\; 0\Big\}\,.$$

The next result is an immediate consequence of the definition and the fact that $L_1(y,\eta)$ is linear in η.

Proposition 2.3.4. $\mathcal{T}(y)$ *is an open convex cone that contains* $(1,0,\dots,0)$. *It is equal to the connected component of* $(1,0,\dots,0)$ *in the noncharacteristic points over* y.

Exercise 2.3.3. *Prove Proposition* 2.3.4.

The linear form $\tau t + \xi x$ is called time-like for the following reason. If one changes to new coordinates with $t' = \tau t + \xi x$, then in the new coordinates the coefficient of $\partial/\partial t'$ is equal to $L_1(y,\tau,\xi)$. It is positive precisely when τ, ξ is time-like. In that case, the system in the new coordinates will be symmetric hyperbolic with the new time variable t'.

More generally a proposed nonlinear change with $t' = t'(t,x)$ leads to a system with coefficient $L_1(y, dt'(t,x))$ in front of $\partial/\partial t'$. Good time functions are those whose differential, dt', belongs to the forward time-like cone.

Examples 2.3.3. 1. The operator $L = \partial_t + \partial_1$ with $\mathcal{T} = \{\tau + \xi_1 > 0\}$ shows that $\mathcal{T}$ need not be a subset of $\{\tau > 0\}$.

2. In the one dimensional constant coefficient case of §2.3.1, the characteristic variety is a finite union of lines given by

$$\operatorname{Char} L \;=\; \bigcup_j \Big\{(\tau,\xi)\ :\ \tau + \lambda_j \xi = 0\Big\},$$

where the λ_j are the eigenvalues of $A_0^{-1/2} A_1 A_0^{-1/2}$ in nondecreasing order. The rays $x = \lambda_j t + \text{const.}$ describe the propagation of traveling waves. The velocity vectors $(1, \lambda_j)$ are orthogonal to the lines

$$\Big\{(\tau,\xi)\ :\ 0 = \psi_j(\tau,\xi)\ :=\ \tau + \lambda_j \xi\Big\},$$

which belong to the characteristic variety. The conormal directions to the line are scalar multiples of the differential

$$d\psi_j \;=\; \Big(\frac{\partial \psi_j}{\partial \tau}, \frac{\partial \psi_j}{\partial \xi}\Big) \;=\; \Big(1, \lambda_j\Big).$$

The lines of the characteristic variety are in the dual space $\mathbb{R}^2_{\tau,\xi}$. The conormals to such lines define directions in the space time $\mathbb{R}^2_{t,x}$.

These relations are illustrated in Figure 2.3.4. where there are two distinct positive eigenvalues, $\lambda_1 < 0 < \lambda_2 < \lambda_3 = -\lambda_1$.

The time-like cone $\mathcal{T}$ is the wedge between the lines labeled 1 and 3. These bounding lines are the steepest lines in the variety on either side of the time-like codirection $(1,0)$. Any line traveling to the right faster than line

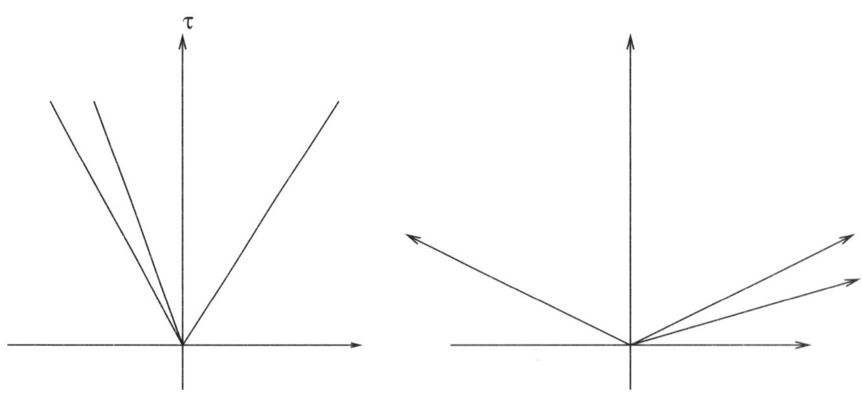

FIGURE 2.3.4. The rays of propagation in the figure on the right are orthogonal to the lines in the characteristic variety on the left.

3 on the right, has conormal that is time-like. Similarly lines whose velocity is more negative than the velocity of line 1 also have time-like conormal.

Exercise 2.3.4. *For a three-speed system with only positive speeds* $0 < \lambda_1 < \lambda_2 < \lambda_3$, *sketch the graphs of the characteristic variety, the rays* $x = t\lambda_j$, *and the forward time-like cone.*

Example 2.3.4. For Maxwell's equations, (2.1.1), $L = L_1$, and

$$\det L_1(\tau, \xi) = \tau^2 \big(\tau^2 - c^2|\xi|^2\big)^2 .$$

The characteristic variety sketched in Figure 2.3.5 is the union of the horizontal hyperplane $\{\tau = 0\}$ and the light cone $\tau^2 = c^2|\xi|^2$. The forward time-like cone is $\{\tau > c\,|\xi|\}$.

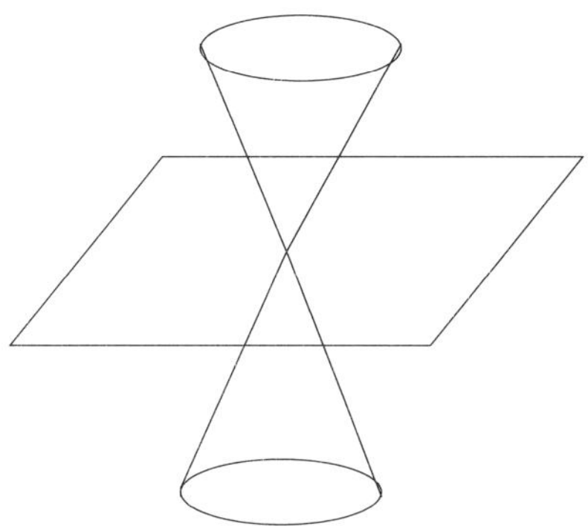

FIGURE 2.3.5. Maxwell's equation characteristic variety

Return to the general development with characteristic polynomial $p(y, \tau, \xi)$.

Definition. *For* $\xi \in \mathbb{R}^d \setminus 0$,

$$\tau_{\max}(y, \xi) \ := \ \max \big\{\tau \in \mathbb{R} \ : \ p(y, \tau, \xi) = 0\big\} . \tag{2.3.10}$$

Then $\tau_{\max}(y, \xi)$ is the largest eigenvalue of the hermitian symmetric matrix $-A_0^{-1/2}(\sum A_j\, \xi_j)A_0^{-1/2}$. Therefore it is uniformly lipschitzean in y, ξ. As a function of ξ, it is positively homogeneous of degree one and convex. The time-like cone and its closure have equations,

$$\mathcal{T}(y) \ = \ \big\{(\tau, \xi) \ : \ \tau > \tau_{\max}(y, \xi)\big\}, \qquad \overline{\mathcal{T}}(y) \ = \ \big\{(\tau, \xi) \ : \ \tau \geq \tau_{\max}(y, \xi)\big\} .$$

Definitions. *The* **closed forward propagation cone** *is the dual of* $\overline{\mathcal{T}}(y)$ *defined by*

$$\Gamma^+(y) \ := \ \Big\{(T, X) \in \mathbb{R}^{1+d} \ : \ \forall (\tau, \xi) \in \overline{\mathcal{T}}(y), \ \ T\tau + X\xi \ \geq \ 0\Big\} . \tag{2.3.11}$$

The section at $T = 1$ *is denoted*

$$\Gamma_1^+(y) \ := \ \Gamma^+ \cap \{T = 1\} .$$

The definition states that $\Gamma^+(y)$ is the set of all points that lie in the future of the origin $(0,0) \in \mathbb{R}^{1+d}$ with respect to each time-like $\tau t + \xi x$, $(\tau, \xi) \in \mathcal{T}(y)$. For points that are not in $\Gamma^+(y)$, there is a time function so that the point is in the past. This suggests that if $(T, X) \notin \Gamma^+(y)$ and $0 < \epsilon \ll 1$, then the point $y + (T, X)\epsilon$ should not be influenced by waves at y. This is verified in the following examples.

Examples 2.3.5. 1. For $\partial_t + c\partial_x$, $\Gamma^+ = \{X = cT\}$, $\Gamma^+_1 = \{c\}$, $\tau_{\max}(\xi) = -c\xi$.

2. For $\partial_t^2 - c^2\Delta$, $\Gamma^+ = \{|x| \le ct\}$, $\Gamma^+_1 = \{|x| \le c\}$, $\tau_{\max}(\xi) = c|\xi|$.

3. For Maxwell's equations, $\tau_{\max}$ is the same as for the second example, so Γ^+ is also the same.

Since $\mathcal{T}(y)$ is convex and contains an open cone about $\mathbb{R}_+(1, 0, \ldots, 0)$, it follows that $\Gamma^+_1(y)$ is a compact convex set. The next proposition gives three more relations between Γ^+ and $\mathcal{T}$.

Proposition 2.3.5. i. *The propagation cone* Γ^+ *has equations*

$$\Gamma^+(y) = \Big\{(T, X) \ : \ T \ge 0 \quad \text{and} \quad \forall \xi \ \ T\,\tau_{\max}(y, \xi) + X\xi \ge 0\Big\}. \tag{2.3.12}$$

ii. *The forward time-like cone is given by the duality*

$$\overline{\mathcal{T}}(y) = \Big\{(\tau, \xi) \ : \ \forall (T, X) \in \Gamma^+(y) \setminus 0\,, \ \ \tau T + X\xi \ \ge \ 0\Big\}. \tag{2.3.13}$$

$\mathcal{T}$ *is given by the same formula with* $\ge$ *replaced by* $>$.

iii.

$$\tau_{\max}(y, \xi) = \max_{X \in \Gamma^+_1(y)} -X\xi\,. \tag{2.3.14}$$

Proof. i. Suppress the y dependence. Take $(\tau, \xi) = (1, 0)$ in (2.3.11) to show that if $(T, X) \in \Gamma^+$, then $T \ge 0$. Γ^+ is defined by $T\tau + x\xi \ge 0$ when $\tau \ge \tau_{\max}$. Since $T \ge 0$ this holds if and only if it holds when $\tau = \tau_{\max}$, proving (2.3.12)

ii. From the definition of Γ^+ it follows that

$$\forall \ (T, X) \in \Gamma^+, \ \ \forall \ (\tau, \xi) \in \overline{\mathcal{T}}, \qquad T\tau + X\xi \ge 0.$$

Therefore,

$$\overline{\mathcal{T}} \subset \Big\{(\tau, \xi) \ : \ \forall (\tau, \xi) \in \overline{\mathcal{T}}, \ \ T\tau + X\xi \ge 0\Big\}. \tag{2.3.15}$$

To prove equality, reason as follows. If $(\underline{\tau}, \underline{\xi}) \notin \overline{\mathcal{T}}$, then $\underline{\tau} < \tau_{\max}(\underline{\xi})$ for some $\underline{\xi}$. The point $(\tau_{\max}(\underline{\xi}), \underline{\xi})$ is a boundary point of the closed convex set $\overline{\mathcal{T}}$, so there is a $(T, X) \ne 0$ so that

$$T\tau_{\max}(\underline{\xi}) + X\underline{\xi} = 0, \qquad \text{and} \qquad \forall \tau, \xi \in \overline{\mathcal{T}}, \ \ T\tau + X\xi \ge 0.$$

For $\tau = 1$ and ξ in a small neighborhood of the origin, $\tau, \xi \in \mathcal{T}$. It follows that $T \neq 0$. Therefore, $T\underline{\tau} + X.\underline{\xi} < 0$ showing that $(\underline{\tau}, \underline{\xi})$ is not in the set on the right of (2.3.15).

iii. In (2.3.12) it suffices to consider (T, X) with $T = 1$ and $X \in \Gamma_1^+$, so

$$\mathcal{T} \;=\; \Big\{(\tau, \xi) \in \mathbb{R}^{1+d} \;:\; \forall X \in \Gamma_1^+, \;\; \tau + X\xi > 0\Big\}.$$

Thus $\mathcal{T}$ has equation $\tau + \min\{X\xi \,:\, X \in \Gamma_1^+\} > 0$. Comparing with $\tau > \tau_{\max}(\xi)$ yields

$$\tau_{\max}(\xi) \;=\; -\min_{X \in \Gamma_1^+} X\xi \;=\; \max_{X \in \Gamma_1^+} -X\xi\,. \qquad \square$$

The cones $\mathcal{T}(y)$ and $\Gamma^+(y)$ are determined by the differential operator at the point y. The results of the preceding section give estimates on propagation that are independent of y. They involve the uniform objects of the next definition.

Definitions.

$$\tau_{\max}^{\mathrm{unif}}(\xi) \;:=\; \sup_{y \in \mathbb{R}^{1+d}} \tau_{\max}(y, \xi)\,,$$

$$\mathcal{T}_{\mathrm{unif}} \;:=\; \Big\{(\tau, \xi) \;:\; \tau \;>\; \tau_{\max}^{\mathrm{unif}}(\xi)\Big\}, \qquad \overline{\mathcal{T}}_{\mathrm{unif}} \;:=\; \Big\{(\tau, \xi) \;:\; \tau \;\geq\; \tau_{\max}^{\mathrm{unif}}(\xi)\Big\},$$

$$\Gamma_{\mathrm{unif}}^+ \;:=\; \Big\{(T, X) \;:\; \forall(\tau, \xi) \in \overline{\mathcal{T}}_{\mathrm{unif}}\,, \quad T\tau + X\xi \geq 0\Big\}.$$

Since $\tau_{\max}(y, \xi)$ is the largest eigenvalue of

$$-A_0(y)^{-1/2}\big(\sum A_j(y)\xi_j\big)A_0(y)^{-1/2},$$

it follows that $\tau_{\max}^{\mathrm{unif}}$ is uniformly lipschitzean. As a function of ξ, it is positive homogeneous of degree one, and convex.

Exercise 2.3.5. *Show that if the coefficients are constant outside a compact set, then* $\mathcal{T}_{\mathrm{unif}} = \bigcap_y \mathcal{T}(y)$. *Show that this need not be the case in general.*

Corollary 2.3.6. *Denote by* u *the solution of* $Lu = 0$, $u\big|_{t=0} = \delta(x - \underline{x})$. *Then for* $t \geq 0$,

$$\operatorname{supp} u \;\subset\; (0, \underline{x}) \;+\; \Gamma_{\mathrm{unif}}^+\,. \tag{2.3.16}$$

For general initial data one has

$$\operatorname{supp} u \;\subset\; \bigcup_{\underline{x} \in \operatorname{supp} u(0)} \Big(\underline{x} \;+\; \Gamma_{\mathrm{unif}}^+\Big)\,. \tag{2.3.17}$$

Remark. We show in the next section that this estimate is sharp in case L_1 has constant coefficients. A comparably sharp result for variable coefficients is proved in §2.5.

Proof. The second assertion follows from the first. Translating coordinates, it suffices treat the first with $\underline{x} = 0$.

The definition (2.3.5) is equivalent to $\Lambda(\xi)$ being the supremum over y of the eigenvalues of $-A_0(y)^{-1/2}\big(\sum A_j(y)\xi_j\big)A_0(y)^{-1/2}$. This in turn is equal to $\tau_{\max}^{\text{unif}}(-\xi)$. Corollary 2.3.3 implies that for $t > 0$

$$\text{supp } u \ \subset \ \Big\{(t,x) \ : \ \forall \xi \quad x\xi \ \le \ t\,\tau_{\max}^{\text{unif}}(-\xi)\Big\}.$$

Therefore, for all ξ, $t\,\tau_{\max}^{\text{unif}}(-\xi) - x\xi \ge 0$. Thus, replacing ξ by $-\xi$ shows that for all ξ, $t\,\tau_{\max}^{\text{unif}}(\xi) + x\xi \ge 0$. Equation (2.3.12) shows that this is equal to the set Γ^+_{unif}. □

Definition. *If Ω_0 is an open subset of $\{t = 0\}$ and Ω is a relatively open subset of $\{t \ge 0\}$, Ω is a* **domain of determinacy** *of Ω_0 when every smooth solution, u, of $Lu = 0$ whose Cauchy data vanish in Ω_0 must vanish in Ω.*

The idea is that the Cauchy data in Ω_0 determine the solution on Ω. Any subset of a domain of determinacy of Ω_0 is also such a domain. The union of a family of domains is one, so there is a largest domain of determinacy. The larger is Ω the more information one has, so the goal is to find large ones.

Definition. *If S_0 is a closed subset of $\{t = 0\}$ and S is a closed subset of $\{t \ge 0\}$, then S is a* **domain of influence** *of S_0 when every smooth solution u of $Lu = 0$ whose Cauchy data is supported in S_0 must be supported in S.*

The idea is that the Cauchy data in S_0 can influence the solution only in S. It does not assert that the data actually does influence in S. In that sense, the name is confusing. Any closed set containing a domain of influence is also such a domain. The intersection of a family of domains of influence of S_0 is such a domain, so there is a smallest one. The smaller is the domain the more information one has.

An immediate consequence of the definitions is that S is a domain of influence of S_0 if and only if $\Omega := \{t \ge 0\} \setminus S$ is a domain of determinacy of $\Omega_0 := \{t = 0\} \setminus S_0$.

The next result rephrases Corollary 2.3.6.

Corollary 2.3.7. i. *For any S_0, the set*

$$S \ := \ \bigcup_{x \in S_0} \ (x + \Gamma^+_{\text{unif}})$$

is a domain of influence.

ii. *For any Ω_0,*

$$\Big\{y \ : \ (y - \Gamma^+_{\text{unif}}) \cap \{t = 0\} \ \subset \ \Omega_0\Big\}$$

is a domain of determinacy.

2.4. Plane waves, group velocity, and phase velocities

Plane wave solutions are the multidimensional analogues of traveling waves $f(x - \lambda t)$ in the $d = 1$ case. They are the backbone of our short wavelength asymptotic expansions. It is not unusual for the analysis of a partial differential equation in science texts to consist only of a calculation of plane wave solutions. Gleaning information this way is part of the tool kit of both scientists and mathematicians. Plane waves will be used to show that Corollary 2.3.6 is precise in the case of operators with constant coefficient principal part.

Functions $f(x - \lambda t)$ generate the general solution of constant coefficient systems without lower order terms when $d = 1$. They are compositions with a linear function of (t, x). The multidimensional analogue is to seek solutions as compositions with real linear functions $t\tau + x\xi = y\eta$.

Definition. Plane waves *are functions which depend only on* $y\eta$ *for some* $0 \neq \eta \in \mathbb{R}^{1+d}$. *That is, functions of the form*

$$u(y) := a(y\eta), \qquad a : \mathbb{R} \to \mathbb{C}^N . \tag{2.4.1}$$

When $L = L_1(\partial)$ is homogeneous with constant coefficients and $u = a(y\eta)$ with $a \in C^1(\mathbb{R})$,

$$\partial_\mu a(y\eta) = a'(y\eta)\,\eta_\mu, \quad \text{so} \quad L(\partial_y)\,u = L_1(\partial_y)\,a(y\eta) = L_1(\eta)\,a'(y\eta)\,.$$[3]

Therefore $L u = 0$ precisely when a' takes values in the kernel of $L_1(\eta)$. There are nontrivial solutions if and only if η belongs to the characteristic variety.

Definition. *For* $(y, \eta) \in \operatorname{Char} L$, $\pi(y, \eta)$ *denotes the* **spectral projection** *of* $L_1(y, \eta)$ *onto its kernel. That is*

$$\pi(y, \eta) := \frac{1}{2\pi i} \oint_{|z|=r} \big(zI - L_1(y, \eta)\big)^{-1}\, dz, \tag{2.4.2}$$

where r *is chosen so small that* 0 *is the only eigenvalue of* $L_1(y, \eta)$ *in the disk* $|z| \leq r$.

If $L = L_1$ has constant coefficients and no lower order terms, then a necessary and sufficient condition for (2.4.1) to define a solution of $Lu = 0$ is that $\pi(\eta)\, a' = a'$. There are nontrivial solutions if and only if $\eta \in \operatorname{Char} L$. Except for an additive constant vector, the equation $\pi\, a' = a'$ is equivalent to

$$\pi(\eta)\, a = a\,. \tag{2.4.3}$$

[3]This formula is valid for arbitrary distributions $a \in \mathcal{D}'(\mathbb{R})$ once $a(y\eta)$ is carefully defined.

This *polarization* for a recurs in all of our formulas from geometric optics.

Exercise 2.4.1. *Compute all plane wave solutions for the following homogeneous constant coefficient operators.*

1. $\prod_{j=1}^{m}\left(\frac{\partial}{\partial t}+\lambda_j\frac{\partial}{\partial x}\right)$ *where x is one dimensional and the λ_j are distinct reals.*

2. $\partial_t + \operatorname{diag}(\lambda_1,\ldots,\lambda_d)\,\partial_x$ *with x and λ_j as in* **1**.

3. $\Box := \partial_t^2 - \Delta_x$.

Example 2.4.1. For $c = 1$ we compute all plane wave solutions of the homogeneous Maxwell's equations. That is, with $\rho = \mathbf{j} = 0$.

The solutions are $a(t\tau + x\xi)$ where a takes values in $\ker L_1(\eta)$. The vector $(\mathbf{e},\mathbf{b})$ belongs to the kernel if and only if $v := e^{iy\eta}(\mathbf{e},\mathbf{b}) := (E,B)$ satisfies $L_1(\partial)v = 0$. On such functions curl $\mapsto i\xi\wedge$ so Maxwell's equations read

$$\tau\mathbf{e} = \xi\wedge\mathbf{b}, \qquad \tau\mathbf{b} = -\xi\wedge\mathbf{e} = \mathbf{e}\wedge\xi\,. \tag{2.4.4}$$

The analysis splits according to $\tau = 0$ and $\tau \neq 0$.

When $\tau = 0$ there is a two dimensional space of solutions. The solutions are the vectors $(\mathbf{e},\mathbf{b})$ with $\mathbf{e}$ and $\mathbf{b}$ parallel to ξ. The dimension of the kernel is equal to the multiplicity of the root τ of the characteristic polynomial $\det L_1(\tau,\xi) = 0$. So $\tau = 0$ is a double root. The corresponding fields E, B have nonzero divergence, so these solutions do not satisfy the divergence free constraints of Maxwell.

When $\tau \neq 0$, the first equation in (2.4.4) implies that $\mathbf{e} \perp \xi$, so automatically, $\operatorname{div} E = 0$ and similarly $\operatorname{div} B = 0$. Multiply the first equation by τ to find

$$\tau^2\mathbf{e} = \xi\wedge\tau\mathbf{b} = \xi\wedge(-\xi\wedge\mathbf{e}) = |\xi|^2\mathbf{e},$$

the last equality using the divergence equation $\xi\cdot\mathbf{e} = 0$. Therefore $\tau^2 = |\xi|^2$. There is a two dimensional space of solutions parametrized by $\mathbf{e}\perp\xi$ as

$$\ker L_1(\tau,\eta) = \Big\{(\mathbf{e}\,,\,\mathbf{e}\wedge(\xi/\tau))\ :\ \mathbf{e}\perp\xi\Big\}\,.$$

When $|\mathbf{e}| = 1$ the three vectors $\mathbf{e}, \mathbf{e}\wedge\xi, \xi/\tau$ form an oriented orthonormal basis for $\mathbb{R}^3$.

For ξ fixed there is a two dimensional kernel for $\tau = 0$ and a two dimensional kernel for each of the roots $\tau = \pm|\xi|$. This accounts for the six roots of $\det L_1(\tau,\xi) = 0$, so the characteristic polynomial is $\tau^2(\tau^2 - |\xi|^2)^2$.

The plane wave solutions of Maxwell's equations, including the divergence constraints, are the functions $a(t\tau + x\xi)$ with $\xi \neq 0$, $\tau = \pm|\xi|$ with a taking values in $\ker L_1(\tau,\xi)$.

Example 2.4.2. *Exact solution of initial value problems with exponential initial data.* Suppose that $L(\partial) = L_1$ is symmetric hyperbolic, has constant coefficients, no lower order terms, and $A_0 = I$. If $\xi \in \mathbb{R}^d$, denote by $\lambda_1 < \lambda_2 < \cdots < \lambda_k$ the eigenvalues, not necessarily simple, of $\sum_j A_j \xi_j$. Then $(\tau, \xi) \in \operatorname{Char} L$ if and only if $\tau = -\lambda_j$ for some j. Then $\sum_j \pi(-\lambda_j, \xi) = I$ is the orthogonal spectral decomposition of $\mathbb{R}^d$ for $\sum_j A_j \xi_j$. The initial value problem

$$Lu = 0, \qquad u(0,x) = r\, e^{ix\xi}, \qquad r \in \mathbb{C}^N,$$

has the explicit solution

$$u = \sum_j e^{i(x\xi - \lambda_j t)}\, \pi(-\lambda_j, \xi)\, r$$

as a finite linear combination of plane waves. Expressing general initial data as a sum of terms of the form $e^{ix\xi} r$ by Fourier decomposition shows that general initial value problems have solutions expressed as combinations of plane waves of the important special form $e^{i(x\xi - \lambda t)}\, \rho$.

Example 2.4.3. *Circular and elliptical polarization for Maxwell's equations.* Consider the exponential plane wave solutions $e^{i(\tau t + x\xi)}\big(\mathbf{e}, \mathbf{e} \wedge (\xi/\tau)\big)$ with $\tau^2 = |\xi|^2$. This is a two dimensional space of solutions parametrized by $\mathbf{e} \perp \xi$. If one is interested in real solutions, for example the real and imaginary parts, one needs to combine with the plane waves associated to the point $(-\tau, -\xi)$ of the characteristic variety generating the four dimensional space of solutions whose electric fields have the form

$$e^{i(\tau t + x\xi)}\, \mathbf{e}_1 + e^{-i(\tau t + x\xi)}\, \mathbf{e}_2, \qquad \mathbf{e}_1 \perp \xi \quad \text{and} \quad \mathbf{e}_2 \perp \xi .$$

This four dimensional space of complex solutions has a real subspace of dimension four with electric fields given by

$$\sin(\tau t + \xi x)\, \mathbf{e}_1 + \cos(\tau t + \xi x)\, \mathbf{e}_2, \qquad \mathbf{e}_1 \perp \xi \quad \text{and} \quad \mathbf{e}_2 \perp \xi .$$

If the $\mathbf{e}_j$ are collinear, then the values of the electric field lie on a line. The solutions are called *linearly polarized.* If not, the values of the electric field lie on an ellipse. The solutions are called *elliptically polarized.* When the $\mathbf{e}_j$ are orthogonal and of equal length then the ellipse is a circle and the solutions are called *circularly polarized.* When the ellipse is not a circle, the direction of the major axis of the ellipse is called the *axis of elliptical polarization.*

For problems with lower order terms, there tend to be few plane wave solutions.

Exercise 2.4.2. *Compute all plane wave solutions for the Klein–Gordon equation* $\Box u + u = 0$. **Discussion.** For each ξ the set of plane wave solutions of the form $f(\tau t + \xi x)$ is finite dimensional in contrast with the wave equation that has solutions $f(\pm|\xi| t + \xi x)$ for arbitrary f.

A plane wave $u = a(\tau t + x\xi)$ has initial value $u_0(x) = u|_{t=0} = a(x\xi)$. Suppose that a' is not identically zero. The solution is said to have velocity $\mathbf{v}$ when it satisfies $u(t,x) = u_0(x - \mathbf{v}t)$. Compute

$$u_0(x - \mathbf{v}t) \;=\; a\big((x - \mathbf{v}t)\xi\big) \;=\; a\big(x\xi - (\mathbf{v}\xi)t\big)\,.$$

Therefore, the solution has velocity $\mathbf{v}$ if and only if $\mathbf{v}$ satisfies

(2.4.5) $$\mathbf{v}\xi \;=\; -\tau\,.$$

These velocities are called *phase velocities.* Note that for $d > 1$ there are many solutions $\mathbf{v}$. They differ by vectors that are orthogonal to ξ.

A remark on units is in order. If t has units of time and x has units of length, then since (τ, ξ) belongs to the dual space, τ (resp., ξ) has units 1/time (resp., 1/length). Therefore the solutions $\mathbf{v}$ of (2.3.4) have dimensions length/time of a velocity.

The amplitude of a plane wave solution is constant on hyperplanes $\{y\eta = \text{const.}\}$. The amplitude seen at $t = 0$ on the hyperplane $x\xi = 0$ appears at $t = 1$ on the hyperplane $\tau + x\xi = 0$. For example when $\xi = (1, 0, \ldots, 0)$, plane waves are functions of $t + x_1$. The amplitude achieved at $t = 0$ on the hyperplane $x_1 = 0$ are achieved at $t = 1$ on the hyperplane $x_1 = -1$.

Any constant vector that translates the first planes to the second is a reasonable candidate velocity. It is traditional in the scientific literature to call the special choice $\mathbf{v} = -\tau\xi/|\xi|^2$ the phase velocity. This choice is always parallel to ξ. This is the unique choice that is orthogonal to the hyperplanes $x\xi = \text{const.}$ with orthogonality measured by the euclidean metric. The reliance on the euclidean metric shows that this notion is not intrinsic. For the simple operator

$$\frac{\partial^2}{\partial t^2} - \frac{\partial^2}{\partial x_1^2} - 4\frac{\partial^2}{\partial x_2^2}$$

or its system analogue

$$\partial_t + \begin{pmatrix} 1 & 0 \\ 0 & -1 \end{pmatrix} \frac{\partial}{\partial x_1} + \begin{pmatrix} 0 & 2 \\ 2 & 0 \end{pmatrix} \frac{\partial}{\partial x_2}\,,$$

the velocity computed in this way does not correspond to the correct propagation velocity associated to the plane wave solutions. The correct velocity is the *group velocity* group velocity from §1.2 and below.

The fact that the phase velocity is not uniquely defined is not well known, so we pause to discuss it a little more. The givens are the space time $\mathbb{R}^{d+1}$ and two linear functions. The first is t which measures the passage of time. The second is the linear function $y\eta$ whose level surfaces are the surfaces of constant amplitude. From these givens by considering the propagation from time $t = 0$ to time $t = 1$, one constructs the pair of hyperplanes in $\mathbb{R}^d_x$ with equations $\tau + x\xi = 0$ and $x\xi = 0$. From two hyperplanes in $\mathbb{R}^d_x$ with $d > 1$, it is impossible to pick a well-defined vector that translates one to the other. The striking exception is the case $d = 1$, where one has two points and the translation is uniquely determined. The conclusion is that the traditional *phase velocity* makes sense only in one dimensional space. It was exactly in the case $d = 1$ that the notion of phase velocity was introduced in the late nineteenth century.

The next result shows that for operators with constant coefficient principal part, the bounds on the support in Corollary 2.3.6 are precise.

Proposition 2.4.1. i. *If $L = L_1(\partial)$ has constant coefficients and is homogeneous, then the fundamental solution u, $Lu = 0$, $u|_{t=0} = \delta(x)$ satisfies for $t \geq 0$,*

$$\operatorname{conv}\big(\operatorname{supp} u\big) \;=\; \Gamma^+ , \tag{2.4.6}$$

where the left-hand side denotes the convex hull.

ii. *If $L = L_1(\partial) + B(y)$ has constant coefficient principal part and v is the fundamental solution, then Γ^+ is the smallest convex cone containing $(\operatorname{supp} v) \cap \{t \geq 0\}$.*

Proof. i. In this case, for each $\xi \in \operatorname{Char} L$ there are plane wave solutions $f(x\xi - t\Lambda(\xi))$. Choosing $f(\sigma)$ vanishing for $s > 0$ and so that $0 \in \operatorname{supp} f$ shows that Theorem 2.3.2 is sharp in the sense that the solution does NOT vanish on any larger set $\{x\xi > t\kappa\}$, $\kappa < \Lambda(\xi)$.

The remainder of the proof of **i** consists of three exercises.

Exercise 2.4.3. *Show that this sharpness implies that the fundamental solution cannot be supported in $\{x\xi \leq t\kappa\}$ for any $\kappa < \Lambda(\xi)$.*

Exercise 2.4.4. *When $L = L_1(\partial)$ is homogeneous with constant coefficients, Γ^+ is the smallest convex cone in $t \geq 0$ that contains the support of the fundamental solution.*

Exercise 2.4.5. *Since $\delta(x)$ is homogeneous of degree $-d$, it follows that when $L = L_1(\partial)$, u is homogeneous of degree $-d$, and (2.4.6) follows.*

ii. Suppose next that $L_1(\partial) + B(y)$ has a fundamental solution v that for $t \geq 0$ is supported in convex cone $\tilde{\Gamma} \subset \{t > 0\}$. Prove that $\Gamma^+ \subset \tilde{\Gamma}$ as follows. Continue to denote by u the fundamental solution of L_1. Define for $\epsilon > 0$, $v^\epsilon(y) := \epsilon^d\, v(\epsilon y)$. Then v^ϵ is the unique solution of

$$\Big(L_1(\partial) + \epsilon\, B(\epsilon y)\Big) v^\epsilon \;=\; 0\,, \qquad v^\epsilon\big|_{t=0} \;=\; \delta(x)\,.$$

By hypothesis, v^ϵ is supported in $\tilde{\Gamma}$.

Exercise 2.4.6. *Prove that as* $\epsilon \to 0$, $v^\epsilon \to u$ *in* $C\big(\mathbb{R}\,;\,H^s(\mathbb{R}^d)\big)$ *for any* $s < -d/2$. **Hints.** Prove uniform bounds for v^ϵ, then extract a weakly convergent subsequence. Use the compactness of $H^s_{\rm comp} \to H^{\tilde s}$ for $s > \tilde s$.

These exercises imply that $\operatorname{supp} u \subset \tilde\Gamma$. Since $\tilde\Gamma$ is convex, part **i** implies that $\Gamma^+ \subset \tilde\Gamma$. □

The characteristic variety is defined by the polynomial equation in η

$$0 \;=\; p(y,\eta) \;:=\; \det L_1(y,\eta)\,.$$

The coefficient of τ^N in p is equal to $\det A_0$ which is real. For ξ real, the remaining coefficients are equal to $\det A_0$ times symmetric functions of the real roots τ. Therefore, the polynomial p has real coefficients.

Consider for y fixed the real algebraic variety $\{\eta \neq 0 : \det L_1(y,\eta) = 0\}$. By definition this is the fiber of $\operatorname{Char} L$ over y. There is a well-defined velocity associated to each point (y,η) of the characteristic variety with the property that the fiber over y is a smooth hypersurface on a neighborhood of $\eta \in \mathbb{R}^{1+d}$.

The next discussion takes place at fixed y and the y dependence of p is suppressed for ease of reading. For each $\xi \in \mathbb{R}^d$, the real algebraic variety $\{p(\tau,\xi) = 0\}$ has only real roots, the variety intersects each line $\tau \to (\tau,\xi)$ in at least one and no more than N points. Therefore the variety has dimension equal to d. The roots are equal to the eigenvalues of $-A_0^{-1/2}\big(\sum_j \xi_j A_j\big)A_0^{-1/2}$.

The fundamental stratification theorem of real algebraic varieties implies that except for a subvariety of dimension at most $d-1$, the fiber is locally a d dimensional real analytic subvariety of $\mathbb{R}^{1+d}_{\tau,\xi}$ (see Proposition 3.2.1.i). Such real analytic points are called *smooth points.*

Proposition 2.4.2. *At smooth points the characteristic variety has conormal vector that is not orthogonal to the time-like codirection* $(1,0,\dots,0)$*. Therefore, the variety is locally a graph*

$$\tau = \tau(\xi)\,, \qquad \tau(\cdot) \in C^\omega\,.$$

Proof. Denote by ν a conormal vector to $\operatorname{Char} L$ at a smooth point $\underline{\eta}$. We need to show that

$$\langle \nu\,,\,(1,0,\dots,0)\rangle \;\neq 0\,. \tag{2.4.7}$$

The proof is by contradiction. If (2.4.7) were not true, changing linear coordinates ξ yields

$$\nu = (0,1,0,\dots,0)\,.$$

Then near $\underline{\tau}, \underline{\eta}$, Char L has an equation

$$\xi_1 = f(\tau, \xi'), \qquad \xi' := (\xi_2, \dots, \xi_n), \tag{2.4.8}$$

with

$$f \in C^\omega, \qquad f(\underline{\tau}, \underline{\xi}') = \underline{\xi}_1, \qquad \partial_{\tau,\xi'} f(\underline{\tau}, \underline{\xi}') = 0. \tag{2.4.9}$$

For ξ' fixed equal to $\underline{\xi}'$ expand f about $\tau = \underline{\tau}$,

$$f(\tau, \underline{\xi}') = \underline{\xi}_1 + a(\tau - \underline{\tau})^r + \text{higher order terms}, \qquad a \in \mathbf{R} \setminus 0.$$

The gradient condition in (2.4.9) implies that the integer $r \geq 2$. Solving (2.4.8) for τ as a function of ξ_1 shows that for ξ_1 near $\underline{\xi}_1$, there are r distinct complex roots $\tau \approx \big[(\xi_1 - \underline{\xi}_1)/a\big]^{1/r}$. Since $r \geq 2$, real values of ξ_1 near $\underline{\xi}_1$ with $(\xi_1 - \underline{\xi}_1)/a < 0$ yield nonreal solutions τ.

The solutions are eigenvalues of a symmetric matrix and therefore real. This contradiction proves (2.4.7). □

At a smooth point $(\tau, \xi) \in \text{Char}\, L$, let H denote the hyperplane in (τ, ξ) space that is tangent to the fiber over y. Proposition 2.4.2 guarantees that H has a normal line $\mathbb{R}(1, \mathbf{v})$. The vector $\mathbf{v}$ is called the *group velocity* associated to (τ, ξ). If the characteristic variety is given locally by the equation $\tau = \tau(\xi)$, then the normal directions are multiples of the differential

$$d_{\tau,\xi}(\tau - \tau(\xi)) = \big(1, -\nabla_\xi \tau(\xi)\big), \qquad \text{therefore} \qquad \mathbf{v} = -\nabla_\xi \tau(\xi). \tag{2.4.10}$$

This agrees with the classic formula for the group velocity in §1.3.

Since $\tau(\xi)$ is homogeneous of degree one in ξ, the Euler homogeneity relation reads $\xi \cdot \nabla_\xi \tau(\xi) = \tau(\xi)$. Thus, the group velocity satisfies $\mathbf{v}\xi = -\tau$ which is the relation (2.4.5) defining the phase velocities. *For $d \geq 2$, at smooth points of the characteristic variety, the group velocity is the correct choice among the infinitely many phase velocities.*

Exercise 2.4.7. *Compute the group velocities for the equation $u_{tt} = u_{x_1x_1} + 4u_{x_2x_2}$ or its system analogue.* **Discussion.** For most ξ, the velocity is not parallel to ξ.

As ξ sweeps out the set of smooth points in the characteristic variety over y, the union of the conormal lines is called the *ray cone* at y. If the characteristic variety at y is given by the *irreducible* equation $q(\tau, \xi) = 0$, then the ray cone is the conic real algebraic variety in (T, X) defined by the polynomial equations

$$\Big\{(T, X) \neq 0 \;:\; \exists (\tau, \xi) \neq 0, \quad q(\tau, \xi) = 0, \quad dq(\tau, \xi) \neq 0, \quad X \frac{\partial q(\tau, \xi)}{\partial \tau} + T\, \nabla_\xi q(\tau, \xi) = 0\Big\}. \tag{2.4.11}$$

In formula (2.4.11), (T, X) are coordinates in the tangent space at y.

General solutions of constant coefficient initial value problems can often be expressed as a Fourier superposition of exponential solutions $e^{i(\tau t + x\xi)}$ with ξ real and $\tau \in \mathbb{C}$. When $\tau \notin \mathbb{R}$, these functions are not plane waves according to our restrictive definition. The reader should be aware that solutions of this exponential form are frequently called plane waves even when τ is not real. For constant coefficient operators and fixed ξ, the solutions of this form come from the (possibly complex) roots τ of the equation

$$\det L(i\tau, i\xi) = 0\,.$$

Exercise 2.4.8. *For each of the following dissipative wave equations show that there are no nonconstant plane wave solutions. Compute all solutions of the form $e^{i(t\tau + x\xi)}$ with real ξ.*

1. *The dissipative wave equation $\Box u + 2u_t = 0$.*
2. *The telegrapher's equation $u_{tt} - u_{xx} + 2u_t + u = 0$.*

Discussion. The roots have imaginary parts greater than zero. The only real root occurs for the dissipative wave equation and $\xi = 0$.

The roots $\tau(\xi)$, when they are nice functions of ξ, define the *dispersion relations* of the equation. Of particular importance is the case of conservative systems for which the roots are real.

Exercise 2.4.9. *Suppose that the constant coefficient $L(\partial_y)$ is symmetric hyperbolic and conservative in the sense that $B = -B^*$. This holds in particular if $B = 0$. Prove that solutions $u \in H^1([0,T] \times \mathbb{R}^d)$ of $Lu = 0$ satisfy the energy law*

$$\int_{\mathbb{R}^d} \langle\, A_0\, u(t,x)\,,\, u(t,x)\,\rangle\; dx \;=\; \textit{independent of time.} \tag{2.4.12}$$

Prove that the roots τ must be real in this case.

Exercise 2.4.10. *Find the dispersion relations for the wave equation, the Klein–Gordon equation $\Box u + u = 0$, and the Schrödinger equation $u_t + i\Delta_x u = 0$.*

For dissipative equations the exponential solutions decay in time which corresponds to roots τ with positive imaginary parts. Hadamard's analysis of well-posedness of initial value problems rests on the observation that one does not have continuous dependence on initial conditions if there exist exponential solutions whose imaginary parts tend to $-\infty$. A systematic use of exponential solutions in the study of initial value problems can be found in [Rauch, 1991, Chap. 3].

Finally, note that the *ellipticity* of a partial differential operator is defined by the absence of plane wave solutions.

Definition. *A first order system of partial differential operators* $L(y, \partial_y)$ *is* **elliptic** *at* $\underline{y}$ *if the constant coefficient homogeneous operator* $L_1(\underline{y}, \partial_y)$ *has no nonconstant plane wave solutions. This is equivalent to the invertibility of* $L_1(\underline{y}, \eta)$ *for all real* $\eta \neq 0$. *The system is elliptic on an open set if this property holds for all points* y *in the set.*

Exercise 2.4.11. *Verify the ellipticity of your favorite elliptic operators. This should include at least the laplacian and the Cauchy–Riemann system.* **Hint.** For the laplacian L_1 must be replaced by L_2 in the definition of ellipticity. In general, ellipticity of an mth order operator is equivalent to the invertibility of $L_m(y, \eta)$ for all real η.

2.5. Precise speed estimate

The last two sections showed that for operators with constant coefficient principal part, the forward propagation cone Γ^+ gives a good bound on the propagation of influence. In the variable coefficient case it is reasonable to expect that $\Gamma^+(y)$ describes the propagation at y. The central concept is that of influence curves which are curves whose tangents lie in the local propagation cones.

Definition. *A lipschitzean curve* $[a, b] \ni t \mapsto (t, \gamma(t))$ *is an* **influence curve** *when for Lebesgue almost all* t, $(1, \gamma'(t)) \in \Gamma^+(t, \gamma(t))$. *The curve* $(-t, \gamma(t))$ *a backward influence curve when* $(-1, \gamma') \in -\Gamma^+(-t, \gamma(t))$ *for almost all* t.

The uniform boundedness of the sets Γ_1^+ implies that influence curves are uniformly lipschitzean. Peano's existence proof, combining Euler's scheme and Ascoli's theorem, implies that influence curves exist with arbitrary initial values.

The convexity of the sets Γ_1^+ implies that uniform limits of influence curves are influence curves. Indeed, if $y^n(t) = (t, \gamma_n(t))$ is such a uniformly convergent sequence, then $\gamma_n' \in \Gamma_1^+(t, \gamma_n(t))$. This uniform bound allows us to pass to a subsequence for which

$$\gamma_n' \quad \to \quad f(t)$$

weak star in $L^\infty([a, b])$. Since the Γ_1^+ are convex, one has $f(t) \in \Gamma_1^+(t, \gamma(t))$ for almost all t.

On the other hand, g_n' converges to g' in the sense of distributions. Therefore $g' = f$ and therefore $(t, \gamma(t))$ is an influence curve.

This fact together with Ascoli's theorem implies that from any sequence of influence curves defined on $[a, b]$ whose initial points lie in a bounded set, one can extract a subsequence uniformly convergent to an influence curve.

Leray (1953) defined emissions as follows.

Definitions. *If $K \subset [0,T] \times \mathbb{R}^d$ is a closed set, the* **forward emission** *of K, denoted $\mathcal{E}^+(K)$, is the union of forward influence curves beginning in K. The* **backward emission**, *defined with backward influence curves, is denoted $\mathcal{E}^-(K)$.*

The compactness result on influence curves implies that emissions are closed subsets of $[0,T] \times \mathbb{R}^d$.

Theorem 2.5.1. *If $u \in C([0,T]\,;\, H^s(\mathbb{R}^d))$ satisfies $L\,u = 0$, then the support of u is contained in $\mathcal{E}^+\big(\operatorname{supp}(u|_{t=0})\big)$.*

Approximating data by smoother data and passing to the limit, it suffices to prove the theorem for s large. We treat $s > 1 + d/2$. In that case $u \in C^1$ and we can integrate by parts in the energy identity.

The proof, taken from [Rauch, 2005], uses fattened propagation cones. The fattening provides a little wiggle room. It also regularizes the boundary of emissions. The propagation cones Γ are fattened by shrinking the dual time-like cones $\mathcal{T}$.

Definitions. *For $\epsilon > 0$, define the shrunken time-like cone,*

$$\mathcal{T}^\epsilon(y) \;:=\; \Big\{(\tau,\xi) \in \mathbb{R}^{1+d} \;:\; \tau \;>\; \tau_{\max}(y,\xi) \;+\; \epsilon|\xi|\Big\}\,.$$

Define the fattened propagation cone, $\Gamma^{+,\epsilon}(y)$, to be the dual cone of $\overline{\mathcal{T}}^\epsilon(y)$. Denote by $\mathcal{E}^\pm_\epsilon$ the emissions defined with the $\Gamma^{\pm,\epsilon}$.

While $\Gamma^+(y)$ can be a lower dimensional cone, $\Gamma^{+,\epsilon}(y)$ has nonempty interior. The fattened cones, $\Gamma^{+,\epsilon}(y)$ are strictly convex, increasing in ϵ and contain $\Gamma^{+,\epsilon/2}(y) \setminus 0$ in their interior. In addition, $\bigcap_{0<\epsilon<1} \Gamma^{+,\epsilon}(y) = \Gamma^+(y)$.

Lemma 2.5.2. *To prove Theorem 2.5.1, it suffices to show that if $\epsilon > 0$, $\underline{y} \in [0,T] \times \mathbb{R}^d$, and $\mathcal{E}^-_\epsilon(\underline{y})$ does not meet $\operatorname{supp}(u|_{t=0})$, then u vanishes on $\mathcal{E}^-_\epsilon(\underline{y})$.*

Proof. One must show that

$$[0,T] \times \mathbb{R}^d \;\ni\; y \;\notin\; \mathcal{E}^+(\operatorname{supp}(u(0,\cdot)) \quad\Longrightarrow\quad y \notin \operatorname{supp}(u)\,.$$

If $y \notin \mathcal{E}^+(\operatorname{supp}(u(0,\cdot))$, then points $\underline{y}$ on a neighborhood of y in $[0,T] \times \mathbb{R}^d$ are also not in $\mathcal{E}^+(\operatorname{supp}(u(0,\cdot))$. Therefore it suffices to show that

$$[0,T] \times \mathbb{R}^d \;\ni\; \underline{y} \;\notin\; \mathcal{E}^+(\operatorname{supp}(u(0,\cdot)) \quad\Longrightarrow\quad u(\underline{y}) = 0\,.$$

From the definitions,

$$\underline{y} \notin \mathcal{E}^+(\operatorname{supp}(u(0,\cdot)) \quad\Longleftrightarrow\quad \mathcal{E}^-(\underline{y}) \,\cap\, \operatorname{supp}(u(0,\cdot)) \;=\; \phi\,.$$

The compact sets $\mathcal{E}_\epsilon^-(\underline{y})$ decrease as ϵ decreases, and

$$\bigcap_{0<\epsilon<1} \mathcal{E}_\epsilon^-(\underline{y}) = \mathcal{E}^-(\underline{y}).$$

Therefore, for ϵ small, $\mathcal{E}_\epsilon^-(\underline{y})$ does not meet $\operatorname{supp}(u|_{t=0})$.

To prove Theorem 2.5.1, it suffices to show that if $\mathcal{E}_\epsilon^-(\underline{y})$ does not meet $\operatorname{supp}(u|_{t=0})$, then $u(\underline{y}) = 0$. This is equivalent to the statement of the lemma. □

The next lemma is an accessibility theorem, in the sense of control theory.

Lemma 2.5.3. *If $\underline{y} = (\underline{t}, \underline{x}) \in [0, T[\times\mathbb{R}^d$ and $\epsilon \in]0,1[$, then there is a $0 < \delta \le T - \underline{t}$ so that*

$$\mathcal{E}_\epsilon^+(\underline{y}) \supset \{\underline{y} + \Gamma^{+,\epsilon/2}(\underline{y})\} \cap \{\underline{t} \le t \le t + \delta\}.$$

δ can be chosen uniformly for $\underline{y} \in [0, T[\times\mathbb{R}^d$. An analogous result holds for backward emissions.

Proof. Continuity of $\Gamma^{+,\epsilon}(y)$ with respect to y implies that there is a δ_0 so that

$$|y - \underline{y}| < \delta_0 \quad \Longrightarrow \quad \Gamma^{+,\epsilon}(y) \supset \Gamma^{+,\epsilon/2}(\underline{y}).$$

Therefore, a curve $(t, \gamma(t))$ with $(1, \gamma') \in \Gamma^{+,\epsilon/2}(\underline{y})$ is an influence curve so long as it stays in $\{|y - \underline{y}| < \delta_0\}$. Choose $0 < \delta \le \delta_0$ so that this holds for $t \in [\underline{t}, \underline{t} + \delta]$ on influence curves starting in $\{|y - \underline{y}| < \delta\}$. This completes the proof for y fixed.

The constants can be chosen uniformly since the $\Gamma^+(y)$ are uniformly continuous. □

Lemma 2.5.4. *For any q and $\epsilon > 0$, the set $\mathcal{E}_\epsilon^-(q)$ has lipschitzean boundary. The boundary has a tangent plane at almost all points. At such points, the conormals belong to $\mathcal{T} \cup -\mathcal{T}$.*

Proof. Suppose that $(\underline{t}, \underline{x}) = \underline{y} \neq q$ belongs to the boundary of $\mathcal{E}_\epsilon^-(q)$. Then for t close to and greater than $\underline{t}$, $\underline{y} + \Gamma^{+,\epsilon/4}(\underline{y})$ belongs to the complement of $\mathcal{E}_\epsilon^-(q)$. To prove this, note that if there were points

$$z = (t, x) \in \underline{y} + \Gamma^{+,\epsilon/4}(\underline{y}) \cap \mathcal{E}_\epsilon^-(q) \quad \text{with} \quad \underline{t} < t < \underline{t} + \delta$$

as in Figure 2.5.1, below, then $\mathcal{E}_\epsilon^-(z) \subset \mathcal{E}_\epsilon^-(q)$. Lemma 2.5.3 implies that $\mathcal{E}_\epsilon^-(z)$ contains a neighborhood of $\underline{y}$ contradicting the fact that $\underline{y}$ is a boundary point.

On the other hand, since $\underline{y}$ belongs to the emission, $\mathcal{E}_\epsilon^-(\underline{y})$ belongs the emission. Lemma 2.5.3 implies that for $\underline{t} > t > t - \delta$, the backward emission

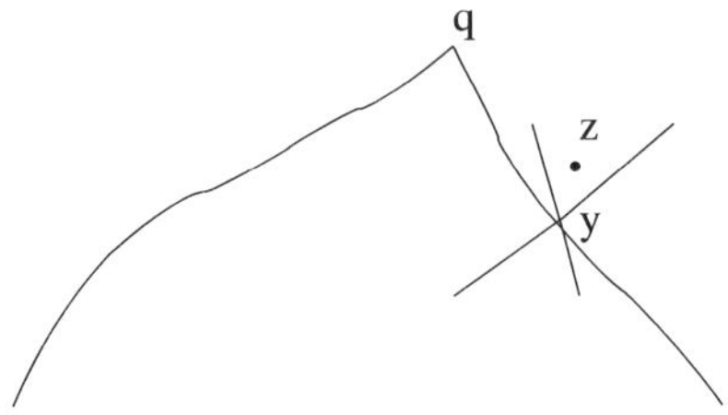

FIGURE 2.5.1.

from $\underline{y}$ contains $\underline{y} - \Gamma_1^{+,\epsilon/2}(\underline{y})$. The interior of that set is thus a subset of the interior of the emission. Thus the boundary of the emission near $\underline{y}$ is sandwiched between $\underline{y} + \Gamma^{+,\epsilon/4}(\underline{y})$ and $\underline{y} - \Gamma^{+,\epsilon/4}(\underline{y})$ as in Figure 2.5.1.

At $\underline{y} = q$, it is also true that the boundary of $\mathcal{E}_\epsilon(q)$ is sandwiched between $\underline{y} + \Gamma^{+,\epsilon/4}(\underline{y})$ and $\underline{y} - \Gamma^{+,\epsilon/4}(\underline{y})$.

This shows that at all points, the boundary satisfies a two sided cone condition with cones $\pm\Gamma^{+,\epsilon/4}(\underline{y})$ that are lipschitzean in their dependence on $\underline{y}$. This proves the desired Lipschitz regularity of the boundary. The differentiability then follows from Rademacher's theorem asserting the almost everywhere differentiability of Lipschitz functions.

At points $\underline{y}$ of differentiability, the tangent plane locally separates the cones $\underline{y} \pm \Gamma^{+,\epsilon/4}(\underline{y})$. Therefore the conormal to the plane that is positive on $(1, 0, \dots, 0)$ is positive on $\Gamma^{+,\epsilon/4}(\underline{y})$, so it is positive on the smaller cone $\underline{y} \pm \Gamma^{+}(\underline{y})$. This implies that the conormal belongs to $\mathcal{T}(\underline{y})$. □

Proof of Theorem 2.5.1. We verify the criterion of Lemma 2.5.2. For $0 < t < \underline{t}$ define

$$\Omega_t := \mathcal{E}_\epsilon^-(\underline{y}) \cap [0, t] \times \mathbb{R}^d, \qquad \mathcal{B}_t := \partial\mathcal{E}_\epsilon^-(\underline{y}) \cap]0, t[\times\mathbb{R}^d .$$

The set $\mathcal{B}_t$ is the lateral boundary of Ω_t.

Define

$$\phi(s) := \int_{\mathcal{E}_\epsilon^-(\underline{y})\cap\{t=s\}} |u(s, x)|^2 \, dx .$$

The energy conservation law (2.3.2) implies that for $0 < t < \underline{t}$, one has

$$0 = \int_{\Omega_t} \partial_t(u, u) + \sum_j \partial_j(A_j u, u) + (Zu, u) \; dx\, dt.$$

Integrating by parts yields

$$\phi(t) - \phi(0) + \int_{\mathcal{B}_t} (\sum_{\mu=0}^{d} \nu_\mu A_\mu u, u) \; d\sigma + \int_{\Omega_t} (Zu, u) \; dx\, dt = 0, \tag{2.5.1}$$

where ν is the unit outward normal to $\mathcal{B}$ and $d\sigma$ is the element of d dimensional surface area in the boundary $\mathcal{B}$. The conormal ν is almost everywhere well defined with respect to surface area. From Lemma 2.5.4, ν belongs to $\mathcal{T} \cup -\mathcal{T}$. Since $\mathcal{E}_\epsilon^-(\underline{y})$ is a backward emission, the outward conormals belong to $\mathcal{T}$. Therefore the matrix $\sum_\mu \nu_\mu A_\mu$ is positive. Using this in (2.5.1) yields

$$\phi(t) \;\le\; \phi(0) + \|Z\|_{L^\infty([0,T]\times\mathbb{R}^d)} \int_0^t \phi(s)\, ds\,.$$

Since $\mathcal{E}^-$ does not meet the support of $u|_{t=0}$, $\phi(0) = 0$. Gronwall's Lemma implies that $\phi(t) = 0$ for $0 \le t \le \underline{t}$. This shows that u vanishes in $\mathcal{E}_\epsilon^-(\underline{y})$, and so completes the proof. □

2.6. Local Cauchy problems

Once finite speed is established, it is not hard to show that the Cauchy problem has unique solutions for data and operators only defined locally. This is needed in Chapter 5 where some operators are defined only where locally defined phases exist.

Assumption. *Suppose that $0 < T < \infty$ and $\mathcal{O}$ is a bounded open subset of $\mathbb{R}^{1+d}$ lying on one side of its smooth compact boundary. Let $\Omega = \mathcal{O} \cap \{0 < t < T\}$. Assume that $\overline{\Omega}$ is a domain of determinacy for $L(y, \partial)$ in the sense that*

$$(2.6.1) \qquad \forall\, y \in \overline{\Omega}, \qquad \mathcal{E}^-(y) \cap \{0 \le t \le T\} \;\subset\; \overline{\Omega}.$$

Denote by $\overline{\Omega}_\sigma := \{(t,x) \in \overline{\Omega} \;:\; t = \sigma\}$ the section at time σ.

Theorem 2.6.1. *Suppose that L is a symmetric hyperbolic operator with coefficients defined on $\overline{\Omega}$ with partial derivatives of all orders bounded. If $g \in C^\infty(\overline{\Omega}_0)$ and $f \in C^\infty(\overline{\Omega})$, then there is one and only one solution $u \in C^\infty(\overline{\Omega})$ of the initial value problem*

$$(2.6.2) \qquad Lu = f \quad \text{on} \quad \overline{\Omega}, \qquad u(0) = g \quad \text{on} \quad \overline{\Omega}_0\,.$$

If $s \in \mathbb{N}$, there is a constant $C = C(L,s)$ so that for all f, g and $0 \le t \le T$,

$$(2.6.3) \quad \sum_{|\alpha|\le s} \|\partial_y^\alpha u(t)\|_{L^2(\Omega_t)}$$
$$\le C\Big(\sum_{|\alpha|\le s} \|\partial_y^\alpha u(0)\|_{L^2(\Omega_t)} + \int_0^t \sum_{|\alpha|\le s} \|\partial_y^\alpha f(\sigma,x)\|_{L^2(\Omega_\sigma)}\, d\sigma \Big)\,.$$

Proof. The first step is to construct a symmetric hyperbolic operator $\widetilde{L}$ defined everywhere on $\mathbb{R}^{1+d}$ and equal to L on $\overline{\Omega}$. To do this, first extend the coefficients A_μ to smooth hermitian valued functions on $\mathbb{R}^{1+d}$ with uniformly bounded derivatives. Extend B similarly but without symmetry requirements.

By continuity, the extended coefficient A_0 is strictly positive definite on a neighborhood of $\overline{\Omega}$. This allows one to construct a possibly different extension with bounded derivatives that is strictly positive everywhere. That completes the construction of $\widetilde{L}$.

Extend f and g to $\widetilde{f} \in C_0^\infty(\mathbb{R}^{1+d})$ and $\widetilde{g} \in C_0^\infty(\mathbb{R}^d)$. Solving the tilde initial value problem on $\mathbb{R}^{1+d}$ constructs a solution.

The energy estimate for $\widetilde{L}$ implies that

$$\sum_{|\alpha|\le s} \|\partial_y^\alpha u(t)\|_{L^2(\Omega_t)}$$
$$\le C(s,\widetilde{L})\Big(\sum_{|\alpha|\le s} \|\partial_y^\alpha \widetilde{g}\|_{L^2(\mathbb{R}^d)} + \int_0^t \sum_{|\alpha|\le s} \|\partial_y^\alpha \widetilde{f}(\sigma,x)\|_{L^2(\mathbb{R}^d)}\, d\sigma \Big).$$

The standard extension process for Sobolev functions shows that the infimum of the right-hand side over extensions $\widetilde{f}$ and $\widetilde{g}$ is a norm equivalent to the right-hand side of (2.6.3).

To prove uniqueness, reason as follows. If u is a solution, choose an extension $\widetilde{u} \in C_0^\infty(\mathbb{R}^{1+d})$. Then $\widetilde{L}\widetilde{u}$ vanishes in $\overline{\Omega}$ and $\widetilde{u}|_{t=0}$ vanishes in $\overline{\Omega}_0$.

The domain of determinacy hypothesis implies that for $y \in \overline{\Omega}$ and $0 \le t \le T$,

$$\mathcal{E}^-(y,\widetilde{L}) \;=\; \mathcal{E}^-(y,L) \;\subset\; \overline{\Omega}\,.$$

Exercise 2.6.1. *Prove this.*

The sharp finite speed result for $\widetilde{L}$ implies that $\widetilde{u}|_{\overline{\Omega}} = 0$. Since $\widetilde{u}$ is equal u on $\overline{\Omega}$, this completes the proof. □

Remark. Solutions with finite regularity can be constructed by an approximation argument using (2.6.3).

Appendix 2.I. Constant coefficient hyperbolic systems

For constant coefficients this section sketches a rough classification of hyperbolic systems (see [Gårding, 1951] and [Hörmander, 1983]).

Consider the Cauchy problem for the differential operator

$$L \;=\; A_0\partial_t \;+\; \sum_{j=1}^{d} A_j\,\partial_j \;+\; B\,,$$

where A_μ and B are constant $N\times N$ matrices. Hyperbolic systems are those for which the initial value problem

$$L\,u = f, \qquad u|_{t=0} = g,$$

has a unique solution for f, g being arbitrary elements of a suitably large family of functions.

The first observation is that $t = 0$ *must be noncharacteristic, that is* A_0 *must be invertible.* In the opposite case, $\operatorname{Rg} A_0$ is a proper subspace of $\mathbb{C}^N$. The differential equation at $t = 0$ then implies that

$$\sum_{j=1}^{d} A_j \partial_j g + Bg - f(0,x) \ \in\ \operatorname{Rg} A_0 \,.$$

This is a nontrivial linear constraint on the data f, g. So to have solvability for reasonably arbitrary data, it is necessary that A_0 is invertible. In that case, multiplying by A_0^{-1} reduces to the case $A_0 = I$, which we assume in the remainder of this appendix.

The Fourier transform yields the solution of the Cauchy problem with $f = 0$,

$$\hat{u}(t,\xi) \ =\ e^{t\left(-i\sum_{j=1}^{d} A_j\xi_j - B\right)}\ \hat{g}(\xi)\,.$$

Hyperbolic systems are those for which this product makes sense for a large class of g. At the very least one would like to solve with g for which the values of g in a neighborhood of a point of x are independent of the values at $\underline{x} \neq x$. This property is not shared by the real analytic data for which solvability is a consequence of the Cauchy–Kovalevskaya theorem. The problem of identifying hyperbolic system at least requires finding solvability in a class of functions without analyticity properties.

Considering $B = 0$, one sees that it is bad if $\sum A_j\xi_j$ has an eigenvalue λ with nonvanishing imaginary part. Replacing ξ by $-\xi$, one may suppose that $\operatorname{Im}\lambda > 0$. Then the matrix $e^{-i\sum A_j\xi_j}$ grows exponentially on a conic neighborhood $a\xi$ with $a \to \infty$. Thus for $e^{-i\sum A_j\xi_j}\,\hat{g}(\xi)$ to be the transform of a nice object, the transform $\hat{g}$ must decay exponentially in such directions. This is a microlocal real analyticity condition on g which shows that such systems must be rejected. The argument works as well when a lower order term $B \neq 0$ is added as the amplification matrix still grows exponentially. The conclusion is that *only systems so that for real* ξ, $\sum_j A_j\xi_j$ *has only real eigenvalues should be called hyperbolic.*

It is not difficult to show that the condition of real spectrum is equivalent to a bound

$$\exists C, \quad \forall \xi \in \mathbb{R}^d, \ 0 \le t \le 1 \quad \left\| e^{t\left(-i\sum A_j\xi_j - B\right)} \right\| \ \le\ C\, e^{(|\xi|^{(N-1)/N})}\,.$$

Thus, for such operators the Cauchy problem is solvable for data whose Fourier transform decays as $e^{-|\xi|^\nu}$ with $1 > \nu > (N-1)/N$. By definition, this is the class of Gevrey data $G^{1/\nu}$. This class is good in the sense that there are Gevrey partitions of unity, and the values at distinct points are entirely independent. A profound result of Bronsteĭn (1980) proves that variable coefficient problems, whose coefficients are $G^{1/\nu}$ smooth and so that $\sum_j A_j(t,x)\xi_j$ has only real eigenvalues, yield good Cauchy problems for $G^{1/\nu}$ data. The result has the weakness that the value of a solution at t, x with $t > 0$ depends on an infinite number of derivatives of the data and coefficients.

The next class of hyperbolic systems, introduced and analyzed by Gårding with earlier contributions of Petrowsky are defined so that the dependence is reduced to a finite number of derivatives. For this it is necessary and sufficient that one has a bound

$$\exists C, m \quad \forall \xi \in \mathbb{R}^d,\ 0 \le t \le 1, \qquad \left\| e^{t\left(-i\sum A_j\xi_j - B\right)} \right\| \le C\,\langle \xi \rangle^m . \tag{2.I.1}$$

When (2.I.1) is satisfied, the Cauchy problem is solvable with loss of no more than m derivatives in the sense that if $g \in H^s(\mathbb{R}^d)$, then there is a solution $u \in C(\mathbb{R}\ ;\ H^{s-m}(\mathbb{R}^d))$.

The bound (2.I.1) is equivalent to the following eigenvalue condition which depends on the lower order term B,

(2.I.2)

$$\exists C,\ \forall \xi \in \mathbb{R}^d, \qquad \text{the eigenvalues of } \Big(\sum_j A_j\xi_j + iB\Big) \text{ satisfy } |\operatorname{Im}\lambda| \le C .$$

This is the *standard definition of hyperbolicity for constant coefficient systems.*

Example 2.I.1. The system

$$\partial_t + \begin{pmatrix} 0 & 1 \\ 0 & 0 \end{pmatrix} \partial_x \tag{2.I.3}$$

satisfies (2.I.1) with $m = 1$, but

$$\partial_t + \begin{pmatrix} 0 & 1 \\ 0 & 0 \end{pmatrix} \partial_x + \begin{pmatrix} 0 & 0 \\ 1 & 0 \end{pmatrix}, \tag{2.I.4}$$

which is obtained by adding a lower order term does not satisfy for any m.

Exercise 2.I.1. *Verify.*

Denote the variables as (u, v). The system (2.I.4) is equivalent to

$$u_t + v_x = 0, \qquad v_t + u = 0 . \tag{2.I.5}$$

Eliminating v yields $u_{tt} + u_x = 0$. The Cauchy problem for this equation (and the original system) is solvable for data in G^μ for $\mu < 2$, and not for data with only a finite number of derivatives. This equation is the sideways heat equation discussed in [Rauch, 1991, §3.9]. The system (2.I.3) yields the hyperbolic equation, $u_{tt} = 0$. The solution of the Cauchy problem

$$u_{tt} \;=\; 0, \qquad u\big|_{t=0} \;=\; g_0, \quad u_t\big|_{t=0} \;=\; g_1,$$

is $u = g_0 + tg_1$. If $g \in H^s$ and $g_1 \in H^{s-1}$, the solution is continuous with values in H^{s-1}. This is a loss of one derivative compared to what one would have for the wave equation with the same initial data. This loss reflects the fact that (2.I.1) is satisfied for $m = 1$ and for no smaller value.

The strongest notion of hyperbolicity corresponds to solvability without loss of derivatives. This requires the bound,

$$\text{(2.I.6)} \qquad \sup_{\xi \in \mathbb{R}^d,\ 0 \le t \le 1} \left\| e^{t\left(-i\sum A_j\xi_j - B\right)} \right\| \;<\; \infty\,.$$

Estimate (2.I.6) is the case $m = 0$ of (2.I.1).

Proposition 2.I.1. *Condition (2.I.6) is equivalent to*

$$\text{(2.I.7)} \qquad \sup_{\xi \in \mathbb{R}^d} \left\| e^{-i\sum A_j\xi_j} \right\| \;<\; \infty\,.$$

Proof. We prove that (2.I.6) implies (2.I.7). The opposite implication is similar. Condition (2.I.6) is a uniform estimate

$$\sup_{0 \le t \le 1,\ \xi \in \mathbb{R}^d} \left\| e^{t(A-B)} \right\| \;\le\; M, \quad \text{where} \qquad A = \sum -iA_j\xi_j\,.$$

We want an estimate $\sup_{\xi \in \mathbb{R}^d} \|u(1)\| < \infty$, where u satisfies $u' = Au$ with unit length initial data. Write the equation for u as

$$u' = (A - B)u \;+\; Bu,$$

so

$$u(t) \;=\; e^{t(A-B)}u(0) \;+\; \int_0^t e^{(t-s)(A-B)}\, B\, u(s)\, ds\,.$$

For $t \le 1$ one has

$$\|u(t)\| \;\le\; M \;+\; \int_0^t M\,\|B\|\,\|u(s)\|\, ds\,.$$

Gronwall's inequality implies that $\|u(1)\| \le M\, e^{M\|B\|}$ proving (2.I.7). □

In particular, the condition (2.I.6) is stable under lower order perturbations.

In order for (2.I.7) to be satisfied, it is necessary that for all real ξ the matrix $\sum_j A_j\xi_j$ is similar to a real diagonal matrix. This follows by considering only the restriction to real multiples of the given ξ. In particular the system (2.I.3) does not satisfy (2.I.6).

When (2.I.7) holds, $f = 0$, and $g \in H^s(\mathbb{R}^d)$, then there is a solution continuous in time with values that lie in H^s. A Gronwall argument shows that this property is also valid if one adds a variable coefficient lower order term, that is for $L = \partial_t + \sum A_j\partial_j + B(t,x)$ with B satisfying (2.1.6).

Kreiss' Matrix Theorem 2.I.2 (Kreiss, 1963). *If V is a complex normed vector space, then $A \in \mathrm{Hom}(V)$ satisfies $\sup_{\sigma\in\mathbb{R}} \|e^{i\sigma A}\| < \infty$ if and only if A is diagonalizable with real eigenvalues. Write $A = \sum_j \lambda_j \pi_j$ with distinct real λ_j and π_j the projector along $\mathrm{Rg}\,(A - \lambda_j I)$ onto $\ker\,(A - \lambda_j I)$. Then*

$$\max_j \|\pi_j\| \;\le\; \sup_{\sigma\in\mathbb{R}} \|e^{i\sigma A}\| \;\le\; \sum_j \|\pi_j\|\,. \tag{2.I.8}$$

Proof. The diagonalizability characterisation is immediate from the Jordan form.

To prove (2.I.8) write

$$e^{i\sigma A} \;=\; \sum_j e^{i\sigma\lambda_j}\,\pi_j\,. \tag{2.I.9}$$

The triangle inequality shows that

$$\forall \sigma \in \mathbb{R}, \qquad \big\|e^{i\sigma A}\big\| \;\le\; \sum_j \|\pi_j\|\,.$$

For the other half of (2.I.8), multiply (2.1.9) by $e^{-\sigma\lambda_k}$ and integrate $d\sigma$ to show that

$$\pi_k \;=\; \lim_{T\to\infty} \frac{1}{T}\int_0^T e^{-i\sigma\lambda_k}\, e^{iA\sigma}\, d\sigma\,.$$

The integral triangle inequality implies that $\big\|\pi_k\big\| \le \sup_{\sigma\in\mathbb{R}} \big\|e^{i\sigma A}\big\|$. □

The result (2.I.8) is often rephrased as follows. The map

$$V \ni u \quad\mapsto\quad K(u) \;:=\; (\pi_1 u\,,\, \pi_2 u\,,\, \dots\,,\, \pi_m u) \;\in\; \bigoplus \ker(A - \lambda_j I)$$

has norm ≤ 1 if the direct sum is normed by the maximum of the norms. Since $u = \sum_j \pi_j u$ and there are at most $N := \dim V$ summands, one has $\max_j \|\pi_j u\| \ge \|u\|/N$ proving that $\|K^{-1}\| \le N$. And KAK^{-1} is diagonal. Thus A is diagonalized by a transformation with $\|K\|\,\|K^{-1}\| \le N \sup_\sigma \|e^{i\sigma A}\|$. The last condition is invariant when K is replaced by $c\,K$.

Therefore, (2.I.7) is satisfied if and only if *there is an invertible matrix valued $K(\xi)$ with K and K^{-1} in $L^\infty(\mathbb{R}^d)$, and so that for all $\xi \in \mathbb{R}^d$, $K(\xi)\,(\sum A_j\xi_j)\,K^{-1}(\xi)$ is diagonal and real.*

Remarks. 1. The condition in italics is satisfied when the A_j are hermitian symmetric in which case K can be chosen unitary and the projectors have norm 1.

2. By homogeneity it suffices to consider ξ with $|\xi| = 1$.

3. The condition is satisfied when for all ξ with $|\xi| = 1$, $\sum_j A_j\xi_j$ is diagonalizable and the multiplicity of its eigenvalues is independent of ξ. In this case $\pi_j(\xi)$ is smooth on $\mathbb{R}^d \setminus 0$.

4. A special case of **3** is when $\sum A_j\xi_j$ has N distinct real eigenvalues for all $\xi \neq 0$. Such systems are called *strictly hyperbolic.*

5. In §5.4.4 we prove that for $d > 1$ and for most symmetric hyperbolic systems, the map $u(0) \mapsto u(\underline{t})$ is unbounded on L^p for $p \neq 2$ unless the A_j are simultaneously diagonalizable. Gués and Rauch (2006) prove the more general result weakening the symmetric hyperbolicity to (2.I.7).

Exercise 2.I.2. i. *Prove* **3.**

ii. *Prove that when* **3** *holds, K can be chosen smooth and homogeneous of degree 0 on $\xi \neq 0$.*

Appendix 2.II. Functional analytic proof of existence

This appendix proves Theorem 2.2.1 from the a priori estimate (2.1.18) by an abstract argument. The idea of using the Sobolev spaces for negative s to give a particularly elegant version dates at least to [Lax, 1955]. The argument uses Lax's duality in the form $L^1([0,T]\,;H^s(\mathbb{R}^d))' = L^\infty([0,T]\,;\,H^{-s}(\mathbb{R}^d))$. We abuse notation in the usual way by writing the duality of H^s and H^{-s} as an integral.

Example 2.II.2. For $\delta' \in H^{-2}(\mathbb{R})$ and $f \in H^2(\mathbb{R})$, $\int \delta'(x)\, f(x)\, dx = f'(0)$ is not an integral.

Proof of Theorem 2.2.1. Step 1. If f, g are as in Theorem 2.2.1, then there is a solution

$$u \in L^\infty([0,T]\,;\,H^s(\mathbb{R}^d)) \,\cap\, C([0,T]\,;\,H^{s-1}(\mathbb{R}^d))\,.$$

Let

$$\Psi \;:=\; \Big\{\psi \in \bigcap_k C^k([0,T]\,;\,H^k(\mathbb{R}^d)) \;:\; \psi(T) = 0\Big\}$$

and

$$L^\dagger w \;:=\; -\partial_t w - \sum \partial_j(A_j w) + B^\dagger w\,,$$

the transposed operator, so

$$\int_{\mathbb{R}^{1+d}} L\phi\;\psi\;dt\,dx \;=\; \int_{\mathbb{R}^{1+d}} \phi\;L^\dagger\psi\;dt\,dx$$

for all smooth ϕ, ψ whose supports intersect in a compact set. Then $-L^\dagger$ is symmetric hyperbolic. Proposition 2.1.1 with initial time T shows that (2.II.1)

$$\forall\,\psi \in \Psi\,, \qquad \sup_{0\le t\le T}\, \|\psi(t)\|_{H^{-s}(\mathbb{R}^d)} \;\le\; C(s,L)\int_0^T \|L^\dagger\psi(\sigma\|_{H^{-s}(\mathbb{R}^d)}\;d\sigma\,.$$

In particular, $L^\dagger$ in injective on Ψ. Then $V := L^\dagger\Psi$ is a linear subspace of $L^1([0,T]\,;\,H^{-s}(\mathbb{R}^d))$. Estimate (2.II.1) asserts that

$$(L^\dagger)^{-1} : V \to C([0,T]\,;\,H^{-s}(\mathbb{R}^d))$$

is continuous. Since $f \in L^1([0,T]\,;\,H^s(\mathbb{R}^d))$, the linear functional $\ell : V \to \mathbb{C}$, defined at $v = L^\dagger\psi$ as

$$\ell(v) \;:=\; \int_0^T \psi\, f\; dt\, dx \;-\; \int \psi(0,x)\, g(x)\; dx,$$

is continuous. The Hahn–Banach theorem[4] implies that there is an extension of ℓ to all of $L^1([0,T]\,;\,H^{-s}(\mathbb{R}^d))$ so there is a $u \in L^\infty([0,T]\,;\,H^s(\mathbb{R}^d))$ so that

$$\ell(v) \;=\; \int_0^T u(t,x)\, v(t,x)\; dt\, dx\,.$$

This proves that for all $\psi \in \Psi$,

$$\text{(2.II.2)}\qquad \int_0^T\!\!\int u(t,x)\; L^\dagger\psi(t,x)\; dt\, dx$$
$$= \int_0^T\!\!\int f(t,x)\;\psi(t,x)\; dt\, dx \;-\; \int \psi(0,x)\, g(x)\; dx\,.$$

Exercise 2.II.1. *Prove that $u \in C([0,T]\,;\,H^{s-1}(\mathbb{R}^d))$. Then prove that (2.II.2) implies that $Lu = f$ and $u|_{t=0} = g$.*

Warning. The x integrals in (2.II.2) are pairings of $H^s(\mathbb{R}^d)$ and $H^{-s}(\mathbb{R}^d)$ not integrals.

This completes the first step.

Step 2. For f, g as in the theorem, choose

$$f_n \in C_0^\infty(\mathbb{R}^{1+d}), \qquad g_n \in C_0^\infty(\mathbb{R}^d),$$

with

$$f_n \to f \quad \text{in} \quad L^1([0,T]\,;\,H^s(\mathbb{R}^d)), \qquad g_n \to g \quad \text{in} \quad H^s(\mathbb{R}^d).$$

Denote by $u_n \in L^\infty([0,T]\,;\,H^{s+1}(\mathbb{R}^d)) \,\cap\, C^1([0,T]\,;\,H^s(\mathbb{R}^d))$ the solution with data f_n, g_n constructed in Step 1. Proposition 2.1.1 applied to $u_n - u_m$ proves that $\{u_n\}$ is a Cauchy sequence in $C([0,T]\,;\,H^s(\mathbb{R}^d))$. The limit $u \in C([0,T]\,;\,H^s(\mathbb{R}^d))$ of this sequence is the desired solution. This proves existence.

Step 3. Uniqueness. Uniqueness is proved as in the earlier proof. $\square$

---

[4]One can avoid the Hahn–Banach theorem (and therefore uncountable choice) by using continuity in $L^2[0,T]\,;\,H^s(\mathbb{R}^d))$. In this Hilbert space choose the unique extension which vanishes on $V^\perp$. This yields a $u \in L^2([0,T]\,;\,H^{-s}(\mathbb{R}^d))$, which requires small modifications in the end of the proof.

Chapter 3

Dispersive Behavior

3.1. Orientation

In this chapter we return to Fourier analysis techniques as in §1.3, §1.4 and Appendix 2.I. The Fourier transform of the solution is written exactly and then analyzed. The results show how the geometry of the characteristic variety of $L = L_1(\partial_y)$ is reflected in qualitative properties of the solutions of $Lu = 0$. The main idea is that when the characteristic variety is curved, the corresponding solutions tend to spread out in space. This dispersive behavior is reflected in solutions becoming smaller in $L^\infty(\mathbb{R}^d)$ in contrast to $L^2(\mathbb{R}^d)$ conservation.

Three simple examples illustrate the theme. The scalar advection operator

$$L := \partial_t + \mathbf{v} \cdot \partial_x , \tag{3.1.1}$$

in dimension d and the system

$$\frac{\partial v}{\partial t} + \begin{pmatrix} 1 & 0 \\ 0 & -1 \end{pmatrix} \frac{\partial v}{\partial x} = 0 \tag{3.1.2}$$

in dimension $d = 1$ have only purely translating modes. The characteristic variety of (3.1.1) is the hyperplane $\tau + \mathbf{v} \cdot \xi = 0$ and for (3.1.2) it is the pair of lines $\tau \pm \xi = 0$. It is not curved at all.

The system analogue of $\Box_{1+2}$,

$$L := \partial_t + \begin{pmatrix} 1 & 0 \\ 0 & -1 \end{pmatrix} \partial_1 + \begin{pmatrix} 0 & 1 \\ 1 & 0 \end{pmatrix} \partial_2 , \tag{3.1.3}$$

behaves differently. Each component satisfies $\Box_{1+2} u = 0$. For smooth compactly supported data, they decay (in sup norm) as $t^{-1/2}$. The characteristic variety is $\tau^2 - |\xi|^2 = 0$. Since characteristic varieties are conic, their Gauss curvatures vanish. The present variety intersects $\tau = 1$ in a strictly convex set. The variety is as curved as a conic set can be. The system is maximally dispersive.

Exercise 3.1.1. *Prove the decay rate for compactly supported solutions of $\Box_{1+2} u = 0$ by expressing solutions as convolutions with the fundamental solution.* **Discussion.** An alternative proof uses the stationary phase inequality. That method is systematically exploited in §3.4.

For all three examples the $L^2(\mathbb{R}^d)$ norm is preserved during the time evolution.

For the solutions of the transport equation (3.1.1), the size of the support of solutions does not change in time. For (3.1.3), solutions spread out over a set whose two dimensional area grows with time. The spread, together with L^2 conservation, explains the $L^\infty(\mathbb{R}^d)$ decay.

In optics, the word *dispersion* is used to mean that the speed of light depends on its wavelength. In that sense, none of the above models is dispersive. The dispersion relations of the first and third models are all positive homogeneous of degree one in ξ. The velocity at $\sigma\xi$ is independent of σ so the standard optical definition classifies them as nondispersive. However for (3.1.3), the velocity depends strongly on ξ, though not on $|\xi|$. The fact that the group velocities point in different directions has the effect of spreading the solution, and for large time the solutions decay.

The variation of the group velocity with ξ is given by the matrix of second derivatives $\nabla^2_\xi \tau$. For our homogeneous operators, $\nabla_\xi \tau$ is homogeneous of degree zero, so ξ belongs to the kernel of $\nabla^2_\xi \tau$. The rank can be at most $d-1$. The D'Alembertian $\Box_{1+d}$ achieves this maximal rank. It is as dispersive as a homogeneous operator can be.

At the extreme opposite is $\nabla^2_\xi \tau \equiv 0$, in which case the dispersion relation is linear in ξ. The graph of τ is a hyperplane that belongs to the characteristic variety. On a hyperplane, $\{\tau = -\mathbf{v} \cdot \xi\}$, contained in the characteristic variety, the group velocity is identically equal to $\mathbf{v}$ so does not depend on ξ. This is the completely nondispersive situation. Solutions translate without spread.

If the variety contains no hyperplanes, the variation of the group velocity spreads wave packets. We will show that as $t \to \infty$, solutions decay in L^∞. These results, presented in §3.2–§3.3, are taken from [Joly, Métivier, and Rauch, Indiana U. Math. J., 1998].

An even stronger notion of uniform dispersion is when the rank of $\nabla^2_\xi \tau$ is everywhere equal to $d-1$. In this case, the sheets of the characteristic variety are uniformly convex cones and smooth compactly supported solutions decay at the rate $t^{-(d-1)/2}$ as $t \to \infty$. This is investigated in §3.4. In §3.4.1 $L^1 \to L^\infty$ decay estimates are proved. These are applied in §6.7 to prove global solvability for nonlinear problems with small initial data and high dimension. In §3.4.3 the $L^1 \to L^\infty$ estimates are used to derive Strichartz estimates. In §6.8, these estimates are applied to prove global solvability of the nonlinear Klein–Gordon equation in the natural energy space.

3.2. Spectral decomposition of solutions

Since $(\tau, 0)$ is noncharacteristic for L, any hyperplane $\{a\tau + \mathbf{b} \cdot \xi = 0\}$ contained in the characteristic variety must have $a \neq 0$. Therefore, it is necessarily a graph $\{\tau = -\mathbf{v} \cdot \xi\}$.

Over each $\xi \in \mathbb{R}^d$ there are at most N points in the characteristic variety. Therefore, the number of distinct hyperplanes in the variety can be no larger than N. Denote by $0 \leq M \leq N$ the number of such hyperplanes, $H_1, \dots, H_M$,

$$H_j = \{\, (\tau, \xi) \,:\, \tau = -\mathbf{v}_j . \xi \,\}, \qquad j = 1, \dots, M \leq N\,. \tag{3.2.1}$$

Examples 3.2.1. 1. When $d = 1$ the characteristic variety is a union of lines so consists only of hyperplanes. There are no curved sheets.

2. The characteristic variety of the operator (3.1.3) is the light cone, $\{\tau^2 = |\xi|^2\}$. There are no hyperplanes.

3. The characteristic varieties of Maxwell's equations and the the linearization at $u = 0$ of the compressible Euler equations are the union of a convex light cone and a single horizontal hyperplane $\tau = 0$.

Convention. *In this chapter we assume that $L(\partial_t, \partial_x)$ is a constant coefficient, homogeneous, symmetric, and $A_0 = I$.*

Definitions. *A $\underline{\xi} \in \mathbb{R}^d \setminus \{0\}$ is* **good wave number** *when there is a neighborhood Ω of $\underline{\xi}$ and a finite number of real valued real analytic functions $\lambda_1(\xi) < \lambda_2(\xi) < \cdots < \lambda_m(\xi)$ so that the spectrum of $\sum_{j=1}^d A_j \xi_j$ is $\{\lambda_1(\xi), \dots, \lambda_m(\xi)\}$ for $\xi \in \Omega$. The complementary set consists of* **bad wave numbers**. *The set of bad (resp., good) wave numbers is denoted $\mathcal{B}(L)$ (resp., $\mathcal{G}$).*

Over a good ξ, the characteristic variety of L contains exactly m nonintersecting sheets $\tau = -\lambda_j(\xi)$. At bad wave numbers, eigenvalues cross and multiplicities change. The examples above have no bad points.

Examples 3.2.2. Consider the characteristic equation $(\tau^2 - |\xi|^2)(\tau - c\xi_1) = 0$ with $c \in \mathbb{R}$. If $|c| < 1$, then the variety is a cone and a hyperplane intersecting only at the origin and all wave numbers are good. If $|c| > 1$, the plane and cone intersect in a conic set whose projection on ξ space is the set of bad wave numbers,

$$\mathcal{B} = \{\, \xi \,:\, (c^2-1)\xi_1^2 = \xi_2^2 + \cdots + \xi_d^2 \,\}\,.$$

When $|c| = 1$, $\mathcal{B}(L)$ degenerates to a line of tangency. □

Proposition 3.2.1. i. *$\mathcal{B}(L)$ is a closed conic set of measure zero in $\mathbb{R}^d \setminus \{0\}$.*

ii. *The complementary set, $\mathbb{R}^d \setminus (\mathcal{B} \cup \{0\})$, is the disjoint union of a finite family of conic connected open sets $\Omega_g \subset \mathbb{R}^d \setminus \{0\}$, $g \in \mathcal{G}$.*

iii. *The multiplicity of $\tau = -\mathbf{v}_j \cdot \xi$ as a root of $\det L(\tau, \xi) = 0$ is independent of $\xi \in \mathbb{R}^d \setminus (\mathcal{B} \cup \{0\})$.*

iv. *If $\lambda(\xi) \in C^\omega(\Omega_g)$ is an eigenvalue of $\sum A_j \xi_j$ depending real analytically on ξ, then either there is $j \in \{1, \ldots, M\}$ such that $\lambda(\xi) = -\mathbf{v}_j \cdot \xi$ or $\nabla^2 \lambda \neq 0$ for almost everywhere on Ω_g.*

Proof. i. The characteristic variety is a conic real algebraic variety in $\mathbb{R}^{1+d} \setminus \{0\}$.

Over each ξ it contains at least one and at most N points. Therefore its projection on $\mathbb{R}^d_\xi$ is the whole space so the variety has dimension at least d. On the other hand it has measure zero by Fubini's theorem so the dimension is at most d, since $d+1$ dimensional algebraic sets contain open sets.

Use the basic stratification theorem of real algebraic geometry (see [Benedetti and Rissler, 1990], [Basu, Pollack, and Roy, 2006]). The singular points are therefore a stratum of dimension at most $d-1$. The bad wave numbers are exactly the projection of this singular locus and so is a real algebraic subvariety of $\mathbb{R}^d_\xi$ of dimension at most $d-1$, and **i** follows.

ii. That there are at most a finite number of components in the complementary set is a classical theorem of Whitney proved in the references cited in **i.**

iii. Denote by m the multiplicity on Ω_g and m' the multiplicity on $\Omega_{g'}$. By definition of multiplicity,

$$\xi \in \Omega_g \quad \text{and} \quad k < m \quad \Longrightarrow \quad \frac{\partial^k \det L(\tau, \xi)}{\partial \tau^k}\bigg|_{\tau = -\mathbf{v}_j \cdot \xi} = 0\,. \tag{3.2.2}$$

Then $\partial_\tau^k L(-\mathbf{v}_j \cdot \xi, \xi)$ is a polynomial in ξ that vanishes on the nonempty open set Ω_g, so must vanish identically. Thus it vanishes on $\Omega_{g'}$ and it follows that $m' \geq m$. By symmetry one has $m \geq m'$.

iv. If λ is a linear function $\lambda = -\mathbf{v} \cdot \xi$ on Ω_g, then $\det L(-\mathbf{v} \cdot \xi, \xi) = 0$ for $\xi \in \Omega_g$ so by analytic continuation, must vanish for all ξ. It follows that the hyperplane $\tau = -\mathbf{v} \cdot \xi$ lies in the characteristic variety and therefore that $\lambda = -\mathbf{v}_j . \xi$ for some j.

If λ is not a linear function, then the matrix $\nabla_\xi^2 \lambda$ is a real analytic function on Ω_g that is not identically zero. It therefore vanishes at most on a set of measure zero in Ω_g. $\square$

Definitions. *Enumerate the roots of* $\det L(\tau, \xi) = 0$ *as follows. Let* $\mathcal{A}_f := \{1, \dots, M\}$ *denote the indices of the* **flat parts**, *and for* $\alpha \in \mathcal{A}_f$, $\tau_\alpha(\xi) := -\mathbf{v}_\alpha \cdot \xi$. *For* $g \in \mathcal{G}$ *and* $\xi \in \Omega_g$, *number the roots other than the* $\{\tau_\alpha : \alpha \in \mathcal{A}_f\}$ *in order* $\tau_{g,1}(\xi) < \tau_{g,2}(\xi) < \cdots < \tau_{g,M(g)}$. *Multiple roots are not repeated in this list. Let* $\mathcal{A}_c$ *denote the indices of the* **curved sheets**

$$\mathcal{A}_c := \left\{ (g, j) \ : \ g \in \mathcal{G} \ \text{ and } \ 1 \le j \le M(g) \right\}. \tag{3.2.3}$$

Let $\mathcal{A} := \mathcal{A}_f \cup \mathcal{A}_c$. *For* $\alpha \in \mathcal{A}_f$ *and* $\xi \in \mathbb{R}^d$ *define* $E_\alpha(\xi) := \pi(-\mathbf{v}_j \cdot \xi, \xi)$. *For* $\alpha \in \mathcal{A}_c$ *define*

$$E_\alpha(\xi) := \begin{cases} \pi\big(\tau_\alpha(\xi), \xi\big) & \text{for } \ \xi \in \Omega_g, \\ 0 & \text{for } \ \xi \notin \Omega_g. \end{cases} \tag{3.2.4}$$

The next proposition decomposes an arbitrary solution of $Lu = 0$ as a finite sum of simpler waves.

Proposition 3.2.2. 1. *For each* $\alpha \in \mathcal{A}$, $E_\alpha(\xi) \in C^\omega(\mathbb{R}^d \setminus (\mathcal{B} \cup \{0\}))$ *is an orthogonal projection valued function positive homogeneous of degree zero.*

2. *For each* $\xi \in \mathbb{R}^d \setminus (\mathcal{B} \cup \{0\})$, $\mathbb{C}^N$ *is equal to the orthogonal direct sum*

$$\mathbb{C}^N = \bigoplus_{\alpha \in \mathcal{A}} \text{Image}\, E_\alpha(\xi). \tag{3.2.5}$$

3. *The operators* $E_\alpha(D_x) := \mathcal{F}^* E(\xi) \mathcal{F}$ *are orthogonal projectors on* $H^s(\mathbb{R}^d)$, *and for each* $s \in \mathbb{R}$. $H^s(\mathbb{R}^d)$ *is equal to the orthogonal direct sum*

$$H^s(\mathbb{R}^d) \ = \ \bigoplus_{\alpha \in \mathcal{A}} \text{Image}\, E_\alpha(D_x). \tag{3.2.6}$$

4. *If* $f \in \mathcal{S}'(\mathbb{R}^d)$ *has Fourier transform equal to a locally integrable function, then the solution of the initial value problem*

$$L(\partial_y)\, u \ = \ 0, \qquad u|_{t=0} \ = \ f, \tag{3.2.7}$$

is given by the formula

$$\hat{u}(t, \xi) \ = \ \sum_{\alpha \in \mathcal{A}} \hat{u}_\alpha(t, \xi) \ := \ \sum_{\alpha \in \mathcal{A}} e^{it\tau_\alpha(\xi)}\, E_\alpha(\xi)\, \hat{f}(\xi). \tag{3.2.8}$$

Remarks. 1. The last decomposition is also written

$$u \ := \ \sum_{\alpha \in \mathcal{A}} u_\alpha \ := \ \sum_{\alpha \in \mathcal{A}} e^{it\tau_\alpha(D_x)}\, E_\alpha(D_x) f.$$

2. Since τ_α is real valued on the support of $E_\alpha(\xi)$, the operator $e^{it\tau_\alpha(D_x)}\, E_\alpha(D_x)$ is a contraction on $H^s(\mathbb{R}^d)$ for all s.

3. If $\alpha \in \mathcal{A}_f$, then $-i\tau_\alpha(D_x) = \mathbf{v}_\alpha \cdot \partial_x$. For $\alpha = (g,j) \in \mathcal{A}_c$, $|\tau_\alpha(\xi)| \le C|\xi|$, so the operator $\tau_\alpha(D_x)f$ is continuous from H^s to H^{s-1}. The mode $u_\alpha = e^{it\tau_\alpha(D_x)} E_\alpha(D_x)f$ satisfies $\partial_t u_\alpha = i\tau_\alpha(D_x)u_\alpha$. For $\alpha \in \mathcal{A}_f$ this is $\big(\partial_t + \mathbf{v}_\alpha \cdot \partial_x\big)u_\alpha = 0$, so

$$u_\alpha \;=\; \Big(E_\alpha(D)f\Big)(x - \mathbf{v}_\alpha t)\,.$$

4. Over $\mathcal{B}(L)$ only the E_α corresponding to the hyperplanes are defined. One does not have a decomposition of $\mathbb{C}^N$. It is important that $\mathcal{B}$ is a negligible set for $\hat f$. The $\hat f \in L^1_{\mathrm{loc}}$ assumption is essential.

3.3. Large time asymptotics

Definition. *Define $\mathbb{A}$ as the set of tempered distributions whose Fourier transforms belong to $L^1(\mathbb{R}^d)$. $\mathbb{A}$ is a Banach space with norm*

$$\|f\|_{\mathbb{A}} \;:=\; (2\pi)^{-d/2} \int_{\mathbb{R}^d} |\hat f(\xi)|\, d\xi\,. \tag{3.3.1}$$

The Fourier inversion formula implies that $\mathbb{A} \subset L^\infty(\mathbb{R}^d)$ and

$$\|f\|_{L^\infty(\mathbb{R}^d)} \;\le\; \|f\|_{\mathbb{A}}\,. \tag{3.3.2}$$

The elements of $\mathbb{A}$ are continuous and tend to zero as $x \to \infty$. Moreover, the Fourier transform of f^2 is a constant multiple of $\hat f * \hat f \in L^1$, so $\mathbb{A}$ is an algebra. It is called the *Wiener algebra.* It is a centerpiece of Wiener's Tauberian theorems.

Theorem 3.3.1 (L^∞ asymptotics for symmetric systems). *Suppose that $f \in \mathbb{A}$ and u is the solution of the initial value problem $L(\partial_x)u = 0$, $u|_{t=0} = f$. Then, with the notations above,*

$$\lim_{t\to\infty} \Big\| u(t) - \sum_{\alpha\in\mathcal{A}_f} \big(E_\alpha(D_x)f\big)\,(x - \mathbf{v}_\alpha t) \Big\|_{L^\infty(\mathbb{R}^d)} = 0\,. \tag{3.3.3}$$

Remarks. 1. This result shows that a general solution of the Cauchy problem is the sum of M rigidly translating waves, one for each hyperplane in the characteristic variety, plus a term that tends to zero in sup norm. The decay comes from the dispersion of waves.

2. The theorem does not extend to f whose Fourier transform is a bounded measure. For example, $u := \big(e^{i(x_1-t)}, 0\big)$ satisfies $Lu = 0$ with L from (3.1.3) and $\hat f$ is equal to a point mass. The characteristic variety contains no hyperplane, so (3.3.3) asserts that solutions with $\hat f \in L^1$ tend to zero in $L^\infty(\mathbb{R}^d)$ while $u(t)$ has sup norm equal to 1 for all t.

Proof of Theorem 3.3.1. Step 1. Approximation-decomposition. Symmetric hyperbolicity implies that for each t, ξ, $\exp\big(it \sum A_j \xi_j\big)$ is unitary on $\mathbb{C}^N$. Therefore $S(t) := \exp\big(-t\sum_j A_j \partial_j\big)$ is isometric on $\mathbb{A}$. Since the family of linear maps

$$f \quad \longmapsto \quad S(t)f - \sum_{\alpha\in\mathcal{A}_f} (E_\alpha(D_x)f)\,(x - \mathbf{v}_\alpha t)$$

is uniformly bounded from $\mathbb{A}$ to $L^\infty(\mathbb{R}^d)$, it suffices to prove (3.3.3) for a set of f dense in $\mathbb{A}$.

For $\alpha \in \mathcal{A}_c$, Proposition 3.2.1.iv shows that the matrix of second derivatives, $\nabla^2_\xi \tau_\alpha$ can vanish at most on a closed set of measure zero. The set of f we choose is that with

$$\hat f \ \in\ C_0^\infty\Big(\mathbb{R}^d \setminus \big\{\mathcal{B}\ \cup\ \{0\}\ \cup\ \bigcup_{\alpha\in\mathcal{A}_c} \{\xi\in\Omega_g\ :\ \nabla^2_\xi \tau_\alpha(\xi) = 0\}\ \big\}\Big).$$

Since the removed set is closed and measures zero, the functions f are dense.

To prove (3.3.3) for such f, decompose

(3.3.4)
$$f = \sum_{\alpha\in\mathcal{A}} f_\alpha := \sum_{\alpha\in\mathcal{A}} E_\alpha(D_x)\, f\,, \qquad u(t) = S(t)f = \sum u_\alpha(t) := \sum S(t)\, f_\alpha\,.$$

For $\alpha \in \mathcal{A}_f$, $u_\alpha(t) = (E_\alpha(D_x)f)\,(x - \mathbf{v}_\alpha t)$ which recovers the summands in (3.15). To prove (3.15), it suffices to show that for $\alpha \in \mathcal{A}_c$,

$$\lim_{t\to\infty} \big\|u_\alpha(t)\big\|_{L^\infty(\mathbb{R}^d)} \ =\ 0\,. \tag{3.3.5}$$

Step 2. Stationary and nonstationary phase. Part 4 of Proposition 3.2.2 shows that for $\alpha \in \mathcal{A}_c$,

$$u_\alpha(t,x) = \int_{\Omega_g} e^{i(\tau_\alpha(\xi)t + x\xi)}\,\hat f_\alpha(\xi)\, d\xi\,, \qquad \hat f_\alpha \ :=\ E_\alpha(\xi)\,\hat f\,. \tag{3.3.6}$$

For each ξ in the support of $\hat f_\alpha$, there is a vector $\mathbf{r} \in \mathbb{R}^d$ so that $\langle \nabla^2_\xi \tau(\xi)\,\mathbf{r}, \mathbf{r}\,\rangle \ne 0$ on a neighborhood of ξ. Using a partition of unity, we can write $\hat f_\alpha$ as a finite sum of functions $h_\mu \in C_0^\infty(\Omega_g)$ so that for each μ there is a $\mathbf{r}_\mu \in \mathbb{C}^N$ so that on an open ball containing the support of h_μ, $\langle \nabla^2_\xi \tau(\xi)\,\mathbf{r}_\mu, \mathbf{r}_\mu\,\rangle \ne 0$. It suffices to show that for each μ

$$\lim_{t\to\infty}\ \sup_{x\in\mathbb{R}^d} \int e^{i(\tau_\alpha(\xi)t + x\xi)}\, h_\mu(\xi)\, d\xi \ =\ 0\,. \tag{3.3.7}$$

For ease of reading, suppress the subscripts. Write $x = tz$. For each $t > 0$, the supremum in x is equal to the supremum in z so it suffices to show that

$$\lim_{t\to\infty}\ \sup_{z\in\mathbb{R}^d} \Big|\int e^{it(\tau(\xi) + z\xi)}\, h(\xi)\, d\xi\Big| \ =\ 0\,.$$

Choose

$$\sigma > \sup_{\xi \in \operatorname{supp} h} \big|\nabla_\xi \tau(\xi)\big| .$$

There is a $\delta > 0$ so that for all $|z| \geq \sigma$,

$$\big|\nabla_\xi(\tau(\xi) + z\xi)\big| \geq \delta .$$

The method of nonstationary phase implies that

$$\forall N > 0,\ \exists C_N,\ \forall |z| \geq \sigma,\ t > 1, \qquad \Big| \int e^{it(\tau(\xi)+z\xi)}\, h(\xi)\, d\xi \Big| \leq C_N\, t^{-N} .$$

It remains to show that

$$\lim_{t\to\infty} \sup_{|z|\leq\sigma} \Big| \int e^{it(\tau(\xi)+z\xi)}\, h(\xi)\, d\xi \Big| = 0 . \tag{3.3.8}$$

Make a linear change of variables in ξ so that $\mathbf{r} = (1, 0, \dots, 0)$ and therefore

$$\frac{\partial^2 \tau}{\partial^2 \xi_1} \neq 0 \qquad \text{on} \qquad \operatorname{supp} h .$$

Choose $R > 0$ so that for $\xi \in \operatorname{supp} h$, $|\xi| \leq R$. Set

$$\Gamma := \{|z_1| \leq \sigma\} \times \{|\xi_2, \dots, \xi_d| \leq R\} \subset \mathbb{R}^1 \times \mathbb{R}^{d-1} .$$

Define

$$\begin{aligned} K(t) &:= \sup_{|z|\leq\sigma,\ |\xi_2,\dots,\xi_d|\leq R} \Big| \int e^{it(\tau(\xi)+z_1\xi_1)}\, h(\xi)\, d\xi_1 \Big| \\ &= \sup_{\Gamma} \Big| \int e^{it(\tau(\xi)+z_1\xi_1)}\, h(\xi)\, d\xi_1 \Big| . \end{aligned}$$

Then

$$\begin{aligned} &\sup_{|z|\leq\sigma} \Big| \int e^{it(\tau(\xi)+z\xi)}\, h(\xi)\, d\xi \Big| \\ &\quad \leq \int_{|\xi_2,\dots,\xi_d|\leq R} e^{i(z_2\xi_2+\cdots+z_d\cdot\xi_d)} \Big(\int e^{it(\tau+z_1\xi_1)}\, h(\xi)\, d\xi_1 \Big) d\xi_2 \cdots d\xi_d \\ &\quad \leq \Big| \{|\xi_2, \dots, \xi_d| \leq R\} \Big|\ K(t) . \end{aligned}$$

It therefore suffices to show that

$$\lim_{t\to\infty} K(t) = 0 . \tag{3.3.9}$$

The points of Γ are split according to whether the phase $\tau(\xi) + z_1\xi_1$ has a stationary point with respect to ξ_1 or not. If $\underline{\gamma} \in \Gamma$ is such that

$$\Big| \frac{\partial \tau}{\partial \xi_1} + z_1 \Big| > \delta > 0 \quad \text{for all} \quad |z_1| \leq \sigma,\ |\xi| \leq R,$$

the same is true on a neighborhood of $\underline{\gamma}$. The principle of nonstationary phase shows that

$$\int e^{it(\tau_\alpha(\xi)+z\xi)}\ \hat{h}_\mu(\xi)\ d\xi_1 \ = \ O(t^{-N})$$

uniformly on such a neighborhood.

On the other hand if for $\underline{\gamma}$ there is a stationary point, then the strict convexity of τ in ξ_1 shows that it is unique and nondegenerate. Therefore for nearby γ there is a nearby unique and nondegenerate stationary point. The inequality of stationary phase (see Appendix 3.II) implies that

$$\int e^{it(\tau_\alpha(\xi)+z\xi)}\ \hat{h}_\mu(\xi)\ d\xi_1 \ = \ O(t^{-1/2})$$

uniformly on a neighborhood of $\underline{\gamma}$.

Covering the compact set Γ by a finite family of neighborhoods proves (3.3.9) and therefore the theorem. □

Definitions. *The operator L is* **purely dispersive** *when its characteristic variety contains no hyperplanes. It is called* **nondispersive** *when its characteristic variety is equal to a union of hyperplanes.*

The nondispersive operators have a discrete set of group velocities. The characteristic variety of purely dispersive operators have only curved sheets. The latter name is justified by the next corollary.

Corollary 3.3.2. *If $L = L_1(\partial_x)$ is a constant coefficient homogeneous symmetric hyperbolic operator, then the following are equivalent.*

1. *The characteristic variety of L contains no hyperplanes (i.e., L is purely dispersive).*

2. *Every solution of $Lu = 0$ with $u\big|_{t=0} \in C_0^\infty(\mathbb{R}^d)$ satisfies*

$$\lim_{t\to\infty}\ \|u(t)\|_{L^\infty(\mathbb{R}^d)} \ \to\ 0\,. \tag{3.3.10}$$

3. *Every solution of $Lu = 0$ with $u\big|_{t=0} \ \in\ \mathbb{A}$ satisfies (3.3.10).*

4. *If $\tau(\xi)$ is a C^∞ solution of $\det L(\tau,\xi) = 0$ defined on a open set of $\xi \in \mathbb{R}^d$, then for every $\mathbf{v} \in \mathbb{R}^d$, $\{\xi \in \mathbb{R}^d : \nabla_\xi \tau = -\mathbf{v}\}$ has measure zero.*

Proof. Theorem 3.3.1 shows that **1** ⇔ **3**. To complete the proof, we show that **3** ⇔ **2** and **1** ⇔ **4**.

The assertions **2** and **3** are equivalent because the family of mappings $u(0) \mapsto u(t)$ is uniformly bounded from $\mathbb{A} \to L^\infty$, and C_0^∞ is dense in $\mathbb{A}$.

That $\sim \mathbf{1} \Longrightarrow \sim \mathbf{4}$ is immediate.

If **4** is violated, there is a smooth solution τ so that $\nabla_\xi \tau = -\mathbf{v}$ on a set of positive measure. It follows from the stratification theorem of real algebraic geometry that $\{\xi : \nabla_\xi \tau = -\mathbf{v}\}$ contains a nonempty conic open in $\mathbb{R}^d \setminus 0$. Integrating shows that $\tau = -\mathbf{v}\cdot\xi$ on this set and we conclude that the polynomial $\det L(-\mathbf{v}\cdot\xi, \xi)$ vanishes on this set and therefore everywhere. Thus the hyperplane $\{\tau = -\mathbf{v}\cdot\xi\}$ is contained in the characteristic variety and **1** is violated. Thus **1** and **4** are equivalent. □

Remarks. 1. Part four of this corollary shows that for any velocity $\mathbf{v}$ the group velocity $-\nabla_\xi \tau$ associated to a curved sheet of the characteristic variety takes the value $\mathbf{v}$ for at most a set of frequencies ξ of measure zero.

2. If $\Omega \subset \mathbb{R}^d$ is a set of finite measure, estimate

$$\int_\Omega |u(t,x)|^2 \, dx \;\leq\; \|u(t)\|^2_{L^\infty(\mathbb{R}^d)} \;\; |\Omega| \,.$$

This implies that for Cauchy data in $\mathbb{A}$ the L^2 norm in any tube of parallel rays tends to zero as $t \to \infty$.

The nondispersive evolutions are described in the next results.

Corollary 3.3.3. *If $L = L_1(\partial_y)$ is a constant coefficient homogeneous symmetric hyperbolic operator with $A_0 = I$, then the following are equivalent:*

1. *The characteristic variety of L is a finite union of hyperplanes.*

2. *The matrices A_j commute* [Motzkin and Tausky, 1952].

3. *If u satisfies $Lu = 0$ with $u(0) \in \mathbb{A}$ and $\|u(t)\|_{L^\infty(\mathbb{R}^d)} \to 0$ as $t \to \infty$, then u is identically equal to zero.*

Proof. 2 ⇒ 3. A unitary change of variable $u = Vv$ replaces the equation $Lu = 0$ with the equivalent equation $\tilde{L}v = 0$ with $\tilde{A}_j := V^* A_j V$. When the A_j commute, V can be chosen so that the $\tilde{A}_j$ are all real diagonal matrices. Property **3** is clear for the tilde equation as each component of the solution rigidly translates as time goes on. The only way its supremum can tend to zero at $t \to \infty$ is for it to vanish.

3 ⇒ 1. This is a consequence of Theorem 3.3.1.

1 ⇒ 2. This result of Motzkin and Tausky is proved next and completes the proof. □

Theorem 3.3.4 (Motzkin and Tausky)**.** *If A and B are hermitian $N \times N$ matrices, then the eigenvalues of $\xi A + \eta B$ are linear functions of ξ, η if and only if A and B commute.*

Proof. We must show that linear eigenvalue implies commutation. The proof is by induction on N. The case $N = 1$ is trivial. We suppose that $N > 1$ and the result is known for dimensions $\leq N - 1$.

Consider the the characteristic variety $\{(\tau, \xi_1, \xi_2) : \det(\tau + \xi_1 A + \xi_2 B) = 0\}$ associated to $\partial_t + A\partial_1 + B\partial_2$. Choose a good wave number $\underline{\xi}$ so that above this point the variety has $k \le N$ real analytic sheets. If $\xi_2 = 0$, leave the spatial coordinates as they are. If $\xi_2 \ne 0$, change orthogonal coordinates in $\mathbb{R}^2$ so that $\underline{\xi}$ is a multiple of dy_1. Thus without loss of generality we may assume that over $\xi_2 = 0$ the variety consists of k real disjoint analytic sheets.

For s small the eigenvalues of $A + sB$ are a real analytic function $\lambda_j(s)$ with $\lambda_j(0) < \lambda_{j+1}(0)$ for $1 \le j < k-1$. Denote by μ_j the multiplicity of $\lambda_j(0)$ and therefore of $\lambda_j(s)$ for s small. By hypothesis the $\lambda_j(s)$ are affine functions of s, so $\lambda'' = 0$. We use this only at $s = 0$.

By a unitary change of variable in $\mathbb{C}^N$ one can arrange that A is block diagonal with diagonal entries $\lambda_j(0) I_{\mu_j \times \mu_j}$.

Corresponding to this block structure, the matrix $A - \lambda_1 I$ has a spectral projector on the kernel and a partial inverse given by

$$\begin{aligned} \pi \;&=\; \mathrm{diag}\Big(I_{\mu_1\times\mu_1}, 0_{\mu_2\times\mu_2}, \ldots, 0_{\mu_k\times\mu_k}\Big), \\ Q \;&=\; \mathrm{diag}\Big(0_{\mu_1\times\mu_1}, \frac{1}{\lambda_2-\lambda_1} I_{\mu_2\times\mu_2}, \ldots, \frac{1}{\lambda_k-\lambda_1} I_{\mu_k\times\mu_k}\Big). \end{aligned} \tag{3.3.11}$$

The matrix B has block structure

$$B \;=\; \begin{pmatrix} B_{1,1} & B_{1,2} & \cdots & B_{1,k} \\ B_{2,1} & B_{2,2} & \cdots & B_{2,k} \\ & & & \\ B_{k,1} & B_{k,2} & \cdots & B_{k,k} \end{pmatrix},$$

with B_{ij} a $\mu_i \times \mu_j$ matrix and $B_{ij} = B_{ji}^*$.

The fundamental formula of second order perturbation theory (3.I.3) from Appendix 3.I yields $\lambda''\pi = 2\pi B Q B \pi$. By hypothesis this is equal to zero.

Straightforward calculation shows that

$$\pi B \;=\; \begin{pmatrix} B_{1,1} & B_{1,2} & \cdots & B_{1,k} \\ 0 & 0 & \cdots & 0 \\ 0 & 0 & & 0 \\ 0 & 0 & \cdots & 0 \end{pmatrix}, \qquad QB\pi \;=\; \begin{pmatrix} 0 & 0 & \cdots & 0 \\ \frac{1}{\lambda_2-\lambda_1} B_{2,1} & 0 & \cdots & 0 \\ & & & \\ \frac{1}{\lambda_k-\lambda_1} B_{k,1} & 0 & \cdots & 0 \end{pmatrix}.$$

Therefore, the $\mu_1 \times \mu_1$ upper left-hand block of $\pi Q B Q \pi$ is equal to

$$\sum_{j=2}^{k} \frac{1}{\lambda_j - \lambda_1} B_{1,j} B_{1,j}^* \,.$$

This sum of positive square matrices vanishes. Thus, for $j \ge 2$, $B_{1,j} = 0$ and $B_{j,1} = 0$.

Thus B and A are reduced by the splitting

$$\mathbb{C}^N = \mathbb{C}^{\mu_1} \times \mathbb{C}^{N-\mu_1}\,.$$

The commutation then follows by the inductive hypothesis applied to the diagonal blocks. This proves the case N and completes the induction. □

Remark. The implication **1** ⇒ **2** is not true without the symmetry hypothesis. For example, the hyperbolic system

$$\partial_t + \begin{pmatrix} 1 & 0 \\ 0 & -1 \end{pmatrix} \partial_1 + \begin{pmatrix} 1 & 1 \\ 0 & 1 \end{pmatrix} \partial_2$$

has flat characteristic variety with equation

$$(\tau + \xi_1 + \xi_2)(\tau - \xi_1 + \xi_2) = 0,$$

and the coefficient matrices do not commute. The implication and the following result extend to the more general hyperbolic systems that generate a semigroup in $L^2(\mathbb{R}^d)$ (see [Guès and Rauch, 2006]).

Theorem 3.3.5 (Brenner). *If $L = L(\partial_y)$ is a constant coefficient homogeneous symmetric hyperbolic operator with $A_0 = I$, then the conditions of Corollary 3.3.3 are equivalent to each of the following.*

i. *For all $t \in \mathbb{R}$ and $p \in [1, \infty]$, the Fourier multiplication operator*

$$S(t) := \mathcal{F}^{-1}\, e^{-it\sum A_j \xi_j}\, \mathcal{F}$$

is a bounded from $L^p(\mathbb{R}^d)$ to itself.

ii. *For some $\underline{t} \in \mathbb{R} \setminus 0$ and $2 \neq \underline{p} \in [1, \infty]$, the operator $S(\underline{t})$ is bounded from $L^{\underline{p}}(\mathbb{R}^d)$ to itself.*

Remarks. The Fourier multiplication operators are unitary on L^2. Property **ii** means that the restriction to $\mathcal{S}(\mathbb{R})$ extends to bounded operators on L^p, equivalently

$$\sup_{f \in \mathcal{S}(\mathbb{R}^d)\setminus 0} \frac{\|S(t)f\|_{L^p(\mathbb{R}^d)}}{\|f\|_{L^p(\mathbb{R}^d)}} < \infty .$$

Proof. The conditions of Corollary 3.3.3 imply that after an orthogonal change of basis, the A_j are all real diagonal matrices. It is then elementary to verify that property **i** is satisfied.

Clearly, property **i** implies **ii**. It remains to show that **ii** implies the conditions of Corollary 3.3.3. Equivalently, if the conditions of the corollary are violated, then **ii** is violated. First observe that **ii** is stronger than it appears. Since $S(\underline{t})$ is unitary on L^2, if **ii** is satisfied then the Riesz–Thorin interpolation theorem implies that $S(t)$ is bounded on L^p for all p between 2 and $\underline{p}$. Thus we may assume that $\underline{p}$ in not equal to 1 or ∞.

For $\sigma \in \mathbb{R}\setminus 0$, $Lu = 0$ if and only if $u^\sigma(t,x) := u(\sigma t, \sigma x)$ satisfies $Lu^\sigma = 0$. It follows that if **ii** is satisfied, then

$$\|S(t)\|_{\mathrm{Hom}(L^{\underline{p}})} = \|S(\underline{t})\|_{\mathrm{Hom}(L^{\underline{p}})} < \infty, \qquad \forall\, t \neq 0. \tag{3.3.12}$$

If $\underline{q}$ is the conjugate index to $\underline{p}$, that is $\frac{1}{\underline{p}} + \frac{1}{\underline{q}} = 1$, then

$$\begin{aligned} \|S(t)\|_{\mathrm{Hom}(L^{\underline{q}}(\mathbb{R}^d))} &= \sup_{f,g\in\mathcal{S}\setminus 0} \frac{(S(t)f\,,\,g)}{\|f\|_{L^{\underline{q}}(\mathbb{R}^d)}\,\|g\|_{L^{\underline{p}}(\mathbb{R}^d)}} \\ &= \sup_{f,g\in\mathcal{S}\setminus 0} \frac{(f\,,\,S(-t)g)}{\|f\|_{L^{\underline{q}}(\mathbb{R}^d)}\,\|g\|_{L^{\underline{p}}(\mathbb{R}^d)}} = \|S(-t)\|_{\mathrm{Hom}(L^{\underline{p}}(\mathbb{R}^d))}\,. \end{aligned}$$

Thus when **ii** is satisfied for $\underline{p}$, it is satisfied for $\underline{q}$ so we may suppose that $\infty > \underline{p} > 2$.

When the conditions of Corollary 3.3.3 are violated, there is a conic set of good wave numbers Ω_g and a sheet $\tau = \tau(\xi)$ over Ω_g with $\nabla^2_{\xi\xi}\tau \neq 0$ for almost all $\xi \in \Omega_g$. Denote by $\pi(\xi)$ the associated spectral projection. Choose an $f \in \mathcal{S}(\mathbb{R}^d)$ with $\hat{f}$ compactly supported in Ω_g. Replacing $\hat{f}$ by $\pi(\xi)\hat{f}$ we may assume that $\pi(D)f = f$. Theorem 3.3.1 implies that

$$\lim_{t\to\infty} \|S(t)\,f\|_{L^\infty(\mathbb{R}^d)} = 0\,.$$

Then

$$\begin{aligned} \|S(t)f\|^{\underline{p}}_{L^{\underline{p}}(\mathbb{R}^d)} &\le \|S(t)f\|^{\underline{p}-2}_{L^\infty(\mathbb{R}^d)}\,\|S(t)f\|^2_{L^2(\mathbb{R}^d)} \\ &= \|S(t)f\|^{\underline{p}-2}_{L^\infty(\mathbb{R}^d)}\,\|f\|^2_{L^2(\mathbb{R}^d)} \to 0\,, \end{aligned}$$

as $t \to \infty$. Therefore,

$$\big\|S(-t)\big\|_{\mathrm{Hom}(L^{\underline{p}})} \ge \frac{\big\|S(-t)\big(S(t)f\big)\big\|_{L^{\underline{p}}}}{\big\|S(t)f\big\|_{L^{\underline{p}}}} = \frac{\big\|f\big\|_{L^{\underline{p}}}}{\big\|S(t)f\big\|_{L^{\underline{p}}}} \to \infty\,.$$

Thus (3.3.12) is violated, and the proof is complete. □

Example 3.3.1. It may seem that (3.3.12) together with $\lim_{t\to 0} S(t)f = f$ might imply that $S(t)$ has norm equal to 1. That this is not true can be seen from the one dimensional example

$$\partial_t + \begin{pmatrix} 0 & 0 \\ 0 & 1 \end{pmatrix} \partial_x$$

and L^p norm defined by

$$\big\|(u_1,u_2)\big\|_p := \Big(\int \|(u_1,u_2)\|^p\,dx\Big)^{1/p}, \qquad \|(u_1,u_2)\| := \Big(|u_1|^2 + |u_2|^2\Big)^{1/2},$$

so that for $p = 2$ one has unitarity. Choosing $u_1(0) = u_2(0) = f \in C_0^\infty(\{|x| \le \rho\})$, one has

$$\|u(0)\|_p^p = (\sqrt{2})^p \|f\|_p^p\,,$$

and for $|t| > \rho$,

$$\|u(t)\|_p^p = 2\,\|f\|_p^p\,.$$

It follows that for all $t \neq 0$ and $p < 2$, $\|S(t)\|^p_{\mathrm{Hom}(L^p)} \geq 2^{1-p/2} > 1$. Reversing time, treats $p > 2$.

3.4. Maximally dispersive systems

3.4.1. The $L^1 \to L^\infty$ decay estimate. If $\tau = \tau(\xi)$ parametrizes a real analytic patch of the characteristic variety of a hyperbolic operator, then τ is homogeneous of degree 1 in ξ. The group velocity $\mathbf{v}(\xi) = -\nabla_\xi \tau(\xi)$ is homogeneous of degree 0. Therefore $\xi \cdot \nabla_\xi \mathbf{v} = 0$ so ξ belongs to the kernel of the symmetric matrix $\nabla_\xi \mathbf{v}(\xi) = -\nabla^2_\xi \tau(\xi)$. Thus the rank of $\nabla^2_\xi \tau$ is at most $d-1$. When the rank is equal to $d-1$, the group velocity depends as strongly on ξ as possible.

Definition. *The homogeneous constant coefficient symmetric hyperbolic operator is* **maximally dispersive** *when*

$$\mathrm{Char}\, L \;=\; \bigcup_{j=1}^{m} \Big\{(\tau,\xi) \;:\; \tau = \tau_j(\xi)\Big\},$$

where for $\xi \in \mathbb{R}^d \setminus 0$,

$$\tau_1(\xi) < \tau_2(\xi) < \cdots < \tau_m(\xi)\,,$$

the τ_j *are real analytic, positive homogeneous of degree one in* ξ, *and*

$$\forall j,\ \ \forall \xi \in \mathbb{R}^d \setminus 0\,, \qquad \mathrm{rank}\, \nabla^2_\xi \tau(\xi) \;=\; d-1\,. \tag{3.4.1}$$

Examples 3.4.1. i. The simplest example is the characteristic equation

$$(\tau^2 - |\xi|^2)\,(\tau^2 - c^2|\xi|^2) \;=\; 0\,, \qquad 0 < c \neq 1\,.$$

The variety contains two sheets $\tau = |\xi|$ and $\tau = c|\xi|$ that have $d-1$ strictly positive principal curvatures. The other two sheets $\tau = -|\xi|$ and $\tau = -c\,|\xi|$ have $d-1$ strictly negative curvatures.

ii. Figure 3.4.1 gives an example with two sheets bounding strictly convex regions for which the functions τ_j change sign. In particular the generator $G = -\sum A_j \partial_j$ is not elliptic since the points where the cone crosses $\tau = 0$ are characteristic for G.

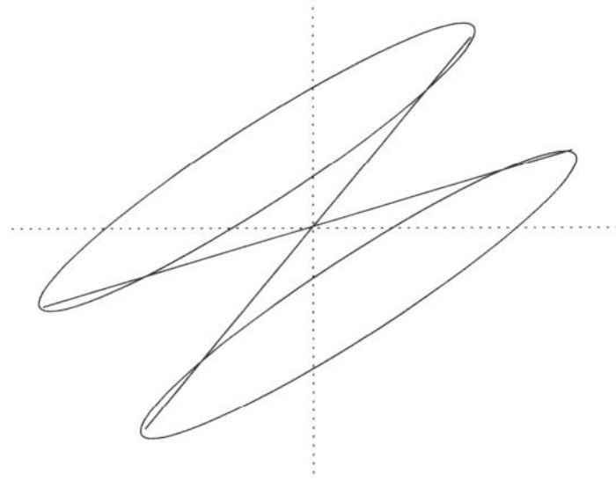

FIGURE 3.4.1.

Proposition 3.4.1. *If $d \geq 2$ and $\tau(\xi)$ is smooth in $\xi \neq 0$, positively homogeneous of degree one and the hessian has rank equal to $d-1$ at all points, then the nonzero eigenvalues of $\nabla_\xi^2 \tau$ have the same sign. When they are positive (resp., negative) τ is convex (resp., concave).*

Proof. Consider the symmetric matrix $\nabla_\xi^2 \tau(\xi)$. By hypothesis it has $d-1$ nonzero eigenvalues and the eigenvalue 0 with eigenspace spanned by ξ.

Denote by $N_+(\xi)$ (resp., N_-) number of strictly positive (resp., strictly negative) eigenvalues. As 0 is an eigenvalue of constant multiplicity and the functions $N\pm$ are continuous integer valued functions on the connected set $\xi \neq 0$, it follows that $N_\pm$ are independent of ξ.

Consider $\tau(\xi)$ as a function on the unit sphere $S^{d-1} = \{|\xi| = 1\}$. At a point $\xi \in S^{d-1}$ where $|\tau(\xi)|$ is maximum, the nonzero eigenvalues must be of one sign. In fact, if $|\tau|$ is attained at a positive maximum (resp., negative minimum) of τ, then the eigenvalues must be strictly negative (resp., positive). Therefore, they are of one sign everywhere. The concavity (resp., convexity) of the cone $\{(\tau(\xi)\,,\,\xi)\}$ follows. □

Examples 3.4.2. The characteristic variety of a maximally dispersive system consists of m disjoint sheets, each the boundary of a strictly convex or concave cone.

Lemma 3.4.2 (Pointwise decay). *If $d \geq 2$, τ is as above, and $k \in C_0^\infty(\mathbb{R}^d \backslash 0)$, then there is a constant C so that*

$$u(t,x) \;:=\; \int e^{it\tau(\xi)}\, e^{ix\xi}\, k(\xi)\, d\xi$$

satisfies

$$\|u(t)\|_{L^\infty(\mathbb{R}^d)} \;\leq\; C\,(1+|t|)^{-(d-1)/2}\,. \tag{3.4.2}$$

Remark. This is the decay rate for solutions of $\square_{1+d} u = 0$ which corresponds to the choice $\tau(\xi) = \pm|\xi|$.

Proof. The easy estimate

$$\|u(t,x)\|_{L^\infty(\mathbb{R}^d)} \leq \int |k(\xi)|\, d\xi$$

shows that only the decay for $|t| \geq 1$ needs to be proved.

Let

$$y := \frac{x}{t}, \qquad x = ty.$$

Then

$$\sup_x \big|u(t,x)\big| = \sup_y \big|u(t,ty)\big| = \sup_y \Big| \int e^{it(\tau(\xi)+y\xi)}\, k(\xi)\, d\xi \Big| .$$

The phase $\tau(\xi) + y\xi$ is stationary when

$$-\nabla_\xi \tau(\xi) = y .$$

The left-hand side is the group velocity.

As in Lemma 3.4.1, denote by $\mathcal{M}$ the set of attained group velocities which is an embedded strictly convex compact $d-1$ manifold.

For any open neighborhood $\mathcal{O}$ of $\mathcal{M}$, the method of nonstationary phase shows that for any N,

$$\sup_{y \in \mathbb{R}^d \setminus \mathcal{O}} \Big| \int e^{it(\tau(\xi)+y\xi)}\, k(\xi)\, d\xi \Big| \leq C_N\, |t|^{-N},$$

as $t \to \infty$.

Choose $0 < r_1 < r_2$ so that

$$\operatorname{supp} k \subset \{r_1 \leq |\xi| \leq r_2\} .$$

Write

$$\int e^{it(\tau(\xi)+y\xi)}\, k(\xi)\, d\xi = \int_{r_1}^{r_2} \Big(\int_{|\xi|=1} e^{it(\tau(\xi)+y\xi)}\, k(r\,\xi)\, d\sigma(\xi) \Big)\, r^{d-1}\, dr .$$

It suffices to show that for any $\underline{y} \in \mathcal{M}$ and $\underline{r} \in [r_1, r_2]$, one has

$$\int_{|\xi|=1} e^{it(\tau(\xi)+y\xi)}\, k(r\,\xi)\, d\sigma(\xi) \leq C\, |t|^{-(d-1)/2}$$

uniformly for r, y in a neighborhood of $\underline{r}, \underline{y}$.

For $\underline{r}, \underline{y}$ fixed, there is a unique $\underline{\xi}$ with $|\underline{\xi}| = \underline{r}$ for which the phase is stationary and the stationary point is nondegenerate because of the rank equal to $d-1$ hypothesis. It follows that for r, y in a neighborhood, there is a unique uniformly nondegenerate stationary point. The desired estimate follows from the inequality of stationary phase proved in Appendix 3.II of this chapter. □

Proposition 3.4.3. *Suppose that* $0 < R_1 < R_2 < \infty$ *and* $\omega := \{\xi \in \mathbb{R}^d : R_1 < |\xi| < R_2\}$. *There is a constant* C *so that for all* $f \in L^1(\mathbb{R}^d_x)$ *with* $\operatorname{supp} \hat{f} \subset \overline{\omega}$,

$$u(t,x) \;:=\; (2\pi)^{-d/2} \int e^{i(t\tau_j(\xi)+x\xi)}\, \hat{f}(\xi)\, d\xi \;:=\; e^{it\tau_j(D_x)}\, f$$

satisfies

$$\|u(t)\|_{L^\infty(\mathbb{R}^d)} \;\leq\; C\,(1+|t|)^{-(d-1)/2}\, \|f\|_{L^1(\mathbb{R}^d)}\,. \tag{3.4.3}$$

The proof is based on a simple idea. The solution u is equal to the convolution of the fundamental solution with f. The Fourier transform of the fundamental solution at $t = 0$ is equal to a constant. To have an analogous but more regular representation, it is sufficient that one convolve with a solution whose initial data has a Fourier transform equal to this constant on the spectrum of f.

Proof. Choose a $k \in C_0^\infty(\mathbb{R}^d\backslash 0)$ with k equal to $(2\pi)^{-d/2}$ on a neighborhood of $\overline{\omega}$. Define G so that $\widehat{G} := k$. Then since $(2\pi)^{d/2}\, k\, \hat{f} = \hat{f}$ one has $G * f = f$. Since $e^{it\tau(D_x)}$ is a Fourier multiplier, one has

$$u(t) := e^{it\tau(D_x)}\, f = e^{it\tau(D_x)}\big(f * G\big) = f * \big(e^{it\tau(D_x)}\, G\big)\,.$$

Then

$$\|u(t)\|_{L^\infty} \;\leq\; \|f\|_{L^1}\, \|e^{it\tau(D_x)}\, G\|_{L^\infty}\,.$$

The preceding lemma shows that

$$\|e^{it\tau(D_x)}\, G\|_{L^\infty} \;\leq\; C\,(1+|t|)^{-(d-1)/2}\,.$$ □

The next subsections consist of two different paths for exploiting the estimates just proved. The first is more elementary and will be used in Chapter 6 to prove, in the spirit of F. John and S. Klainerman (see [Klainerman, 1998, part III]), that in high dimensions there is global solvability for maximally dispersive nonlinear problems with small data. The second is devoted to Strichartz estimates that serve, among other things, to analyze existence for low regularity data. That problem is important in trying to pass from local solvability to global solvability for nonlinear problems for which the natural a priori estimates control few derivatives.

3.4.2. Fixed time dispersive Sobolev estimates.[1]

First find decay estimates for $\|u(t)\|_{L^\infty}$ for sources with Fourier transform supported in $\lambda\overline{\omega}$ for $0 < \lambda$. The starting point is

$$\|u(t)\|_{L^\infty(\mathbb{R}^d)} \;\leq\; C\,|t|^{-(d-1)/2}\, \|f\|_{L^1(\mathbb{R}^d)}\,, \qquad \operatorname{supp} \hat{f} \subset \omega\,. \tag{3.4.4}$$

[1]The material in this subsection is not needed for the Strichartz estimates in the next subsection.

The idea is to apply this identity to the function $(t,x)\mapsto u(\lambda t\,,\,\lambda x)$. The estimate (3.4.4) is not invariant under such scalings. To overcome this, note that since the spectrum of f is contained in ω, the norms

$$\big\||D|^\gamma f\big\|_{L^1(\mathbb{R}^d)}$$

are all equivalent. The index γ is chosen so that the resulting inequality is scale invariant.

Proposition 3.4.4. *There is a constant C so that for all $\lambda>0$ and $f\in L^1$ with $\operatorname{supp}\hat f\subset\lambda\omega$, the solution of*

$$L\,u\;=\;0\,,\qquad u\big|_{t=0}\;=\;f\,,$$

satisfies

$$\|u(t)\|_{L^\infty(\mathbb{R}^d)}\;\le\;C\,|t|^{-(d-1)/2}\,\|\,|D|^{(d+1)/2}f\|_{L^1(\mathbb{R}^d)}\,. \tag{3.4.5}$$

Check that the two sides of (3.4.5) have the same dimensions. With t,x having the dimensions of a length ℓ, the factor $|t|^{(d-1)/2}$ has dimension $\ell^{(d-1)/2}$. On the other hand, in

$$\|\,|D|^\gamma f\|_{L^1(\mathbb{R}^d)}\;=\;\int\big||D|^\gamma f\big|\;dx,$$

the integrand has dimension $\ell^{-\gamma}$ and dx has dimension ℓ^d. In total the right-hand side of (3.4.5) has dimension $\ell^{d-\gamma-(d-1)/2}$. It is dimensionless as is the left-hand side exactly when $\gamma=(d+1)/2$.

Proof of Proposition 3.4.4. Choose $\psi_\pm\in C_0^\infty(\mathbb{R}^d_\xi)$ so that $\psi_\pm\;=\;|\xi|^{\pm\gamma}$ on $\overline{\omega}$. Then

$$|D|^\gamma f\;=\;C\,\hat\psi_+ * f\qquad\text{and}\qquad f\;=\;C\,\hat\psi_- * (|D|^\gamma f)\,.$$

Young's inequality implies that when $\operatorname{supp}\hat f\,\subset\,\omega$, $\|\,|D|^\gamma f\|_{L^1}$ is a norm equivalent to that on the right in (3.4.4). Therefore,

$$\|u(t)\|_{L^\infty(\mathbb{R}^d)}\;\le\;C\,|t|^{-(d-1)/2}\,\|\,|D|^\gamma f\|_{L^1(\mathbb{R}^d)}\,,\qquad \operatorname{supp}\hat f\subset\omega\,.$$

If $u_\lambda(t,x)\;:=\;u(\lambda t,\lambda x)$, then $Lu_\lambda\;=\;0$ if and only if $Lu\;=\;0$. Since $\hat u_\lambda(\lambda t,\xi)=\lambda^{-d}\hat u(t,\xi/\lambda)$, the spectrum of u is contained in ω if and only if the spectrum of u_λ is contained in $\lambda\omega$.

Exercise 3.4.1. *Show that if $f_\lambda(x):=f(\lambda x)$, then*

$$(\mathcal{F}f_\lambda)(\xi)=\lambda^{-d}(\mathcal{F}f)(\xi/\lambda),\quad (|D|^\gamma f_\lambda)(x)=\lambda^\gamma\big(|D|^\gamma f\big)(\lambda x)$$

and

$$\||D|^\gamma f_\lambda\|_{L^1}=\lambda^{\gamma-d}\||D|^\gamma f\|_{L^1}.$$

Then (3.4.3) yields

$$\begin{aligned}\big\|u_\lambda(t)\big\|_{L^\infty} &= \|u(\lambda t)\|_{L^\infty} \le C\,|\lambda t|^{-(d-1)/2}\,\|\,|D|^\gamma f\|_{L^1(\mathbb{R}^d)}\\ &= C\,\lambda^{-(d-1)/2-\gamma+d}|t|^{-(d-1)/2}\|\,|D|^\gamma f_\lambda\|_{L^1(\mathbb{R}^d)}\,.\end{aligned}$$

With $\gamma = (d+1)/2$, the exponent of λ is equal to zero. □

Since $\hat{u}$ and $\hat{f}$ are locally integrable functions, the point $\xi = 0$ is negligible so we have the diadic Littlewood–Paley decompositions

$$u = \sum_{j=-\infty}^{\infty} \chi(2^{-j}D_x)\,u := \sum_{j=-\infty}^{\infty} u_j, \qquad f = \sum_{j=-\infty}^{\infty} \chi(2^{-j}D)\,f := \sum_{j=-\infty}^{\infty} f_j,$$

as in Lemma 3.II.2. This expresses a solution of $Lu = 0$ as a sum of spectrally localized solutions. The next exercise shows that $|D|^\sigma$ acts like multiplication by $2^{\sigma j}$ on f_j.

Exercise 3.4.2. *Show that there is an integer k and a constant C depending on σ and χ so that for $p \in [1,\infty]$*

$$\big\|\,|D|^\sigma f_j\big\|_{L^p} \le C\,2^{\sigma j} \sum_{|n-j|\le k} \|f_n\|_{L^p}\,, \tag{3.4.6}$$

$$\big\|\,2^{\sigma j}\,f_j\big\|_{L^p} \le C \sum_{|n-j|\le k} \|\,|D|^\sigma f_n\|_{L^p}\,. \tag{3.4.7}$$

Theorem 3.4.5. i. *If $Lu = 0$ and $u\big|_{t=0} = f$, then*

$$\|u\|_{L^\infty} \le C\,|t|^{-(d-1)/2} \sum_{j=-\infty}^{\infty} \|\,|D|^\gamma f_j\|_{L^1}\,, \qquad \gamma = \frac{d+1}{2}\,. \tag{3.4.8}$$

ii. *If $0 < \delta < \gamma$, there is a constant $C(\gamma,\delta)$ so that*

$$\sum_{j=-\infty}^{\infty} \|\,|D|^\gamma f_j\|_{L^1} \le C\Big(\big\|\,|D|^{\gamma-\delta} f\big\|_{L^1(\mathbb{R}^d)} + \big\|\,|D|^{\gamma+\delta} f\big\|_{L^1(\mathbb{R}^d)}\Big)\,. \tag{3.4.9}$$

Remarks. 1. The sum on the right of (3.4.8) is the definition of the norm in the homogeneous Besov space $\dot{B}^\gamma_{1,1}$. Estimate (3.4.9) yields a bound that is not as sharp but avoids these spaces.

2. A slightly weaker estimate than (3.4.8)–(3.4.9) is proved by Lucente and Ziliotti (1999).

3. It is impossible to have a decay estimate of the form

$$\|u(t)\|_{L^\infty} \le g(t)\|f\|_{H^s}\,, \qquad \lim_{t\to\infty} g(t) = 0\,,$$

with a conserved norm on the right-hand side. If there were such an estimate, apply it to $v(t) = u(t-T)$ at $t = T \to \infty$ to find

$$\|u(0)\|_{L^\infty} = \|v(T)\|_{L^\infty} \le g(T)\,\|v(0)\|_{H^s} = g(T)\,\|f\|_{H^s} \to 0\,.$$

The appearance of norms that are not propagated by the equation is crucial.

4. To see that estimate (3.4.4) is essentially optimal, consider for $t \gg 1$, $u(t)$ with $\operatorname{supp} \hat{u}(t) \subset \omega$ and $u(0)$ concentrated on $\{t-\rho \le |x| \le t+\rho\}$ with ρ independent of t. Then since $\|u(t)\|_{L^2}^2 = \|u(0)\|_{L^2}^2$, a typical amplitude a of $u(0)$ satisfies

$$a^2 \, \big|\{t-\rho \le |x| \le t+\rho\}\big| \sim 1\,, \qquad \text{so} \qquad a \sim t^{-(d-1)/2}\,.$$

Therefore,

$$\|u(0)\|_{L^1} \sim \big|\{t-\rho \le |x| \le t+\rho\}\big| a \sim 1$$

and

$$\frac{\|u(t)\|_{L^\infty}}{\|u(0)\|_{L^1}\, t^{-(d-1)/2}} \sim 1\,,$$

saturating (3.4.4).

Proof of Theorem 3.4.5. i. Estimate (3.4.5) implies

$$\|u_j(t)\|_{L^\infty} \le C\,|t|^{-(d-1)}\, \|\,|D|^\gamma f_j\|_{L^1}\,.$$

Summing yields

$$\|u\|_{L^\infty} \le \sum \|u_j\|_{L^\infty} \le C\,|t|^{-(d-1)} \sum \|\,|D|^\gamma f_j\|_{L^1}\,.$$

ii. For $j \ge 0$, estimate (3.4.6) implies

$$\big\|\,|D|^\gamma f_j\big\|_{L^1} \le C\, 2^{\gamma j} \sum_{|n-j|\le k} \big\|f_n\big\|_{L^1}\,.$$

Estimate (3.4.7) implies

$$\big\|f_n\big\|_{L^1} \le C\, 2^{-\sigma n} \sum_{|m-n|\le k} \big\|\,|D|^\sigma f_m\big\|_{L^1}\,.$$

Finally,

$$\big\|\,|D|^\sigma f_m\big\|_{L^1} \le C\,\big\|\,|D|^\sigma f\big\|_{L^1}\,.$$

Combining yields

$$\sum_{j\ge 0} \|\,|D|^\gamma f_j\|_{L^1} \le C\,\big\|\,|D|^\sigma f\big\|_{L^1} \sum_{j\ge 0}\ \sum_{|n-j|\le k} 2^{\gamma j-\sigma n}\,.$$

With $\sigma = \gamma + \delta$, the sum is finite, so

$$\sum_{j\ge 0} \|\,|D|^\gamma f_j\|_{L^1} \le C\,\big\|\,|D|^{\gamma+\delta} f\big\|_{L^1}\,.$$

Exercise 3.4.3. *Prove the complementary low frequency estimate*

$$\sum_{j< 0} \|\,|D|^\gamma f_j\|_{L^1} \le C\,\big\|\,|D|^{\gamma-\delta} f\big\|_{L^1}\,.$$

This completes the proof. □

Corollary 3.4.6. *For any $d/2 > \delta > 0$ there is a constant C so that if $Lu = 0$, then*

$$
\begin{aligned}
\|u(t)\|_{L^\infty(\mathbb{R}^d)} \;\le\; & C\,\langle t\rangle^{-(d-1)/2}\Big(\|f\|_{H^{d/2+\delta}(\mathbb{R}^d)} \\
& + \|\,|D|^{(d+1)/2+\delta} f\|_{L^1(\mathbb{R}^d)} \;+\; \|\,|D|^{(d+1)/2-\delta} f\|_{L^1(\mathbb{R}^d)}\Big)\,. \qquad (3.4.10)
\end{aligned}
$$

Remark. The smaller $\delta > 0$ is, the stronger the conclusion.

Proof. Sobolev's inequality yields

$$\|u(t)\|_{L^\infty(\mathbb{R}^d)} \;\le\; C\,\|u(t)\|_{H^{\delta+d/2}(\mathbb{R}^d)} \;=\; C\,\|f\|_{H^{\delta+d/2}(\mathbb{R}^d)}\,.$$

This yields (3.4.10) for $|t| \le 1$.

For $|t| \ge 1$ use Theorem 3.4.5.ii. □

3.4.3. Strichartz estimates.

Strichartz bounds involve norms

$$\|u\|_{L^q_t L^r_x} \;:=\; \Big(\int_0^\infty \|u(t)\|^q_{L^r(\mathbb{R}^d_x)}\,dt\Big)^{1/q}$$

that integrate over space and time. If such a norm is finite, then the integrand must be small for large times. This requires $r > 2$. The estimates express time decay because of dispersion.

For maximally dispersive systems, the group velocities lie on the strictly convex manifold $\mathcal{M}$. The Fourier transform lives on a set of positive measure so the ignited group velocities include a set of positive $d-1$ area on $\mathcal{M}$. The method of nonstationary phase suggests that the solution is expected to live on the union of the rays with these speeds and with feet in the support of the initial data. As t grows, the volume of the set swept out by these rays grows like t^{d-1}. An example is the growing annulus of constant thickness $\rho_1 < |x| - t < \rho_2$, or, the part of the annulus subtending a fixed solid angle.

Conservation of $L^2(\mathbb{R}^d)$ and also Lemma 3.4.2 show that the expected amplitude is $O(t^{-(d-1)/2})$. Then

$$\|u(t)\|^r_{L^r} \;\sim\; t^{-r(d-1)/2}\, t^{d-1}\,,$$

so

$$\|u\|^q_{L^q_t L^r_x} \;\sim\; \int_1^\infty \Big(t^{-r(d-1)/2}\, t^{d-1}\Big)^{q/r}\,dt\,.$$

The limiting indices are those for which the power of t is equal to -1. With

$$\sigma \;:=\; d-1\,,$$

$$\Big(\frac{-r\sigma}{2} + \sigma\Big)\frac{q}{r} \;=\; -1\,, \qquad \text{equivalently,} \qquad \frac{-\sigma}{2} + \frac{\sigma}{r} \;=\; \frac{-1}{q}\,.$$

The admissible indices are those for which the power is less than or equal to -1,

$$\frac{-\sigma}{2} + \frac{\sigma}{r} \le \frac{-1}{q}.$$

Definitions. *The pair* $2 < q, r < \infty$ *is* **σ-admissible** *if*

$$\frac{1}{q} + \frac{\sigma}{r} \le \frac{\sigma}{2}.$$

It is **sharp σ-admissible** *when equality holds.*

The estimates involve the homogeneous Sobolev norms

$$\| \, |D|^\gamma f \|_{L^2} := \Big(\int \big| |\xi|^\gamma \hat f(\xi) \big|^2 \, d\xi \Big)^{1/2}.$$

Theorem 3.4.7 (Strichartz inequality)**.** *Suppose that* $L(\partial)$ *is maximally dispersive,* $\sigma = d - 1$, q, r *is* σ*-admissible, and* γ *is the solution of*

$$\frac{1}{q} + \frac{d}{r} = \frac{d}{2} - \gamma.$$

There is a constant C *so that for* $f \in L^2$ *with* $\| \, |D|^\gamma f \|_{L^2} < \infty$, *the solution of* $Lu = 0$, $u|_{t=0} = f$ *satisfies*

$$\big\| u \big\|_{L^q_t L^r_x} \le C \, \big\| \, |D|^\gamma f \big\|_{L^2(\mathbb{R}^d)} . \tag{3.4.11}$$

There are two complicated relations in this assertion. The first is the definition of admissibility. It encodes the rate of decay of solutions. The second is the definition of γ. Once admissible q, r are chosen, γ is forced so that the two sides of (3.4.11) scale the same under the dilation, $(t, x) \mapsto (at, ax)$. The admissibility encodes the $t^{-(d-1)/2}$ decay, and the γ is from dimensional analysis.

There is an alternative perspective. Start from the scaling relation that is independent of the dispersion. For example if you are obliged to work with a specific γ (e.g., when we treat the energy space in §6.8), then the scaling restricts $1/q, 1/r$ to lie on a line. The admissability chooses an interval on that line. Changing the dispersion, for example considering a problem with the same scaling but weaker dispersion, leaves the line fixed but constrains the $1/q, 1/r$ to lie on a smaller subinterval.

We follow the proof of Keel and Tao (1998); Ginibre and Velo (1995) is a second standard reference. The limit point case (not discussed here) is treated in the first reference. The key step is an estimate for spectrally localized data.

Lemma 3.4.8. *Suppose* q, r *is* σ*-admissible for* $\sigma := d - 1$, *and* ω *is as in Proposition 3.4.3. There is a constant* C *so that for all* $f \in L^2(\mathbb{R}^d)$ *with*

$\operatorname{supp}\hat{f} \subset \overline{\omega}$,

$$u(t) \;:=\; e^{it\tau_j(D_x)}f \;:=\; U(t)f\,, \qquad U(t)^* = U(-t)\,,$$

satisfies

(3.4.12)
$$\big\|u\big\|_{L^q_t L^r_x} \;\le\; C\,\big\|f\big\|_{L^2}\,.$$

Furthermore, for all $F \in L^{q'}_t L^{r'}_x$ *with* $\operatorname{supp}\hat{F}(t,\,\cdot\,) \subset \overline{\omega}$,

(3.4.13)
$$\Big\|\int_0^\infty U(s)^*\,F(s)\;ds\Big\|_{L^2(\mathbb{R}^d)} \;\le\; C\,\big\|F\big\|_{L^{q'}_t L^{r'}_x}\,.$$

Remark. The estimate is true in the sharp admissible case even though for the heuristics given before the definition, the integral diverged. It is not possible to achieve the concentration suggested in the heuristics with data that has spectrum with support in an annulus. For example, if one considers the wave operator $\square$ on $\mathbb{R}^{1+3}$ with data supported in $|x| \le 1$, the solutions are supported in $|x| - t \le 1$ and decay along with their derivatives exactly as in the heuristic. Thus one gets divergent integrals. However, compact support and compactly supported Fourier transform are not compatible, and the compact spectrum is enough to overcome the logarithmic divergence of $\int^\infty 1/t\,dt$ that appears in the heuristics.

Proof. Denote by $(\ ,\ )$ the $L^2(\mathbb{R}^d)$ scalar product. Since

$$\int_0^\infty (U(t)\,f\,,\,F(t))\;dt \;=\; \int_0^\infty (f\,,\,U(t)^*\,F(t))\;dt \;=\; \Big(f\,,\,\int_0^\infty U(t)^*\,F(t)\;dt\Big)\,,$$

estimates (3.4.12) and (3.4.13) are equivalent thanks to the pair of duality representations of norms,

$$\Big\|\int_0^\infty U(t)^*\,F(t)\;dt\Big\|_{L^2(\mathbb{R}^d)} = \sup\Big\{\Big(f\,,\,\int_0^\infty U(t)^*\,F(t)\;dt\Big) \;:\; \hat{f} \in C_0^\infty(\omega),\;\; \|f\|_{L^2} = 1\Big\}\,,$$

$$\big\|U(t)f\big\|_{L^qL^r} \;=\; \sup\Big\{\int_0^\infty (U(t)\,f\,,\,F(t))\;dt \;:\; \hat{F} \in C_0^\infty(]0,\infty[\times\omega),\;\; \|F\|_{L^{q'}L^{r'}} = 1\Big\}\,.$$

Estimate (3.4.13) holds if and only if

$$\Big(\int_0^\infty (U(t)^*F(t))\;dt\;,\;\int_0^\infty (U(s)^*G(s))\;ds\Big)$$

is a continuous bilinear form on $L^{q'}L^{r'}$, that is

(3.4.14)
$$\Big|\int_0^\infty\!\!\int_0^\infty \big(U(s)^*\,F(s)\,,\,U(t)^*\,G(t)\big)\;ds\,dt\Big| \;\le\; C\,\|F\|_{L^{q'}_t L^{r'}_x}\|G\|_{L^{q'}_t L^{r'}_x}\,.$$

Unitarity implies that

$$\forall s,t\,,\quad B\ :=\ U(t)U^*(s) = U(t-s)\quad \text{satisfies}\quad \|Bf\|_{L^2}\ \le\ \|f\|_{L^2}\,.$$

The dispersive estimate (3.4.3) is

$$\forall s,t\,,\quad \|Bf\|_{L^\infty}\ \le\ C\langle t-s\rangle^{-\sigma}\|f\|_{L^1}\,.$$

With $r'\in\,]1,2[$ the dual index to r, choose $\theta\in\,]0,1[$ so that

$$(3.4.15)\qquad \frac{1}{r'}\ =\ \theta\,\frac{1}{1}\ +\ (1-\theta)\,\frac{1}{2}\,,\qquad \text{then}\qquad \theta = \frac{2-r'}{r'} = \frac{r-2}{r}\,.$$

The Riesz–Thorin theorem implies that

$$\|Bf\|_{L^r}\ \le\ C^\theta\,\langle t-s\rangle^{-\sigma\theta}\,\|f\|_{L^{r'}}\,.$$

With Hölder's inequality, this yields the interpolated bilinear estimate,

$$\Big|\big(U(s)^*\,F(s)\,,\,U(t)^*\,G(t)\big)\Big|\ \le\ C^\theta\,\langle t-s\rangle^{-\sigma\theta}\,\|F(s)\|_{L^{r'}}\|G(t)\|_{L^{r'}}.$$

Admissibility implies that

$$\frac{1}{q}\ \le\ \sigma\Big(\frac{1}{2}-\frac{1}{r}\Big)\ =\ \sigma\Big(\frac{r-2}{2r}\Big)\ =\ \frac{\sigma\,\theta}{2}\,.$$

When strict inequality holds in the definition of admissibility, $\langle t-s\rangle^{-\sigma\theta}\in L^{q/2}(\mathbb{R}_t)$. The hypothesis $q>2$ is used here. For the limiting case, it is nearly so. The Hardy–Littlewood inequality shows that convolution with $|t|^{-2/q}$ has the L^p mapping properties that convolution with an element of $L^{q/2}(\mathbb{R})$ would have.

The Hausdorff–Young inequality shows that

$$(3.4.16)\qquad L^{p_1}\,*\,L^{p_2}\,\subset\,L^{p_3}\,,\qquad \text{provided}\qquad \frac{1}{p_1}\,+\,\frac{1}{p_2}\,=\,1\,+\,\frac{1}{p_3}\,.$$

The Hardy–Littlewood inequality asserts that when $1<p_1,p_2,p_3<\infty$,

$$(3.4.17)\quad \frac{1}{\langle t\rangle^{1/p_1}}\,*L^{p_2}(\mathbb{R})\,\subset\,L^{p_3}(\mathbb{R})\,,\qquad \text{provided}\qquad \frac{1}{p_1}+\frac{1}{p_2}\,=\,1+\frac{1}{p_3}\,.$$

Set

$$(3.4.18)\qquad p_1\ =\ \frac{q}{2}\,,\qquad p_2\ =\ q'\,,\qquad \text{and}\qquad p_3\ =\ q\,.$$

The index conditions in (3.4.16)–(3.4.17) becomes

$$\frac{2}{q}\ +\ \frac{1}{q'}\ =\ 1+\frac{1}{q}.$$

It is satisfied thanks to the definition of q'. Then (3.4.16) in the admissible case and (3.4.17) in the sharp admissible case imply that

$$(3.4.19)\qquad \Big\|\int_{-\infty}^{\infty}\langle t-s\rangle^{-\sigma\theta}\,\|F(s)\|_{L^{r'}}\,ds\Big\|_{L^q(\mathbb{R}_t)}\ \le\ C\,\big\|F\big\|_{L_t^{q'}L_x^{r'}}\,.$$

Hölder's inequality yields

$$\int_0^\infty \Big(\int_0^\infty \langle t-s\rangle^{-\sigma\theta} \, \big\|F(s)\big\|_{L^{r'}} \, ds \Big) \, \big\|G(t)\big\|_{L^{r'}} \, dt \;\le\; C \, \big\|F\big\|_{L_t^{q'} L_x^{r'}} \, \big\|G\big\|_{L_t^{q'} L_x^{r'}} \,.$$

This proves the desired estimate (3.4.14). □

A scaling yields estimates for sources with Fourier transform supported in $\lambda\overline{\omega}$ for $0 < \lambda$.

Lemma 3.4.9. *With q, r, ω, σ as in the previous lemma and γ as in the theorem, there is a C so that for all $0 < \lambda$ and $f \in L^2$ with $\operatorname{supp} \hat f \subset \lambda\overline{\omega}$,*

$$u(t) \;:=\; e^{it\tau_j(D_x)} f \;:=\; U(t) f$$

satisfies

$$\big\|u\big\|_{L_t^q L_x^r} \;\le\; C \big\| \, |D|^\gamma f \big\|_{L^2} \,. \tag{3.4.20}$$

Proof of Lemma 3.4.9. If $u_\lambda(t,x) := u(\lambda t, \lambda x)$, then $L u_\lambda = 0$ and the spectrum of u_λ is contained in $\overline{\omega}$.

The two sides of (3.4.12) scale differently. We verify that with the γ in the lemma, the two sides (3.4.20) have the same scaling properties. Compute

$$\big\|u_\lambda(t)\big\|_{L^r} \;=\; \Big(\int |u_\lambda(t,x)|^r \, dx \Big)^{1/r} \;=\; \Big(\int |u(\lambda t\,,\, \lambda x)|^r \, dx \Big)^{1/r} .$$

The substitution $y = \lambda x$, $dx = \lambda^{-d} dx$ yields

$$= \; \lambda^{-d/r} \Big(\int |u(\lambda t\,,\, y)|^r \, dy \Big)^{1/r} \;=\; \lambda^{-d/r} \big\|u(\lambda t)\big\|_{L^r} \,.$$

A similar change of variable for the time integral shows that

$$\big\|u_\lambda\big\|_{L_t^q L_x^r} \;=\; \lambda^{-1/q - d/r} \big\|u\big\|_{L_t^q L_x^r} \,.$$

For any γ, $\| \, |D|^\gamma f\|_{L^2}$ is a norm equivalent to the norm on the right-hand side for sources with spectrum in $\overline{\omega}$. Exercise 3.4.1 show that $\| \, |D|^\gamma f_\lambda\|_{L^2} = \lambda^{\gamma - d/2} \| \, |D|^\gamma f\|_{L^2}$.

Given q, r, the γ of the theorem is the unique value so that the two norms scale the same. Therefore the estimate of the present lemma follows from the preceding lemma. □

Proof of Theorem 3.4.7. With χ from the dyadic partition of unity for $\mathbb{R}^d_\xi \setminus 0$ from Lemma 3.II.2, introduce the Littlewood–Paley decomposition of tempered distributions with $\hat g \in L^1_{\rm loc}(\mathbb{R}^d)$

$$g \;=\; \sum_{J \in \mathbb{Z}} g_j \,, \qquad g_j \;:=\; \chi(D/2^j)\, g \;:=\; (2\pi)^{-d/2} \int e^{ix\xi} \, \chi(\xi/2^j) \, \hat g(\xi) \, d\xi \,.$$

Then for $1 < r < \infty$ the square function estimate (see [Stein, 1970]) asserts that there is a $C = C(p) > 1$ so that

$$C^{-1}\|g\|_{L^r} \;\le\; \Big\|\Big(\sum_{j\in\mathbb{Z}} |g_j|^2\Big)^{1/2}\Big\|_{L^r} \;\le\; C\,\|g\|_{L^r}\,.$$

Lemma 3.4.10. *If $2 \le q, r < \infty$, there is a constant C so that*

$$\big\|F\big\|^2_{L^q_tL^r_x} \;\le\; C\sum_{j\in\mathbb{Z}} \|F_j\|^2_{L^q_tL^r_x}\,, \tag{3.4.21}$$

where $F(t) = \sum_j F_j(t)$ is the Littlewood–Paley decomposition in x.

Proof of Lemma 3.4.10. The square function estimate yields

$$\big\|F(t)\big\|^2_{L^r_x} \;\le\; C\Big(\int\Big(\sum_j |F_j(t)|^2\Big)^{r/2}\,dx\Big)^{2/r} \;=\; C\,\Big\|\sum_j |F_j(t)|^2\Big\|_{L^{r/2}}\,.$$

Minkowski's inequality in $L^{r/2}$ shows that this is

$$\le\; C\sum_j \big\|F_j(t)^2\big\|_{L^{r/2}} \;=\; C\sum_j \big\|F_j(t)\big\|^2_{L^r}\,.$$

Using this yields

$$\begin{aligned}\big\|F\big\|^2_{L^q_tL^r_x} &\le C\Big(\int_0^\infty\Big(\sum_j \|F_j(t)\|^2_{L^r}\Big)^{q/2}\,dt\Big)^{2/q}\\ &= C\,\Big\|\sum_j \|F_j(t)\|^2_{L^r(\mathbb{R}^d_x)}\Big\|_{L^{q/2}(\mathbb{R}_t)}\,.\end{aligned}$$

Minkowski's inequality in $L^{q/2}(\mathbb{R}_t)$ shows this is

$$\le\; C\sum_j \Big\|\|F_j(t)\|^2_{L^r(\mathbb{R}^d_x)}\Big\|_{L^{q/2}(\mathbb{R}_t)} \;=\; C\sum_j \|F_j(t)\|^2_{L^q_tL^r_x}\,.$$

□

Return now to the proof of the theorem. Associate to the sheet $\tau = \tau_k(\xi)$ the projector $\pi_k(\xi) := \pi(\tau_k(\xi), \xi)$ from §3.2. The π_k are real analytic on $\xi \ne 0$ and homogeneous of degree 0 in ξ. In addition $\sum_k \pi_k = I$. The solution u satisfies

$$u \;=\; \sum_k e^{it\tau_k(D)}\,\pi_k(D)\,f \;:=\; \sum_k u_k\,.$$

Apply (3.4.21) to u_k using (3.4.20) to find

$$\begin{aligned}\big\|u_k\big\|^2_{L^q_tL^r_x} &\le C\sum_j \|u_{k,j}\|^2_{L^q_tL^r_x}\\ &\le\; C'\sum_j \|\,|D|^\gamma\pi_k(D)f_j\|^2_{L^2} \;\le\; C'\,\|\,|D|^\gamma f\|^2_{L^2}\,.\end{aligned}$$

The finite sum on k completes the proof of the theorem. □

Corollary 3.4.11. *Denote by $S(t)$ the L^2 unitary mapping $u(0) \mapsto u(t)$ for solutions of $Lu = 0$. With the indices of the theorem one has*

$$(3.4.22) \qquad \Big\| \int_0^\infty S(s)^*\, F(s)\, ds \Big\|_{L^2(\mathbb{R}^d)} \leq C \, \big\| |D_x|^\gamma \, F \big\|_{L_t^{q'} L_x^{r'}} .$$

Proof. Estimate (3.4.22) is equivalent to the Strichartz estimate (3.4.11) by a duality like that used to establish the equivalence of (3.4.12) and (3.4.13). □

Exercise 3.4.4. *Prove the following complement to (3.4.21) that comes from the other side of the square function inequality. If $1 < p \leq 2$ and $1 \leq r \leq 2$, then there is a C so that*

$$(3.4.23) \qquad \sum_{j=-\infty}^{\infty} \|F_j\|^2_{L_t^r L_x^p} \leq C \, \|F\|^2_{L_t^r L_x^p} .$$

Appendix 3.I. Perturbation theory for semisimple eigenvalues

Formulas from the perturbation theory of eigenvalues are needed in the proof of Theorem 3.3.4 and are also central elements in geometric optics. These results for multiple eigenvalues that are semisimple are not that well known. The key idea is that one should not make a choice of basis of eigenfunctions, but work systematically with the spectral projections.

Definitions. *An eigenvalue λ of a matrix A is* **semisimple** *when the kernel and range of $A - \lambda I$ are complementary subspaces. In this case denote by π the* **spectral projection** *onto the kernel of $A - \lambda I$ along its range and by Q the* **partial inverse** *defined by*

$$(3.I.1) \qquad Q\,\pi = 0 , \qquad Q\big(A - \lambda I\big) = I - \pi .$$

Examples 3.I.1. i. Every eigenvalue of a hermitian symmetric or normal matrix is semisimple.

ii. More generally, a matrix is similar to a diagonal matrix if and only if each of its eigenvalues is semisimple.

Theorem 3.I.1. *Suppose that $\Omega \subset \mathbb{R}^m$ is open and*

$$M(Y) \in C^\infty(\Omega, \mathrm{Hom}\,(\mathbb{C}^N))$$

is a matrix valued function. Suppose that there is a disk $\mathbb{D} \subset \mathbb{C}$ so that for every $Y \in \Omega$ there is exactly one eigenvalue, $\lambda(Y)$ of $M(Y)$ in $\overline{\mathbb{D}}$ and that eigenvalue is semisimple. Denote by $\pi(Y)$ the projection along the range of

$M - \lambda I$ onto the kernel of $M - \lambda I$ and by $Q(Y)$ the partial inverse defined by

$$Q(Y)\,\pi(Y) \;=\; 0, \qquad Q(Y)\,(M(Y) - \lambda I) \;=\; I - \pi(Y).$$

Then λ, π, and Q are C^∞ functions of Y.

Proof. It suffices to prove smoothness at an arbitrary $\underline{Y}$. Suppose that $\mathbb{D} = \{|z - \underline{z}| < r\}$. Choose $\epsilon > 0$ so that for $|Y - \underline{Y}| < \epsilon$ the disk $|z - \underline{z}| \le r + \epsilon$ contains only one eigenvalue. The regularity of $\pi(Y)$ for those Y follows from the contour integral representation,[2]

$$\pi(Y) \;=\; \frac{1}{2\pi i}\oint_{|z-\underline{z}|=r+\epsilon} \Big(z - (M(Y) - \lambda(Y) I)\Big)^{-1}\, dz\,.$$

The regularity of Q and λ follow from the identities,

$$Q(Y) = (I - \pi(Y))\,\big(\pi(Y) + M(Y)\big)^{-1}, \qquad \lambda(Y) \;=\; \frac{\operatorname{trace}(M(Y)\,\pi(Y)}{\operatorname{trace}\pi(Y)}\,. \qquad \square$$

Theorem 3.I.2. *Suppose that $]a,b[\ \ni s \to A(s)$ is a smooth family of complex matrices so that for all s the disk $\overline{\mathbb{D}} \subset \mathbb{C}$ contains exactly one eigenvalue, $\lambda(s)$ of $A(s)$ in $\overline{\mathbb{D}}$, and that eigenvalue is semisimple. Denoting d/ds by $'$, one has the following perturbation formulas,*

$$\lambda'(s)\;\pi(s) \;=\; \pi(s)\;A'(s)\;\pi(s)\,, \tag{3.I.2}$$

$$\lambda''\,\pi \;=\; \pi\,A''\,\pi \;-\; 2\,\pi\,A'\,Q\,A'\,\pi\,, \tag{3.I.3}$$

$$\pi' \;=\; -\pi\,A'\,Q - QA'\pi\,. \tag{3.I.4}$$

Example 3.I.2. If A is hermitian symmetric, semisimplicity is automatic and π is hermitian. If $\mathbf{v}$ is a unit eigenvector, then multiplying (3.I.2) on the right by $\mathbf{v}$ and then taking the scalar product with $\mathbf{v}$ yields the standard formula, $\lambda' = \langle \mathbf{v}\,,\, A'\,\mathbf{v}\rangle$.

Proof. The smoothness of $\pi(s), Q(s), \lambda(s)$ follows from Theorem 3.I.1.

The formulas (3.I.2)–(3.I.4) are proved by differentiating the identity $(A - \lambda)\pi = \pi(A - \lambda) = 0$ with respect to s. The equation for each d^j/ds^j is analysed by considering its projections π and $I - \pi$. Equivalently, each equation is multiplied first by π, and then by Q.

Differentiating $(A - \lambda)\pi$ yields

$$(A - \lambda)'\;\pi \;+\; (A - \lambda)\;\pi' \;=\; 0\,. \tag{3.I.5}$$

Multiplying on the left by π eliminates the second term to yield

$$\pi\;(A - \lambda)'\;\pi \;\;=\; 0\,. \tag{3.I.6}$$

This is (3.I.2).

[2]A short proof of the representation is to evaluate the right-hand sides in a basis whose first k elements form a basis for $\ker(M - \lambda I)$ and whose last elements are a basis for the range.

Multiply equation (3.I.5) on the left by Q to find

$$(I-\pi)\pi' \;=\; -Q\,\big(A-\lambda\big)'\,\pi\,.$$

Since $Q\,\pi=0$, this simplifies to

$$(I-\pi)\pi' \;=\; -QA'\pi\,. \tag{3.I.7}$$

Equation (3.I.5) is exhausted, and we take a second derivative,

$$(A-\lambda)''\pi \;+\; 2(A-\lambda)'\pi' \;+\; (A-\lambda)\pi'' \;=\; 0\,.$$

Multiply on the left by π to eliminate the last term,

$$\pi(A-\lambda)''\pi + 2\pi(A-\lambda)'\pi' \;=\; 0\,.$$

Subtract $2(\pi(A-\lambda)'\pi)\pi'=0$ to find,

$$\pi(A-\lambda)''\pi + 2\pi(A-\lambda)'(I-\pi)\pi' \;=\; 0\,.$$

Then (3.I.7) yields

$$\pi(A-\lambda)''\pi + 2\pi(A-\lambda)'(-QA'\pi) \;=\; 0\,. \tag{3.I.8}$$

Since $\pi\,Q=0$, one has

$$2\,\pi\lambda'(-QA'\pi) \;=\; 0\,. \tag{3.I.9}$$

Adding (3.I.8) and (3.I.9) yields (3.I.3).

To prove (3.I.4) knowing (3.I.7), what is needed is $\pi\,\pi'$. Differentiate $\pi^2=\pi$ to find

$$\pi\,\pi' + \pi'\,\pi \;=\; \pi'\,, \qquad \text{whence} \qquad \pi\,\pi' \;=\; \pi'(I-\pi)\,. \tag{3.I.10}$$

Differentiate $\pi\,(A-\lambda)=0$ to find

$$\pi'(A-\lambda) + \pi\,(A-\lambda)' \;=\; 0\,.$$

Multiply on the right by Q to find

$$\pi'(I-\pi) \;=\; -\pi\,(A'-\lambda')\,Q\,.$$

Use (3.I.10) and simplify using $\pi\,Q=0$ to find

$$\pi\,\pi' \;=\; \pi'(I-\pi) \;=\; -\pi\,A'\,Q\,.$$

Adding this to (3.I.7) completes the proof. □

Appendix 3.II. The stationary phase inequality

Definition. *A point $\underline{x}$ in an open subset $\Omega \subset \mathbb{R}^d$ is a* **stationary point** *of $\phi \in C^\infty(\Omega\,;\,\mathbb{R})$ when $\nabla_x \phi(\underline{x}) = 0$. It is a* **nondegenerate stationary point** *when the matrix of second derivatives at $\underline{x}$ is nonsingular.*

When $\underline{x}$ is a nondegenerate stationary point, the map $x \mapsto \nabla_x \phi(x)$ has nonsingular jacobian at $\underline{x}$. It follows that the map is a local diffeomorphism and in particular the stationary point is isolated.

Taylor's theorem shows that

$$\nabla_x \phi(x) \;=\; \frac{1}{2}\,\nabla_x^2 \phi(\underline{x})\,(x - \underline{x}) \;+\; O(|x - \underline{x}|^2)\,.$$

Therefore if $\omega \Subset \Omega$ contains $\underline{x}$ and no other stationary point, nondegeneracy implies that there is a constant $C > 0$, so

$$\text{(3.II.1)} \qquad \forall \underline{x} \in \omega\,, \qquad \big|\nabla_x \phi(x)\big| \;\geq\; C\,|x - \underline{x}|\,.$$

We estimate the size of oscillatory integrals whose phase has a single nondegenerate stationary point. These integrals have a complete asymptotic expansion. I learned the following proof from G. Métivier; see [Stein, 1993] for an alternate proof.

Theorem 3.II.1. *Suppose that $\phi \in C^\infty(\Omega\,;\,\mathbb{R})$ has a unique stationary point $\underline{x} \in \Omega$. Suppose that $\underline{x}$ is nondegenerate, and let m denote the smallest integer strictly larger than $d/2$. Then for any $\omega \Subset \Omega$ there is a constant C so that for all $f \in C_0^m(\omega)$, and $0 < \epsilon < 1$,*

$$\text{(3.II.2)} \qquad \Big|\int e^{i\phi/\epsilon} f(x)\; dx\Big| \;\leq\; C\,\epsilon^{d/2} \sup_{|\alpha| \leq m}\; \|\partial^\alpha f(x)\|_{L^\infty(\omega)}\,.$$

Lemma 3.II.2. *There is a nonnegative $\chi \in C_0^\infty(\mathbb{R}^d \backslash 0)$ so that for all $x \neq 0$, $\sum_{k=-\infty}^{\infty} \chi(2^k\,x) = 1$.*

Proof of Lemma 3.II.2. Choose nonnegative $g \in C_0^\infty(\mathbb{R}^d \backslash 0)$ so that $g \geq 1$ on $\{1 \leq |x| \leq 2\}$. Define the locally finite sum

$$G(x) \;:=\; \sum_{k=-\infty}^{\infty} g(2^k x)\,, \qquad \text{so} \qquad G(2^k x) = G(x)\,.$$

Then $G \in C^\infty(\mathbb{R}^d \setminus 0)$ and $G \geq 1$. The function $\chi := g/G$ has the desired properties. □

Proof of Theorem 3.II.1. Translating coordinates, we may suppose that $\underline{x} = 0$. Choose χ as in the lemma and write

$$\int e^{i\phi/\epsilon} f(x)\; dx \;=\; \sum_{k=-\infty}^{\infty} \int \chi(2^k\,x)\; e^{i\phi/\epsilon}\; f(x)\; dx \;:=\; \sum_{k=-\infty}^{\infty} I(k)\,.$$

The half sum $\sum_{k<0} \chi(2^k x)$ is a smooth function on $\mathbb{R}^d$ that vanishes on a neighborhood of the origin and is identically equal to 1 outside a large ball. The nonstationary phase Lemma 1.2.2 implies that for all m,

$$\Big| \int e^{i\phi/\epsilon} \Big(\sum_{k<0} \chi(2^k\, x) \Big)\ f(x)\ dx \Big| \ \le\ C(m)\, \epsilon^m \sup_{|\alpha|\le m} \|\partial^\alpha f(x)\|_{L^1(\omega)}\,.$$

The sum $\sum_{2^k\epsilon^{1/2}\ge 1} \chi(2^k\, x)$ is a bounded function supported in a ball $|x| \le C\epsilon^{1/2}$ so

$$\Big| \int e^{i\phi/\epsilon} \Big(\sum_{2^k\epsilon^{1/2}\ge 1} \chi(2^k\, x) \Big)\ f(x)\ dx \Big| \ \le\ C\, \epsilon^{d/2}\, \|f(x)\|_{L^\infty(\omega)}\,.$$

There remains the sum over $1 \le 2^k < \epsilon^{-1/2}$. The change of variable $y = 2^k\, x$ yields

$$I(k)\ =\ 2^{-kd} \int \chi(y)\, e^{i\phi_k(y)/(2^{2k}\epsilon)}\ f(2^{-k}y)\ dy\,, \qquad \phi_k(y)\ :=\ 2^{2k}\, \phi(2^{-k}\, y)\,.$$

It follows from (3.II.1) that there is a constant $c > 0$, so that on the support of χ,

$$c^{-1} \le \big|\nabla \phi_k\big| \le c\,.$$

In addition there are constants $C(\alpha)$ independent of $k \ge 0$ so that $|\partial^\alpha \phi_k| \le C_\alpha$. The method of nonstationary phase shows that there is a constant independent of $k \ge 0$ so that

$$\Big| \int \chi(y)\ e^{i\phi_k(y)/(2^{2k}\epsilon)}\ f(2^{-k}y)\ dy \Big| \ \le\ C\, (2^{2k}\, \epsilon)^m \sup_{|\alpha|\le m} \|\partial^\alpha f(x)\|_{L^1(\omega)}\,.$$

Therefore,

$$\sum_{1\le 2^k<\epsilon^{-1/2}} |I(k)| \ \le\ C\, \epsilon^m \sum_{1\le 2^k<\epsilon^{-1/2}} 2^{-kd}\, 2^{2km} \sup_{|\alpha|\le m} \|\partial^\alpha f(x)\|_{L^1(\omega)}\,.$$

The finite geometric sum has ratio $r = 2^{2m-d} > 1$. If K is the largest index, then

$$r^K \ \le\ 1 + r + r^2 + \cdots + r^K \ =\ \frac{r^{K+1} - 1}{r - 1} \ <\ \frac{r}{r-1}\, r^K \ :=\ C(r)\, r^K\,.$$

The sum is comparable to the last term. Therefore, with $C = C(m,d) = r/(r-1)$,

$$\epsilon^m \sum_{1\le 2^k<\epsilon^{-1/2}} 2^{-kd}\, 2^{2km} \ \le\ C\, \epsilon^m\, (2^K)^{2m-d} \ \le\ C\, \epsilon^m\, \big(\epsilon^{-1/2}\big)^{2m-d} \ =\ C\, \epsilon^{d/2}\,.$$

This completes the proof. □

Corollary 3.II.3. *Suppose that $\phi(x,\zeta)$ is a family of phases depending smoothly on ζ on a neighborhood of $0 \in \mathbb{R}^q$ and that $\phi(x,0)$ satisfies the hypotheses of the preceding theorem. Then there is a neighborhood $0 \in \mathcal{O}$ so that the hypotheses are satisfied for $\zeta \in \mathcal{O}$ and the estimate (3.II.1) holds with a constant independent of $\zeta \in \mathcal{O}$.*

Proof. The first assertion follows from the implicit function theorem applied to the system of equations $\nabla_x \phi(x,\zeta) = 0$. The estimates of the proof are all locally uniform, proving the second assertion. $\square$

Chapter 4

Linear Elliptic Geometric Optics

The study of oscillatory solutions of elliptic equations is easier than the corresponding hyperbolic theory. The reason is simple. Oscillations propagate in the hyperbolic case and have only local effects in the elliptic case. Nevertheless, the elliptic case is a good starting point for several reasons. First, it is easier to introduce some of the basic notions in this case. Second, the elliptic results are needed in the proofs of nonlinear hyperbolic results.

For the analysis of this section there is no need for symmetry or any other hypothesis of hyperbolicity. Similarly, the independent variable y is not split into space and time. The partial differential operator

$$L(y, \partial_y) = \sum A_\mu(y) \frac{\partial}{\partial y_\mu} + B(y)$$

is assumed to have smooth matrix valued coefficients on an open set $\mathcal{O} \subset \mathbb{R}^n$.

The fundamental dichotomy is between oscillations with phases ϕ such that $(y, d\phi(y))$ is characteristic or not. For elliptic operators only the second possibility occurs while for hyperbolic problems both are possible.

When the phase is noncharacteristic, one can have oscillatory solutions only when there are oscillatory sources.

4.1. Euler's method and elliptic geometric optics with constant coefficients

This section begins the discovery of short wavelength asymptotic expansions of WKB type starting from the elementary theory of constant coefficient ordinary differential operators

$$L(d/dt) \;:=\; p_m \frac{d^m}{dt^m} + p_{m-1}\frac{d^{m-1}}{dt^{m-1}} + \cdots + p_1 \frac{d^1}{dt^1} + p_0\,, \qquad p_m \neq 0\,.$$

The key identity is $L(d/dt)e^{i\tau t} = L(i\tau)e^{i\tau t}$. Therefore, a solution of

$$Lu = b\,e^{i\tau t}$$

is given by

$$u = L(i\tau)^{-1}\, b\, e^{i\tau t}, \qquad \text{provided that} \qquad L(i\tau) \neq 0\,. \tag{4.1.1}$$

Since

$$|L(\tau)| \;\geq\; |p_m\,\tau^m| - \sum_0^{m-1} |p_j\,\tau|^j\,, \qquad p_m \neq 0\,,$$

it follows that $L(i\tau) \neq 0$ for large τ, so this method suffices for rapidly oscillating sources.

Consider a localized strongly oscillatory source,

$$L(d/dt)u \;=\; b(t)\,e^{i\phi(t)/\epsilon}\,, \qquad b \in C_0^\infty(\mathbb{R})\,, \qquad 0 < \epsilon \ll 1\,.$$

To guarantee oscillation, suppose that φ is real valued with $\varphi' \neq 0$ on $\operatorname{supp} b$. For t outside of $\operatorname{supp} b$, u satisfies $Lu = 0$. The general solution of this homogeneous equation is a linear combination of m solutions of the form $q(t)\,e^{rt}$ with polynomial q and roots r of $L(r) = 0$. In particular it does not oscillate on scale ϵ for small ϵ. Oscillations do not spread beyond the support of b. Elliptic partial differential equations behave like this. In the hyperbolic case, some oscillations can propagate beyond the support of the sources.

For a constant coefficient partial differential operator, Euler's identity is $L(\partial_y)e^{iy\eta} = L(i\eta)e^{iy\eta}$, so a particular solution of a constant coefficient system of partial differential equations,

$$L(\partial_y)\,u = b\,e^{iy\eta}\,, \qquad b \in \mathbb{C}^N\,, \tag{4.1.2}$$

is given by

$$u = L(i\eta)^{-1}\,b\,e^{iy\eta}, \qquad \text{provided that} \qquad \det L(i\eta) \neq 0\,.$$

To study the case of a first order system in the limit of small wavelength, use η/ϵ in place of η, and consider $\epsilon \to 0$. Since

$$L(i\eta/\epsilon) \;=\; L_1(i\eta/\epsilon) + L_0 \;=\; \frac{1}{\epsilon}\,\big(L_1(i\eta) + \epsilon L_0\big)\,,$$

it follows that if η is not characteristic, then $L(i\eta/\epsilon)$ is invertible for ϵ small. For such η, ϵ, an explicit solution of (4.2) is given by

$$u = e^{iy\eta/\epsilon}\,\big(L_1(i\eta/\epsilon) + L_0\big)^{-1} b = \epsilon\, e^{iy\eta/\epsilon}\,\big(L_1(i\eta) + \epsilon\,L_0\big)^{-1} b\,.$$

Write

$$L_1(i\eta) + \epsilon\, L_0 \;=\; L_1(i\eta)\big(I \;+\; \epsilon\, L_1(i\eta)^{-1}\, L_0\big)$$

to show that for ϵ small, the inverse is given by a convergent Neumann series,

$$u(y) = \epsilon\, e^{iy\eta/\epsilon} \sum_{n=0}^{\infty} \big(- \epsilon\, L_1(i\eta)^{-1}\, L_0\big)^n\, L_1(i\eta)^{-1}\, b$$

$$= \epsilon\, e^{iy\eta/\epsilon}\, \big(L_1(i\eta)^{-1}\, b + \text{higher order terms}\big)\,.$$

The form of the solution is a series

$$e^{iy\eta/\epsilon}\left(\epsilon\, a_1 + \epsilon^2 a_2 + \cdots\right),$$

where the vector valued summands are each obtained by multiplying b by a finite product of matrices.

Multiplying the source and the solution by $e^{ic/\epsilon}$ with real c shows that the computation above works for the affine phase $\phi = c + y\eta$, in which case $\eta = d\phi$.

A key feature of this solution is that the leading term depends only on the principal symbol L_1. The reason is simple. For highly oscillatory solutions, the derivatives are of order $1/\epsilon$ larger than u. So long as $L_1(i\eta)$ is invertible, the sum of those terms is dominant. The general principle here is that for noncharacteristic short wavelength oscillations, the principal symbol dominates. In contrast when $L_1(i\eta)$ is not invertible, the lower order terms play an important role (see §1.4 and Chapter 5).

4.2. Iterative improvement for variable coefficients and nonlinear phases

The next step is a key insight that passes from finding exact solutions to constant coefficient problems to approximate solutions of variable coefficient problems. Consider a source term that is rapidly oscillating with possibly nonlinear phase and with amplitude that depends on y. Consider the differential equation with possibly variable coefficients

$$L(y, \partial_y)\, u \;=\; b(y)\, e^{i\phi(y)/\epsilon}\,. \tag{4.2.1}$$

The phase ϕ is a smooth real valued function whose gradient is assumed to be nonvanishing on the support of $b(y)$. The coefficients of L are assumed to be smooth on a neighborhood of this support.

Imagine an observer who looks at u near a point $\underline{y}$. Suppose that the observation region is large compared to ϵ but small compared to the scale on which b, the coefficients of L, and $d\phi$ vary. To such an observer, these quantities appear constant and the differential equation looks like

$$L(\underline{y}, \partial_y)u \;=\; b(\underline{y})\, e^{i\left(\phi(\underline{y}) + d\phi(\underline{y})(y-\underline{y}))/\epsilon\right)}\,. \tag{4.2.2}$$

If $(\underline{y}, d\phi(\underline{y}))$ is not in the characteristic variety of L, the previous analysis shows that for ϵ small an approximate solution on this region is given by

$$\begin{aligned} u_{\text{approx}} \sim\ & \epsilon\, e^{i\big(\phi(\underline{y})+d\phi(\underline{y})(y-\underline{y})\big)/\epsilon}\, L_1(\underline{y}, id\phi(\underline{y}))^{-1}\, b(\underline{y}) \\ \approx\ & \epsilon\, e^{i\phi(y)/\epsilon}\, L_1(\underline{y}, id\phi(\underline{y}))^{-1}\, b(\underline{y})\,. \end{aligned}$$

These computations suggest that

$$u(y) \;=\; \epsilon\, e^{i\phi(y)/\epsilon}\, a_1(y)\,, \qquad a_1(y) := L_1(y, id\phi(y))^{-1}\, b(y)\,, \tag{4.2.3}$$

defines a reasonable approximate solution.

The idea leading to this guess was that in the limit of very small wavelength the problem can be replaced by an approximate problem with constant coefficients, a source with constant amplitude, and an affine phase. To assess the accuracy of the approximation, take u as defined in (4.2.3) and apply $L(y, \partial)$. The largest $O(1/\epsilon)$ arise when a derivative falls on the exponential factor where $L_1(y, \partial)e^{i\phi/\epsilon} = L(y, id\phi/\epsilon)\, e^{i\phi/\epsilon}$, yielding

$$\begin{aligned} L(y, \partial_y)\big(\epsilon\, e^{i\phi/\epsilon} a_1(y)\big) &= e^{i\phi/\epsilon}\big(L_1(y, id\phi)\, a_1 + L(y, \partial)a_1\big) \\ &= e^{i\phi/\epsilon}\big(b(y) + \epsilon b_1(y)\big), \end{aligned} \tag{4.2.4}$$

where

$$b_1(y) := L(y, \partial_y)\, a_1(y)\,. \tag{4.2.5}$$

The error on the right-hand side is of the same order of magnitude as the approximate solution. This is not an auspicious beginning. The good news is that the previous computation tells us a corrector. It suffices to subtract from the approximate solution an approximate solution, v, to $L\, v = \epsilon b_1\, e^{i\phi/\epsilon)}$. Thus, take

$$u \;:=\; e^{i\phi(y)/\epsilon}\Big(\epsilon\, a_1(y) + \epsilon^2\, a_2(y)\Big)\,, \qquad a_2(y) \;:=\; -L_1(y, id\phi(y))^{-1}\, b_1(y)\,, \tag{4.2.6}$$

to find

$$L(y, \partial_y)\, u = e^{i\phi(y)/\epsilon}\Big(b(y) + \epsilon^2 b_2(y)\Big)\,, \tag{4.2.7}$$

where

$$b_2(y) := L(y, \partial_y)\, a_2(y)\,. \tag{4.2.8}$$

This process, by induction on m, yields the following theorem.

Theorem 4.2.1. *Suppose that $m \geq 1$ is an integer, $\Omega \subset \mathbb{R}^n$ is a bounded open set, $b(y)$ is a smooth amplitude on Ω, and ϕ is a smooth real valued phase such that for all $y \in \Omega$, $d\phi(y) \neq 0$ and $(y, d\phi(y)) \notin \operatorname{Char} L$. Then, there are smooth amplitudes a_j on Ω so that*

$$u^\epsilon := e^{i\phi(y)/\epsilon}\Big(\epsilon\, a_1(y) + \epsilon^2\, a_2(y) + \cdots + \epsilon^m\, a_m(y)\Big) \tag{4.2.9}$$

satisfies

$$Lu^\epsilon = e^{i\phi(y)/\epsilon}\, b(y) + \epsilon^m\, e^{i\phi(y)/\epsilon}\, r^\epsilon(y)\,, \tag{4.2.10}$$

with

$$\forall \alpha \qquad \sup_{(\epsilon,y)\in]0,1]\times\Omega} \left|\partial_y^\alpha r^\epsilon(y)\right| \;<\; \infty\,.$$

The principal amplitude is given by $a_1 = L(y, id\phi(y))^{-1}\, b(y)$.

This ends a path leading from the method of Euler for constant coefficient ordinary differential equations to these expansions (4.2.9) of WKB type after Wentzel, Kramers, and Brillouin. Such expansions date to the early nineteenth century. They were used independently by these three authors in 1926 in quasiclassical expansions of Schrödinger's equation as Plank's constant $\hbar \to 0$ (see §5.2.2).

4.3. Formal asymptotics approach

Once the form of the expansion (4.2.9) is known, the exact coefficients can be computed without going through the above recursion. We treat a more general situation where a sequence of smooth amplitudes b_j are given and we seek amplitudes a_j so that

(4.3.1)

$$L(y,\partial_y)\left(e^{i\phi(y)/\epsilon}\left(\epsilon\, a_1(y) + \epsilon^2 a_2(y) + \cdots\right)\right) \;\sim\; e^{i\phi(y)/\epsilon}\left(b_0(y) + \epsilon^1 b_1(y) + \cdots\right).$$

The precise interpretation of (4.3.1) is left for later to show that the method is so natural that it works almost by itself. The previous analysis was the case where the $b_j = 0$ for $j \ge 1$. The source on the right-hand side is $O(1)$ while the response in $O(\epsilon)$ is as it was in the preceding sections. To compute the left-hand side of (4.3.1), compute as in (4.2.4)

$$L(y,\partial_y)\left(e^{i\phi(y)/\epsilon}\left(\epsilon\, a\right)\right) \;=\; e^{i\phi(y)/\epsilon}\Big(L_1(y, id\phi(y))a \;+\; \epsilon L(y,\partial_y)a\Big).$$

Plug in $a \sim \sum \epsilon^j a_j$ to find

(4.3.2)

$$e^{i\phi(y)/\epsilon}\left(L_1(y,id\phi(y))a_1(y) + \sum_{j=1}^{\infty} \epsilon^j\left(L_1(y,id\phi(y))a_{j+1}(y) + L(y,\partial_y)a_j(y)\right)\right).$$

The equations determining the a_j are then read off to be

$$L_1(y, id\phi(y))\, a_1(y) = b_0(y)\,, \tag{4.3.3}$$

and for $j > 1$,

$$L_1(y, id\phi(y))\, a_j(y) = -L(y,\partial_y)\, a_{j-1}(y) + b_{j-1}\,. \tag{4.3.4}$$

They are uniquely solvable provided that $L_1(y, d\phi(y))$ is invertible or equivalently $(y, d\phi(y))$ is everywhere noncharacteristic.

To put meat on the bones of the formal series, one has to give meaning to the sums in (4.3.1). These sums do not usually converge but represent asymptotic expansions as $\epsilon \to 0$. The interpretation is like that of Taylor expansions of smooth but not analytic functions.

Definitions. 1. *If $\mathcal{O}$ is an open set in $\mathbb{R}^n$, $b_j(y)$ is a sequence of smooth functions on $\mathcal{O}$, and $b \in C^\infty(]0,1[\, \times \mathcal{O}\,;\, \mathbb{C}^N)$, then the asymptotic relation*

$$b(\epsilon, y) \sim b_0(y) + \epsilon\, b_1(y) + \epsilon^2\, b_2(y) + \cdots \quad \text{in} \quad C^\infty(\mathcal{O}) \tag{4.3.5}$$

means that for every integer $m \geq 0$, every multi-index $\alpha \in \mathbb{N}^n$, and every compact subset $K \subset \mathcal{O}$,

$$\sup_K \left| \partial_y^\alpha \Big(b - \sum_{j=0}^{m} \epsilon^j\, b_j(y) \Big) \right| = O(\epsilon^{m+1}) \qquad \text{as } \epsilon \to 0\,. \tag{4.3.6}$$

If $c \in C^\infty(]0,1[\, \times \mathcal{O})$, the asymptotic relation

$$b(\epsilon, y) \sim c(\epsilon, y)$$

means that $b - c \sim \sum \epsilon^j\, 0$.

2. *If $\mathcal{O}$ is a bounded open set, then $b \sim \sum \epsilon^j b_j$ in $C^\infty(\overline{\mathcal{O}})$ is defined similarly by replacing the supremum over compact subsets K by the supremum over $\overline{\mathcal{O}}$.*

Remarks. 1. Instead of $\sim \sum \epsilon^j\, 0$, we write ~ 0.

2. If b is smooth on a neighborhood of $\epsilon = 0$, then Taylor's theorem shows that (4.3.5) is equivalent to

$$b_j(y) = \frac{1}{j!}\, \frac{\partial^j b(0,y)}{\partial \epsilon^j}\,.$$

The definition still leaves the interpretation of (4.3.1) unclear since the construction went from b_j to a_j with no mention of functions of $a(\epsilon, y)$ and $b(\epsilon, y)$. The key link is Borel's theorem.

Borel's Theorem 4.3.1. *Given a sequence b_j of smooth functions on the open set $\mathcal{O} \subset \mathbb{R}^n$, there is a smooth function $b(\epsilon, y)$ defined on $\mathbb{R} \times \mathcal{O}$ so that*

$$b(\epsilon, y) \sim b_0(y) + \epsilon b_1(y) + \epsilon^2 b_2(y) + \cdots\,.$$

Remarks. 1. If $\tilde{b}(\epsilon, y)$ is a second such function, then $\tilde{b} \sim b$.

2. Suppose that b and b_j are as in the definition above. Borel's Theorem implies that there is a $c(\epsilon, y) \in C^\infty(\mathbb{R} \times \mathcal{O})$ with $c \sim \sum \epsilon^j\, b_j(y)$. Then $b \sim c$ and $j!\, b_j = \partial^j c(0,y)/\partial \epsilon^j$. This shows that the smooth in ϵ case of Remark 2 preceding the theorem is the general case.

The proof of Borel's theorem 4.3.1 is a direct generalization of the proof of the following seemingly much more special result.

Borel's Theorem 4.3.2. *Given a sequence b_j, $0 \le j$, of complex numbers, there is a smooth function $b(\epsilon)$ on $\mathbb{R}$ whose Taylor series at the origin is $\sum \epsilon^j\, b_j$.*

Proof. The idea is to set $b(\epsilon) = \sum \epsilon^j\, b_j$. However, this series has no reason to converge since the b_j may grow arbitrarily rapidly. The clever idea is to cut off the summands so that they live only where $|\epsilon|$ is so small that the ϵ^j compensate the b_j.

Choose a function $\chi \in C_0^\infty(]-1,1[)$ such that $\chi(\epsilon) = 1$ for $|\epsilon| \le 1/2$.

The summand $\epsilon^j b_j$ is replaced by $\epsilon^j\, \chi(M_j \epsilon) b_j$ where the sequence of positive numbers M_j is chosen as follows.

Set $M_0 = 1$. For $j \ge 1$, choose $M_j \ge 1$ so that for $m = 0, 1, 2, \dots, j-1$ and all $\epsilon \in \mathbb{R}$,

$$\left| \frac{d^m}{d\epsilon^m} \big(\chi(M_j \epsilon)\, \epsilon^j\, b_j \big) \right| \le \frac{1}{2^j}\,. \tag{4.3.7}$$

This is possible since when the derivatives are expanded there are a finite number of terms. Each term is a bounded function of ϵ times

$$b_j\, \epsilon^{j-l}\, M_j^k\, \frac{d^k \chi}{d\epsilon^k}(M_j \epsilon)\,, \qquad k + l = m \le j - 1\,. \tag{4.3.8}$$

In the support of the $\chi^{(k)}$ term, $\epsilon < 1/M_j$. Thus the term (4.3.8) is bounded by

$$\frac{c(\chi, j)|b_j|}{M_j^{j-k-l}} \le \frac{c(\chi, j)|b_j|}{M_j}\,,$$

so (4.3.7) can be achieved by choosing M_j sufficiently large.

Then

$$\sum \epsilon^j\, \chi(M_j \epsilon)\, b_j$$

converges uniformly with all of its derivatives to a function $b(\epsilon)$. That it satisfies the conditions of the theorem is immediately verified by differentiating term by term and setting $\epsilon = 0$. □

Exercise 4.3.1. *Prove Borel's Theorem 4.3.1.*

In the construction ellipticity was used only to ensure that $L_1(y, d\phi)$ was invertible. The next result is stated for possibly nonelliptic operators for which this is true.

Theorem 4.3.3. *Suppose that $\Omega \subset \mathbb{R}^n$ is an open set,*

$$b(\epsilon, y) \sim \sum_{j=0}^{\infty} \epsilon^j\, b_j(y)$$

with amplitudes $b_j \in C^\infty(\Omega)$, *and* ϕ *is a smooth real valued phase such that for all* $y \in \Omega$ $d\phi(y) \neq 0$ *and* $(y, d\phi(y)) \notin \operatorname{Char} L$. *Then, there is an* $a \in C^\infty(\mathbb{R} \times \Omega)$ *with*

$$a(\epsilon, y) \sim \sum_{j=1}^{\infty} \epsilon^j \, a_j(y)$$

and

$$L(y, \partial_y) \left(e^{i\phi(y)/\epsilon} \, a(\epsilon, y) \right) \; - \; b(\epsilon, y) e^{i\phi(y)/\epsilon} \sim 0 \,.$$

The $a_j(y)$ *are uniquely determined from (4.3.3) and (4.3.4). The leading amplitude is given by (4.2.3).*

Remark. One can take both $a(\epsilon, y)$ and $b(\epsilon, y)$ as smooth functions on $[0, \infty[\times \Omega$. However, neither the source $e^{i\phi(y)/\epsilon} \, b(\epsilon, y)$ nor the response $e^{i\phi(y)/\epsilon} \, a(\epsilon, y)$ is smooth up to $\epsilon = 0$. This follows from

$$\frac{d^k}{d\epsilon^k} \, e^{i\phi(y)/\epsilon} \, b(\epsilon, y) = \frac{(-1)^{k-1}(k-1)!}{\epsilon^k} \, \phi(y) \, e^{i\phi(y)/\epsilon} \, b_0(y) + O(\epsilon^{1-k}) \,,$$

and a similar expression differing by a power of ϵ for the response. Derivatives of order ≥ 1 of the source and ≥ 2 of the response diverge to infinity as $\epsilon \to 0$.

Exercise 4.3.2. *Compute two terms of an asymptotic solution of*

$$\frac{\partial^2 u}{\partial x_1^2} + \frac{\partial^2 u}{\partial x_2^2} + \frac{\partial u}{\partial x_1} = e^{ix.\xi/\epsilon} \,.$$

It is interesting to ask whether the asymptotic solution $u(\epsilon, y) := a(\epsilon, y) e^{i\phi(y)/\epsilon}$ can be corrected by a term $c(\epsilon, y) \sim 0$ so that

$$L(u + c) \; = \; e^{i\phi/\epsilon} \, a \,.$$

Define the residual by

$$r(\epsilon, y) \; := \; L(y, \partial_y) \, u(\epsilon, y) - e^{i\phi/\epsilon} \, b \; \sim \; 0 \,.$$

One needs to solve

$$L(y, \partial_y) \, c(\epsilon, y) \; = \; -r(\epsilon, y) \,, \qquad \text{with} \qquad c \sim 0 \,.$$

Under a variety of conditions this is possible. For example, if L is symmetric hyperbolic, it suffices to supplement the equation for c with the initial condition $c|_{t=0} = 0$. A similar argument works for parabolic equations determining c in $t \geq 0$.

If L is elliptic, then inhomogeneous equations like that for c are solvable on sufficiently small neighborhoods of arbitrary points. For example one can solve the Dirichlet problem on small balls for the strongly elliptic equation $L(y, \partial)^\dagger L(y, \partial) c = -L(y, \partial)^\dagger r$.

If L has constant coefficients, one can choose a fundamental solution E and a plateau cutoff χ and set $c := E * (\chi r)$. This works on compact subsets of space-time.

Levy showed (in 1957) that linear partial differential equations with variable coefficients, even with polynomial coefficients, need not be locally solvable. In such cases the equation for c need not have solutions, and the construction of an asymptotic solution is the best that one can do.

4.4. Perturbation approach

The fundamental equations, (4.3.3), (4.3.4), have now been derived two different ways, one inductive and one by plugging in the right *ansatz*. Here is a third derivation with the feel of perturbation theory. It is useful to have a variety of approaches for at least three reasons. First one sees that they are all versions of the same thing. In much of the mathematical and scientific literature these different computations are confused as fundamentally different. Second, in extending these ideas sometimes one or the other point of view is more easily adaptable. Finally, different arguments appeal to different people.

Suppose that as $\epsilon \to 0$,

$$b(\epsilon, y) \;\sim\; b_0(y) + \epsilon b_1(y) + \epsilon^2 b_2(y) + \cdots . \tag{4.4.1}$$

Seek $u(\epsilon, y)$ solving

$$L(y, \partial_y)\, u \;\sim\; e^{i\phi(y)/\epsilon}\, b . \tag{4.4.2}$$

Motivated by the case of constant coefficient ordinary differential equations, try

$$u \;=\; e^{i\phi(y)/\epsilon}\, a(\epsilon, y) . \tag{4.4.3}$$

As in §4.2 start with

$$L(y, \partial_y)\,\big(e^{i\phi(y)/\epsilon}\, a\big) = e^{i\phi(y)/\epsilon}\,\Big(\frac{1}{\epsilon}\, L_1(y, id\phi(y)) + L(y, \partial_y)\Big)\, a . \tag{4.4.4}$$

Thus (4.4.2) holds if and only if

$$\Big(L_1(y, id\phi(y)) + \epsilon L(y, \partial_y)\Big)\, a \;\sim\; \epsilon b . \tag{4.4.5}$$

If there is a solution a whose derivatives are $O(1)$ as $\epsilon \to 0$, then there are two terms on the left, one of order 1 and the other of order ϵ. Neglecting the latter yields a first approximation that is identical to (4.2.3). There are at least two natural ways to proceed from here. One is to seek a as an asymptotic (a.k.a. Taylor series) in ϵ

$$a(\epsilon, y) \;\sim\; \epsilon\, a_1(y) + \epsilon^2\, a_2(y) + \cdots .$$

Plugging this into (4.4.5) is the method of §4.3.

An alternative is to write (4.4.5) as a fixed point equation

$$a(y) = L(y, id\phi(y))^{-1}\Big(b - \epsilon L(y, \partial_y)\, a\Big),$$

and to perform fixed point iteration. This generates a sequence of approximations

$$a^\nu(y) = L(y, id\phi(y))^{-1}\Big(b - \epsilon L(y, \partial_y)\, a^{\nu-1}\Big). \tag{4.4.6}$$

The first approximation is found by dropping the $\epsilon\, L$ term from the right-hand side to find

$$a^1 = \epsilon\, L(y, id\phi(y))^{-1}\, b. \tag{4.4.7}$$

Iteration then yields

$$a^{\nu+1} - a^\nu = -\epsilon\, L(y, id\phi(y))^{-1}\, L(y, \partial_y)\, \big(a^\nu - a^{\nu-1}\big). \tag{4.4.8}$$

This implies that

$$a^\nu = \epsilon\, a_1 + \epsilon^2\, a_2 + \cdots + \epsilon^\nu\, a_\nu + O(\epsilon^{\nu+1}), \tag{4.4.9}$$

with the a_j from (4.3.3), (4.3.4).

With three distinct approaches to solving (4.4.5) all leading to the same answer, you may have the misimpression that the solution of this equation is simple. The differential operator on the left of (4.4.5) is a first order operator with the property that the differentiation terms have a coefficients of size ϵ. The derivative terms that are normally the main terms have a small coefficient and so end up playing the role of corrections. One consequence is that the successive correction terms are generated by applying operators of high order in $\epsilon\partial_y$. This is all to say that the approximation just produced is subtle, and also that convergence of the series in ϵ is unlikely except when the operators and sources satisfy real analyticity hypotheses. Such hypotheses are unnatural in many scientific applications since they imply that knowledge of sources in one neighborhood determines them everywhere.

4.5. Elliptic regularity

A striking application of Theorem 4.3.3 is a proof of the interior elliptic regularity theorem. The proof is modified in §4.6 to give the microlocal version that is one of the two central results in linear microlocal analysis. The other is proved in Chapter 5.

The most familiar elliptic regularity theorems assert that harmonic functions and solutions of the Cauchy–Riemann equations are real analytic. More generally if Δu is real analytic, then so is u. Such results extend to elliptic equations with real analytic coefficients.

We will treat sources that are smooth or only finitely differentiable. Elliptic regularity asserts that if L is an mth order elliptic operator and

Lu has k derivatives in an appropriate sense, then u has $m+k$ derivatives. In dimension greater than one, it is false that if $Lu \in C^k$, then $u \in C^{k+m}$. That solutions of $\Delta u = \rho$ with ρ in the Hölder space $C^{k+\alpha}$, $\alpha \in]0,1[$, satisfy $u \in C^{2+k+\alpha}$ is a classical regularity theorem for Newtonian potentials. This Hölder version extends to general elliptic equations. The version we give is for Sobolev spaces.

Whenever there is elliptic regularity, there is a corresponding estimate. For example, if L is first order on $\overline{\Omega}$ and elliptic on $\omega \Subset \Omega$, Theorem 4.5.1 implies that if u and Lu belong to $H^s(\Omega)$, then $u \in H^{s+1}(\omega)$. Thus there is an inclusion

$$\Big\{ u \in H^s(\Omega) \ : \ Lu \in H^s(\Omega) \Big\} \quad \hookrightarrow \quad H^{s+1}(\omega) \,.$$

Both sides are Hilbert spaces. The norm for the left-hand side is equal to

$$\Big(\|u\|^2_{H^s(\Omega)} \ + \ \|L\,u\|^2_{H^s(\Omega)} \Big)^{1/2} \,.$$

Exercise 4.5.1. *Prove that the inclusion has closed graph so is continuous by the Closed Graph Theorem.*

The continuity proved in the exercise implies that there is a constant $C = C(L, \omega, \Omega, s)$ so that

$$\text{(4.5.1)} \qquad \|u\|_{H^{s+1}(\omega)} \ \le \ C \left(\|Lu\|_{H^s(\Omega)} + \|u\|_{H^{s-1}(\Omega)} \right).$$

Such closed graph arguments showing that qualitative results imply quantitative estimates were invented by Banach. They show that in practice to prove regularity you must prove the estimate. In some cases, like the proof below, this is done but is not emphasized.

If $L = L_1(\partial)$ is homogeneous with constant coefficients, (4.5.1) cannot hold for nonelliptic operators. In the nonelliptic case, there would be a point $\eta \in \operatorname{Char} L$ and associated plane wave solutions $Lv^\epsilon = 0$,

$$v^\epsilon \ := \ e^{iy\eta/\epsilon}\, a\,, \qquad a \in \ker L_1(\eta)\,.$$

The functions

$$u^\epsilon \ := \ \psi \, v^\epsilon$$

with $\psi \in C_0^\infty(\Omega)$ violate (4.5.1) in the limit $\epsilon \to 0$. This construction can be lifted to the variable coefficient case.

Exercise 4.5.2. *Show that in the variable coefficient case, if there is a point* $(\underline{y}, \underline{\eta}) \in \operatorname{Char} L$, *then (4.5.1) cannot be satisfied for any neighborhood* ω *of* $\underline{y}$. **Hint.** Use plane waves for the operator $L_1(\underline{y}, \partial)$ and a cutoff of the form $\psi((y-\underline{y})/\epsilon^\mu))$ for suitable $\mu > 0$ and $\psi \in C_0^\infty(\{|y| < 1\})$.

Definitions. *If $\Omega \subset \mathbb{R}^n$ is open, $\underline{y} \in \Omega$, and u is a distribution on Ω, we say that u* **is in** H^s **at** $\underline{y}$ *and write $u \in H^s(\underline{y})$, if and only if there is a $\psi \in C_0^\infty(\Omega)$ with $\psi(\underline{y}) \neq 0$ and $\psi u \in H^s(\Omega)$.*

Elliptic Regularity Theorem 4.5.1. *Suppose that $\underline{y} \in \Omega$, $L(y, \partial_y)$ is an elliptic operator of order 1 on Ω, u is a distribution on Ω, and, $Lu \in H^s(\underline{y})$. Then $u \in H^{s+1}(\underline{y})$.*

Proof. The distribution $f := Lu$ is H^s on a neighborhood of $\underline{y}$. Choose a smooth $\tilde{\psi}$, compactly supported in this neighborhood, and identically equal to 1 on a smaller neighborhood of $\underline{y}$ so that L is elliptic on a neighborhood of the support of $\tilde{\psi}$. Choose a second such function, ψ supported in the set where $\tilde{\psi} = 1$

Denote by ω the points of the unit sphere, $S^{n-1} \subset \mathbb{R}^n$.

The strategy is to prove that $\psi u \in H^{s+1}$ by studying the Fourier transform $\widehat{\psi u}(\eta)$ for $k = |\eta| \to \infty$. Let $k := 1/\epsilon$ and $\omega := \eta/|\eta|$. Compute

$$\widehat{\psi u}(\eta) = \widehat{\psi u}\,(k\omega) = \big\langle \psi u, e^{-ik\omega.y} \big\rangle = \big\langle u\,,\, e^{-i\omega.y/\epsilon} \psi(y) \big\rangle\,.$$

The differential equation $Lu = f$ asserts that for $v \in C_0^\infty$,

$$\langle u, L^\dagger v \rangle \;=\; \langle f, v \rangle\,. \tag{4.5.2}$$

If $v(\epsilon, \omega, y)$ is a good approximate solution of $L^\dagger(y, \partial_y)\, v = \psi\, e^{-iy.\omega/\epsilon}$, then

$$\widehat{\psi\, u}(\omega/\epsilon) \;=\; \big\langle u, \psi\, e^{-ix.\omega/\epsilon} \big\rangle \;\approx\; \big\langle u, L^\dagger v \big\rangle \;=\; \big\langle f, v(\epsilon, \omega, \cdot) \big\rangle\,.$$

Since L is elliptic, every ω is noncharacteristic. The same is therefore true for the transposed operator $L(y, \partial_y)^\dagger$ since the principal symbol is equal to the transpose of $-L_1(y, \eta)$. Thus we have constructed, for each ω, asymptotic solutions $v(\epsilon, \omega, y)$ to

$$L(y, \partial_y)^\dagger\, v(\epsilon, \omega, y) \;-\; \psi(y) e^{-iy.\omega/\epsilon} \;\sim\; 0\,. \tag{4.5.3}$$

The construction is uniform in the parameters in the sense that it yields $a_j(\omega, y) \in C^\infty(S^{n-1} \times \mathbb{R}^n)$ that vanish for y outside the support of ψ. Borel's theorem yields

(4.5.4)

$$C^\infty([0,1] \times S^{n-1} \times \mathbb{R}^n) \ni a(\epsilon, \omega, y) \;\sim\; \sum_{j=0}^{\infty} \epsilon^j\, a_j(\omega, y) \quad \text{in} \quad C^\infty(S^{n-1} \times \mathbb{R}^n)\,,$$

with a vanishing for y outside the support of ψ.

Setting

$$v(\epsilon, \omega, y) \;:=\; \epsilon\, a(\epsilon, \omega, y)\, e^{iy.\omega/\epsilon}\,, \tag{4.5.5}$$

(4.5.3) holds in $C^\infty(S^{n-1} \times \mathbb{R}^n)$. Then

$$\widehat{\psi u}(\eta) \;=\; \big\langle u\,,\, L^\dagger v \big\rangle + \big\langle u\,,\, \psi e^{ix.\omega/\epsilon} - L^\dagger v \big\rangle \tag{4.5.6}$$

and

$$\forall M, \quad \langle u, \psi e^{ix.\omega/\epsilon} - L^\dagger v\rangle = O(|\eta|^{-M})\,.$$

The proof is completed by showing that $\langle\eta\rangle^{s+1}\widehat{\psi u}(\eta) \in L^2$. It suffices to show that $\langle\eta\rangle^{s+1}$ times each of the two summands on the right of (4.5.6) belongs to $L^2(\mathbb{R}^n_\eta)$.

For the second summand it suffices to choose M with $2M - 2(s+1) > n$ so that

$$\int_{|\eta|>1} \frac{\langle\eta\rangle^{2(s+1)}}{|\eta|^{2M}}\, d\eta \;<\; \infty\,.$$

The approximate solution v vanishes for y outside the support of ψ. Therefore, $\widetilde{\psi}\, v = v$. This with (4.5.2) shows that $\langle u\,,\, L^\dagger v\rangle \;=\; \langle \tilde{\psi} f\,,\, v\rangle$. Formula (4.5.5) shows that the right-hand side is equal to

$$\int \tilde{\psi}\; f\; a\; e^{-iy.\omega/\epsilon}\; dy \;=\; \epsilon\, \mathcal{F}\Big(a(\epsilon,\omega,\cdot)\,\tilde{\psi}(\cdot)\, f(\cdot)\Big)(\eta)\,, \qquad \eta = \omega/\epsilon\,. \tag{4.5.7}$$

Expressing the Fourier transform of a product as a convolution yields

$$\langle u\,,\, L^\dagger v\rangle \;=\; \epsilon\; (2\pi)^{d/2} \int \mathcal{F}(\tilde{\psi} f)(\zeta)\; \hat{a}(\epsilon,\omega,\zeta-\eta)\; d\zeta\,. \tag{4.5.8}$$

The smoothness and compact support of a implies that

$$\forall N,\ \exists C_N,\ \forall\,(\epsilon,\omega) \in [0,1]\times S^{n-1}, \qquad \big|\hat{a}(\epsilon,\omega,\zeta)\big| \;\le\; C_N\, \langle\zeta\rangle^{-N}. \tag{4.5.9}$$

Since $\tilde{\psi} f \in H^s$,

$$\big|\mathcal{F}(\tilde{\psi} f)(\zeta)\big| \;=\; \langle\zeta\rangle^{-s}\, g\,, \qquad \text{with} \quad g(\zeta) \in L^2(\mathbb{R}^n)\,. \tag{4.5.10}$$

Estimates (4.5.8)–(4.5.10) imply that for $|\eta| > 1$

$$\Big|\langle u\,,\, L^\dagger v\rangle\Big| \;\le\; \int \frac{C\; g(\zeta)}{\langle\zeta\rangle^s\; |\eta|\; \langle\zeta-\eta\rangle^N}\; d\zeta\,,$$

where the key is the factor $\epsilon = |\eta|^{-1}$. Multiplying by $\langle\eta\rangle^{s+1}$ and using the fact that $|\eta| \ge 1$, yields the bound

$$\langle\eta\rangle^{s+1}\, \Big|\langle u\,,\, L^\dagger v\rangle\Big| \;\le\; C \int \frac{\langle\eta\rangle^s}{\langle\zeta\rangle^s\; \langle\zeta-\eta\rangle^N}\; g(\zeta)\; d\zeta\,. \tag{4.5.11}$$

Suppose that $s \ge 0$. Since $(\zeta - \eta) + \zeta \;=\; \eta$, either $|\zeta - \eta| \;>\; |\eta|/2$ or $|\zeta| > |\eta/2|$, so the integrand is bounded by

$$\frac{\langle\eta\rangle^s}{\langle\zeta\rangle^s\; \langle\zeta-\eta\rangle^s}\; \frac{g(\zeta)}{\langle\zeta-\eta\rangle^{N-s}} \;\le\; C\; \frac{g(\zeta)}{\langle\zeta-\eta\rangle^{N-s}}\,.$$

Choose $N > n + s$ so $\langle\zeta\rangle^{-N+s} \in L^1(\mathbb{R}^n_\zeta)$. Young's inequality implies that

$$\begin{aligned}\Big\| \langle\eta\rangle^{s+1} \, \langle u\,,\, L^\dagger v\rangle \Big\|_{L^2(\mathbb{R}^n_\eta)} &\leq C \, \big\| \langle\zeta\rangle^{N-s} * g \big\|_{L^2(\mathbb{R}^n_\zeta)} \\ &\leq C \, \|\langle\zeta\rangle^{-N+s}\|_{L^1(\mathbb{R}^n_\zeta)} \, \|g\|_{L^2(\mathbb{R}^n)} < \infty\,.\end{aligned}$$

This completes the proof when $s \geq 0$.

When $s < 0$ the integrand is bounded by

$$\frac{g(\zeta)\,\langle\zeta\rangle^{|s|}}{\langle\eta\rangle^{|s|}\,\langle\zeta-\eta\rangle^{N}} \leq \frac{\langle\zeta\rangle^{|s|}}{\langle\eta\rangle^{|s|}\,\langle\zeta-\eta\rangle^{|s|}} \, \frac{g(\zeta)}{\langle\zeta-\eta\rangle^{N-|s|}} \leq C \, \frac{g(\zeta)}{\langle\zeta-\eta\rangle^{N-|s|}}\,,$$

since either $|\eta| \geq |\zeta|/2$ or $|\zeta-\eta| \geq |\zeta|/2$. Young's inequality completes the proof. □

Remarks. 1. In the heart of the proof the gain of one derivative came from the factor ϵ in the approximate solution. Elliptic regularity is a reflection of the fact that a right-hand side oscillating with wavelength ϵ yields a response oscillating in the same way whose amplitude is smaller by a factor ϵ.

2. A standard proof of the regularity theorem is to construct a pseudodifferential operator $P(y,\partial)$ of order -1 so that $LP = I +$ smoothing. Then $P\big(b(y)e^{iy\omega/\epsilon}\big)$ is an infinitely accurate approximate solution of $Lu = b(y)\,e^{iy\omega/\epsilon}$. The computation of the full symbol of P is the same analytic problem as computing the full asymptotic expansion $a(\epsilon,y)\,e^{iy\omega/\epsilon}$. From this perspective the computation in §4.2 resembles the Levi construction of elliptic parametrices, while the computations in §4.3 and §4.4 resemble more closely the symbol calculus for pseudodifferential operators.

3. The heart of the proof is an explicit formula for $\widehat{\psi u}(\eta)$ with error bound $O(|\eta|^{-\infty})$. This is an impressive achievement for a variable coefficient problem.

Corollary 4.5.2. *If $L(y,\partial_y)$ is elliptic on the open set Ω and $u \in \mathcal{D}'(\Omega)$ satisfies $Lu \in C^\infty(\Omega)$, then $u \in C^\infty(\Omega)$.*

Proof. If $y \in \Omega$, then u is $H^s(y)$ for some possibly very negative s. The theorem implies that $u \in H^{s+1}(y)$. A second application implies that $u \in H^{s+2}(y)$. An induction shows that u belongs to $H^{s+m}(y)$ for all integers m. Sobolev's embedding theorem implies that for every k, $u \in C^k$ on a neighborhood of y. □

4.6. The Microlocal Elliptic Regularity Theorem

This is one of the two basic theorems in microlocal analysis. What needs to be added to the ideas of the last section are two definitions.

Definitions. *If u is a distribution defined on an open set Ω, $\underline{y} \in \Omega$, $s \in \mathbb{R}$, and $\underline{\eta} \in \mathbb{R}^n \setminus 0$, we say that u* **is in H^s microlocally at** $(\underline{y}, \underline{\eta})$ *and write $u \in H^s(\underline{y}, \underline{\eta})$ if and only if there is a $\psi \in C_0^\infty(\Omega)$ with $\psi(\underline{y}) \neq 0$ and a $\chi \in C^\infty(\mathbb{R}^n \setminus 0)$ homogeneous of degree 0 with $\chi(\underline{\eta}) \neq 0$ so that*

$$\int_\Gamma \chi(\eta)\, \left|\widehat{\psi u}(\eta)\right|^2\, \langle\eta\rangle^{2s}\, d\eta \;<\; \infty\,. \tag{4.6.1}$$

The set of (y, η) so that $u \notin H^s(y, \eta)$ is called the H^s **wave front set** *and is denoted $WF_s(u)$.*

Remarks. i. $u \in H^s(y, \eta)$ if and only if $u \in H^s(y, r\eta)$ for all $r > 0$. Therefore, $WF_s(u)$ is a conic subset of $\Omega \times (\mathbb{R}^d \setminus 0)$.

ii. By definition, the complement of $WF_s(u)$ is an open subset of $\Omega \times (\mathbb{R}^d \setminus 0)$, so $WF_s(u)$ is a closed subset of $\Omega \times \mathbb{R}^n \setminus 0$.

iii. If $u \in \mathcal{D}'(\Omega)$ and $y \in \Omega$, then $u \in H^s(y)$ if and only $u \in H^s(y, \eta)$ for all $\eta \neq 0$ if and only if $u \in H^s(y, \eta)$ for all $|\eta| = 1$. **Discussion.** The definition of microlocal regularity takes care of $\mathcal{F}(\psi u)(\eta)$ for large η. For η in bounded sets, use the fact that $\mathcal{F}(\psi u)$ is analytic so surely $L^2_{\rm loc}$.

Examples 4.6.1. i. If $u = \delta(x)$, then for $s \geq -d/2$, $WF_s(\delta) = \{(0, \eta) : \eta \neq 0\}$. For $s < -d/2$, $u \in H^s$ and WF_s is empty.

ii. Consider $u(y_1, y_2) = h(y_2)$ with h denoting Heaviside's function. The singular support of u is equal to $\{y_2 = 0\}$. Thus if $y_2 \neq 0$, $u \in H^s(y, \eta)$ for all s, η.

For points with $y_2 = 0$, take a cutoff function $\psi = \phi_1(y_1)\phi_2(y_2)$ so

$$\widehat{\psi u}(\eta_1, \eta_2) \;=\; \widehat{\phi}_1(\eta_1)\; \mathcal{F}(\phi_2(y_2)\, h(y_2))\,.$$

Since $\widehat{\phi}_1(\eta_1)$ is rapidly decreasing, this proves that for any η with $\eta_1 \neq 0$, $u \in H^\infty(y, \eta)$, that is in $H^s(y, \eta)$ for all s.

There remain the points with $y_2 = 0$ and $\eta_1 = 0$. There one has $|\mathcal{F}(\phi_2\, h)(\eta_2)| \sim c/|\eta_2|$. On a conic neighborhood one has

$$\widehat{\psi u}(\eta_1, \eta_2) \;\leq\; \frac{|\hat{\phi}_1(\eta_1)|}{\langle\eta_2\rangle}\,.$$

Using Fubini's theorem shows that u is microlocally $H^s(y, \eta)$ if and only if $s < 1/2$.

Exercise 4.6.1. *Prove that $WF_{s-1}(\partial_j u) \subset WF_s(u)$.*

Exercise 4.6.2. *Prove that if $u \in H^s(\underline{y}, \underline{\eta})$ and $\psi \in C^\infty(\Omega)$, then $\psi u \in H^s(\underline{y}, \underline{\eta})$.* **Hints.** Cut off in space by $\phi \in C_0^\infty(\Omega)$, and decompose the resulting function as $u_1 + u_2$ with $\widehat{u_1} = O(\langle\xi\rangle^M)$ and vanishes on a conic neighborhood of $\underline{\eta}$ while $u_2 \in H^s(\mathbb{R}^d)$. Write the transform of the product as a convolution. Consider only η in a small conic neighborhood of $\underline{\eta}$. For

$\widehat{\psi} * \widehat{u}_1$ show that the argument of $\widehat{\psi}$ is $\geq c|\underline{\eta}|$. **Discussion.** Exercises 4.6.1 and 4.6.2 imply that $WF_s(Lu) \subset WF_{s+1}(u)$.

Microlocal Elliptic Regularity Theorem 4.6.1. *Suppose that on a neighborhood of $\underline{y}$, $L(y, \partial_y)$ is a system of differential operators of order 1 with smooth coefficients and that L is noncharacteristic at $(\underline{y}, \underline{\eta})$. If u is a distribution with $Lu \in H^s(\underline{y}, \underline{\eta})$, then $u \in H^{s+1}(\underline{y}, \underline{\eta})$.*

Remark. Together with the preceding discussion this shows that

$$WF_s(Lu) \setminus \operatorname{Char} L \;=\; WF_{s+1}u \setminus \operatorname{Char} L\,.$$

Proof of Theorem 4.6.1. The proof follows the proof of Theorem 4.5.1. The change comes in the construction of the approximate solution (4.5.3). In the elliptic case this was done for all ω using the fact that $L_1(y, \omega)$ is invertible for all (y, ω).

In the present context we choose the cutoffs ψ and χ with sufficiently small support so that that (y, η) is noncharacteristic on $\operatorname{supp}\psi \times \operatorname{supp}\chi$.

Then $v(\epsilon, y, \eta)$ is defined for $(y, \eta) \in \operatorname{supp}\psi \times \operatorname{supp}\chi$. Estimates are uniform on the corresponding compact set of (y, ω).

The estimate for the integral (4.5.8) must be changed because it is no longer assumed that $\widetilde{\psi} f \in H^s$. What is true is that one can choose $\widetilde{\psi}$ so that $\mathcal{F}(\widetilde{\psi} f)$ is the sum of two terms. One is as in the right-hand side of (4.5.10) and is treated as before. The second term, $h(\zeta)$, vanishes on a conic neighborhood of $\underline{\eta}$ and for some σ, possibly very large,

$$\frac{h(\zeta)}{\langle\zeta\rangle^\sigma} \;\in\; L^2(\mathbb{R}^n_\zeta)\,.$$

For this h part of (4.5.8) we must show that

$$\int \frac{h(\zeta)\;\langle\eta\rangle^s}{\langle\zeta\rangle^s\;\langle\zeta-\eta\rangle^N}\;d\zeta$$

is square integrable on a small closed conic neighborhood of $\underline{\eta}$. Choose that neighborhood small enough that it will be disjoint from a closed cone containing the support of h. Then, in the support of the integrand, $|\zeta - \eta| \geq C\max\{|\zeta|, |\eta|\}$. For $s \geq 0$ and ζ in the support of the integrand,

$$\frac{h(\zeta)\;\langle\eta\rangle^s}{\langle\zeta\rangle^s\;\langle\zeta-\eta\rangle^N} \;\leq\; \frac{h(\zeta)}{\langle\zeta\rangle^\sigma}\;\frac{C}{\langle\zeta-\eta\rangle^{N-\sigma-s}}\,.$$

Choosing $N > \sigma + s + n$, Young's inequality finishes the proof.

For $s < 0$, the integrand is dominated by

$$\frac{h(\zeta)\;\langle\zeta\rangle^{|s|}}{\langle\eta\rangle^{|s|}\;\langle\zeta-\eta\rangle^N} \;\leq\; \frac{h(\zeta)}{\langle\zeta\rangle^\sigma}\;\frac{C}{\langle\zeta-\eta\rangle^{N-\sigma-|s|}}\,,$$

and choosing $N > \sigma + |s| + n$, Young's inequality completes the proof. □

A candidate for a microlocal smoothness set is the set of points (y, η) such that $u \in H^s(y, \eta)$ for all s. This is the complement of $\bigcup_s WF_s(u)$. There is a stronger notion which leads to a possibly smaller set.

Definition. *If u is a distribution on Ω, the* **wavefront set** *of u, denoted $WF(u)$, is a set of points $(y, \eta) \in \Omega \times \mathbb{R}^n \setminus 0$ so that $(y, \eta) \notin WF(u)$ if and only if there is a $\psi \in C_0^\infty(\Omega)$ with $\psi(y) \neq 0$ and a $\chi \in C^\infty(\mathbb{R}^n \setminus 0)$ positive homogeneous of degree 0 with $\chi(\eta) \neq 0$ so that for all $N \in \mathbb{Z}_+$,*

$$\sup_{\eta \in \mathbb{R}^n} |\eta|^N \mathcal{F}(\chi u)(\eta) < \infty .$$

Exercise 4.6.3. *Prove that if $\psi \in C^\infty(\Omega)$, then $WF(\psi u) \subset WF(u)$.* **Hint.** Read the hint in Exercise 4.6.2.

Exercise 4.6.4. *Prove that $WF(\partial^\alpha u) \subset WF(u)$.* **Discussion.** Combining this with Exercise 4.6.3 proves that $WF(p(x, \partial)u) \subset WF(u)$ for partial differential operators p with smooth coefficients. More general operators that do not increase wavefront sets are called *microlocal.* Classical pseudodifferential operators are microlocal.

Example 4.6.2. Define $u = \mathcal{F}^{-1}k$ where $k(\eta) \in C^\infty(\mathbb{R}^n)$ is positive homogeneous of degree σ on $|\eta| \geq 1$. That is, there is a positive homogeneous function K so that $k = K$ on $|\eta| \geq 1$. Combining the results of the next three exercises shows that $WFu = \{0\} \times \operatorname{supp} K$.

Exercise 4.6.5. *Prove that the singular support of u is equal to the origin by showing that for any k there is an N so that $|y|^{2N} u \in C^k$.*

Exercise 4.6.6. *Show that for any η disjoint from the support of K, $(0, \eta) \notin WF(u)$.* **Hint.** Estimate products as in the proof of Theorem 4.6.1.

Exercise 4.6.7. *If $\psi \in C_0^\infty(\mathbb{R}^2)$ with $\psi(0) = 1$, then*

$$\mathcal{F}(\psi u) - K = O(|\eta|^{\sigma - 1})$$

as $\eta \to \infty$. **Hint.** Write the transform of the product as a convolution. Then use the fact that $\widehat{\psi}$ is rapidly decreasing.

Microlocal Elliptic Regularity Theorem 4.6.2. *If L is as in Theorem 4.6.1, $(y, \eta) \notin \operatorname{Char} L$, and $(y, \eta) \notin WF(Lu)$, then $(y, \eta) \notin WF(u)$.*

Exercise 4.6.8. *Prove this by suitably modifying the proof of Theorem 4.6.1.*

Next analyze the wavefront set of a piecewise smooth function by using two methods.

Definition. *If M is an embedded hypersurface, $m \in M$, and u is a distribution defined on a neighborhood of M, then u is* **piecewise smooth**

at m if the restriction of u to each side of M has a smooth extension to a neighborhood of m.

Exercise 4.6.9. *If u is piecewise smooth at m prove using the definition of WF that on a neighborhood of m, $WF(u)$ is contained in the conormal bundle of M.* **Hint.** By subtracting a function smooth near m, reduce to the case of u vanishing on one side. Estimate $\mathcal{F}(\psi u)$ in directions not conormal by the method of nonstationary phase. Compute the boundary terms that appear in the integration by parts. Show that they are small by the method of nonstationary phase.

Remark. For the same u one can prove the slightly weaker result that for all s, $WF_s(u)$ is contained in the conormal variety using only the microlocal elliptic regularity theorem. If η is not conormal, at m one can find a smooth vector field $V(y, \partial)$ on a neighborhood of m so that V is tangent to M and $V(m, \eta) \neq 0$. Then for any k, $V^k u \in L^2_{\text{loc}}$. Microlocal elliptic regularity applied for the operator V^k implies that $u \in H^j(m, \eta)$ for all k.

Chapter 5

Linear Hyperbolic Geometric Optics

5.1. Introduction

The mathematical subject of geometric optics is devoted to the construction and analysis of asymptotic solutions of partial differential equations that are accurate when wavelengths are small compared to other natural lengths in the problem. Since the wavelength of visible light is of the order of magnitude 5×10^{-5}cm, a great deal of what one sees falls into this category. The description of optical phenomena was and is a great impetus to study short wavelength problems.

The key feature of geometric optics, propagation along rays, is not present in the elliptic case of the last section. Rays occur for hyperbolic problems, and in the same spirit, for singular elliptic problems that arise when looking for high frequency time periodic solutions of a hyperbolic equation. An example of the latter is that solutions of the form

$$u(t,x) = e^{i\tau t}\, w(x)\,, \qquad \tau \gg 1, \tag{5.1.1}$$

to D'Alembert's equation

$$\Box u = 0 \tag{5.1.2}$$

must satisfy the reduced wave equation

$$\tau^2\, w + \Delta\, w = 0\,, \tag{5.1.3}$$

which is elliptic. The singularity in this elliptic equation is that one is interested in solutions in the limit $\tau \to \infty$. A coefficient is tending to infinity.

The key dichotomy is that in the (nonsingular) elliptic case, rapid oscillations have only local effects. For the asymptotic solutions, the values of the amplitudes a_j in the neighborhood of a point are determined by the values of the b_j in the same neighborhood. For hyperbolic equations (and their related singular elliptic problems, such as (5.1.3)) the oscillations may and usually do propagate to distant parts of space-time. This is why they are the main carriers of information in both the communications industry and in the universe.

The starting point for the construction of the asymptotic solutions of hyperbolic geometric optics is the observation that if $L = L_1(\partial_y)$ has constant coefficients and no lower order terms, then there are plane wave solutions

$$u(y) \ := \ a(y\eta)\,, \qquad y\eta \ := \ \sum_{\mu=0}^{d} y_\mu \eta_\mu\,.$$

In §2.4 we showed that u satisfies $L\,u = 0$ if and only if $\eta \in \operatorname{Char} L$ and

$$\forall \sigma\,, \quad a'(\sigma) \in \ker L_1(\eta)\,. \tag{5.1.4}$$

For $\eta \in \operatorname{Char} L$, the plane wave solutions (modulo constants) are parametrized by functions $a : \mathbb{R} \to \ker L(\eta)$. With the notation

$$\eta \ = \ (\eta_0, \eta_1, \ldots, \eta_d) \ := \ (\tau, \xi_1, \ldots, \xi_d)\,, \tag{5.1.5}$$

plane wave solutions have the form

$$u = a(\tau t + x\xi). \tag{5.1.6}$$

They translate at velocity $\mathbf{v}$ in the sense that $u(t,x) = u(0, x - \mathbf{v}t)$ if and only if $\mathbf{v}$ satisfies $\mathbf{v}\xi = -\tau$ (see §2.4).

The heart of the present chapter is the construction of a generalization of the plane wave solutions. The corresponding solutions can be localized in space and the method applies to systems with variable coefficients and lower order terms.

Among the plane wave solutions described above are those with short wavelength sinusoidal oscillatory behavior,

$$u \ = \ e^{iy\eta/\epsilon}\, a(y\eta), \tag{5.1.7}$$

where a is smooth and ϵ is small compared to the typical distances on which a varies. Equation (5.1.7) represents wave packets with wavelength $\epsilon \ll 1$. They have derivatives of order $1/\epsilon \gg 1$. A zero order term applied to (5.1.7) is bounded so much smaller. When the coefficients do not vary much on the scale ϵ of a wavelength, this suggests that at least locally, replacing an operator by its constant coefficient part without lower order terms is a reasonable approximation for high frequency solutions. This turns out to be not quite true since in the solutions the large contributions of the highest order terms nearly cancel and the lower order terms can be important as in the example of §1.5. In addition, the derivatives of the coefficients affect the local propagation. Only a detailed computation reveals the exact laws.

The constant coefficient problem of §1.4 is a first such calculation based on an exact solution formula using the Fourier transform. In the next section we treat the case of scalar second order equations with constant coefficient principal part. These equations are interesting in their own right. The scalar case is algebraically a little more straightforward than the systems treated afterward. They are a natural starting point. For many applications, systems are essential. An impatient or experienced reader can skip directly to §5.3.

5.2. Second order scalar constant coefficient principal part

Section 5.2.1 presents the case of scalar hyperbolic equations with constant coefficient principal part. It serves to introduce the hyperbolic theory with a simple case. Section 5.2.2 is a detour through the semiclassical limit of quantum mechanics. It can be postponed for later reading. It is included because it is so similar to the first and is of extraordinary scientific importance.

5.2.1. Hyperbolic problems. Begin with the example of scalar operators with constant principal part,

$$L := \sum_{\mu,\nu=0}^{d} a_{\mu\nu}\partial_\mu\partial_\nu + \sum_{\mu=0}^{d} b_\mu(y)\partial_\mu + c(y), \qquad a_{\mu\nu} = a_{\nu\mu}. \tag{5.2.1}$$

The operator need not be hyperbolic. Suppose that the coefficients $a_{\mu\nu}$ of second order derivatives are real and do not depend on y. Otherwise, linear phases would be unrealistic. For variable coefficient operators, surfaces of constant phase are unlikely to be planar.

In §1.4 we found short wavelength solutions of the form

$$e^{iy\eta/\epsilon}\, a(\epsilon, y), \qquad a(\epsilon, y) \sim \sum_j a_j(y)\, \epsilon^j .$$

Similar expansions were successful in Chapter 4. Seek solutions of that form. To compute $L(y,\partial)\big(e^{iy\eta/\epsilon}\, a\big)$, use

$$\frac{\partial}{\partial y_j}\Big(e^{iy\eta/\epsilon}\, a\Big) = e^{iy\eta/\epsilon}\Big(\frac{i\eta_j}{\epsilon} + \frac{\partial}{\partial y_j}\Big) a$$

to find

$$L(y,\partial)\,\big(e^{iy\eta/\epsilon}\, a\big) = e^{iy\eta/\epsilon}\, L\Big(y\,,\ \frac{i\eta_j}{\epsilon} + \frac{\partial}{\partial y_j}\Big) a\,.$$

Suppress the y dependence of $L(y,\partial)$ for ease of reading. And denote by

$$L_2(\eta,\zeta) \;=\; \sum_{\mu\nu} a_{\mu\nu}\,\eta_\mu\,\zeta_\nu$$

the symmetric bilinear form associated to the principal symbol. Ordering by power of ϵ yields

$$L\left(e^{iy\eta/\epsilon}\,a\right) \;=\; e^{iy\eta/\epsilon}\Big(\epsilon^{-2}L_2(i\eta,i\eta) + \epsilon^{-1}\big(2\,L_2(i\eta,\partial_y) + L_1(i\eta)\big) + L(\partial)\Big)a\,.$$

The most singular term is

$$\epsilon^{-2}L_2(i\eta,i\eta)a \;=\; \epsilon^{-2}\sum a_{\mu\nu}\,(i\eta_\mu)(i\eta_\nu)\,a\,.$$

In order to find solutions, one must have

$$\sum a_{\mu\nu}\,\eta_\mu\eta_\nu = 0\,. \tag{5.2.2}$$

This asserts that $\eta\in\operatorname{Char}L$. Recall that for an mth order differential operator, the characteristic variety is defined by the equation $\det L_m(y,\eta) = 0$. The equation (5.2.2) has a solution $0\neq\eta\in\mathbb{R}^{1+d}$ if and only if L is not elliptic.

Given (5.2.2) one has

$$L(\partial)\left(e^{iy\eta/\epsilon}\,a(\epsilon,y)\right) = \epsilon^{-1}e^{iy\eta/\epsilon}\left(i\left(2\sum_{\mu,\nu} a_{\mu,\nu}\,\eta_\nu\,\partial_\mu + \sum_\mu b_\mu\eta_\mu\right) + \epsilon\,L(\partial)\right)a\,.$$

Seek $a(\epsilon,y)$ satisfying

$$\left(i\left(2\sum_{\mu,\nu} a_{\mu,\nu}\,\eta_\nu\,\partial_\mu + \sum_\mu b_\mu\eta_\mu\right) + \epsilon\,L(\partial)\right)a \;=\; 0 \tag{5.2.3}$$

at least at the level of formal power series in ϵ.

Injecting the expansion of $a(\epsilon,y)$ and setting the coefficient of ϵ^{j-1} equal to zero yields the recurrence,

$$i\Big(2\sum_{\mu,\nu} a_{\mu\nu}\eta_\nu\partial_\mu + b\eta\Big)a_j + L(\partial)\,a_{j-1} = 0\,, \qquad a_{-1} \;:=\; 0\,. \tag{5.2.4}$$

Define the vector field V and scalar c by

$$V \;:=\; \sum a_{\mu\nu}\,\eta_\nu\,\partial_\mu, \qquad \gamma \;:=\; \frac{1}{2}\,b\eta \;=\; \frac{1}{2}\sum_\mu b_\mu\eta_\mu\,. \tag{5.2.5}$$

The case $j=0$ of (5.2.4) shows that the leading amplitude $a_0(y)$ must satisfy the transport equation

$$\big(V+\gamma\big)\,a_0 = 0\,. \tag{5.2.6}$$

The integral curves of V are straight lines called *rays*. Equation (5.2.6) shows that the restriction of a_0 to a ray satisfies a first order linear ordinary differential equation. If $\gamma = 0$, then a_0 is constant on rays. The eikonal equation (5.2.2) shows that the phase $\phi(y) = y\eta$ satisfies $V\phi = 0$ so ϕ is constant on rays.

Example of phases and rays. For the operator $\partial_t^2 - c^2\Delta$ and $\eta = (\tau, \xi)$ with $\tau = \pm|c\xi|$, the phase is

$$\phi(y) = \phi(t,x) \;=\; y\eta \;=\; (t,x)\cdot(\tau,\xi) \;=\; \tau t + x\xi\,.$$

The vector field V is given by

$$V \;=\; \tau\,\partial_t - c^2\,\xi\nabla_x \;=\; \pm|c\xi|\,\Big(\partial_t \;\mp\; c\,\frac{\xi}{|\xi|}\,\nabla_x\Big).$$

The integral curves of V move with speed c in the direction $\mp\xi/|\xi|$. This is equal to the group velocity associated to (τ,ξ) in §2.4. In the present computation there is no hint of the stationary phase argument used to introduce the group velocity in §1.3.

The eikonal equation implies that this velocity satisfies $\mathbf{v}\xi = -\tau$, the condition defining the phase velocities. Usually (as in Exercise 5.2.2 below), $\mathbf{v}$ is not parallel to ξ.

Exercise 5.2.1. *Suppose that the principal symbol defines a nondegenerate symmetric bilinear form $b(\eta,\zeta) = \sum a_{\mu\nu}\eta_\mu\zeta_\nu$. Define an isomorphism $\eta \mapsto y$ from $\mathbb{R}^{d+1}_\eta$ to $\mathbb{R}^{d+1}_y$ by*

$$\forall\zeta, \qquad b(\eta,\zeta) \;=\; y\zeta\,.$$

Prove that the vector V is the image by this isomorphism of the covector $\eta \in \operatorname{Char} L$.

Exercise 5.2.2. *Suppose that $L_2 = \partial_t^2 - c^2\partial_1^2 - \partial_2^2$. Compute the characteristic variety and the velocity of transport for every $\eta \in \operatorname{Char} L$.* **Discussion.** For $c \neq 1$ the velocity of transport is not parallel to ξ except when $\xi_1\xi_2 = 0$.

Initial conditions for a_0 can be prescribed, for example, on a hyperplane P transverse to V. The other a_j can be similarly determined from their values on P. For $j > 0$ the transport equation for a_j has a source term depending on a_{j-1}.

Proposition 5.2.1. *Suppose that η is characteristic, P is a hyperplane transverse to V in (5.2.5), and $f_j \in C^\infty(P)$ for $j \geq 0$. Then there are uniquely determined a_j with $a_j\big|_P = f_j$ so that (5.2.4) is satisfied. In this case $L\big(e^{iy\eta/\epsilon}\,a\big) \sim 0$. If the f_j are supported in a fixed subset $E \subset P$, then the a_j are supported in the tube of rays with feet in E.*

In sharp contrast to the elliptic case, there are oscillatory asymptotic solutions without oscillatory terms on the right-hand side of the equation.

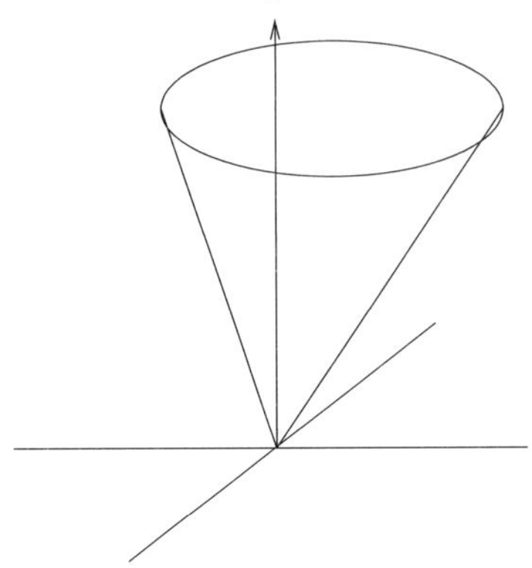

FIGURE 5.2.1. The time-like cone

Definitions. *The operator (5.2.1) is* **strictly hyperbolic with time variable** *t if and only if $L_2(1, 0, \ldots, 0) \neq 0$ and for every $\xi \in \mathbb{R}^d \setminus 0$ the equation $L_2(\tau, \xi) = 0$ has two distinct real roots. The* **time-like cone** *is the connected component in $\{\eta : L_2(\eta) \neq 0\}$ containing $(1, 0, \ldots, 0)$. A hyperplane is* **space like** *if it has a conormal belonging to the time-like cone.*

Exercise 5.2.3. *For what real values γ is the operator $\partial_t^2 + \gamma\partial_t\partial_x - \partial_x^2$ strictly hyperbolic with time t?*

Exercise 5.2.4. *Show that $L = \partial_t^2 - \sum_{ij} a_{ij}\partial_i\partial_j$ is strictly hyperbolic with time t if and only if $\sum a_{ij}\xi_i\xi_j$ is a strictly positive definite symmetric bilinear form.*

Exercise 5.2.5. *Show that if L of form (5.2.1) with real symmetric $a_{\mu\nu}$ is strictly hyperbolic with time t, then the same is true for real symmetric $\widetilde{a}_{\mu\nu}$ sufficiently close to $a_{\mu\nu}$.* **Discussion.** The set of such operators is open.

Exercise 5.2.6. *Carry out the construction of asymptotic solutions of the Klein–Gordon operator $L = \partial_t^2 - \sum_{ij} a_{ij}\partial_i\partial_j + m^2$, with a_{ij} real and strictly positive. In particular show that the equations for the leading amplitude do not involve m.* **Discussion.** The propagation of short wavelength oscillations is to leading order unaffected by the m^2u term. This is consistent with the idea that for highly oscillatory solutions, lower order terms are less important. Note however that the first order terms $b\partial_y$ would affect the leading amplitude a_0. These results are closely related to the fact (see §1.2) that the jump discontinuities in the fundamental solution of $u_{tt} - u_{xx} + m^2u = 0$ are not affected by m.

Exercise 5.2.7. *Prove that if L is strictly hyperbolic with time t, H is a space-like hyperplane and η is characteristic, then the transport vector field is transverse to H.*

The structure of the leading approximation $e^{iy\eta/\epsilon}a_0(y)$ is visualized as follows. The surfaces of constant phase are the hyperplanes $y\eta = \text{const}$. The eikonal equation (5.2.2) implies that the transport vector field V is tangent to these surfaces. In the absence of lower order terms, the leading amplitude a_0 is constant on the rays that are integral curves of the constant vector field V. If $a_0\big|_{t=0}$ is supported in a set $E \subset \mathbb{R}^d_x$, then a_0 is supported in the tube of rays

$$\mathcal{T} \;=\; \big\{(0,x) + tV \;:\; x \in E,\; t \in \mathbb{R}\big\}\,.$$

The velocity associated to V is called the *group velocity*. From the definition of V one has that

$$\text{group velocity} \;:=\; -\frac{\sum_{j=1}^{d}\sum_{\mu} a_{j\mu}\,\eta_\mu\,\partial_j}{\sum_\mu a_{0\mu}\eta_\mu}.$$

Exercise 5.2.8. *Show that in the strictly hyperbolic case, the characteristic variety is parametrized as a graph $\tau = \tau(\xi)$ and the group velocity is equal to $-\nabla_\xi \tau$.* **Discussion.** This is the same formula for group velocity found in §1.3 and §2.4.

Examples of ray-like solutions. For $\partial_t^2 - c^2\Delta_x$, and $\eta = (\tau, \xi)$ with $\tau = \pm|c\xi|$, if $a_0(0,x) = f(x)$, then $a_0(t,x) = f(x \mp ct\xi/|\xi|)$. A particularly interesting case is when f has support in a small ball centered at a point $\underline{x}$. In this case the principal term in the approximate solution is supported in a cylinder in space-time about the ray starting at $\underline{x}$. If one takes initial data $a_j(0,x) = 0$ for $j \geq 1$, then Theorem 5.2.1 yields approximate solutions u^ϵ that are supported in such a narrow cylinder and whose residual is infinitely small in ϵ.

This recovers, in a different way, the approximate solutions derived in §1.4 by Fourier transform. The latter method is restricted to constant coefficient problems.

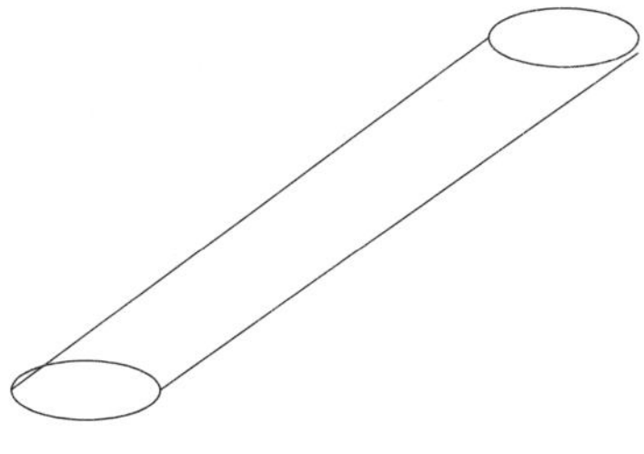

FIGURE 5.2.2. The cylindrical support of the amplitudes of a ray-like solution

The reader is encouraged to think of these as flashlight beams or ray-like approximate solutions to the wave equation. It is worth noting that the only exact solution of the multidimensional wave equation that is supported in such a cylinder is the solution $u = 0$.

Exercise 5.2.9. *Prove that if $d > 1$ and $u \in C^\infty(\mathbb{R}^{1+d})$ satisfies $\Box u = 0$ and is supported in a tube*

$$\Big\{(t,x)\ :\ \big|(x_1 - ct, x_2, \ldots, x_d)\big| \le R\Big\},$$

then $u = 0$. **Hint.** This follows from the fact that the energy in the tube tends to zero. In fact, $\nabla_{t,x} u = O(t^{(1-d)/2})$. This sharp decay rate can be proved using the fundamental solution. **Alternative Hint 1.** Derive the decay as in the proof of Theorem 3.4.5. One only needs $o(1)$, not the sharp rate. The weaker decay can be obtained as in Corollary 3.3.1. **Alternative Hint 2.** The boundary of the tube is characteristic where $x_2 = x_3 = \cdots = x_d = 0$. It is noncharacteristic at the other points. Apply Fritz John's Global Hölmgren Uniqueness Theorem to prove that u vanishes in the cylinder.

Exercise 5.2.9 completes an analysis begun in Warning 1 following (1.4.12). We next perform the estimate used in Warning 2 at that same place, assuming that γ is constant. It shows that in this hyperbolic context, the series $\sum \epsilon^j a_j$ is usually divergent even for real analytic f_j, and always divergent for compactly supported f_j. Equation (5.2.4) implies that

$$(V+\gamma)a_j \;=\; \frac{i\,L(\partial)}{2}\,a_{j-1}\,.$$

An exponential integrating factor eliminates the γ term,

$$V(e^{\gamma t}\,a_j) \;=\; e^{\gamma t}(V+\gamma)a_j \;=\; e^{\gamma t}\,\frac{i\,L(\partial)}{2}\,a_{j-1} \;=\; \frac{i\,L(\partial_t-\gamma\,,\,\partial_x)}{2}\,\big(e^{\gamma t}\,a_{j-1}\big)\,.$$

Iterate j times and insert the explicit solution a_0 to find

$$V^j(e^{\gamma t}\,a_j) \;=\; \Big(\frac{i\,L(\partial_t-\gamma\,,\,\partial_x)}{2}\Big)^j\big(e^{\gamma t}\,a_0\big) \;=\; \Big(\frac{i\,L(\partial_t-\gamma,\partial_x)}{2}\Big)^j f_0(x-\mathbf{v}t),$$

where $\mathbf{v}$ denotes the group velocity. Using the chain rule for the time derivatives yields

$$V^j(e^{\gamma t}\,a_j) \;=\; \Big(\frac{i\,L(-\mathbf{v}\partial_x-\gamma\,,\,\partial_x)}{2}\Big)^j(f_0)\,(x-\mathbf{v}t)\,. \tag{5.2.7}$$

For simplicity suppose that $f_j = 0$ for $j \ge 1$. The profile $e^{\gamma t}a_j = b_1 + b_2$ is the sum of two terms. The first term is the solution with homogeneous initial data

$$V^j(b_1) \;=\; \Big(\frac{i\,L(-\mathbf{v}\partial_x-\gamma\,,\,\partial_x)}{2}\Big)^j(f_0)\,(x-\mathbf{v}t)\,,$$

$$V^m b_1\big|_{t=0} \;=\; 0 \quad \text{for} \quad 0 \le m \le j-1\,.$$

Since L is second order one sees that for fixed $T > 0$, b_1 is bounded by

$$\|b_1\|_{L^\infty([0,T]\times\mathbb{R}^d)} \;\le\; \frac{(Ct)^j}{j!} \sum_{|\beta|\le 2j} \|\partial_x^\beta f_0\|_{L^\infty}\,,$$

with constant C independent of j. For a real analytic function f_0 whose derivatives of order α are $\sim C^{|\alpha|}\,\alpha!$, one has $b_1 \sim C_1^{2j}\,(2j)!/j!$.

The term b_2 satisfies $V^j b_2 = 0$ and at $t = 0$ has $V^m b_2 = V^m(e^{\gamma t} a_j)$ for $0 \le m \le j-1$. It is given by

$$b_2 \;=\; \sum_{0\le m\le j-1} \frac{t^m}{m!}\, V^m(e^{\gamma t} a_j)(0, x - \mathbf{v}t)\,.$$

Estimate the worst V^{j-1} term. Use

$$\frac{V^{j-1}(e^{\gamma t}\, a_j)}{(j-1)!} \;=\; \frac{1}{(j-1)!}\left(\frac{i\, L(\partial_t - \gamma, \partial_x)}{2}\right)^{j-1} \left(e^{\gamma t}\, a_1\right) \;\sim\; \frac{D^{2j-2} f_0}{(j-1)!}\,,$$

which is comparable to the growth estimate for b_1. This yields a growth rate $a_j \sim D^{\le 2j} f_0/j!$, the qualitative behavior used in Warning 2.

5.2.2. The quasiclassical limit of quantum mechanics. This section is not used in the sequel, so it can be skipped or postponed. It uses Hamilton–Jacobi theory discussed in the next sections and Appendix 5.I. The computation belongs here since it is of second order with constant coefficient leading part. The quasiclassical limit, also known as the *correspondence principal*, was important in the development of quantum mechanics. It remains fundamental in understanding that theory.

The dynamical equation of a nonrelativistic quantum mechanical particle corresponding to a classical particle of mass m moving in a force field with real valued potential energy $V(x)$ is Schrödinger's equation

(5.2.8)

$$0 \;=\; i\,\hbar\, u_t \;-\; \frac{\hbar^2}{2m}\,\Delta u + V(x) u \;:=\; L(x, \hbar\partial) u\,, \quad L(x,\partial) := i\partial_t - \frac{\Delta}{2m} + V(x)\,.$$

Planck's constant $\hbar$, equal to 6.6×10^{-34} in MKS units, is very small. It is an invitation to perform an asymptotic analysis as $\hbar \to 0$.

The symmetry of the operator $-(\hbar^2/2m)\Delta + V$ implies that solutions suitably small as $x \to \infty$ satisfy the conservation law

$$\partial_t \int_{\mathbb{R}^d} |u(t,x)|^2\, dx \;=\; 0\,.$$

The solutions of interest satisfy $\int_{\mathbb{R}^d} |u(t,x)|^2\, dx = 1$, in which case the probability of finding the particle in Ω at time t is $\int_\Omega |u(t,x)|^2\, dx$.

When $V = 0$, the equation is invariant under the parabolic scaling $(t,x) \mapsto (\lambda^2 t, \lambda x)$. The scaling $(t', x') = (\hbar t, \hbar x)$ transforms the equation to the case $\hbar = 1$. The $V = 0$ equation has the exponential plane waves $e^{i(\tau t+\xi x)}$ with dispersion relation

$$\hbar\tau \;=\; \frac{-|\hbar\xi|^2}{2m}$$

and group velocity

$$\mathbf{v} = -\nabla_\xi \tau = \frac{\hbar\xi}{m}.$$

The last relation suggests that $\hbar\xi$ is related to momentum, and in fact $\int_\omega |\hat{u}(t,\xi)|^2\, d\xi$ is the probability that the particle has momentum in $\omega/\hbar$ at time t. The energy of a free particle of momentum p is $|p|^2/2m$, so the dispersion relation suggests correctly that $\hbar\tau$ is related to energy.

For nonzero V if u varies on scale $\hbar$ in both t and x, then the three terms of (5.2.8) are of comparable size. It is this class of short wavelength solutions that we construct. When $V = 0$, exponential solutions have form $e^{i(\sigma t+\zeta x)/h}$ with $\sigma = -|\zeta|^2/2m$. Seek asymptotic solutions of similar form,

$$u^\hbar \;\sim\; e^{iS(t,x)/\hbar}\Big[a_0(t,x) \;+\hbar\, a_1(t,x) \;+\; \hbar^2 a_2(t,x) \;+\; \cdots\;\Big]. \tag{5.2.9}$$

In quantum mechanics the phase is commonly denoted S and is called *action*. Compute

$$(\hbar\partial)\big(e^{iS/\hbar}a\big) = e^{iS/\hbar}\big(\hbar\partial a + i(\partial S)a\big),$$
$$\text{so}\quad L(x,\hbar\partial)\big(e^{iS/\hbar}a\big) = e^{iS/\hbar}\, L\big(x,\hbar\partial + i\partial S\big)a.$$

Here

$$L\big(x,\hbar\partial + i\partial S\big) \;=\; i(iS_t + \hbar\partial_t) \;-\; \frac{1}{2m}(i\nabla_x S + \hbar\partial)(i\nabla_x S + \hbar\partial) \;+\; V(x).$$

Expanding in powers of $\hbar$ yields

$$\begin{aligned} L\big(x,\hbar\partial + i\partial S\big) = \;& \Big(-S_t + \frac{|\nabla_x S|^2}{2m} + V(x)\Big) \\ & +\; h\,i\Big(\partial_t - \frac{1}{m}\nabla_x S\partial_x - \frac{\Delta S}{2m}\Big) \;-\; h^2\Big(\frac{\Delta}{2m}\Big). \end{aligned}$$

Injecting (5.2.9) yields

$$\begin{aligned} \Big[\Big(-S_t + \frac{|\nabla_x S|^2}{2m} + V(x)\Big) + h\,i\Big(\partial_t - \frac{1}{m}\nabla_x S\partial_x - \frac{\Delta S}{2m}\Big) - h^2\Big(\frac{\Delta}{2m}\Big)\Big] \\ \times \Big[a_0(t,x) + \hbar a_1(t,x) + \cdots\;\Big] \sim 0. \end{aligned}$$

In order to have solutions with $a_0 \neq 0$, the action S must satisfy the eikonal equation

$$S_t \;=\; H(x,\nabla_x S), \qquad H(x,p) \;:=\; \frac{|p|^2}{2m} \;+\; V(x). \tag{5.2.10}$$

The function $H(x,p)$ is the *hamiltonian* of the classical mechanical system. The function S is uniquely determined locally from arbitrary smooth initial data $S(0,x)$ chosen so that $\nabla_x S \neq 0$. The eikonal equation eliminates one third of the terms on the left-hand side. The term of order h yields a transport equation determining a_0 from arbitrary initial data

$$\left(\partial_t - \frac{1}{m}\nabla_x S\,\partial_x - \frac{\Delta S}{2m}\right) a_0 \;=\; 0\,. \tag{5.2.11}$$

For $j \geq 1$ the coefficients a_j are determined inductively from arbitrary initial data by the transport equations

$$i\left(\partial_t - \frac{1}{m}\nabla_x S\partial_x - \frac{\Delta S}{2m}\right) a_j \;=\; \frac{\Delta a_{j-1}}{2m}\,. \tag{5.2.12}$$

Borel's theorem yields $a(h,t,x) \sim \sum_j \hbar^j a_j$. Then $u^h := e^{iS/h}a$ satisfies $Lu^h \sim 0$.

Hamilton–Jacobi theory (see Propagation Lemma 5.I.2 or Tangency Lemma 5.I.5) identifies the integral curve of $\partial_t - m^{-1}\nabla_x S\,\partial_x$ starting at $(0,\underline{x})$ as the projection on (t,x) space of the orbit $(x(t),p(t))$ of classical mechanics with initial conditions $x(0) = \underline{x}$ and $p(0) = \nabla_x S(0,\underline{x})$.

Example 5.2.1. Suppose that S is fixed and $a_j(0,\,\cdot\,)$ has support in a small neighborhood of $\underline{x}$. Then the family u^h can be chosen supported in the narrow tube of classical orbits with initial momenta given by $-\nabla_x S(0,\cdot)$. The quantum mechanical approximate solutions shadows that classical orbit. The hyperbolic solutions that follow rays have quantum mechanical analogues that follow classical mechanical paths. The classical orbits with hamiltonian H play the role that bicharacteristics play in the hyperbolic context. The accuracy of the approximation is briefly discussed in the last remark of the next section.

5.3. Symmetric hyperbolic systems

Convention. *From here on, the underlying operator $L(y,\partial_y)$ is assumed to be a symmetric hyperbolic first order system and $y = (t,x)$.*

With the experience from the wave equation derived from the Fourier integral representation in §1.4, the elliptic case from Chapter 4, and the second order case with constant coefficient principal symbol in §5.2, it is natural to seek solutions

$$L(y,\partial_y)\left(a(\epsilon,y)\,e^{i\phi(y)/\epsilon}\right) \;\sim\; 0 \tag{5.3.1}$$

with vector valued

$$a(\epsilon,y) \;\sim\; \sum_{j=0}^{\infty} \epsilon^j\, a_j(y) \qquad \text{as } \epsilon \to 0\,. \tag{5.3.2}$$

Here ϕ is assumed to be a smooth real valued function with $d\phi \neq 0$ on the domain of interest. This guarantees that the solutions are rapidly oscillating as $\epsilon \to 0$. The surfaces of constant phase in space-time have conormal vectors proportional to $d\phi$.

Computing as in the derivation of (4.3.2) yields

$$\begin{aligned} L(y, \partial_y) \left(e^{i\phi(y)/\epsilon} a \right) &= e^{i\phi(y)/\epsilon} L\Big(y\,,\,\partial_y + \frac{id\phi(y)}{\epsilon}\Big) a \\ &= e^{i\phi(y)/\epsilon} \left(\frac{1}{\epsilon} L_1(y, id\phi(y))\, a \;+\; L(y, \partial_y) a \right) \\ &\sim e^{i\phi(y)/\epsilon} \bigg(\frac{1}{\epsilon} L_1(y, id\phi(y))\, a_0 \\ &\qquad + \sum_{j=0}^{\infty} \epsilon^j \left[L_1(y, id\phi(y))\, a_{j+1}(y) + L(y, \partial_y)\, a_j(y) \right] \bigg). \end{aligned}$$

The convention $a_{-1} := 0$ yields

$$L(y, \partial_y) \left(e^{i\phi(y)/\epsilon} a \right) = e^{i\phi(y)/\epsilon} \sum_{j=-1}^{\infty} \epsilon^j \left[L_1(y, id\phi(y))\, a_{j+1}(y) + L(y, \partial_y)\, a_j(y) \right]. \tag{5.3.3}$$

Equation (5.3.1) holds if and only if

$$L_1\big(y, id\phi(y)\big)\, a_j(y) + L(y, \partial_y)\, a_{j-1}(y) \;=\; 0\,, \qquad \text{for} \quad j = 0, 1, 2, \dots . \tag{5.3.4}$$

The special case $j = 0$ is

$$L_1(y, id\phi(y))\, a_0 \;=\; 0\,. \tag{5.3.5}$$

Proposition 5.3.1. *If*

$$a(\epsilon, y) \;\sim\; \sum_{j=0}^{\infty} \epsilon^j\, a_j(y) \quad \textit{in} \quad C^\infty(\Omega),$$

then

$$L(y, \partial_y) \left(a(\epsilon, y)\, e^{i\phi(y)/\epsilon} \right) \;\sim\; 0 \quad \textit{in} \quad C^\infty(\Omega)$$

if and only if the coefficients $a_j(y)$ and phase ϕ satisfy (5.3.4).

In contrast to the scalar case of the previous section, it is not at all obvious how or whether the equations (5.3.4) determine the amplitudes a_j and phase ϕ. The rest of this section is devoted to that question.

In order for (5.3.5) to admit nonzero solutions, the matrix valued function $L_1(y, d\phi(y))$ must be singular,

$$\det L_1(y, d\phi(y)) \;=\; 0. \tag{5.3.6}$$

This is a nonlinear first order partial differential equation for the real valued phase ϕ. It is called the *eikonal equation.* It asserts that the graph of $d\phi$ belongs to the characteristic variety of L. Such equations are the subject of Hamilton–Jacobi theory that is recalled in Appendix 5.I.

Example 5.3.1. If L is the Maxwell system, the equation (5.3.6) becomes

$$(\partial_t \phi)^2 \left((\partial_t \phi)^2 - c^2 |\nabla_x \phi|^2 \right)^2 \;=\; 0\,.$$

The phases that give solutions satisfying the divergence constraints are those that satisfy

$$(\partial_t \phi)^2 \;=\; c^2 |\nabla_x \phi|^2\,. \tag{5.3.7}$$

For the macroscopic Maxwell's equations in matter with $\epsilon(y)$ and $\mu(y)$, one obtains the same result with the speed of light $c = c(y)$ depending on y.

Examples 5.3.2. *Examples of solutions of (5.3.7) with* $c =$ const.

1. We have already encountered the linear phases $\phi = t\tau + x\xi$, with $\tau^2 = c^2|\xi|^2$, that occur in plane waves.

2. Spherically symmetric solutions satisfy $\phi_t^2 = c^2\phi_r^2$, so long as $d\phi \neq 0$, $\phi_t \neq 0$, and they satisfy one of the two equations $\phi_t = \pm c\phi_r$. The general solution is $\phi = f(ct \pm |x|)$, $f \in C^\infty$ with $f' \neq 0$. For the plus sign, the surfaces of constant phase are incoming spheres. The sphere that starts at radius R degenerates at time $t = R/c$, showing that solutions of the eikonal equation will normally exist only locally in time.

3. An interesting class of solutions to (5.3.7) are those of the form

$$\phi(t,x) := ct \pm \psi(x)\,, \qquad \text{with} \qquad |\nabla \psi| = 1\,.$$

Discussion. If ψ is a smooth solution of $|\nabla\psi| = 1$ on a neighborhood of $\underline{x}$, then on a neighborhood of $\underline{x}$, $\psi - \psi(\underline{x})$ is equal to the signed distance from x to the level set $M := \{\psi(x) = \psi(\underline{x})\}$.

Proof. Replace ψ by $\psi - \psi(\underline{x})$ to reduce to the case $\psi(\underline{x}) = 0$. Denote by ϕ the signed distance which is positive on the same side of M as ψ. Then, $|\nabla\phi| = |\nabla\psi| = 1$ and $\phi|_M = \psi|_M$. It follows from the sign condition that $d\phi = d\psi$ on M. The uniqueness for solutions of the Cauchy problem from Hamilton–Jacobi theory applied to the equation $|\nabla\psi| = 1$ completes the proof. □

If the level set $\{\psi = \psi(\underline{x})\}$ in the last example is curved, then the distance function will not be smooth, developing singularities at centers of curvature. Theorem 6.6.3.ii proves that for $d \geq 2$ the only C^2 solutions of $|\nabla\psi| = 1$ everywhere defined on $\mathbb{R}^d$ are affine functions.

There are essentially two strategies for finding solutions of the eikonal equation. The first is to look for exact solutions. The simplest are linear functions $\phi(t,x) = t\tau + x\xi$ when L_1 has constant coefficients and (τ, ξ) is characteristic. The second strategy is to appeal to Hamilton–Jacobi theory to obtain local solutions. To apply the results of Appendix 5.I, let $F(y,\eta) := \det L_1(y,\eta)$ so the eikonal equation reads $F(y, d\phi(y)) = 0$. Seek to determine ϕ from initial values $\phi(0,x) = g(x)$. To determine $\phi_t(0,x)$ from its values at a single point $(0,\underline{x})$, the implicit function theorem requires

$$\frac{\partial F(0,\underline{y},\tau,dg(x))}{\partial\tau}\Big|_{\tau=\phi_t(0,\underline{x})} \neq 0.$$

This is equivalent to τ being a simple root of the polynomial equation $F(0,\underline{x},\tau,dg(\underline{x})) = 0$ and is also equivalent to $\dim\ker L_1(y,d\phi(y)) = 1$. When these conditions are satisfied, Hamilton–Jacobi theory yields a local solution ϕ.

For the example of Maxwell's equations (see §2.5), the roots are double. However the double roots result from the fact that the factors τ and $\tau^2 - c^2|\xi|^2$ appear squared. To apply Hamilton–Jacobi theory, one solves (5.3.7) and not $(\phi_t^2 - c^2|\nabla_x\phi|^2)^2 = 0$.

This phenomenon of multiple roots which appear because of repeated factors is so common that we give a general treatment. Suppose that

$$F(0,\underline{x},\phi_t(0,\underline{x}),dg(\underline{x})) = 0$$

and that on a neighborhood of

$$(\underline{x},\underline{\eta}) := (0,\underline{x},\phi_t(0,\underline{x}),dg(\underline{x})),$$

one has with integer $p > 1$,

$$F(y,\eta) \;=\; G(y,\eta)^p\, K(y,\eta)$$

with

$$\frac{\partial G(0,\underline{x},\tau,dg(\underline{x}))}{\partial\tau}\Big|_{\tau=\phi_t(0,\underline{x})} \neq 0, \qquad \text{and} \qquad K(\underline{y},\underline{x}) \neq 0.$$

In this case, one applies Hamilton–Jacobi theory to the reduced equation

$$G(y,d\phi(y)) \;=\; 0,$$

that satisfies the simple root condition. The root has multiplicity $p > 1$ for F. This shows that $-\tau$ is an eigenvalue of multiplicity p of the hermitian matrix $A_0^{-1/2}\big(\sum A_j\xi_j\big)A_0^{-1/2}$. Therefore, $\dim\ker L_1(y,d\phi(y)) = p$.

All the above strategies lead to phases ϕ satisfying the following hypothesis.

Constant rank hypothesis. *On an open connected subset* $\Omega \subset \mathbb{R}^{1+d}$, $\phi \in C^\infty(\Omega)$, $\partial_t \phi \neq 0$, *and* $\ker L_1(y, d\phi(y))$ *has strictly positive dimension independent of* y.

Example 5.3.3. If $L = L(\partial_y)$ has constant coefficients and $\phi = y\eta$ is linear with $\eta \in \operatorname{Char} L$, then $d\phi$ is constant, so $\ker L_1(d\phi)$ is independent of y and the constant rank hypothesis is automatic. This is so even when η is a singular point of the characteristic variety.

Recall from (2.4.4) that $\pi(y, d\phi(y))$ denotes the orthogonal projection of $\mathbb{C}^N$ onto the kernel of $L_1(y, d\phi(y))$. Introduce the partial inverse, $Q(y)$ of the singular symmetric matrix valued function $L_1(y, d\phi(y))$ by

$$(5.3.8)\quad Q(y)\,\pi(y, d\phi(y)) \;=\; 0\,, \qquad Q(y)\,L_1(y, d\phi(y)) \;=\; I - \pi(y, d\phi(y))\,.$$

When the constant rank hypothesis is satisfied, Theorem 3.I.1 implies that these are smooth matrix valued functions.

Remark. The example of linear phase and constant coefficient L shows that while it is true that the points $\nabla_y \phi \in \operatorname{Char} L$ are smooth in y, the characteristic variety near the points $(y, d\phi(y))$ can be singular. Conical refraction in crystals is an important example (see [Ludwig, 1961], [Joly, Métiver, and Rauch, Ann. Inst. Fourier, 1994] or [Métivier, 2009]).

Equation (5.3.5) holds if and only if

$$a_0(y) \;\in\; \ker L_1(y, d\phi(y))\,.$$

Equation (5.3.6) implies that $L_1(y, d\phi(y))$ is not surjective. Thus the case $j = 1$ of (5.3.4) has information about a_0, namely

$$(5.3.9)\qquad L(y, \partial)\, a_0 \;\in\; \operatorname{range} L_1(y, d\phi(y))\,.$$

The last two displayed equations are sufficient to determine a_0 from its initial data, though this is by no means obvious. In the scalar second order analysis of §5.2.1, at the analogous stage one immediately found a transport equation determining a_0 from its initial data.

For ease of reading, we suppose that ϕ is fixed and write $\pi(y)$ for $\pi(y, d\phi(y))$. Equation (5.3.5) is equivalent to

$$(5.3.10)\qquad \pi(y)\, a_0(y) = a_0(y)\,.$$

The condition (5.3.9) is satisfied if and only if,

$$(5.3.11)\qquad \pi(y)\, L(y, \partial_y)\, a_0 = 0\,.$$

Equations (5.3.10) and (5.3.11) are our second formulation of the equations that determine a_0.

First check that the number of equations is equal to the number of unknowns. The polarization equation (5.3.10) shows that a takes values in a linear space of dimension equal to $k := \dim\big(\ker L_1(y, \phi(y)\big)$. Thus there k unknown functions. Equation (5.3.11) asserts that $L(y, \partial_y)\, a_0$ takes values in range $L_1(y, d\phi(y))$, which has codimension k. Thus (5.3.11) is k partial differential equations for the k unknowns a_0.

There are other equivalent forms of (5.3.9). In the science literature, the usual procedure is to take linear combinations of the equations given by the rows of the system $L(y, \partial)a_0 + iL_1(y, d\phi(y))a_1 = 0$. Combinations that eliminate the $L_1 a_1$ terms are chosen. The constant rank hypothesis implies that the annihilator of $\mathrm{rg}\, L_1(y, d\phi(y))$ is a smoothly varying subspace of dimension k. If $\ell_1(y), \dots, \ell_k(y)$ form a basis, then (5.3.9) holds if and only if

$$\big\langle \ell_j(y)\,,\, L(y,\partial)a_0 \big\rangle \;=\; 0\,, \qquad 1 \le j \le k\,.$$

Equation (5.3.11) is of this form with the ℓ_j chosen to be k linearly independent rows of the matrix of the projector $\pi(y)$. More generally, if $K(y)$ is an $N \times N$ matrix satisfying for all y,

$$\mathrm{rank}\, K(y) \;=\; k, \qquad K(y)\big(\mathrm{rg}\, L(y, d\phi(y))\big) \;=\; 0,$$

then the equation (3.4.9) is equivalent to $K(y)\, L(y, \partial)\, a_0 = 0$. See [Métivier and Rauch, 2011] for more on this circle of ideas, especially if you like vector bundles.

A consequence of the nonlinearity of the eikonal equation is that phases ϕ are usually only defined locally. As a result, the amplitudes a_j are also only constructed locally, at best where the ϕ are defined. For the next result the local existence theorem for symmetric hyperbolic systems is used. In most concrete situations, the equations for the a_j are simple and sharper results are available. This theme is investigated in §5.4.

Theorem 5.3.2. *Suppose that $\Omega \subset \{0 < t < T < \infty\}$ is a domain of determinacy for L as defined in the assumption at the start of §2.6 and that the phase $\phi \in C^\infty(\overline{\Omega})$ satisfies the constant rank hypothesis and the eikonal equation on $\overline{\Omega}$. Given $g_j(x) \in C^\infty(\overline{\Omega}_0)$ satisfying $\pi(0,x)g_j = g_j$, there is one and only one sequence $a_j \in C^\infty(\overline{\Omega})$ satisfying (5.3.4) and the initial conditions*

$$\pi(0,x)\, a_j(0,x) = g_j(x) \qquad j = 0, 1, \dots\,. \tag{5.3.12}$$

As initial data, what is needed is the projections $\pi(y)\, a_j(0,x)$. For $j = 0$ this is equal to a_0. For $j \ge 1$, it is only part of the values of $a_j|_{t=0}$. The amplitudes $a_1, a_2, \dots$ are usually not polarized.

Proof of Theorem 5.3.2. For $j \ge 0$ denote by (5.3.4j) the case j of (5.3.4), with the convention $a_{-1} = 0$. Each equation (5.3.4j) is equivalent to the pair of equations arising from projecting on the kernel and range of $L_1(y, d\phi)$,

$$(5.3.4j) \qquad \Longleftrightarrow \qquad \Big\{\pi(y)(5.3.4j) \quad \text{and} \quad \big(I - \pi(y)\big)(5.3.4j)\Big\}.$$

Since Q is an isomorphism on $\text{Range}\,(I - \pi(y))$,

$$\big(I - \pi(y)\big)(5.3.4j) \qquad \Longleftrightarrow \qquad Q(y)(5.3.4j)\,.$$

We prove that for each $J \geq 0$, there are uniquely determined a_j for $j \leq J$ satisfying, for $j = 0, 1, \dots, J$,

$$\pi(y)\,(5.3.4j)\,, \quad \pi(y)a_j|_{t=0} = g_j\,, \quad \text{and,} \quad (I - \pi(y))(5.3.4(j+1)).$$

Since $a_{-1} = 0$, the case $J = 0$ reduces to showing that a_0 is uniquely determined by the three equations

$$\pi(y)\,a_0 \;=\; a_0, \qquad \pi(y)\,L(y, \partial_y)\,a_0 \;=\; 0, \qquad \text{and,} \qquad a_0\big|_{t=0} \;=\; g_0\,.$$

This is a consequence of the following lemma whose proof is postponed.

Lemma 5.3.3. *For any $f \in C^\infty(\overline{\Omega})$ and $g \in C^\infty(\overline{\Omega}_0)$ satisfying*

$$\pi(y)\,g = g\,, \qquad \pi(y)\,f \;=\; f\,,$$

there is a unique $w \in C^\infty(\overline{\Omega})$ satisfying

$$\pi(y)\,L(y, \partial_y)\,w \;=\; f\,, \qquad \pi(y)\,w \;=\; w\,, \qquad w\big|_{t=0} \;=\; g\,.$$

Assuming the lemma, the the proof of the theorem is by induction on J. The case $J = 0$ has just been verified. Suppose that $j \geq 0$ and that a_k are known for $k \leq j-1$ satisfying the profile equations up to $(I - \pi(y))(5.3.4j)$.

Multiplying (5.3.4j) by Q shows that

$$(5.3.13) \quad \big(I - \pi(y)\big)(5.3.4j) \qquad \Longleftrightarrow \qquad \big(I - \pi\big)a_j \;=\; -\,Q\,L(y, \partial_y)\,a_{j-1}\,.$$

Thus, $(I - \pi)a_j$ is determined from a_{j-1}.

Express

$$a_j \;=\; \pi\,a_j + (I - \pi)\,a_j \;=\; \pi\,a_j \;-\; Q\,L(y, \partial_y)\,a_{j-1}\,.$$

Inject this into $\pi(y)(5.3.4(j+1))$ to find

(5.3.14)
$$\pi(y)\,L(y, \partial_y)\,\pi(y)\,a_j \;=\; f_j\,, \qquad f_j \;:=\; \pi(y)\,L(y, \partial_y)\,Q\,L(y, \partial_y)\,a_{j-1}\,.$$

Reversing the steps shows that (5.3.13) and (5.3.14) imply the equations $(I - \pi(y))(5.3.4j)$ and $\pi(y)(5.3.4(j+1))$.

Thus, to prove the inductive step, it suffices to show that the equation (5.3.14) is uniquely solvable for $\pi(y)a_j$ for arbitrary initial data $\pi(y)a_j(0, y) = g_j$. That is guaranteed by the lemma. □

Proof of Lemma 5.3.3. For a solution, $\pi\, w = w$. Inject in the differential equation to find $\pi\, L\, \pi\, w = 0$. Since $(I-\pi)w = 0$, one has $(I-\pi)\, L\,(I-\pi)w = 0$. Adding yields

$$\pi\, L \pi\, w \;+\; (I-\pi)\, L\,(I-\pi)\, w \;=\; f\,. \tag{5.3.15}$$

The result is proved by showing that the differential operator $\widetilde{L} := \pi\, L\, \pi + (I-\pi)\, L\,(I-\pi)$ is symmetric hyperbolic and for all y the time-like cones satisfy $\overline{\mathcal{T}}(y, L) \supset \overline{\mathcal{T}}(y, \widetilde{L})$. The inclusion of time-like cones implies that the propagation cones of $\widetilde{L}$ are contained in the propagation cones of L. Therefore, since $\overline{\Omega}$ is a domain of determinacy for L, it follows that it is a domain of determinacy for $\widetilde{L}$.

The coefficient matrices of $\widetilde{L}$ are

$$\widetilde{A}_\mu \;:=\; \pi\, A_\mu \pi + (I-\pi)\, A_\mu\,(I-\pi).$$

They are symmetric since A_μ and π are. The coefficient of ∂_t is

$$\widetilde{A}_0 \;:=\; \pi\, A_0 \pi + (I-\pi)\, A_0\,(I-\pi)\,.$$

Since A_0 is positive definite, one estimates with $c > 0$ that

$$\begin{aligned}
\langle \widetilde{A}_0 v\,,\, v\rangle &= \langle \pi\, A_0\, \pi\, v\,,\, v\rangle + \langle (I-\pi)\, A_0\,(I-\pi)\, v\,,\, v\rangle \\
&= \langle A_0\, \pi\, v\,,\, \pi\, v\rangle + \langle A_0\,(I-\pi) v\,,\, (I-\pi) v\rangle \\
&\ge c\left(\|\pi\, v\|^2 + \|(I-\pi)v\|^2\right) = c\|v\|^2.
\end{aligned}$$

Thus, $\widetilde{A}_0$ is positive definite, so $\widetilde{L}$ is symmetric hyperbolic on $\overline{\Omega}$.

The principal symbols satisfy

$$\widetilde{L}_1(y,\eta) \;=\; \pi L_1(y,\eta)\pi \;+\; (I-\pi) L_1(y,\eta)(I-\pi)\,.$$

The same argument showing that $\widetilde{A}_0$ is strictly positive shows that if $L_1(y,\eta) \ge cI > 0$, then $\widetilde{L}_1(y,\eta) \ge cI$. This implies the comparison of time-like cones,

Theorem 2.6.1 then implies that for a given initial data $w(0,x)$, equation (5.3.16) has a unique solution w on $\overline{\Omega}$. This proves uniqueness of w.

To prove existence, we show that the function w so constructed solves the equations of the lemma. Multiplying (5.3.16) by $\pi(y)$ shows that the solution w satisfies

$$\pi(y)\, L(y, \partial_y)\, \pi(y)\, w \;=\; f\,. \tag{5.3.16}$$

Thus, if w satisfies (5.3.16), then so does $v := \pi w$. Since $v = w$ at $t = 0$, it follows that v satisfies the same symmetric hyperbolic initial value problem as w so

$$\pi(y)\, w = w\,. \tag{5.3.17}$$

This together with (5.3.16) yields the equations of the lemma completing the existence proof. □

Remark. An alternate approach to the lemma is to consider g as a $\ker L_1(y, d\phi(y))$ valued function. The operator $g \mapsto \pi(y)\, L(y, \partial_y)\, g$ maps such functions to themselves. Choosing a smooth orthonormal basis $\mathbf{e}_j(y)$, $1 \le j \le k$ for $\ker L_1(y, d\phi(y))$ and expanding $w = \sum w_j(y)\mathbf{e}_j(y)$ yields a symmetric hyperbolic system for the w_j. Such bases exist locally. To avoid the choice of bases, one can consider symmetric hyperbolic systems on hermitian vector bundles. The proof above avoids appealing to that abstract framework.

The next result proves that the asymptotic solutions differ by an infinitely small quantity from exact solutions.

Theorem 5.3.4 (Lax, Duke Math. J., 1957). *Suppose that Ω, ϕ, and a_j are as in Theorem 5.3.2 and that*

$$a(\epsilon, y) \sim \sum_{j=0}^{\infty} \epsilon^j\, a_j(y) \qquad \text{in} \quad C^\infty(\overline{\Omega})\,.$$

Suppose that $u(\epsilon, y) \in C^\infty(\overline{\Omega})$ is the exact solution of the initial value problem

$$\text{(5.3.18)} \qquad L(y, \partial_y)\, u(\epsilon, y) \;=\; 0\,, \qquad u(\epsilon, 0, x) \;=\; a(\epsilon, 0, x)\, e^{i\phi(0,x)/\epsilon}\,.$$

Then

$$\text{(5.3.19)} \qquad u(\epsilon, y) \;-\; e^{i\phi(y)/\epsilon}\, a(\epsilon, y) \;\sim\; 0, \qquad \text{in} \quad C^\infty(\overline{\Omega})\,.$$

Remarks. i. The utility of these results is twofold. First from the solution of one initial value problem without small parameter, one has an approximation of the *family* of solutions with short wavelength. The elements of the family with $\epsilon \ll 1$ are inaccessible by direct computer simulation. Second, in most cases the equations determining a_0 are significantly simpler than the original equations. This is not clear in the analysis above, and it is the content of §5.4.

ii. As in Theorem 4.3.3, neither the family of exact solution u^ϵ nor the family of approximation $e^{i\phi(y)/\epsilon}\, a(\epsilon, y)$ is smooth at $\epsilon = 0$.

iii. Once one knows that there is such an infinitely accurate approximation, one usually studies only the leading term, which for ϵ small is dominant. Though one does not compute the correctors in practice, their existence is crucial for the error analysis.

iv. $L(y, \partial_y)(a_0\, e^{i\phi/\epsilon}) = O(1)$. The leading term alone does not lead to a small residual. This explains why in the science literature the slowly varying envelope approximation is never checked by injecting it into the original equations.

Proof. For any $m, s \in \mathbb{N}$ there is a constant C so that

$$\|L(y, \partial_y)\big(u(\epsilon, y) - e^{i\phi(y)/\epsilon}\, a(\epsilon, y)\big)\|_{H^s(\overline{\Omega})} \;\leq\; C\,\epsilon^m$$

and

$$\|\, u(\epsilon, 0, x) - e^{i\phi(0,x)/\epsilon}\, a(0, x)\,\|_{H^s(\overline{\Omega}_0)} \;\leq\; C\,\epsilon^m\,.$$

The basic linear H^s energy estimate from §2 implies that

$$\|u(\epsilon, y) - e^{i\phi(y)/\epsilon}\, a(\epsilon, y)\|_{H^s(\overline{\Omega})} \;\leq\; C'(T, m, s)\,\epsilon^m\,.$$

Since this is true for all m, s, the result follows from the Sobolev embedding theorem. □

For the study of nonlinear equations, it is important to understand the effect of oscillatory source terms. The case of nowhere characteristic phase is treated in Chapter 4. The case of an everywhere characteristic phase is analyzed exactly as above. The result is the following.

Theorem 5.3.5 (Lax, Duke Math. J., 1957). *Suppose that the domain of determinacy Ω and the real phase ϕ satisfying the eikonal equation are as above. Given smooth functions $b_j \in C^\infty(\overline{\Omega})$, there are uniquely determined amplitudes $\underline{a}_j \in C^\infty(\overline{\Omega})$ satisfying $\pi(y)\underline{a}_j(0, x) = 0$ and $\underline{a}_0(0, x) = 0$ and so that if*

$$\underline{a}(\epsilon, y) \;\sim\; \sum_{j=0}^{\infty} \epsilon^j\, \underline{a}_j(y) \quad \text{and} \quad b(\epsilon, y) \;\sim\; \sum_{j=0}^{\infty} \epsilon^j\, b_j(y),$$

then

$$L(y, \partial_y)\big(a(\epsilon, y)\, e^{i\phi(y)/\epsilon}\big) \;-\; b(\epsilon, y)\, e^{i\phi(y)/\epsilon} \;\sim\; 0 \quad in \quad C^\infty(\Omega_T).$$

The principal amplitude is determined by the pair of equations

$$\pi(y)\, \underline{a}_0 = \underline{a}_0\,, \qquad \pi(y)\, L(y, \partial_y)\, \pi(y)\, \underline{a}_0 = b_0\,, \tag{5.3.20}$$

with the initial condition $\underline{a}_0(0, x) = 0$.

Remark. This result shows that a source of size one with characteristic phase yields waves of size one. This contrasts with the case of noncharacteristic phases in Chapter 4 where the response is order ϵ.

Exercise 5.3.1. *Prove Theorem 5.3.5.* **Discussion.** The proof of the error estimate is the same as Theorem 5.3.4.

Remarks on quasiclassical quantum mechanics. An analog of Theorem 5.3.4 is that the approximate solutions of Schrödinger's equation from §5.2.2 have error $O(\hbar^\infty)$. One needs a good smooth existence theorem. That is not hard if, for example, $\partial^\alpha V \in L^\infty$ for all α. The key a priori estimate is

$$\sup_{0 \leq t \leq T} \|u\|_{L^2(\mathbb{R}^d)} \;\leq\; \|u(0)\|_{L^2(\mathbb{R}^d)} \;+\; \frac{1}{\hbar}\int_0^T \|(L\,u)(t)\|_{L^2(\mathbb{R}^d)}\, dt\,.$$

This comes from the differential form of the L^2 conservation law. The factor $1/\hbar$ comes from the $\hbar\partial_t$ in the equation. The conservation law expresses $\partial_t(\hbar|u|^2)$ in terms of Lu. Since there is no finite speed, the approximate solution must be defined on all of $\mathbb{R}^d$. This is achieved by constructing u^h supported in the domain of definition of the phase S and extending to vanish outside that domain. The h^{-1} factor is not a problem as one applies the estimate where $L\,u^h = O(h^\infty)$.

5.4. Rays and transport

5.4.1. The smooth variety hypothesis. One of the key ideas in geometric optics is transport along rays. The equation

$$\pi\,L\,\pi\,a_0 \;=\; 0 \tag{5.4.1}$$

determines $a_0 = \pi(y)a_0$ from its polarized initial data. The key and not obvious fact is that under the smooth variety hypothesis that is satisfied in the vast majority of applications, *the system of differential operators* $\pi(y)\,L(y,\partial_y)\,\pi(y)$ *has first order part that is a directional derivative.*

Using the product rule for the derivatives $\partial_\mu(\pi(y)a_j)$ yields

$$\pi(y)\,L(y,\partial_y)\,\pi(y) \;=\; \sum_\mu \pi A_\mu \pi\,\partial_\mu \;+\; \sum_\mu \pi(A_\mu(\partial_\mu\pi) + B)\pi\,. \tag{5.4.2}$$

Each matrix $\pi(y)A_\mu(y)\pi(y)$ defines a linear transformation from $\ker L_1(y,d\phi(y))$ to itself. Where the variety is smooth, it is true but not obvious that each of these transformations is a scalar multiple of $\pi(y)A_0(y)\pi(y)$, so the differential operator is essentially a directional derivative.

Example 5.4.1. When $\ker L_1(y,d\phi(y))$ is one dimensional, it is easy to see that one gets a directional derivative. In that case, $\pi(y)$ is a projector of rank 1, and the polarization $\pi\,a_0 = a_0$ determines a_0 up to a scalar multiple. Since $\ker L_1(y,d\phi(y))$ is one dimensional, there are uniquely determined scalars $v_\mu(y)$ such that

$$\pi(y)A_\mu(y)\pi(y) \;=\; v_\mu(y)\,\pi(y)\,.$$

Similarly there is a unique scalar value $\gamma(y)$ such that

$$\pi(y)\sum_\mu \big(A_\mu(\partial_\mu\pi) + B\big)\,\pi(y) \;=\; \gamma(y)\,\pi(y)\,.$$

Define a vector field $V(y,\partial_y)$ by

$$V(y,\partial_y) \;:=\; \sum_{\mu=0}^{3} v_\mu(y)\,\frac{\partial}{\partial y_\mu}\,.$$

Then $\pi\, L \pi a_0 = 0$ is equivalent to the transport equation

$$V(y, \partial_y)\, a_0 + \gamma(y)\, a_0 \;=\; 0$$

along the integral curves of V.

Example 5.4.2. Consider

$$L \;=\; \partial_t \;+\; \begin{pmatrix} c_1(y) & 0 \\ 0 & c_2(y) \end{pmatrix} \frac{\partial}{\partial x} \;+\; \begin{pmatrix} B_{11} & B_{12} \\ B_{21} & B_{22} \end{pmatrix}, \qquad c_1 \;>\; c_2\,,$$

where $x \in \mathbb{R}^1$. The eikonal equation is

$$\big(\phi_t + c_1(y)\phi_x)\big)\,\big(\phi_t + c_2(y)\phi_x)\big) \;=\; 0\,.$$

Phases satisfy one of the two linear equations $(\partial_t + c_j \partial_x)\phi = 0$. Consider $j = 1$. The projector π and principal profile are

$$\pi \;=\; \begin{pmatrix} 1 & 0 \\ 0 & 0 \end{pmatrix}, \qquad a_0 \;=\; (a(y)\,,\,0)\,.$$

The transport equation is

$$\Big(\partial_t + c_1(y)\partial_x + B_{11}(y)\Big)a \;=\; 0\,.$$

A particularly interesting case is when $L = \partial_t + G$ with G antiselfadjoint, precisely

$$L \;=\; \partial_t + \begin{pmatrix} c_1(y) & 0 \\ 0 & c_2(y) \end{pmatrix} \frac{\partial}{\partial x} + \begin{pmatrix} \partial_x c_1(y) & 0 \\ 0 & \partial_x c_2(y) \end{pmatrix} + B\,, \qquad B = -B^*\,.$$

In this case solutions of $Lu = 0$ suitably small as $x \to \infty$ satisfy

$$\partial_t \int |u(t,x)|^2\, dx \;=\; 0\,.$$

The transport equation is

$$\Big(\partial_t + c_1(y)\partial_x + \partial_x c_1 + B_{11}\Big)a \;=\; 0\,, \qquad B_{11}(y) \in i\mathbb{R}\,.$$

Where $\partial_x c > 0$ (resp., < 0), the amplitude a^2 decreases (resp., increases) along rays. Where $c_x > 0$ (resp., < 0) neighboring rays spread apart (resp., approach). The energy between neighboring rays is spread over a larger region in the first case, so the amplitude decreases to compensate. In the second case the energy is compressed and the amplitude increases. These results are illustrated in Figure 5.4.1. More general results are presented in the next sections.

FIGURE 5.4.1. Compression on the left and expansion on the right

The operator $\pi(y)\, L(\partial)\, \pi(y)$ is a transport operator under the next hypothesis.

Smooth characteristic variety hypothesis. *The smooth characteristic variety hypothesis is satisfied at* $(\underline{y}, \underline{\eta}) = (\underline{\tau}, \underline{\xi}) \in \operatorname{Char} L$ *if there is a neighborhood of* $(\underline{y}, \underline{\eta})$ *so that in that neighborhood, the characteristic variety is a smooth graph* $\tau = \tau(y, \xi)$.[1]

Theorem 3.I.1, applied to $L_1(y, \tau(y,\xi), \xi)$ which has 0 as an isolated eigenvalue, implies that $\pi(y, \tau(y,\xi), \xi)$ and $Q(y, \tau(y,\xi), \xi)$ are smooth functions of y, ξ.

Examples 5.4.3. 1. When $\ker L_1(\underline{\tau}, \underline{\xi})$ is one dimensional, the eigenvalue τ of $A_0^{-1/2}\big(\sum_j A_j \xi_j\big) A_0^{-1/2}$ is simple. Thus for (y, ξ) near $(\underline{y}, \underline{\xi})$, there is a unique simple eigenvalue near $\underline{\tau}$ and the smooth variety hypothesis is satisfied.

2. If $L_1 = L_1(\partial)$ has constant coefficients, it was remarked in §2.4 that the stratification theorem of real algebraic geometry implies that the set of points where the smooth variety hypothesis is violated is at most a codimension one subvariety if $\operatorname{Char} L$. Therefore, the smooth variety hypothesis is satisfied for generic linear phases, $\phi(y) = y\eta$, $\eta \in \operatorname{Char} L$.

3. The argument in **2** extends to $L_1(y, \partial)$ with real analytic coefficients showing in that case that with the exception of a codimension one set in $\operatorname{Char} L$, the smooth variety hypothesis is satisfied.

Definitions. *If* (y, τ, ξ) *belongs to the characteristic variety and satisfies the smooth variety hypothesis, define the group velocity* $\mathbf{v}(y, \tau, \xi)$ *by*

$$\mathbf{v}(y, \tau, \xi)\, \partial_x \;:=\; -\sum_{j=1}^{d} \frac{\partial \tau(y, \xi)}{\partial \xi_j}\, \frac{\partial}{\partial x_j}.$$

If $\phi(t, x)$ *is a solution of the eikonal equation and the points* $(y, d_y\phi(y)) \in \operatorname{Char} L$ *satisfy the smooth variety hypothesis with associated function*

[1] It is sufficient to know that the variety is locally a continuous graph. The smoothness then follows from Theorem 3.I.1.

$\tau(y,\xi)$, *define the associated transport operator by*

$$(5.4.3)\ V(y,\partial_y)\ :=\ \partial_t\ -\ \sum_{j=1}^{d}\frac{\partial\tau}{\partial\xi_j}(y,d\phi(y))\,\frac{\partial}{\partial x_j}\ =\ \partial_t\ +\ \mathbf{v}(y,d\phi(y))\partial_x\,.$$

A geometric construction leading to (5.4.3) was given in §2.4. The velocity also appeared in the nonstationary phase calculation in §1.3. To show that this same velocity is hidden in the leading profile equation requires an algebraic identity.

Proposition 5.4.1. *At characteristic points* (y,τ,ξ) *where the smooth variety hypothesis is satisfied, the following fundamental algebraic identities hold for* $1\le j\le d$*:*

$$\begin{aligned}&\pi\big(y,\tau(y,\xi)\big)\,A_j(y)\,\pi\big(y,\tau(y,\xi)\big)\\ &\quad=\ -\,\frac{\partial\tau(y,\xi)}{\partial\xi_j}\ \pi\big(y,\tau(y,\xi)\big)\,A_0(y)\,\pi\big(y,\tau(y,\xi)\big).\end{aligned}$$

Example 5.4.4. When $\ker L_1$ is not one dimensional and $A_0=I$, the operators $\pi\,A_j\pi$ act as scalars on $\ker L_1(y,d\phi(y))$ in spite of their appearance as typical symmetric linear transformations on that space.

Proof. The variable y acts purely as a parameter and is suppressed. Consider the map

$$\xi_j\quad\mapsto\quad L_1\big(\tau(\xi),\xi\big)\ \pi\big(\tau(\xi),\xi\big)\,.$$

The matrix on the right has $\lambda=0$ as an isolated eigenvalue of constant multiplicity thanks to the smooth variety hypothesis.

The perturbation equation (3.I.2) implies that

$$\pi(\tau(\xi),\xi)\,\big(\frac{\partial\tau(\xi)}{\partial\xi_j}\,A_0+A_j\big)\,\pi(\tau(\xi),\xi)=0\,,$$

where the term in the middle is the derivative with respect to ξ_j of L_1. This identity is the desired relation. □

Using these identities yields

$$\pi(y)\,A_j(y)\,\pi(y)\,\frac{\partial}{\partial x_j}\ =\ \pi(y)\,A_0(y)\pi(y)\,\Big(-\,\frac{\partial\tau(y,d\phi(y))}{\partial\xi_j}\Big)\,\frac{\partial}{\partial x_j}\,.$$

Using this in (5.4.2) together with the definition of the group velocity $\mathbf{v}(y,d\phi(y))$ yields

$$\pi\,L\pi\ =\ \pi\,A_0\,\pi\Big(\partial_t\ +\ \mathbf{v}\partial_x\Big)\ +\ \pi\Big(B(y)\ +\ \sum_\mu A_\mu\partial_\mu\pi\Big)\pi\,.$$

The matrix $\pi\big(B + \sum_\mu A_\mu \partial_\mu \pi\big)\pi$ annihilates $\ker \pi$ and maps the image of π to itself. Since $\pi\, A_0\, \pi$ is an isomorphism of the image of π to itself, there is a unique smooth matrix value $\gamma(y)$ such that

$$\gamma(y)\,(I - \pi(y)) = (I - \pi(y))\,\gamma(y) = 0$$

and

$$\pi\Big(B + \sum_\mu A_\mu \partial_\mu \pi\Big)\pi = \pi\, A_0\, \pi\, \gamma\,.$$

Therefore,

$$\pi\, L\, \pi = \pi(y)\, A_0(y)\, \pi(y)\Big(\partial_t + \mathbf{v}\partial x + \gamma(y)\Big)\,,$$

and equation (5.4.1) is equivalent to the homogeneous transport equation,

$$\Big(\partial_t + \mathbf{v}\partial x + \gamma(y)\Big)a_0 = 0\,. \tag{5.4.4}$$

Definitions. *The integral curves of $\partial_t + \mathbf{v}\partial_x$ are called* **rays** *associated to ϕ. Equation (5.4.4) is called the* **transport equation for** a_0.

Solving equation (5.4.4) amounts to solving ordinary differential equations. When the smooth variety hypothesis is satisfied, the operator $\pi\, L\, \pi$ is essentially a linear transport operator, and the existence theorem, Theorem 5.3.2, can be strengthened.

Theorem 5.4.2. *Suppose that ϕ satisfies the eikonal equation on a set Ω, the smooth variety hypothesis is satisfied at the points $(y, d\phi(y))$, and for each point in Ω the backward ray can be continued in Ω until it reaches $t = 0$. Then for the asymptotic solutions of $L(a\, e^{i\phi/\epsilon}) \sim 0$, the determination of the a_j from the initial data $(\pi\, a_j)|_{t=0} \in C^\infty(\Omega \cap \{t = 0\})$ reduces to the solution of inhomogeneous transport equations*

$$\Big(\partial_t + \mathbf{v}\,\partial_x + \gamma(y)\Big)a_j = f_j\,,$$

where $f_0 = 0$ and $f_j = \pi(y)\, f_j$ is determined from a_{j-1}. If for all j, the support of $\pi(y)\, a_j|_{t=0}$ is contained in a set E, then the a_j are supported in the tube of rays with feet in E.

When the smooth variety hypothesis is satisfied, let p denote the dimension of $\ker L_1(\underline{y}, \underline{\eta}))$. Then p is also the dimension for nearby characteristic points since $\pi(y, \tau(y,\xi), \xi)$ is smooth. Division of polynomials in τ depending smoothly on y, ξ shows that

$$\det L(y,\eta) = (\tau - \tau(y,\xi))^p\, K(y,\tau,\xi), \qquad K \in C^\infty, \qquad K(\underline{y}, \underline{\tau}, \underline{\xi}) \neq 0.$$

Therefore phases can be determined from their initial data applying Hamilton–Jacobi theory to the *reduced eikonal equation*

$$\phi_t(y) = \tau(y, \nabla_x \phi(t,x)), \tag{5.4.5}$$

as in the discussion before the constant rank hypothesis with $G := (\tau - \tau(y, \xi))$.

Introduce the hamiltonian,

$$H(y, \eta) \;=\; H(t, x, \tau, \xi) \;:=\; \tau - \tau(y, \xi) \;=\; \tau - \tau(t, x, \xi)\,,$$

and its Hamilton vector field

$$\mathcal{X}_H \;:=\; \sum_{\mu=0}^{d} \Big(\frac{\partial H}{\partial \eta_\mu} \frac{\partial}{\partial y_\mu} - \frac{\partial H}{\partial y_\mu} \frac{\partial}{\partial \eta_\mu} \Big). \tag{5.4.6}$$

Hamilton–Jacobi theory shows that the graph of $d\phi$, that is $\{(y, d\phi(y))\}$, is generated from the initial points $(0, x, \tau(0, x, \nabla_x \phi(0, x)), \nabla_x \phi(0, x))$ by flowing along the the integral curves of $\mathcal{X}_H$.

The function H is constant on integral curves. The integral curves, along which $H = 0$, are curves in (y, η) space that lie in the characteristic variety and are called *bicharacteristics* or *null bicharacteristics*. The graph of $d\phi$ is foliated by a family of bicharacteristics parametrized by initial points over $\{t = 0\}$. The definitions of $\mathcal{X}_H$ and V yield the following result.

Theorem 5.4.3. *Suppose that ϕ satisfies the eikonal equation and the points $(y, d\phi(y))$ satisfy the smooth characteristic variety hypothesis so (5.4.5) holds. Then, the projection on space-time of the bicharacteristics foliating the graph of $d\phi$ are exactly the rays. Equivalently,*

$$\sum_{\mu=0}^{d} \frac{\partial H}{\partial \eta_\mu}(y, d\phi(y))\, \frac{\partial}{\partial y_\mu} \;=\; \partial_t \;+\; \mathbf{v}\, \partial_x\,. \tag{5.4.7}$$

Remark on numerics. If you find ϕ by solving the Hamilton–Jacobi equation using bicharacteristics, you will have computed integral curves of the vector field (5.4.6). The amplitude a_0 satisfies a transport equation along the space-time projections of these curves. The additional computational cost required to determine a_0 is negligible. This is true theoretically and also when the theory is used for numerical simulations. This method has numerical defects when rays grow far apart, since then the phase ϕ and amplitudes are determined at a sparse set of points. There is a well developed computational art of inserting new rays to help overcome this weakness. The methods are called *ray tracing* algorithms.

5.4.2. Transport for $L = L_1(\partial)$. When $L = L_1(\partial)$ has constant coefficients and no lower order terms, the transport equation for phases with $(y, d\phi(y))$ satisfying the smooth variety hypothesis can be understood in purely geometric terms. To simplify the formulas, we suppose that $A_0 = I$, which can always be achieved by the change of variable $u = A_0^{-1/2}\tilde{u}$.

Theorem 5.4.4. *Suppose that $L = L_1(\partial)$ has constant coefficients, no lower order terms, and $A_0 = I$. Suppose ϕ solves the eikonal equation and the points $(y, d\phi(y))$ satisfy the smooth characteristic variety hypothesis. Then when $\pi(y)w = w$, one has*

$$\pi(y)\, L(\partial)\, w \;=\; \Big(\partial_t + \mathbf{v}(y)\partial_x + \frac{1}{2}\,\mathrm{div}\,\mathbf{v}\Big) w\,, \tag{5.4.8}$$

where $\mathbf{v}$ *denotes the group velocity determined by* ϕ*. The transport equation determining* a_0 *is*

$$\Big(\partial_t + \mathbf{v}(y)\partial_x + \frac{1}{2}\,\mathrm{div}\,\mathbf{v}\Big) a_0 \;=\; 0\,. \tag{5.4.9}$$

The divergence of $\mathbf{v}$ involves second derivatives of τ. The proof, from [Guès and Rauch, 2006], uses second order perturbation theory from Theorem 3.I.2.

Proof. In $\pi\, L(\partial)\, w$, write the spatial derivatives as

$$\pi\, A_j \partial_j w \;=\; \pi\, A_j \partial_j\, (\pi w) \;=\; \pi\, A_j\, \pi\, \partial_j w + \pi\, A_j (\partial_j \pi)\, \pi\, w\,.$$

The eikonal equation satisfied by ϕ is (5.4.5). Consider the eigenvalue $-\tau(\xi)$ and eigenprojection $\pi(\xi)$ of the matrix $A(\xi) := \sum_j A_j \xi_j$ as functions of the parameter ξ_j. Formula (3.I.2) implies that

$$\pi\, A_j\, \pi \;=\; -\frac{\partial \tau}{\partial \xi_j}\, \pi \;=\; \mathbf{v}_j \pi\,. \tag{5.4.10}$$

Consider $-\tau(\underline{\xi}+s\xi)$ as an eigenvalue of $\sum_j A_j(\underline{\xi}_j+s\xi_j)$. Since the second derivative of the matrix with respect to s vanishes, formula (3.I.3) implies that

$$\begin{aligned}
-\pi\, \frac{d^2}{ds^2} \tau(\underline{\xi} + s\xi) \;&=\; \sum_{i,j} \pi\, \frac{\partial^2 \tau}{\partial \xi_i \partial \xi_j}\; \xi_i\; \xi_j \\
&=\; -2\pi \frac{d}{ds}\Big(\sum_j A_j(\underline{\xi}_j + s\xi_j)\Big)\, Q\, \frac{d}{ds}\Big(\sum_i A_i(\underline{\xi}_i + s\xi_i)\Big)\, \pi \\
&=\; -2\pi \Big(\sum_j A_j \xi_j\Big)\, Q\, \Big(\sum_i A_i \xi_i\Big)\, \pi \\
&=\; -2 \sum_{i,j} \pi\, A_j Q A_i \xi_j \xi_i\, \pi \\
&=\; -\sum_{i,j} \pi\, \big(A_j Q A_i \xi_j \xi_i \;+\; A_i Q A_j \xi_i \xi_j\big)\, \pi\,,
\end{aligned}$$

the last step by symmetrization. Therefore

$$\frac{\partial^2 \tau}{\partial \xi_j \partial \xi_k}\, \pi \;=\; -\pi A_j Q A_k \pi \;-\; \pi A_k Q A_j \pi. \tag{5.4.11}$$

Next consider the eigenvalue $-\tau(\nabla_x\phi)$ and eigenprojections $\pi(y)$ of $M(y) := \sum_k A_k\partial_k\phi(y)$ as functions of the parameter x_j. The formula (3.I.4) yields

$$(\partial_j\pi)\,\pi \;=\; -Q\,\partial_j M\,\pi \;=\; -Q\sum_k A_k\,\pi\,\frac{\partial^2\phi}{\partial x_k\partial x_j}\,.$$

Using (5.4.11) yields

$$\pi\,A_j\,(\partial_j\pi)\pi \;=\; -\sum_k \pi A_j Q A_k\pi\,\frac{\partial^2\phi}{\partial x_k\partial x_j} \;=\; \sum_k \frac{1}{2}\,\frac{\partial^2\tau}{\partial\xi_j\partial\xi_k}\,\frac{\partial^2\phi}{\partial x_k\partial x_j}\,\pi\,.$$

Differentiating $\mathbf{v} = -\nabla_\xi\tau\big(\nabla_x\phi(y)\big)$ shows that

$$\operatorname{div}\mathbf{v} \;=\; -\sum_j \frac{\partial}{\partial x_j}\,\frac{\partial\tau(\nabla_x\phi)}{\partial\xi_j} \;=\; \sum_{j,k}\frac{\partial^2\tau}{\partial\xi_j\partial\xi_k}\,\frac{\partial^2\phi}{\partial x_k\partial x_j}\,.$$

Therefore,

$$\sum_j \pi\,A_j\,(\partial_j\pi)\pi \;=\; \frac{1}{2}\big(\operatorname{div}\mathbf{v}\big)\,\pi\,. \tag{5.4.12}$$

Combining (5.4.10), (5.4.11), and (5.4.12) with the first identity of the proof yields, when $w = \pi\,w$,

$$\pi\,L(\partial)w \;=\; \pi\,\big(\partial_t + \mathbf{v}\partial_x\big)w + \frac{1}{2}\,\big(\operatorname{div}\mathbf{v}\big)w\,.$$

Since $\nabla_x\phi$ is constant along integral curves of $\partial_t + \mathbf{v}\partial_x$ it follows that $\pi(y)$ is also constant, so

$$\pi\,\big(\partial_t + \mathbf{v}\partial_x\big)w \;=\; \big(\partial_t + \mathbf{v}\partial_x\big)\,\pi w \;=\; \big(\partial_t + \mathbf{v}\partial_x\big)w\,,$$

and the proof is complete. □

Exercise 5.4.1. *The corresponding computation in the scalar case is significantly simpler. For a homogeneous constant coefficient second order scalar strictly hyperbolic operator without lower order terms, modify the computation of §5.2 to treat the case of nonlinear phases $\phi(y)$. In particular find a formula for the associated group velocity $\mathbf{v}$ and show that the leading amplitude a_0 satisfies (5.4.9).* **Example.** D'Alembert's wave equation.

Both ϕ and $\nabla_x\phi$ are constant on the ray that start at point $(0,x)$ and has velocity given by $\big(1, -\tau_\xi(\nabla_x\phi_0(x))\big) = (1,\mathbf{v})$. This ray has equation

$$t \;\mapsto\; \Big(t\,,\,x + t\,\mathbf{v}\big(\nabla_x\phi(0,x)\big)\Big) \;:=\; \big(t,\Phi(t,x)\big)\,. \tag{5.4.13}$$

The rays $t\mapsto \big(t,\Phi(t,x)\big)$ are integral curves of the vector field $\partial_t + \mathbf{v}\partial_x$. Equivalently,

$$\frac{d}{dt}\,\big(t,\Phi(t,x)\big) \;=\; \Big(1\,,\,\mathbf{v}(\Phi(t,x))\Big)\,, \qquad \Phi(0,x) = x\,.$$

This equation shows that Φ is the flow of the time dependent vector field $\mathbf{v}$,

$$\frac{d}{dt}\,\Phi(t,x) \;=\; \mathbf{v}(\Phi(t,x))\,, \qquad \Phi(0,x) = x\,.$$

The jacobian determinant,

$$J(t,x) \;:=\; \det\big(D_x\Phi(t,x)\big),$$

describes the infinitesimal deformation of d dimensional volumes as in (5.4.20) below.

Differentiating with respect to x yields the evolution of $D_x\Phi(t,x)$,

$$\frac{d}{dt}\,D_x\Phi(t,x) = \big(D_x\mathbf{v}\big)(\Phi(t,x))\;D_x\Phi(t,x)\,.$$

For x fixed this is an equation of the form

$$\frac{dM}{dt} \;=\; A(t)\,M(t),$$

where A and M are smooth $N\times N$ matrix valued functions. It follows that

$$\frac{d\det M}{dt} \;=\; \big(\mathrm{trace}A(t)\big)\;\det M\,.$$

This follows from the following three estimates as $\Delta t \to 0$:

$$\begin{aligned} M(t+\Delta t) \;&=\; (I + A(t)\Delta t)M(t) \;+\; O((\Delta t)^2),\\ \det M(t+\Delta t) \;&=\; \det(I + A(t)\Delta t)\,\det M(t) + \;O((\Delta t)^2)\,,\\ \det(I + A(t)\Delta t) \;&=\; 1 \;+\; \mathrm{trace}A(t)\,\Delta t \;+\; O((\Delta t)^2)\,. \end{aligned}$$

Applying the formula for $(\det M)'$ yields

(5.4.14) $$\frac{d}{dt}\,J(t,x) \;=\; \mathrm{trace}\big(D_x\mathbf{v}\big(\Phi(t,x)\big)\;J \;=\; \big(\mathrm{div}\,\mathbf{v}(\Phi(t,x))\big)\,J\,.$$

Corollary 5.4.5. *The amplitude $a_0 = \pi(y)\,a_0$ satisfies the transport equation (5.4.9) if and only if the function*

(5.4.15) $$a_0(t,\Phi(t,x))\,\sqrt{J(t,x)}$$

does not depend on t.

Proof. Equation (5.4.14) implies that $\sqrt{J(t,x)}$ satisfies

$$\partial_t\sqrt{J(t,x)} \;=\; \frac{1}{2\,\sqrt{J(t,x)}}\;\partial_t J \;=\; \frac{1}{2}\,\big(\mathrm{div}\,\mathbf{v})(\Phi(t,x)\big)\;\sqrt{J(t,x)}\,.$$

Therefore

$$\begin{aligned} &\partial_t\Big(a_0(t,\Phi(t,x))\,\sqrt{J(t,x)}\Big)\\ &\quad= \;\big(\partial_t\sqrt{J(t,x)}\big)\,a_0(t,\Phi(t,x)) \;+\; \sqrt{J(t,x)}\,\Big(\partial_t a_0 + \mathbf{v}\partial_x a_0\Big)\,(t,\Phi(t,x))\,. \end{aligned}$$

Using the formula for $\partial_t\sqrt{J(t,x)}$ yields

$$\partial_t\Big(a_0(t,\Phi(t,x))\,\sqrt{J(t,x)}\Big)$$
$$= \sqrt{J(t,x)}\,\Big(\partial_t a_0 + \mathbf{v}\partial_x a_0 + \frac{1}{2}(\operatorname{div}\mathbf{v})a_0\Big)(t,\Phi(t,x))\,. \qquad \square$$

For fixed t denote by $X = \Phi(t,x)$ the point on the ray whose initial point is x. Then the infinitesimal volumes satisfy

$$dX = \left|\det\frac{\partial X}{\partial x}\right| dx = J(t,x)\,dx\,. \tag{5.4.16}$$

Equation (5.4.15) implies that

$$|a_0(t,X)|^2\,J(t,x) = |a_0(0,x)|^2\,,$$

$$\text{equivalently} \qquad |a_0(t,X)|^2\,dX = |a_0(0,x)^2|\,dx\,.$$

This is an *infinitesimal conservation of energy law.*

If $\omega \subset \{t=0\}$ is a nice bounded open set, the family of rays starting in ω is called a *tube of rays* and is denoted $\mathcal{T}$. Its section at time t is denoted $\omega(t)$. In particular, $\omega=\omega(0)$. Then,

$$\int_\omega |a_0(0,x)^2|\,dx = \int_{\omega(t)} |a_0(t,X)|^2\,dX\,. \tag{5.4.17}$$

Since $A_0 = I$, the energy density for solutions of $Lu=0$ is $\langle u, A_0 u\rangle = \langle u,u\rangle$. Equation (5.4.17) shows that to leading order, the energy in any tube of rays is conserved.

Example. Linear phases. Suppose that $\phi(y) = t\tau + x\xi$ is linear and satisfies the smooth variety hypothesis. The group velocity $\mathbf{v} = -\nabla_\xi\tau(\xi)$ is then constant and the rays are lines in space-time with this velocity. The divergence of $\mathbf{v}$ vanishes, so a_0 is constant on rays, so $a_0(t,x) = g(x - t\,\mathbf{v})$, where $g(x) := a_0(0,x)$ is the initial value of a_0. The leading approximation is

$$e^{i(t\tau+x\xi)/\epsilon}\,g(x-t\mathbf{v})\,, \qquad \pi\,g = g\,.$$

This generalizes the result obtained in §1.4.

The transport equation and its solution in the last two results depend only the phase ϕ and its associated group velocity. Two constant coefficient homogeneous systems leading to the same eikonal equation (5.4.5) lead to the same profile equation. For example Maxwell's equations, Dirac equations, and the wave equation all have the same eikonal equation $\partial_t\phi^2 = |\nabla_x\phi|^2$, so their principal profiles satisfy the same transport equations. A third example is the operator

$$L := \partial_t + \begin{pmatrix}1 & 0\\ 0 & -1\end{pmatrix}\frac{\partial}{\partial x_1} + \begin{pmatrix}0 & 1\\ 1 & 0\end{pmatrix}\frac{\partial}{\partial x_2}\,. \tag{5.4.18}$$

The characteristic polynomial is $\det L_1(\tau,\xi) = \tau^2 - |\xi|^2$, and the eikonal equation is $\phi_t^2 = |\nabla_x\phi|^2$.

The computation that follows is performed in dimension d. The case $d = 3$ applies to Maxwell's equations. The computation for any d applies to any Lorentz invariant field equation without lower order terms, e.g., the wave equation.

Example. Outgoing spherical solutions of $\phi_t^2 = |\nabla_x\phi|^2$. A particular solution of $\phi_t^2 = |\nabla_x\phi|^2$ for $|x| \neq 0$ is

$$\phi(y) := t - |x|, \qquad \phi_t = |\nabla_x\phi|. \tag{5.4.19}$$

Here $\tau(\xi) = |\xi|$, and the group velocity is

$$\mathbf{v} = -\nabla_\xi\tau(\nabla_x\phi) = -\frac{\xi}{|\xi|}\bigg|_{\xi=\nabla_x\phi} = \frac{x}{|x|}.$$

Since

$$\partial_j\frac{x_j}{|x|} = \frac{1}{|x|} + x_j\partial_j\frac{1}{|x|}, \qquad \operatorname{div}\mathbf{v} = \frac{d}{|x|} + r\partial_r(r^{-1}) = \frac{d}{|x|} - \frac{r}{r^2} = \frac{d-1}{|x|}.$$

The rays move radially away from the origin with speed equal to one. The flow is given by

$$\Phi(t,x) = x + t\frac{x}{|x|}.$$

The annulus $\rho < r < \rho+\delta\rho$ is mapped to the annulus $\rho+t < r < \rho+t+\delta\rho$ as in Figure 5.4.2.

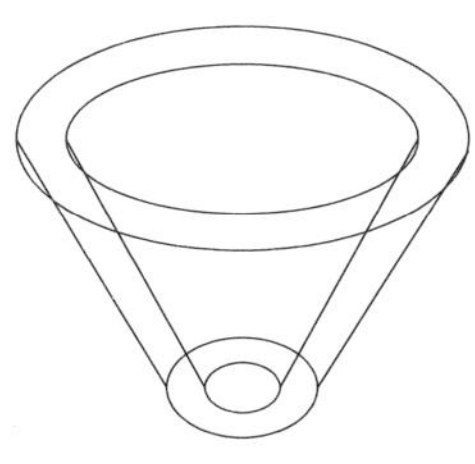

FIGURE 5.4.2. Outgoing annulus has growing volume

Considering $\delta\rho \ll 1$ shows that the volume is amplified by $\big((\rho+t)/\rho\big)^{d-1}$. Therefore the jacobian, that depends only on $|x|$, is given by

$$J(t,x) = \left(\frac{|x|+t}{|x|}\right)^{d-1}. \tag{5.4.20}$$

Given initial values $a_0(0,x) = g(x) \in C_0^\infty(\mathbb{R}^d \setminus 0)$, the amplitude a_0 is defined for all $t \geq 0$ and has support in the set of outgoing rays with feet in the initial support. Equation (5.4.15) yields

$$a_0(t, x + tx/|x|)\ J(t,x)^{1/2} \;=\; a_0(0,x),$$

$$\text{so}\quad a_0(t,x) \;=\; g(x - tx/|x|)\left(\frac{|x|}{|x|+t}\right)^{(d-1)/2}.$$

The leading term in the geometric optics approximation is

$$e^{i(t-|x|)/\epsilon}\, g(x - tx/|x|)\left(\frac{|x|}{|x|+t}\right)^{(d-1)/2}. \tag{5.4.21}$$

Exercise 5.4.2. *Compute formula (5.4.15) for the jacobian directly using the definition of $J(t,x)$ and/or (5.4.14).*

Example. Incoming spherical solutions of $\phi_t^2 = |\nabla_x\phi|^2$. The preceding example has an important sibling. If one considers the outgoing example for times $t < 0$, the level sets $\phi =$ const. are incoming spherical shells, that focus to a point. If the closest point to the origin in the support of $a_0(t,x)$ is at distance r, then the equation for the amplitude becomes singular at $t = -r$.

Equivalently, had we considered the phase $\phi(t,x) = t + |x|$, the group velocity would have been $\mathbf{v} = -x/|x|$ and the wavefronts would be incoming as in Figure 5.4.3.

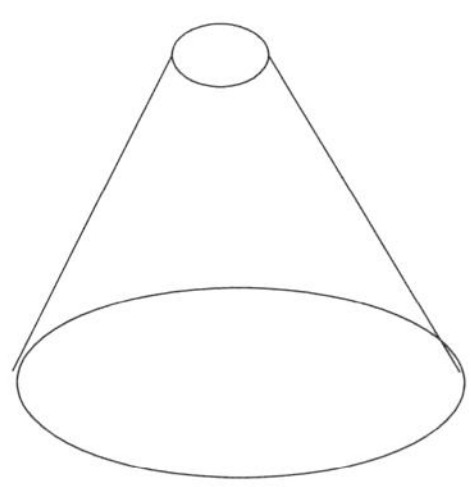

FIGURE 5.4.3. Incoming or focusing wavefronts

The leading approximation of geometric optics for such focusing spherical wavefronts is

$$e^{i(t+|x|)/\epsilon}\, g(x + tx/|x|)\left(\frac{|x|}{|x|-t}\right)^{(d-1)/2}. \tag{5.4.22}$$

The amplitudes become infinitely large where $x = t$. If you follow a ray approaching the origin, the amplitudes grow to compensate for the volume compression. As a spherical front focuses to a point, the amplitude diverges to infinity. For ϵ fixed, the initial data are smooth and the exact solution of the initial value problem is smooth. *But* the approximation of geometric optics becomes infinitely large at focal points. The conclusion is that in a small neighborhood of the focal point, the approximation is inaccurate. What is surprising is that after the focus one finds that, in the linear case, the approximation becomes accurate again, with the phase changed by an additive constant called the Guoy shift (see [Hörmander, Chapter 2, 1983]) after the physicist whose two mirror experiments verified the phenomenon for $d = 3$.

The geometric optics approximation is valid until the first ray along which a_0 is nonzero touches the origin. For example so long as the rays lie in $|x| \geq \delta > 0$, choosing δ small shows that solutions do grow as they focus. They just do not grow infinitely large.

Exercise 5.4.3. *Prove that the solutions of the the wave equation in $d = 3$ with oscillatory radial initial data*

$$u^\epsilon(0,r) = a(r)\, e^{ir/\epsilon}, \qquad u^\epsilon_t(0,r) = \epsilon^{-1}\, b(r)\, e^{ir/\epsilon}$$

with smooth compactly supported radial a, b vanishing for $r \leq R$, has maximum value that grows no faster than $1/\epsilon$ as $\epsilon \to 0$. **Hint.** Use the formula for the general radial solution of the $d = 3$ wave equation,

$$u \;=\; \frac{f(t+r) - f(t-r)}{r} \quad \text{for} \quad r \neq 0, \qquad u(t,0) = 2f'(t).$$

Find f in terms of a, b. **Discussion. i.** Each solution u^ϵ is a sum of an outgoing and an incoming spherical solution. For $\delta > 0$ and $0 \leq t \leq R - \delta$, the solution is supported in $r \geq \delta$ and the approximations of geometric optics are accurate. When $t = R$, the incoming wave can arrive at the origin, the phase loses smoothness, and the approximation breaks down. **ii.** On $0 \leq t \leq R - \delta$, the family $(\epsilon\partial)^\alpha u^\epsilon$ is bounded as $\epsilon \to 0$. The exercise shows that for typical a, b, this is not true on $[0, R+1] \times \{|x| \leq 1\}$.

5.4.3. Energy transport with variable coefficients.

The energy identity (2.3.1) implies that when $Lu = 0$,

$$\frac{d}{dt}\int \big\langle A_0(y)u(t,x), u(t,x)\big\rangle\, dx \\ + \int \big\langle (B + B^* - \sum \frac{\partial A_\mu}{\partial y_\mu}) u(t,x), u(t,x)\big\rangle\, dx = 0.$$

Denote,

$$Z(y) \;:=\; B + B^* - \sum \partial_\mu A_\mu\,.$$

Definition. *A symmetric hyperbolic system is* **conservative** *when*

$$B + B^* - \sum \frac{\partial A_\mu}{\partial y_\mu} \;\equiv\; 0\,. \tag{5.4.23}$$

Since Cauchy data at time t are arbitrary, one has the following equivalence.

Proposition 5.4.6. *A system is conservative if and only if*

$$\int_{\mathbb{R}^d} \langle A_0 u(t,x), u(t,x)\rangle \, dx$$

is independent of time for all solutions of $Lu = 0$ whose Cauchy data are compactly supported in x.

Theorem 5.4.7. *Suppose that L is conservative and ϕ is a phase satisfying the smooth characteristic variety hypothesis at all points $\big(y, d\phi(y)\big)$ over a tube of rays $\mathcal{T}$ with sections $\omega(t)$. If $a(\epsilon, y)e^{i\phi(y)/\epsilon}$ is an asymptotic solution of $Lu \sim 0$, then at leading order the energy in the tube is conserved, that is*

$$\int_{\omega(t)} \big\langle A_0(y)\, a_0(t,x), a_0(t,x)\big\rangle \, dx \tag{5.4.24}$$

is independent of t.

Proof. For $0 < \delta \ll 1$, choose a cutoff function $0 \le \chi_\delta(x) \le 1$ such that χ is equal to one on ω_0 and $\chi(x) = 0$ when $\operatorname{dist}(x, \omega_0) > \delta$. Construct a Lax solution $\tilde{a}(\epsilon, y)e^{i\phi(y)/\epsilon}$ with

$$\tilde{a}_0(0,x) \;=\; \chi_\delta(x)\, a_0(x)\,. \tag{5.4.25}$$

Then Lax's theorem together with conservation of energy implies that for all m and t

$$\int_{\mathbb{R}^d} \big\langle A_0\, \tilde{a}(\epsilon,t,x), \tilde{a}(\epsilon,t,x)\big\rangle \, dx \;-\; \int_{\mathbb{R}^d} \big\langle A_0\, \tilde{a}(0,x), \tilde{a}(0,x)\big\rangle \, dx \;=\; O(\epsilon^m)\,.$$

In addition the quantity on the left is controlled by its principal term, so

$$\int_{\mathbb{R}^d} \big\langle A_0\, \tilde{a}_0(t,x), \tilde{a}_0(t,x)\big\rangle \, dx \;-\; \int_{\mathbb{R}^d} \big\langle A_0\, \tilde{a}_0(0,x), \tilde{a}_0(0,x)\big\rangle \, dx \;=\; O(\epsilon)\,. \tag{5.4.26}$$

Since the left-hand side is independent of ϵ it must vanish identically.

The amplitudes $\tilde{a}$ are uniformly bounded for bounded t and for $\delta < 1$. They differ from a on a set of measure $O(\delta)$. Therefore, for t in bounded sets

$$\int_{\mathbb{R}^d} \big\langle A_0\, \tilde{a}_0(t,x), \tilde{a}_0(t,x)\big\rangle \, dx \;-\; \int_{\omega(t)} \big\langle A_0\, a_0(t,x), a_0(t,x)\big\rangle \, dx \;=\; O(\delta)\,. \tag{5.4.27}$$

Combining the last two results yields

$$\begin{aligned}\int_{\omega(t)} \big\langle A_0\, a_0(t,x), a_0(t,x)\big\rangle\, dx \;&=\; \int_{\mathbb{R}^d} \big\langle A_0\, \tilde a_0(t,x), \tilde a_0(t,x)\big\rangle\, dx \;+\; O(\delta)\\ &=\; \int_{\mathbb{R}^d} \big\langle A_0\, \tilde a_0(0,x), \tilde a_0(0,x)\big\rangle\, dx \;+\; O(\delta)\\ &=\; \int_{\omega(0)} \big\langle A_0\, a_0(0,x), a_0(0,x)\big\rangle\, dx \;+\; O(\delta)\,.\end{aligned}$$

Letting δ tend to zero proves the theorem. □

Consider $\omega(0)$ shrinking to a point $\underline{x}$ so the tube converges to the ray through $\underline{x}$. Then

$$\frac{\text{vol}\,(\omega(t))}{\text{vol}\,(\omega(0))} \;\to\; J(t,\underline{x}), \tag{5.4.28}$$

where $J(t,\underline{x})$ is the jacobian $\det d\Phi/dx$ of the flow $\Phi(t,x)$ generated by the vector field $\mathbf{v}(y) = -\nabla_\xi \tau(y, \nabla_x\phi(y))$.

The law of conservation of energy applied to a tube of diameter δ implies that

$$\begin{aligned}&\text{vol}\,(\omega(t))\,\big\langle A_0(\Phi(t,\underline{x}))\, a_0(\Phi(t,\underline{x}))\,, a_0(\Phi(t,\underline{x}))\,\big\rangle\\ &\qquad = \text{vol}\,(\omega(0))\,\big\langle A_0(\Phi(0,\underline{x}))\, a_0(\Phi(0,\underline{x}))\,,\, a_0(\Phi(0,\underline{x}))\big\rangle\,(1+O(\delta))\,.\end{aligned}$$

Dividing by $\text{vol}\,(\omega(0))$ and passing to the limit $\delta \to 0$ implies that the quantity

$$\big\langle A_0(\Phi(t,\underline{x}))\, a_0(\Phi(t,\underline{x}))\,,\, a_0(\Phi(t,\underline{x}))\,\big\rangle J(t,\underline{x})$$

is independent of t, proving the next result.

Corollary 5.4.8. *Suppose that the the smooth characteristic variety hypothesis is satisfied at all points $\big(y, d\phi(y)\big)$ over a tube of rays $\mathcal{T}$, and that the system L is conservative. If $a(\epsilon, y)\, e^{i\phi(y)/\epsilon}$ is an asymptotic solution of $Lu \sim 0$, then the quantity*

$$\big\langle A_0(y)\, a_0(y)\,,\, a_0(y)\,\big\rangle\; J(y)$$

is constant on the rays associated to ϕ.

Remark. This shows that for conservative problems the size of the leading amplitude a_0 is determined from its initial size entirely by conservation and volume deformation. This is weaker than the results for $L = L_1(\partial)$ where volume deformation alone determined the exact values of a_0.

The above results show that there is negligible energy flux into or out of the tube of rays. We give an alternate proof that allows us to generalize the results to nonconservative problems. The point of departure is the energy law (2.3.1). For a smooth w, the energy flux per unit area across an element of hypersurface $d\sigma$ with unit outward conormal ν is given by

$$\Big\langle \sum_\mu \nu_\mu\, A_\mu\, w\,,\, w \Big\rangle .$$

Proposition 5.4.9. *Suppose that ϕ satisfies the eikonal equation and $\big(y, d\phi(y)\big)$ satisfies the smooth variety hypothesis. Then transport is along $(1,\mathbf{v}) = (1, -\nabla_\xi \tau(y, \nabla_x \phi(y)))$. If the conormal to an infinitesimal hypersurface element $d\sigma$ is orthogonal to $(1,\mathbf{v})$, that is,*

$$\nu_0 \;+\; \sum_j \nu_j\, \mathbf{v}_j \;=\; 0\,, \tag{5.4.29}$$

then, for any polarized $w = \pi(y, d\phi(y))\, w$, the flux through $d\sigma$ vanishes,

$$\Big\langle \sum_\mu \nu_\mu\, A_\mu\, w\,,\, w \Big\rangle \;=\; 0\,. \tag{5.4.30}$$

Proof. Using the polarization, the flux is equal to

$$\Big\langle \sum_\mu \nu_\mu\, A_\mu\, w\,,\, w \Big\rangle \;=\; \Big\langle \sum_\mu \nu_\mu\, A_\mu\, \pi\, w\,,\, \pi\, w \Big\rangle \;=\; \Big\langle \sum_\mu \nu_\mu\, \pi\, A_\mu\, \pi\, w\,,\, w \Big\rangle .$$

The identity of Proposition 5.4.1 implies that this is equal to

$$\Big\langle \big(\nu_0 + \sum_j \nu_j\, \mathbf{v}_j\big)\, \pi\, A_0\, \pi\, w\,,\, w \Big\rangle .$$

This vanishes thanks to (5.4.29). □

Corollary 5.4.10. *If ϕ satisfies the eikonal equation and the smooth variety hypothesis on a tube of rays, and $w = \pi\, w$ satisfies the profile equation $\pi\, L\, \pi\, w = 0$, then along each ray $(t, \Phi(t,x))$ of $\mathcal{T}$ the energy density satisfies*

$$\frac{\partial}{\partial t}\bigg(\Big\langle A_0\, w\,,\, w \Big\rangle_{(t,\Phi(t,x))}\, J(t,x) \bigg) + \Big\langle Z\, w\,,\, w \Big\rangle_{(t,\Phi(t,x))}\, J(t,x) \;=\; 0\,. \tag{5.4.31}$$

Remark. In the conservative case, $Z = 0$, one recovers the result of Corollary 5.4.8.

Proof. Using the polarization and transport equation yields

$$\Big\langle L\, w\,,\, w \Big\rangle \;=\; \Big\langle L\, \pi\, w\,,\, \pi\, w \Big\rangle \;=\; \Big\langle \pi\, L\, \pi\, w\,,\, w \Big\rangle \;=\; 0\,.$$

The energy identity (2.3.1) yields

$$\sum_\mu \partial_\mu \Big\langle A_\mu\, w\,,\, w \Big\rangle \;+\; \Big\langle Z\, w\,,\, w \Big\rangle \;=\; 0\,.$$

Integrate over the tube from $t = 0$ to t. The lateral boundaries of the tube are foliated by rays. Therefore the conormal to the lateral boundaries are orthogonal to the transport direction. Proposition 5.4.9 shows that the flux through the lateral boundaries vanishes. Therefore,

$$\int_{\omega(t)} \Big\langle A_0\, w\,,\, w\Big\rangle\, dx \;+\; \int_0^t \int_{\omega(s)} \Big\langle Z\, w\,,\, w\Big\rangle\, dx\, ds \;=\; \int_{\omega(0)} \Big\langle A_0\, w\,,\, w\Big\rangle\, dx\,.$$

Dividing by the volume of $\omega(0)$ and shrinking $\omega(0)$ to the single point x, the tube contracts to the ray $(t, \Phi(t,x))$, and one finds using (5.4.27) that

$$\Big\langle A_0\, w\,,\, w\Big\rangle_{(t,\Phi(t,x))} J(t,x) \;+\; \int_0^t \Big\langle Z\, w\,,\, w\Big\rangle_{(s,\Phi(s,x))} J(s,x)\, ds \;=\; \Big\langle A_0\, w\,,\, w\Big\rangle_{(0,x)}\,.$$

This is equivalent to (5.4.30). □

5.5. The Lax parametrix and propagation of singularities

The seminal paper [Lax, Duke Math. J., 1957] made several crucial advances. It systematized the formal aspects of the high frequency asymptotic solutions in the strictly hyperbolic case, and it showed how to prove their accuracy using energy estimates. Taking a very large step toward the creation of Fourier integral operators, it used these solutions to solve the the Cauchy problem with distribution initial data up to a smooth error, at least for small time. The necessity of small time comes from the fact that the nonlinear eikonal equations are solvable only locally in time.

The locality in time was removed as a hypothesis by Ludwig [Ludwig, 1960] by piecing together the local solutions and making a nonstationary phase argument that is a special case of general results of Hörmander on the composition of Fourier integral operators.

In the late 1960s, Hörmander introduced the wavefront set. This refined the notion of singular support motivated by work of Sato in the analytic category. In this section we show that using the notion of wavefront set, the local construction of Lax gives the global result of Ludwig and a microlocal version of Hörmander that is the second of our two basic theorems of microlocal analysis.

5.5.1. The Lax parametrix. The point of departure is a representation of the solution of the initial value problem,

$$Lu \;=\; 0\,, \qquad u(0,x) \;=\; f(x) \;\subset\; \mathcal{S}'(\mathbb{R}^d)\,, \tag{5.5.1}$$

using the Fourier integral representation of the initial data,

$$f(x) \;=\; (2\pi)^{-d/2} \int e^{ix\xi}\, \hat f(\xi)\, d\xi\,. \tag{5.5.2}$$

From the point of view of singularities, only large ξ are important. The solution of initial value problems with data $e^{ix\xi}\hat{f}(\xi)$ with $|\xi| \gg 1$ are the key ingredients.

As in the treatment of elliptic regularity, introduce $\omega := \xi/|\xi|$ and $\epsilon = 1/|\xi|$. The initial data of interest are $e^{ix\omega/\epsilon}\hat{f}(\xi)$ with $0 < \epsilon \ll 1$. This is a family of short wavelength data parametrized by $\{|\omega| = 1\}$.

Convention. *By a linear change of dependent variable we may, without loss of generality, assume that $A_0 = I$ for the remainder of §5.5.*

Hypothesis. *Suppose that the smooth characteristic variety hypothesis holds at every point of the characteristic variety. In addition, assume that the distance between the sheets of the variety is bounded below by $C|\eta|$ uniformly in y.*

Remarks. 1. Since there is finite speed of propagation, the uniform separation hypothesis as $|y| \to \infty$ is just for convenience and simplicity of statements.

2. If one assumes the smooth characteristic variety hypothesis only along a neighborhood of a bicharacteristic, there is an analogous construction microlocally along that curve.

This hypothesis implies that the eigenvalues $\lambda_\nu(t,x,\xi)$ of $-\sum_j \xi_j A_j$ are defined and smooth in $\xi \neq 0$ and are uniquely determined by the ordering $\lambda_\nu < \lambda_{\nu+1}$ and

$$\operatorname{Char} L \;=\; \bigcup_\nu \{\tau = \lambda_\nu(t,x,\xi)\}\,.$$

Define

$$\pi_\nu(t,x,\xi) \;:=\; \pi_\mu(t,x,\lambda_\nu(t,x,\xi),\xi) \tag{5.5.3}$$

to be the projection along the range and onto the kernel of $L(t,x,\lambda_\nu(t,x,\xi),\xi)$. Since $A_0 = I$, these are also the orthogonal projections onto the eigenspace associated to λ_ν. The rank of $\pi_\nu(t,x,\lambda_\nu(t,x,\xi),\xi)$ is constant on each connected sheet of the variety. One says that L has *constant multiplicity*. In addition,

$$\sum_\nu \pi_\nu(t,x,\xi) \;=\; I\,.$$

Define $\phi_\nu(t,t',x,\xi)$ as the solution of the eikonal equation

$$\partial_t\phi_\nu = \lambda_\nu(t,x,\nabla_x\phi_\mu)\,, \qquad \phi_\nu(t,t',x,\xi)\big|_{t=t'} \;=\; x\xi\,. \tag{5.5.4}$$

Since ϕ_ν is positive homogeneous of degree one in ξ, it suffices to consider $|\xi| = 1$. Choose $\underline{T} > 0$ so that for each $\nu, t', |\xi| = 1$ the solution of the eikonal equation (5.5.4) exists and is smooth on $]t' - 2\underline{T}, t' + 2\underline{T}[\, \times \mathbb{R}^d$.

Begin with the choice $t' = 0$ and for ease of reading, suppress the t' dependence of ϕ_ν. Seek matrix valued asymptotic solutions

$$(5.5.5)\quad \mathbf{U}_\nu \;=\; e^{i\phi_\nu(y,\omega)/\epsilon}\,\mathbf{A}_\nu(\epsilon,y,\omega)\,, \qquad \mathbf{A}_\nu(\epsilon,y,\xi)\;\sim\;\sum_{j=0}^{\infty}\epsilon^j\,\mathbf{A}_{\nu,j}(t,x,\omega).$$

For the matrix valued solution $\mathbf{U}_\nu$, one either repeats the derivation of the equations for the profiles, or reasons column by column. From either point of view it is no harder to consider matrix valued asymptotic solutions than the vector valued case.

Seek $\mathbf{U}_\nu$ so that

$$(5.5.6)\qquad L\,\mathbf{U}_\nu\;\sim\;0\,,\qquad \sum_{\nu,j}\epsilon^j\,\mathbf{A}_{\nu,j}(0,x)\;\sim\;I\,.$$

The Lax parametrix for the initial value problem (5.5.1) is then

$$(5.5.7)\quad u_{\text{approx}}\;:=\;\sum_\nu u_\nu\,,\qquad u_\nu\;:=\;\int \mathbf{A}_\nu(\epsilon,y,\xi)\,e^{i\phi_\nu(t,x,\omega)/\epsilon}\,\chi(\xi)\,\hat f(\xi)\,d\xi\,,$$

$\chi \in C^\infty$ with support in $|\xi| > 1$ with $\chi(\xi) = 1$ for $|\xi| \geq 2$. The expression emphasizes the parameter ϵ and the origin in short wavelength asymptotics.

Next determine smooth initial values of $\mathbf{A}_{\nu,j}(0,x)$ to achieve the second relation in (5.5.6). The leading symbols $\mathbf{A}_{\nu,0}$ must satisfy

$$\sum_\nu \mathbf{A}_{\nu,0}\;=\;I\,,\qquad \pi_\nu\,\mathbf{A}_{\nu,0} = \mathbf{A}_{\nu,0}\,.$$

Multiplying the first by π_ν shows that these two equations uniquely determine

$$\mathbf{A}_{\nu,0}(0,x,\omega)\;=\;\pi_\nu(0,x,\omega)\,.$$

The transport equations

$$\Big(\partial_t + \mathbf{v}(y,\omega).\partial_x + \pi_\nu\big(B + (L\,\pi)\big)\pi_\nu\Big)\mathbf{A}_{\nu,0}\;=\;0\,,\qquad \pi_\nu = \pi_\nu(y,\nabla_x\phi(y,\omega)),$$

then determine $\mathbf{A}_{\nu,0}$. Since π_ν and $\mathbf{v}_\nu$ are smooth and homogeneous in $\xi \neq 0$, it follows that the transport equations are solvable with uniform estimates on $\{|t| < 2\underline{T}\} \times \mathbb{R}^d \times \{|\omega| = 1\}$.

The components $(I - \pi_\nu)\mathbf{A}_{\nu,1}$ are determined from $\mathbf{A}_{\nu,0}$ by

$$(I-\pi_\nu)\,\mathbf{A}_{\nu,1}\;=\;-Q_\nu\,L(y,\partial_y)\,\mathbf{A}_{\nu,0}\,,$$

where $Q_\nu(y)$ is the partial inverse of $L(y,\lambda_\nu(y,\xi),\xi)|_{\xi=\nabla_x\phi(y)}$.

To achieve (5.5.6), the symbols $\mathbf{A}_{\nu,1}$ must satisfy

$$\sum_\nu \mathbf{A}_{\nu,1}\big|_{t=0} = 0\,.$$

Decomposing $\mathbf{A}_{\nu,1} = \pi_\nu \mathbf{A}_{\nu,1} + (I - \pi_\nu)\mathbf{A}_{\nu,1}$ yields

$$\Big(\sum_\nu \pi_\nu \mathbf{A}_{\nu,1} \;-\; \sum_\nu Q_\nu \; L \; \mathbf{A}_{\nu,0} \Big)_{t=0} \;=\; 0\,.$$

Multiplying by π_ν shows that this uniquely determines

$$\pi_\nu \mathbf{A}_{\nu,1}(0,x,\omega) \;=\; \sum_{\nu' \neq \nu} \pi_\nu \; Q_{\nu'} \; L \; \mathbf{A}_{\nu,0}(0,x,\omega)\,.$$

Then transport equations determine $\pi_\nu \, \mathbf{A}_{\nu,1}$ in $\{|t| \le 2\underline{T}\}$. Once these are known, algebraic equations determine $(I - \pi_\nu)\mathbf{A}_{\nu,2}$, and so on.

Choose

$$\mathbf{A}_\nu(\epsilon,y,\omega) \;\sim\; \sum_j \mathbf{A}_{\nu,j}(\epsilon,y,\omega)\, \epsilon^j \quad \text{in} \quad \{|\omega| = 1\}.$$

For $|\xi| > 1$ set $\epsilon = 1/|\xi|, \omega = \xi/|\xi|$ and define $\mathbf{A}_\nu(y,\xi) := \mathbf{A}_\nu(\epsilon,y,\omega)$ and $\mathbf{A}_{\nu,j}(y,\xi) := \mathbf{A}_{\nu,j}(\epsilon,y,\omega)$ to find

$$\mathbf{A}_\nu(y,\xi) \;\sim\; \sum_j \mathbf{A}_{\nu,j}(y,\omega)\, |\xi|^{-j} \quad \text{in} \quad \{|\xi| \ge 1\}\,.$$

Since the π_ν and $\mathbf{A}_\nu$ are bounded on $[-2T, 2T] \times \mathbb{R}^d$, the integral $d\xi$ in (5.5.7) is absolutely convergent as soon as $\hat f \in L^1(\mathbb{R}^d)$. The formula

$$u_\nu(t,x) \;:=\; \int \mathbf{A}_\nu(t,x,\xi)\; e^{i(\phi_\nu(t,0,x,\xi) - w\xi)}\; \chi(\xi)\; f(w)\; dw\, d\xi\,, \tag{5.5.8}$$

for u_ν in terms of f is not absolutely convergent. The linear map $f \;\mapsto\; u$ has expression $\int K(t,x,w)\, f(w)\, dw$ with distribution kernel $K(t,x,w)$ given by

$$K(t,x,w) \;:=\; \int \mathbf{A}_\nu(t,x,\xi)\; e^{i(\phi_\nu(t,0,x,\xi) - w\xi)}\; \chi(\xi)\; d\xi\,. \tag{5.5.9}$$

This oscillatory integral has integrand that is not L^1. The next section introduces the technique used to analyze expressions such as (5.5.8) and (5.5.9).

5.5.2. Oscillatory integrals and Fourier integral operators. In this section, the method of nonstationary phase, introduced in §1.3, is used to study oscillatory integrals. We present the key definitions and two fundamental results. For a more complete treatment of Fourier integral operators, see [Duistermaat, 1973] and [Hörmander, 1971]. The two theorems show that (5.5.8) and (5.5.9) define well defined distribution for any $f \in \mathcal{E}'(\mathbb{R}^d)$ and compute their wavefront sets in terms of the wavefront set of f.

The method of oscillatory integrals is a generalization of the definition of the Fourier transform of distributions. For example, the expression

$$\delta(x) \;=\; (2\pi)^{-d} \int e^{-ix\xi}\, d\xi$$

is a familiar oscillatory integral. The interpretation is that for $\gamma \in C_0^\infty(\mathbb{R}^d)$ that is equal to one on a neighborhood of the origin,

$$\lim_{\epsilon \to 0}\ (2\pi)^{-d} \int \gamma(\epsilon\xi)\, e^{-ix\xi}\, d\xi \;=\; \delta(x)\,,$$

in the sense of distributions. Equivalently, for test functions $\psi(x)$,

$$\lim_{\epsilon \to 0}\ (2\pi)^{-d} \int \psi(x)\, \gamma(\epsilon\xi)\, e^{-ix\xi}\, d\xi\, dx \;=\; \psi(0)\,.$$

In (5.5.8) one can fix t in which case the output $u_\mu(t)$ is a distribution on $\mathbb{R}^d_x$ or one can leave t as a variable in which case the output in a distribution on $\mathbb{R}^{1+d}_{t,x}$. In the first case the kernel is a distribution in x, w depending on a parameter t while in the second case the kernel is a distribution in t, x, w. In (5.5.8) the phase, amplitude, and oscillatory integral are

$$\phi(y,\xi) \;:=\; \phi_\mu(t,0,x,\xi) - w\xi\,, \quad a(y,\xi) = \mathbf{U}_\mu\, \pi_\mu\, \chi\,, \quad \int a(y,\xi)\, e^{i\phi(y,\xi)}\, d\xi\,.$$

The phase is homogeneous of degree 1 in ξ. Showing that u_μ defines a well defined distribution requires showing that the expression

$$\int \psi(y)\, a(y,\xi)\, e^{i\phi(y,\xi)}\, d\xi\, dy$$

is a continuous linear function of the test function ψ. As in the case of the δ function the analysis is by cutting off in ξ and analysing the limit.

A typical symbol a behaves like a function homogeneous in ξ perhaps of positive degree. The integrals are not absolutely convergent. The integral is finite only because of cancellations. Conditionally convergent integrals are often dangerous for analysis. The theory of oscillatory integrals gives a notable exception.

Definition. *If Ω is an open subset of $\mathbb{R}^M$, then the symbol class $S^m(\Omega\times\mathbb{R}^N)$ consists of functions $a(y,\xi) \in C^\infty(\Omega \times \mathbb{R}^N)$ such that*

$$\forall \omega \Subset \Omega,\ \ \forall (\alpha,\beta) \in \mathbb{N}^{M+N},\ \ \exists C,\ \ \forall (y,\xi) \in \omega \times \mathbb{R}^N,$$
$$\big|\partial_y^\alpha\, \partial_\xi^\beta\, a(y,\xi)\big| \;\le\; C\, \langle\xi\rangle^{m-|\beta|}\,.$$

Examples 5.5.1. 1. Functions a that are everywhere smooth and also positively homogeneous of degree m in ξ for $|\xi| \ge 1$.

2. Finite sums of functions as in **1** with degrees of homogeneity less than or equal to m.

3. The function $\mathbf{A}_\mu(y,\xi)\,\chi(\xi)$ with

$$\mathbf{A}_\mu \;\sim\; \sum_j |\xi|^{-j}\,\mathbf{A}_{\mu,j}(y,\xi/|\xi|)$$

belongs to $S^0(([0,T]\times\mathbb{R}^d)\times\mathbb{R}^d)$ with symbol estimates $|\partial_y^\alpha\,\partial_\xi^\beta\,a| \le C\langle\xi\rangle^{-|\beta|}$ uniform on $y\in[0,T]\times\mathbb{R}^d$.

4. The function $\mathbf{A}_\mu$ is an asymptotic sum of homogeneous terms. Such symbols are called *polyhomogeneous* and are the most common examples. To avoid singularities at $\xi=0$, they must be multiplied by a cutoff like $\chi(\xi)$ to place them in the classes S^m. For a polyhomogeneous symbol

$$a(y,\xi) \;\sim\; \sum_{j\le m} a_j(y,\xi)$$

with a_j homogeneous of degree j, one has $\chi(\xi)\,a\in S^m$.

Exercise 5.5.1. *Prove the assertions in the example.*

The results of the next exercise are used in defining oscillatory integrals by passing to the limit in symbols compactly supported in ξ.

Exercise 5.5.2. *Prove the following assertions.* **i.** *S^m is a Fréchet space.* **ii.** *If $\gamma\in C_0^\infty(\mathbb{R}^d_\xi)$ and $\gamma=1$ on a neighborhood of $\xi=0$, then the symbols $\gamma(\epsilon\xi)\,a$, $0<\epsilon\le 1$ are bounded in S^m.* **iii.** *For any $\mu<m$, $\gamma(\epsilon\xi)\,a$ converges to a in S^μ as $\epsilon\to 0$.*

Definition. *If Ω is an open subset of $\mathbb{R}^M$, then a real valued $\phi(y,\xi)\in C^\infty(\Omega\times(\mathbb{R}^N\setminus 0))$ is a* **nondegenerate phase function** *if it is positive homogeneous of degree one in ξ and has nowhere vanishing gradient $\nabla_{y,\xi}\phi$.*

Examples 5.5.2. The phases $-x\xi$ and the phase in the Lax parametrix (5.5.8).

Proposition 5.5.1. *If ϕ is a nondegenerate phase function, then the map*

$$C_0^\infty(\Omega)\ni\psi \quad\mapsto\quad \int \psi(y)\;a(y,\xi)\;e^{i\phi(y,\xi)}\;d\xi\,dy$$

is well defined for smooth a with compact support in ξ. It extends uniquely by continuity to $a\in S^m(\Omega\times\mathbb{R}^N)$ for any m. For such symbols, it defines a distribution of order $k\in\mathbb{N}$ provided $k<-m-N$.

Proof. For a smooth $\chi(\xi)$ vanishing on a neighborhood of $\xi=0$ and identically equal to one outside a compact set, write $a=\chi a+(1-\chi)a$. The second term is compactly supported and the associated map defines a distribution of order $-\infty$. So, it suffices to construct the extension for symbols that vanish on a neighborhood of $\xi=0$. That is done as follows.

Since ϕ is nondegenerate, we can introduce the first order differential operator in $\xi \neq 0$,

$$\mathcal{L} := \frac{-i}{|\nabla_y\phi|^2 + |\xi|^2\,|\nabla_\xi\phi|^2}\Big(\nabla_y\phi\partial_y + |\xi|^2\nabla_\xi\phi\partial_\xi\Big), \quad \text{so that} \quad \mathcal{L}e^{i\phi} = e^{i\phi}.$$

The denominator is nonvanishing and homogeneous of degree 2 in ξ. Therefore the coefficient of ∂_y (resp., ∂_ξ) is homogeneous of degree -1 (resp., 0) in ξ. Away from $\xi = 0$, the coefficient of ∂_y belongs to S^{-1} and the coefficient of ∂_ξ belongs to S^0.

If a is compactly supported and vanishes on a neighborhood of $\xi = 0$, one has for any k

$$\int \psi(y)\; a(y,\xi)\; e^{i\phi(y,\xi)}\; d\xi\, dy \;=\; \int \psi(y)\; a(y,\xi)\; \mathcal{L}^k\, e^{i\phi(y,\xi)}\; d\xi\, dy\,.$$

Denote by $\mathcal{L}^\dagger$ the transposed differential operator. Integrating by parts yields

$$\int \psi(y)\; a(y,\xi)\; e^{i\phi(y,\xi)}\; d\xi\, dy \;=\; \int e^{i\phi(y,\xi)}\; (\mathcal{L}^\dagger)^k\big(\psi(y)\,a(y,\xi)\big)\; d\xi\, dy\,.$$

Since $\mathcal{L} = S^{-1}\partial_x + S^0\partial_\xi$ one has away from $\xi = 0$,

$$\mathcal{L}^\dagger \;=\; S^{-1}\partial_x \;+\; S^0\partial_\xi \;+\; (\partial_x S^{-1}) \;+\; (\partial_\xi S^0) \;=\; S^{-1}\partial_x \;+\; S^0\partial_\xi \;+\; S^{-1}\,.$$

Each summand maps S^r to S^{r-1}. For $a \in S^m$ vanishing on a neighborhood of $\xi = 0$, it follows that $(\mathcal{L}^\dagger)^k(\psi\, a) \in S^{m-k}$. The integrand in $\int e^{i\phi}(\mathcal{L}^\dagger)^k(\psi\, a) d\xi dy$ is $O(\langle\xi\rangle^{m-k})$. When k is so large that $m - k < -N$, it is absolutely integrable. The integral for those values of k yields the extension by continuity for symbols $a \in S^m$ vanishing on a neighborhood of $\xi = 0$.

For those k, the integral is bounded by $C(\omega, k)\|\psi\|_{C^k}$ proving that the distribution has order k. □

Exercise 5.5.3. *Use Proposition 5.5.1 to estimate the order of* $\delta(x) = (2\pi)^{-d}\int e^{-ix\xi}\, d\xi$.

Remarks. 1. If $\gamma(\xi) \in C_0^\infty(\mathbb{R}^N)$ is identically equal to one on a neighborhood of $\xi = 0$, Proposition 5.5.1 shows that

$$\lim_{\epsilon\to 0} \int \gamma(\epsilon\xi)\; e^{i\phi(y,\xi)}\; a(y,\xi)\; d\xi$$

exists in the topology of distributions of order k. Following the lead of Hörmander and the tradition of the Fourier transform, the oscillatory integral is simply written as $\int e^{i\phi}\, a\, d\xi$ as if it were an integral.

2. In case m is very negative, the formula yields negative values of k. This corresponds to $u \in C^{-k} = C^{|k|}$. That conclusion is correct and is justified by direct differentiation under the integral sign defining the oscillatory integral. In particular, if $a \in \bigcap_m S^m$, then the distribution is a C^∞ function. Therefore, if $a \sim 0$ is polyhomogeneous, the associated distribution is smooth.

3. The proposition allows one to manipulate oscillatory integrals as if they were integrals. For example, the integration by parts formula $\int \psi a e^{i\phi} d\xi dy = \int e^{i\phi} (\mathcal{L}^\dagger)^k (\psi a) d\xi dy$ is true for all $a \in S^m$, since the two sides are continuous and are equal on symbols with compact support.

4. If the test function ψ and symbol a depend in a continuous fashion on a parameter, then the associated oscillatory integral also depends continuously on the parameter since it is the uniform limit of the cutoff oscillatory integrals. Similarly, one justifies differentiation under the (oscillatory) integral sign.

Exercise 5.5.4. *Prove* **2.**

The next definition uses the notion of the *distribution kernel of a linear map*. Formally, the operator with distribution kernel $A(y_1, y_2) \in \mathcal{D}'(\Omega_1 \times \Omega_2)$ is given by

$$(Au)(y_2) = \int A(y_1, y_2)\, u(y_1)\, dy_1 .$$

The precise version is for $u \in C_0^\infty(\Omega_1)$ and test function $\zeta(y_2) \in C_0^\infty(\Omega_2)$,

$$\langle \zeta(y_2)\, ,\, Au\rangle := \langle \zeta(y_2)\, u(y_1)\, ,\, A\rangle .$$

Any distribution A defines a continuous linear map $A : C_0^\infty(\Omega_1) \to \mathcal{D}'(\Omega_2)$.[2]

Definition. *If $\Omega_j \in \mathbb{R}^{N_j}$ are open subsets, a continuous linear map $A : C_0^\infty(\Omega_1) \to \mathcal{D}'(\Omega_2)$ is a* **Fourier integral operator** *when it is given by a kernel, $A \in \mathcal{D}'(\Omega_1 \times \Omega_2)$, defined by on oscillatory integral. Equivalently, the operator is given by a formula*

$$Au(y_2) = \int a(y_1, y_2, \xi)\; e^{i\phi(y_1, y_2, \xi)}\; u(y_1)\; d\xi\, dy_1 , \tag{5.5.10}$$

with amplitude $a(y, \xi) \in S^m$ for some m and nondegenerate phase $\phi(y, \xi)$.

Examples 5.5.3. 1. With $\Omega_1 = \Omega_2$ and $D = \frac{1}{i}\partial$, the identity map and the differential operator $a_\alpha(y) D^\alpha$ have kernels

$$\delta(y_2 - y_1) = (2\pi)^{-d} \int e^{i(y_2 - y_1).\xi}\, d\xi , \qquad (2\pi)^{-d} \int a_\alpha(y_2)\, \xi^\alpha e^{i(y_2 - y_1).\xi}\, d\xi ,$$

with $\phi = (y_2 - y_1)\xi$. For a general differential operator, $a(y, D)$, one inserts the the symbol of $a(y, \xi)$. Pseudodifferential operators, about which we say

[2]The converse, called the Schwartz kernel theorem, is true and is not needed below.

very little, are also Fourier integral operators with this same phase. The pseudodifferential operator with symbol $a(y,\xi) \in S^m$ has kernel

$$(2\pi)^{-d} \int a(y_2,\xi)\ e^{i(y_2-y_1)\xi}\ d\xi,$$

so,

$$a(y,D)u \ = \ \int a(y,\xi)\ e^{iy\xi}\ \hat{u}(\xi)\ d\xi.$$

2. The operator appearing in the Lax parametrix has the special structure

$$\int a(y_2,\xi)\ e^{i(\phi(y_2,\xi)-y_1\xi)}\ u(y_1)\ dy_1\, d\xi\,, \qquad \nabla_{y_2,\xi}\phi \neq 0\,. \tag{5.5.11}$$

The phase is $\Phi(y_1,y_2,\xi) := \phi(y_2,\xi) - y_1\xi$.

Proposition 5.5.2. Suppose that A is a Fourier integral operator with phase ϕ and amplitude a.

i. *If $\nabla_{y_1,\xi}\phi$ is nowhere zero on the $\{\xi \neq 0\} \cap \operatorname{supp} a$, then A maps $C_0^\infty(\Omega_1) \to C^\infty(\Omega_2)$.*

ii. *If $\nabla_{y_2,\xi}\phi$ is nowhere zero on $\{\xi \neq 0\} \cap \operatorname{supp} a$, then A extends uniquely to a continuous linear map $\mathcal{E}'(\Omega_1) \to \mathcal{D}'(\Omega_2)$.*

Examples 5.5.4. All of the preceding examples satisfy Proposition 5.5.2.i and ii.

Proof of Proposition 5.5.2. i. One has

$$\langle Au\,,\,\psi\rangle \ = \ \int \psi(y_2)\ u(y_1)\ e^{i\phi(y_1,y_2,\xi)}\ a(y_1,y_2,\xi)\ d\xi\, dy_1\, dy_2. \tag{5.5.12}$$

The hypothesis **i** implies that the $d\xi\, dy_1$ integral is an oscillatory integral. The nondegenerate phase and the amplitude depend smoothly on y_2. Remark **4** shows that

$$\int u(y_1)\ e^{i\phi(y_1,y_2,\xi)}\ a(y_1,y_2,\xi)\ d\xi\, dy_1 \ \in\ C^\infty(\Omega_2)\,.$$

That this smooth function is equal to Au follows by $\gamma(\epsilon\xi)$ truncation and passage to the limit.

ii. To show that the operator extends to a continuous linear map from $\mathcal{E}'(\Omega_1) \to \mathcal{D}'(\Omega_2)$, one must show that

$$\int \psi(y_2)\ a(y_1,y_2,\xi)\ e^{i\phi(y_1,y_2,\xi)}\ u(y_1)\ d\xi\, dy_1\, dy_2$$

is a continuous functional of $\psi \in C_0^\infty(\Omega_2)$ and $u \in \mathcal{E}'(\Omega_1)$. This is so if (and only if) for every test function $\psi \in C_0^\infty(\Omega_2)$ one has

$$\int \psi(y_2)\ a(y_1,y_2,\xi)\ e^{i\phi(y_1,y_2,\xi)}\ d\xi\, dy_2 \ \in\ C^\infty(\Omega_1)\,. \tag{5.5.13}$$

Hypothesis **ii** guarantees that the phase in (5.5.13) is nondegenerate for the $d\xi\, dy_2$ integration. This implies (5.5.13) and **ii** follows. □

The next result estimates the action on wavefront sets of the class of Fourier integral operators (5.5.11) that arise in the Lax construction. The idea of treating only this class follows [Taylor, 1981]. For more general results see [Hörmander, 1971], [Duistermaat, 1973], and [Gabor, 1972].

Proposition 5.5.3. *For a Fourier integral operator of the form (5.5.11) with phase $\phi(y_2, \xi) - y_1\xi$, $\nabla_{y_2,\xi}\phi \neq 0$, and, $u \in \mathcal{E}'(\Omega_1)$,*
(5.5.14)
$$WF(Au) \subset \Big\{(y_2, \nabla_{y_2}\phi(y_2,\xi)) \,:\, \exists (y_2,\xi) \in \operatorname{supp} a,\ \big(\nabla_\xi \phi(y_2,\xi),\, \xi\big) \in WF\,u\Big\}.$$

Examples 5.5.5. 1. For t fixed, consider solutions of the wave equation that are superpositions of the plane waves $e^{i(-|\xi|t + x\xi)} := e^{i\phi(t,x,\xi)}$,

$$\int e^{i(\phi(t,x,\xi) - w\xi)}\, \chi(\xi)\, f(w)\, dw\, d\xi\,, \qquad \phi(t,x,\xi) \;=\; -t|\xi| + x\xi\,.$$

This has form (5.5.11). The variable w plays the role of y_1 and x the role of y_2. We consider t as a parameter. One has

$$\phi_\xi = x - t\frac{\xi}{|\xi|} \qquad \text{and} \qquad \phi_x = \xi\,.$$

A point $(x - t\xi/|\xi|, \xi) \in WF\,f$ may produce a point (x,ξ) in $WFu(t)$. The frequency ξ is unchanged. The position has moved t units in the direction $\xi/|\xi|$. Singularities associated with the plane waves $e^{i(-t|\xi|+x\xi)}$ propagate at the group velocity $\xi/|\xi|$. The opposite sign choice, $+t|\xi$, in the phase yields the velocity $-\xi/|\xi|$.

2. The identity map and pseudodifferential operators are Fourier integral operators with phase $(y_2 - y_1)\xi$. The proposition shows that for this phase, $WF(Au) \subset WF(u)$. This is the *pseudolocal property* of pseudodifferential operators.

Remark. Statement (5.5.14) asserts that $WF(Au)$ is contained in the set of (y_2, η_2) such that there is a point $(y,\xi) \in WF(u)$ so that
(5.5.15)
$$(y_2,\xi) \in \operatorname{supp} a, \qquad \eta_2 = \nabla_{y_2}\phi(y_2,\xi), \qquad \text{and} \qquad y = \nabla_\xi\phi(y_2,\xi)\,.$$
The last two equations assert that $\nabla_{y_2,\xi}\big(\phi(y_2,\xi) - y_2\eta_2 - y_1\xi\big) = 0$.

Proof of Proposition 5.5.3. It suffices to show that if $(\underline{y}_2, \underline{\eta}_2)$ is not in the set on the right of (5.5.14), then it does not belong to $WF(Au)$. To show that $(\underline{y}_2, \underline{\eta}_2) \notin WF(Au)$, it suffices to show that for $\zeta \in C^\infty$ supported near $\underline{y}_2$ and η_2 in a small conic neighborhood of $\underline{\eta}_2$,

$$\int \zeta(y_2)\, e^{-iy_2\eta_2}\, a(y_2,\xi)\, e^{i\phi(y_2,\xi)}\, e^{-iy_1\xi}\, u(y_1)\, dy_1\, d\xi\, dy_2$$

is rapidly decreasing as $\eta_2 \to \infty$.

Choose a finite smooth partition of unity $\{\psi_\nu(y_1)\}$ of $\operatorname{supp} u$, and $\theta_\mu(\xi)$ of $\mathbb{R}^N \setminus 0$ with θ_μ homogeneous of degree zero in ξ. It suffices to consider individual summands,

(5.5.16)

$$\int \zeta(y_2)\ e^{-iy_2\eta_2}\ a(y_2,\xi)\ e^{i\phi(y_2,\xi)}\ e^{-iy_1\xi}\ \psi_\nu(y_1)\ \theta_\mu(\xi)\ u(y_1)\ dy_1\, d\xi\, dy_2\,.$$

If the support of $\psi_\nu\,\theta_\mu$ does not meet $WF\,u$, then Exercise 4.6.3 implies that $\mathbb{R}^{N_1} \times \operatorname{supp}\theta_\mu$ does not meet $WF(\psi_\nu u)$, so

$$b(\xi)\ :=\ \int\ e^{-iy_1\xi}\ \psi_\nu(y_1)\ \theta_\mu(\xi)\ u(y_1)\ dy_1$$

is rapidly decreasing. Then differentiation under the integral shows that

$$g(y_2)\ :=\ \int\ a(y_2,\xi)\ b(\xi)\ e^{-i\phi}\ d\xi\ \in\ C^\infty(\Omega_2)\,.$$

Then

$$(5.5.16)\ =\ \int \zeta(y_2)\ g(y_2)\ e^{-iy_2\eta_2}\ dy_2\ =\ \mathcal{F}\big(\zeta(y_2)g(y_2)\big)\ \in\ \mathcal{S}(\mathbb{R}^{N_2}_{\eta_2})$$

is rapidly decreasing.

Therefore, it suffices to consider μ,ν so that the support of $\psi_\nu(y_1)\,\theta_\mu(\xi)$ belongs to a small conic neighborhood of a point $(\underline{y}_1,\underline{\xi}) \in WFu$.

The remark after the proposition shows that when $(\underline{y}_2,\underline{\eta}_2)$ is not in the set on the right of (5.5.14), the integral (5.5.16) for the important μ,ν has phase satisfying

$$\nabla_{y_2,\xi}\Phi(y_1,y_2,\xi,\eta)\ :=\ \nabla_{y_2,\xi}\Big(\phi(y_2,\xi) - y_2\eta_2 - y_1\xi\Big)\ \neq\ 0$$

on the support of the integrand of

$$\int \zeta(y_2)\ a(y_2,\xi)\ e^{i\Phi}\ \psi_\nu(y_1)\ \theta_\mu(\xi)\ d\xi\, dy_2\, dy_1\,.$$

By homogeneity,

$$|\nabla_{y_2}\Phi|^2\ +\ |\xi|^2\,|\nabla_\xi\Phi|^2\ \geq\ C\,(|\xi|^2 + |\eta|^2) \tag{5.5.17}$$

in the support of the integrand. Introduce

$$\mathcal{L}\ :=\ \frac{-i}{|\nabla_{y_2}\Phi|^2\ +\ |\xi|^2\,|\nabla_\xi\Phi|^2}\ \Big(\nabla_{y_2}\Phi.\partial_{y_2} + |\xi|^2\nabla_\xi\Phi.\partial_\xi\Big)\,,\quad \text{so}\quad \mathcal{L}e^{i\Phi} = e^{i\Phi}\,.$$

Then

$$\int \zeta(y_2)\ a(y_2,\xi)\ e^{i\Phi}\ \psi_\nu(y_1)\ \theta_\mu(\xi)\ d\xi\, dy_2\, dy_1$$

$$=\ \int e^{i\Phi}\ (\mathcal{L}^\dagger)^k\big[\zeta(y_2)\ a(y_2,\xi)\ \psi_\nu(y_1)\ \theta_\mu(\xi)\big]\ d\xi\, dy_2\, dy_1$$

since the identity is true for compactly supported a and so extends by continuity for arbitrary a.

Use $a \in S^m$ and (5.5.17) to see that each application of $\mathcal{L}^\dagger$ gains a factor of $(|\xi|^2 + |\eta|^2)^{-1/2}$. Only one term requires a remark, namely

$$\frac{O(|\xi|^2)\ \partial_\xi S^\mu}{O(|\xi|^2 + |\eta|^2)} \sim \frac{O(|\xi|^2)\ O(1/|\xi|)\ S^\mu}{O(|\xi|^2 + |\eta|^2)} \leq \frac{S^\mu}{O((|\xi|^2 + |\eta|^2)^{1/2})}\,.$$

Choosing k large, it follows that the integral on the right is no larger than

$$C_k \int_{|y_1, y_2| \leq R} \int_{|\xi| \geq r} \frac{|\xi|^m}{(|\xi|^2 + |\eta|^2)^{k/2}}\ d\xi\, dy_2\, dy_1\,,$$

implying the desired rapid decrease. □

5.5.3. Small time propagation of singularities.

Small Time Propagation of Singularities Theorem 5.5.4 (Lax and Hörmander). *There is a $0 < T_1 \leq \underline{T}$ so that $u_{\rm approx} = \sum u_\mu$ from (5.5.7) satisfies the following:*

i. *If $f \in \bigcup_s H^s(\mathbb{R}^d)$ and u the solution of (5.5.1), then $u - u_{\rm approx} \in C^\infty([0, T_1] \times \mathbb{R}^d)$.*

ii. *For $t \in [0, T_1]$ the wavefront set of $u_\mu(t)$ in $T^*(\mathbb{R}^d_x)$ is the image of the wavefront set of $u_\mu(0)$ by the symplectic map on $T^*(\mathbb{R}^d_x)$ that is the time t flow of the hamiltonian field with time dependent hamiltonian $-\lambda_\mu(t, x, \xi)$.*

iii. *The wavefront set of u_μ in $T^*(\mathbb{R}^{1+d})$ is contained in the set $\tau = \lambda_\mu(t, x, \xi)$ and is invariant under the hamiltonian flow of $\tau - \lambda_\mu(t, x, \xi)$. $WF\, u_\mu$ contains only integral curves that pass over the wavefront set of f.*

Remarks. 1. Reversing time $t' := -t$ yields the analogous result on $-T_1 \leq t \leq 0$.

2. If one had started at a time T_0, one would obtain an analogous result on $|t - T_0| \leq T_1$. It is important to note that one can find T_1 independent of T_0. The interval is determined by the fact that the eikonal equation is solvable and the maps C_μ stay close to the identity. These intervals cannot shrink because of the uniform bounds on the derivatives of the coefficients of L together with the uniform smooth variety hypothesis at the beginning of §5.5. If one were making such hypotheses only locally, one would obtain a T_1 uniformly bounded on compact sets.

Proof. i. It suffices to verify that

$$L(u - u_{\rm approx}) \in C^\infty\big([0, \underline{T}] \times \mathbb{R}^d\big) \qquad \text{and} \qquad u\big|_{t=0} - u_{\rm approx}\big|_{t=0} \in C^\infty(\mathbb{R}^d).$$

These are consequences of (5.5.6). We show how the first part of (5.5.6) yields the first of the two conclusions.

Since $Lu = 0$, it is sufficient to show that $Lu_\mu \in C^\infty([0,\underline{T}] \times \mathbb{R}^d)$ for each μ. Differentiating under the oscillatory integral sign in (5.5.8) shows that

$$Lu_\mu := \int L\big(\mathbf{A}_\mu(t,x,\xi)\ e^{i(\phi_\mu(t,0,x,\xi)-w\xi)}\big)\ \chi(\xi)\ f(w)\ dw\, d\xi\,.$$

This identity would be true if $\mathbf{A}_\mu$ had compact support so extends to general $\mathbf{A}_\mu$ by continuity.

By construction,

$$L\big(\mathbf{A}_\mu(t,x,\xi)\ e^{i(\phi_\mu(t,0,x,\xi))}\big) = b(t,x,\xi)\, e^{i(\phi_\mu(t,0,x,\xi)},$$
$$b \in S^{-\infty}(([0,\underline{T}] \times \mathbb{R}^d_x) \times \mathbb{R}^d_\xi).$$

Therefore,

$$Lu_\mu = \int b(t,x,\xi)\ e^{i(\phi_\mu(t,0,x,\xi)-w\xi)}\ \chi(\xi)\ f(w)\ dw\, d\xi\,.$$

Since the amplitude b is of order $-\infty$, this oscillatory integral is smooth by Proposition 5.5.1. □

Exercise 5.5.5. *Prove that* $u(0,x) - u_{\text{approx}}(0,x) \in C^\infty(\mathbb{R}^d)$.

To prove Theorem 5.5.4.ii, analyze the oscillatory integral defining $u_\mu(t)$,

$$\int \mathbf{A}_\mu(t,x,\xi)\ e^{i[\phi_\mu(t,0,x,\xi)-w\xi]}\ \chi(\xi)\ f(w)\ dw\, d\xi\,. \tag{5.5.18}$$

For ease of reading, omit the subscript μ and the $s = 0$ argument from ϕ. The proposition estimating $WF(Au)$ implies that

$$WF\,u(t) \subset \Big\{\big(x, \nabla_x\phi(t,x,\xi)\big)\ :\ \big(\nabla_\xi\phi(t,x,\xi),\xi\big) \in WF\,u_\mu(0)\Big\}.$$

The mapping transforming the initial wavefront to that at time t is given by

$$\big(\nabla_\xi\phi(t,x,\xi)\,,\,\xi\big) \ \mapsto\ \big(x\,,\,\nabla_x\phi(t,x,\xi)\big)\,. \tag{5.5.19}$$

At $t = 0$, $\phi(0,x,\xi) = x\xi$. Thus at $t = 0$, the transformation is equal to the identity. It follows from uniform smoothness and the implicit function theorem that there is a $\underline{T}_1 > 0$ so that for $0 \le t \le \underline{T}_1$ both of the maps,

$$(x,\xi) \mapsto (x, \nabla_\xi\phi(t,x,\xi)) \qquad \text{and} \qquad (x,\xi) \mapsto (\nabla_x\phi(t,x,\xi),\xi)\,,$$

are invertible maps close to the identity. It follows that for the same t, (5.5.19) defines a diffeomorphism of $T^*(\mathbb{R}^d)$ to itself that is close to the identity. It is called the transformation with generating function ϕ and the next lemma is a classical result in mechanics.

Lemma 5.5.5. *Denote by* $C_\mu(t)$ *the flow on* $T^*(\mathbb{R}^d\times)\setminus 0$ *of the time dependent Hamilton field with hamiltonian* $-\lambda_\mu(t,x,\xi)$. *Then for* $t\in[0,T_1]$, $C_\mu(t)$ *is equal to the diffeomorphism defined by (5.5.19).*

Proof of Lemma 5.5.5. Suppress the subscripts μ. Denote by $(x(t),\xi(t))$ the curve traced by the diffeomorphism (5.5.19) so that $(x(0),\xi(0))=(x_0,\xi_0)$. The formula (5.5.19) means that to compute $(x(t),\xi(t))$ one must find a pair (x,ξ) so that

$$\big(\nabla_\xi\phi(t,x,\xi)\,,\,\xi\big)\;=\;(x_0,\xi_0)\,,\quad\text{then}\quad\big(x\,,\,\nabla_x\phi(t,x,\xi)\big)\;=\;(x(t),\xi(t))\,.$$

Equivalently, $\big(x(t),\xi(t)\big)$ is determined by

$$x_0\;=\;\nabla_\xi\phi(t,x(t),\xi_0),\qquad \xi(t)\;=\;\nabla_x\phi(t,x(t),\xi_0)\,. \tag{5.5.20}$$

It suffices to show that

$$x'=\nabla_\xi(-\lambda(t,x(t),\xi(t)))\,,\qquad \xi'=-\nabla_x(-\lambda(t,x(t),\xi(t)))\,. \tag{5.5.21}$$

Determine x' by differentiating the first equation in (5.5.20) with respect to t to find

$$0\;=\;\nabla_\xi\,\phi_t+\nabla^2_{x\xi}\phi\big|_{(t,x(t),\xi_0)}\,x'\,. \tag{5.5.22}$$

Differentiating the eikonal equation $\phi_t(t,x,\xi)=\lambda(y,\nabla_x\phi(t,x,\xi))$ with respect to ξ and evaluating at $(t,x(t),\xi_0)$ yields

$$0\;=\;\nabla_\xi\,\phi_t\;-\;\nabla_\xi\lambda(t,x(t),\xi(t))\,\nabla^2_{\xi x}\phi\big|_{(t,x(t),\xi_0)}\,. \tag{5.5.23}$$

The linear equation (5.5.22) satisfied by $x'(t)$ is identical to the equation (5.5.23) satisfied by $-\nabla_\xi\lambda(t,x(t),\xi(t))$. By the choice of T_1, the matrices $\nabla^2_{x\xi}\phi$ are invertible and close to the identity. Therefore, using (5.5.20),

$$x'\;=\;-\nabla_\xi\lambda(t,x(t),\xi(t))\,,$$

verifying half of (5.5.21).

Determine ξ' by differentiating the second equation in (5.5.20) with respect to time to find

$$\xi'\;=\;\nabla_x\phi_t\;+\;\nabla^2_{xx}\phi\;x'\;=\;\nabla_x\phi_t\;-\;\nabla^2_{xx}\phi\,\nabla_\xi\lambda\,. \tag{5.5.24}$$

Differentiating $\phi_t=\lambda(y,\nabla_x\phi)$ with respect to x yields

$$\nabla_x\phi_t\;=\;\nabla_\xi\lambda\,\nabla^2_{xx}\phi+\nabla_x\lambda\,. \tag{5.5.25}$$

Equations (5.5.24), (5.5.25) imply that $\xi'=-\nabla_x\lambda$, completing the proof of the lemma. □

It follows that

$$WF(u_\mu(t))\;\subset\;C_\mu(t)WF(u_\mu(0))\,.$$

Using this same conclusion for the Lax parametrix for the Cauchy problem with initial time at t shows that

$$WF(u_\mu(0)) \subset C_\mu(-t)WF(u_\mu(t)), \qquad \text{equivalently} \quad C_\mu(t)\,WF(u_\mu(0)) \subset WF(u_\mu(t)).$$

Combining implies that

$$WF(u_\mu(t)) = C_\mu(t)WF(u_\mu(0)),$$

completing the proof of Theorem 5.5.4.ii.

The proof of Theorem 5.5.4.iii is similar. The result on $WF(Au)$ shows that the wavefront set of Au is a set of points $\big(t, x, \phi_t(t,x), \nabla_x\phi(t,x)\big)$. Since $\phi_t = \lambda_\mu(t, x, \nabla_x\phi)$, it follows that the wavefront set is a subset of $\tau = \lambda_\mu(t,x,\xi)$.

The formula for $WF(Au)$ together with the formula from Theorem 5.5.4.ii shows that

$$(5.5.26)\quad WF_{\mathbb{R}^{1+d}}\, u_\mu(t,x) \subset \Big\{\big(t, x, \lambda_\mu(t,x,\xi), \xi\big) \;:\; (x,\xi) \in WF_{\mathbb{R}^d}\, u_\mu(t)\Big\}.$$

Next prove that there is equality in (5.5.26). If (t, x, λ_μ, ξ) on the right were not in the wavefront set, then for every real τ, (t,x,τ,ξ) would not be in the wavefront set. The limit point $(t,x,1,0)$ is also not in the wavefront set by the microlocal elliptic regularity theorem.

Using a finite covering by cones outside the wavefront set, it follows that there is a $\zeta \in C_0^\infty$ supported near (t,x) and nonvanishing at (t,x), and a conic neighborhood Γ of ξ so that

$$\forall n, \quad \exists C_n, \quad \forall \gamma \in \Gamma, \quad \big|\mathcal{F}_{\mathbb{R}^{1+d}}(\zeta u)\big| \le C_n \langle \tau, \gamma\rangle^{-n}.$$

The spatial Fourier transform is equal to

$$\mathcal{F}_{\mathbb{R}^d}\big(\zeta(t,\cdot)u(t)\big)(\gamma) = c\int \mathcal{F}_{\mathbb{R}^{1+d}}(\zeta u)(\tau,\gamma)\; e^{it\tau}\; d\tau,$$

so is rapidly decreasing. Therefore, $(x,\xi) \notin WF(u_\mu(t))$.

Hence,

$$(5.5.27)\quad WF_{\mathbb{R}^{1+d}}\, u(t,x) = \Big\{\big(t, x, \lambda_\mu(t,x,\xi), \xi\big) \;:\; (x,\xi) \in WF_{\mathbb{R}^d}\, u_\mu(t)\Big\}.$$

To prove that this is equivalent to Theorem 5.5.4.iii, reason as follows. For an integral curve $(x(s), \xi(s)) \in T^*(\mathbb{R}^d) \setminus 0$ of the Hamilton field with time dependent hamiltonian $-\lambda_\mu(t,x,\xi)$, there is a unique lift to an integral curve of the Hamilton field on $T^*(\mathbb{R}^{1+d}) \setminus 0$ with hamiltonian $\tau - \lambda_\mu(t,x,\xi)$ along which $\tau = \lambda_\mu$. The lift is given by

$$(t, x, \tau, \xi)(s) = \big(s, x(s), \lambda_\mu(s, x(s), \xi(s)), \xi(s)\big). \qquad \square$$

5.5.4. Global propagation of singularities. This subsection shows that the analysis restricted to $0 \le t \le T_1$ from §5.5.3 implies the analogous global in time result.

First show that the approximate solution $\sum_\mu u_\mu$ has a natural extension to all times, which is beyond the domain where the Lax parametrix construction applies. For each μ, choose a function $g_\mu \in C^\infty(\mathbb{R}^{1+d})$ so that

$$g_\mu \;=\; L\,u_\mu \quad \text{for} \quad 0 \le t \le T_1\,. \tag{5.5.28}$$

Extend u_μ beyond $0 \le t \le T_1$ by solving

$$L\,u_\mu \;=\; g_\mu\,, \qquad u_\mu\big|_{0<t<T_1} \;=\; \text{given } u_\mu\,. \tag{5.5.29}$$

If one makes a different choice, $\tilde g_\mu$, the resulting function $\tilde u_\mu$ satisfies $u_\mu - \tilde u_\mu \in C^\infty(\mathbb{R}^{1+d})$ since

$$L(u_\mu - \tilde u_\mu) \;\in\; C^\infty(\mathbb{R}^{1+d}) \qquad \text{and} \qquad u_\mu - \tilde u_\mu \;=\; 0 \;\text{ on }\;]0,T_1[\,\times\,\mathbb{R}^d\,.$$

Thus u_μ and $u_{\text{approx}} = \sum_\mu u_\mu$ are well defined on $\mathbb{R}^{1+d}$ modulo $C^\infty(\mathbb{R}^{1+d})$. One has

$$u - \sum u_\mu \;\in\; C^\infty(\mathbb{R}^{1+d})\,, \tag{5.5.30}$$

since $L(u - \sum u_\mu) \in C^\infty$ and $u - \sum u_\mu \in C^\infty(]0,T[\,\times\,\mathbb{R}^d)$. *The singularities of the extended u_μ determine those of u.*

Global Propagation of Singularities Theorem 5.5.6 (Ludwig and Hörmander). **i.** *The wavefront set of u_μ is invariant under the Hamilton flow of $\tau - \lambda_\mu(t,x,\xi)$. Precisely, a μ-bicharacteristic belongs to $WF\,u_\mu$ if and only if it does so at $\{t = 0\}$.*

ii. *If $0 \le t_1 < t_2$, then $WF\,u_\mu(t_2)$ is the image of $WF\,u_\mu(t_1)$ by the symplectic map that is the flow of the Hamilton field with time dependent hamiltonian $-\lambda_\mu(t,x,\xi)$.*

Proof. Denote by $\Gamma(t) = (t, \underline{x}(t), \underline{\xi}(t), \lambda_\mu(t,\underline{x}(t),\underline{\xi}(t)), \underline{\xi}(t))$ a μ-bicharacteristic. It suffices to show that $\Gamma \in WF\,u_\mu$ if and only if $\Gamma(0) \in WF\,u_\mu$. It suffices to prove this for times $t \le T$ for arbitrary $T \in\,]0,\infty[$.

Consider the set

$$\Big\{\underline{t} \in [0,T] \;:\; \Gamma(t) \in WF\,u_\mu \quad \text{for} \quad 0 \le t \le \underline{t}\Big\}\,.$$

By definition of the wavefront set, this is a closed set. It suffices to prove that this set is open since, once that is known, we know that the set is either empty or the entire interval. It is empty when $\Gamma(0) \notin WF\,u_\mu(0)$, and it is the entire interval when $\Gamma(0) \in WF\,u_\mu(0)$.

To prove that Γ is open, it suffices to show that if $\Gamma([0,\underline{t}]) \subset WF\,u_\mu$ and $\underline{t} < T$, then $\Gamma\big([0,\underline{t}+\delta[\big) \subset WF\,u_\mu$ for small positive δ. If $\underline{t} = 0$ this follows from Theorem 5.5.4.

If $0 < \underline{t} < T$, choose $0 < T_1$ from the local in time result and choose $0 < \delta \le T_1/2$ satisfying $\delta < \underline{t}$ and $\underline{t} + \delta \le T$. Define $0 < t' := \underline{t} - \delta$. The Lax parametrix construction yields a solution v_μ defined on $\{|t - t'| \le 2\delta\}$ to

$$Lv_\mu \in C^\infty([\underline{t} - 2\delta, \underline{t} + 2\delta] \times \mathbb{R}^d)\,, \qquad v_\mu(t') - u_\mu(t') \in C^\infty(\mathbb{R}^d)\,.$$

Then, $v_\mu - u_\mu \in C^\infty([\underline{t} - 2\delta, \underline{t} + 2\delta] \times \mathbb{R}^d)$.

The local in time result implies that the wavefront set of v_μ for $\underline{t} - 2\delta \le t \le \underline{t} + 2\delta$ is a union of bicharacteristics. We know that for $0 < t < \underline{t}$ it is a union of only μ-bicharacteristics so it follows that the wavefront set of v_μ is invariant under the $\tau - \lambda_\mu$ Hamilton flow.

For $t \in \,]t', \underline{t}[$, the wavefront set agrees with that of u_μ. It follows that the wavefront set of v_μ consists exactly of the continuations to $\underline{t} < t < t + 2\delta$ of the μ-bicharacteristics in $WF(u_\mu)$ for $t < \underline{t}$.

Since $u_\mu - v_\mu \in C^\infty$, this implies that $\Gamma\big([0, \underline{t} + \delta[\big) \subset WF\, u_\mu$. This completes the proof of Theorem 5.5.4.i.

Denote by $C(t_2, t_1)$ from $T^*(\mathbb{R}^d \backslash 0)$ to itself the flow from time t_1 to t_2 by the flow the Hamilton field with time dependent hamiltonian $-\lambda_\mu(t, x, \xi)$. The Lax parametrix construction shows that if $0 < t_1 < t_2 < t_1 + T_1$, then

$$WF\, u_\mu(t_2) \;=\; C(t_2, t_1)\, WF\, u_\mu(t_1)\,.$$

A finite number of applications of this result proves Theorem 5.5.4.ii. □

Exercise 5.5.6. *Denote by Γ_μ the bicharacteristics passing over $(0, \underline{x}, \underline{\xi})$. Prove that when $Lu \in C^\infty$,*

$$\Big(\bigcup_\mu \Gamma_\mu\Big) \cap WF(u) \;=\; \bigcup_\mu \big(\Gamma_\mu \cap WF(u_\mu)\big)\,.$$

Exercise 5.5.7. *Under the same hypotheses, prove that*

$$(\underline{x}, \underline{\xi}) \notin WF(u(0)) \quad \Longleftrightarrow \quad \Big(\bigcup_\mu \Gamma_\mu\Big) \cap WF(u) \;=\; \phi\,.$$

The first part of the next result restates Theorem 5.5.6 without reference to the decomposition as $\sum u_\mu$. The second part gives an H^s version.

Theorem 5.5.7. i. *If $Lu \in C^\infty(\mathbb{R}^{1+d})$, then $WF(u)$ is contained in the characteristic variety of L and is invariant under the Hamilton flow with hamiltonian equal to $\Pi_\mu\big(\tau - \lambda_\mu(t, x, \xi)\big)$.*

ii. *The same conclusion is valid with $WF(u)$ replaced by $WF_s(u)$.*

Remarks. 1. The hamiltonian field restricted to $\tau = \lambda_\mu(t, x, \xi)$ is parallel to the field with hamiltonian $\tau - \lambda_\mu(t, x, \xi)$.

2. It is the reduced hamiltonian and not $\det L(t, x, \tau, \xi)$ that appears on sheets in the characteristic variety with multiplicity greater than one. The hamiltonian vector field associated to $\det L$ vanishes at such points.

3. Theorems 5.5.6 and 5.5.7 are global in time. They are not restricted to domains where the eikonal equation is solvable. They go beyond caustics and focusing.

Example 5.5.6. Suppose that $\mathcal{O} \subset \mathbb{R}^d$ is a smoothly bounded open subset and $u(0,x)$ is a function smooth on $\overline{\mathcal{O}}$ and vanishing on $\mathbb{R}^d \setminus \overline{\mathcal{O}}$. The function and its derivatives may jump at the boundary. Exercise 4.6.9 shows that $WF(u(0))$ is contained in the conormal variety $N^*(\partial\mathcal{O})$. Therefore the $\mathbb{R}^{1+d}$ wavefront set of u_ν at time $t = 0$ is contained in the set of points $\big(0, x, \lambda_\nu(0,x,\xi), \xi\big)$ with (x,ξ) conormal to $\partial\mathcal{O}$. $WF(u_\nu)$ is then a subset of the flowout by the Hamilton flow of $\tau - \lambda_\nu$ of the points $(0, x, \lambda_\nu(0,x,\xi), \xi)$ with $(x,\xi) \in N^*(\partial\mathcal{O})$.

This same set appears in Hamilton–Jacobi theory as follows. Choose a function g smooth on a neighborhood of $\partial\mathcal{O}$ with $g = 0$ and $dg \neq 0$ at all points of $\partial\mathcal{O}$. Solve the initial value problem

$$\psi_t = \lambda(t, x, \partial_x \psi), \qquad \psi|_{t=0} = g\,.$$

The set containing $WF(u)$ is exactly equal to the set of points

$$\Big\{ \big(t, x, \sigma(\psi_t(t,x), \partial_x\psi(t,x))\big) \ : \ 0 \neq \sigma \in \mathbb{R}\,, \quad \text{and} \quad \psi(t,x) = 0 \Big\}\,.$$

This set is the conormal variety of $\{\psi = 0\}$. Thus, as long as the eikonal equation is smoothly solvable, $WF(u_\nu) \subset N^*(\{\psi = 0\})$.

The propagation of singularities theorem is global in time so is not limited by the local solvability of the eikonal equation. The result applies after caustics and focusing. On the other hand, locally in time one can show (see the discussion of progressing waves in [Lax, 2006]) that the solution remains piecewise smooth with singularities along $\{\psi = 0\}$. For such small times, this example is a generalization of the piecewise smooth solutions in §1.1.

Proof of Theorem 5.5.7. Part i is just a restatement.

For ii compute for $\mu \neq \nu$ that on $\tau = \lambda_\mu$,

$$\tau - \lambda_\nu \ = \ \tau - \lambda_\mu + (\lambda_\mu - \lambda_\nu) \ = \ \lambda_\mu - \lambda_\nu \ \neq \ 0.$$

Therefore if Γ_μ is a $\tau - \lambda_\mu$ bicharacteristic along which $\tau = \lambda_\mu$, one has for $\mu \neq \nu$, $\Gamma_\mu \cap WF(u_\nu) = \phi$. Therefore, **ii** is equivalent to the assertion that $WF_s(u_\mu)$ is invariant under the Hamilton flow of $\tau - \lambda_\mu(t,x,\xi)$. It is sufficient to prove that if $\Gamma_\mu(\underline{t}) \notin WF_s(u_\mu)$, then $\Gamma_\mu \cap WF_s(u_\mu) = \phi$.

Since the curve γ_μ and WF_s are both closed, it follows that

$$\mathcal{B} \ := \ \big\{ t \ : \ \Gamma_\mu(t) \notin WF_s(u_\mu) \big\} \quad \text{is open and nonempty.}$$

Suppose that t_2 belongs to the closure $\overline{\mathcal{B}}$. It suffices to show that $\Gamma_\mu(t_2) \notin WF_s(u_\mu))$.

Choose $\mathcal{B} \ni t_1 \neq t_2$ with $|t_1 - t_2| < T_1$ with T_1 from Theorem 5.5.4. We treat the case $t_1 < t_2$, the other being similar. Choose $\beta(t) \in C^\infty(\mathbb{R})$ with β identically equal to one on a neighborhood of t_2 and vanishing for $t < t_1$. Choose the support of $d\beta/dt$ so close to t_1 so that $\Gamma(t) \notin WF_s(u(t))$ when $t \in \operatorname{supp} d\beta/dt$. Then

$$L(\beta u_\mu) \;=\; \beta\, L\, u_\mu \;+\; [L, \beta] u_\mu \;=\; \beta\, L\, u_\mu \;+\; A_0\, \frac{d\beta}{dt}\, u_\mu\,.$$

The first term is smooth and $WF_s([L,\beta]u_\mu) \subset WF_s(u_\mu) \cap \operatorname{supp} d\beta/dt$ by Exercise 4.6.1. The choice of β guarantees that $\Gamma_\mu \cap WF_s(L(\beta u_\mu)) = \phi$.

Write $[L,\beta]u_\mu = f_1 + f_2$ with supports in $t \geq t_1$. The summand f_1 is supported very close to t_1 and satisfies $f_1 \in H^s_{\mathrm{loc}}$. The summand f_2 satisfies $\Gamma_\mu \cap WF(f_2) = \phi$. Denote by v_j the solutions of $Lv_j = f_j$ that vanish for $t \leq t_1$. Then $v_1 \in H^s_{\mathrm{loc}}$ so $\Gamma_\mu(t_2) \notin WF_s(v_1)$.

We are only interested in v_2 near the space-time projection of $\Gamma_\mu(t_2)$. Using finite speed, we may cut off f_2 to have compact support without modifying the solution v_2 on a neighborhood of the projection of $\Gamma_\mu(t_2)$.

Use the Lax parametrix and Duhamel's representation to write for t near t_2, $v_2 = \sum_\nu v_{2,\nu} + C^\infty$,

$$v_{2,\nu}(t,x) \;:=\; \int a_\nu(t,\sigma,x,\xi)\; e^{i(\phi_\nu(t,\sigma,x,\xi) - x\xi)}\; \chi(\xi)\; f_2(\sigma,x)\; dx\, d\sigma\; d\xi\,.$$

As in the proof of Theorem 5.5.4, these oscillatory integrals are analyzed using Proposition 5.5.3. First, $WF(v_{2,\nu}) \subset \{\tau = \lambda_\nu\}$ so $\Gamma_\mu(t_2) \notin WF(v_{2,\nu})$ for $\nu \neq \mu$. The same proposition shows that the behavior of $v_{2,\mu}$ microlocally on Γ_μ is determined by f_μ microlocally on Γ_μ. By construction we have $\Gamma_\mu \cap WF(f_2) = \phi$, and it follows that $\Gamma_\mu \cap WF(v_{2,\mu}) = \phi$.

Exercise 5.5.8. *Give the details of these two applications of Proposition 5.5.3.*

This completes the proof that $\Gamma_\mu(t_2) \notin WF_s(v_j)$, and thereby proves the theorem. □

5.6. An application to stabilization

This section shows how the results on microlocal analysis yield a complete solution to a nontrivial problem of stabilization. By duality these yield results on controlability, an important subject both in engineering and mathematics. An alternative treatment of the same problem using defect measures is presented in the book manuscript of Evans and Zworski.

This section requires familiarity with Riemannian geometry, specifically with the geodesic flow. Suppose that M is a compact connected Riemannian manifold without boundary. The metric is

$$g_{ij}(x)\, dx^i\, dx^j\,. \tag{5.6.1}$$

The induced metric on one forms is denoted $\langle\ \ \rangle$ and in local coordinates is

$$g^{ij}(x)\, d\xi_i\, d\xi_j\,. \tag{5.6.2}$$

The volume form is

$$dv(x) \;=\; (\det g_{ij}(x))^{1/2}\, dx\,.$$

The Dirichlet integral is

$$D(u,u) \;:=\; \int_M \big\langle du(x), du(x)\big\rangle\, dv(x) \;:=\; \int_M |du(x)|^2\, dv(x)\,. \tag{5.6.3}$$

For functions u supported in a single coordinate patch, this is equal to

$$\int \sum_{i,j} g^{ij}(x)\, \partial_i u(x)\, \partial_j u(x)\, (\det g_{ij}(x))^{1/2}\, dx\,.$$

The Laplace–Beltrami operator, $\Delta = \Delta_g$, is defined by
(5.6.4)

$$\big\langle \Delta u, \psi\big\rangle \;:=\; \frac{-1}{2}\, \frac{dD(u+s\psi, u+s\psi)}{ds}\bigg|_{s=0} \;=\; -\int_M \langle du(x), d\psi(x)\rangle\, dv\,.$$

In local coordinates,

$$-\int_M \langle du(x), d\psi(x)\rangle dv(x) = -\int g^{ij}(x)\partial_i u(x)\partial_j\psi(x)(\det g_{ij}(x))^{1/2}dx$$

$$= \int (\det g_{ij}(x))^{-1/2}\partial_j\Big(g^{ij}(x)(\det g_{ij}(x))^{1/2}\partial_i u(x)\Big)\psi(x)(\det g_{ij}(x))^{1/2}dx,$$

so

$$\Delta u \;=\; (\det g_{ij}(x))^{-1/2}\, \partial_j\Big(g^{ij}(x)\, (\det g_{ij}(x))^{1/2}\, \partial_i u(x)\Big)\,. \tag{5.6.5}$$

Consider the damped Klein–Gordon equation,

$$Lu \;:=\; u_{tt} \;-\; \Delta u \;+\; u \;+\; a(x)\, u_t \;=\; 0\,, \qquad C^\infty(M) \ni a \geq 0\,. \tag{5.6.6}$$

One could treat the damped wave equation in the same way with the inconvenience of working modulo the constants.

The principal symbol of the operator L is

$$h(t,x,\tau,\xi) \;:=\; \tau^2 - g^{ij}(x)\, \xi_i\, \xi_j \;=\; \tau^2 - \langle \xi_i dx^i, \xi_j dx^j\rangle \;:=\; \tau^2 - |\xi|^2\,,$$

with $|\cdot|$ denoting the Riemannian length. The characteristic variety lies in $\tau \neq 0$ and consists of two smooth sheets,

$$\tau \;=\; \pm|\xi| \;=\; \pm\Big(g^{ij}(x)\, \xi_i\, \xi_j\Big)^{1/2}\,.$$

The roots are simple, and the smooth variety hypothesis is everywhere satisfied.

The crucial energy identity is

$$\frac{d}{dt}\int_M |u_t|^2 \ + \ |du(x)^2| \ + \ |u|^2 \ dv \ = \ 2\text{Re} \int_M \overline{u}_t \ \big(u_{tt} \ - \ \Delta\, u \ + \ u\big) \ dv\,.$$

Therefore, for solutions of $Lu = 0$, energy is dissipated according to

$$\frac{d}{dt}\int_M |u_t|^2 \ + \ |du(x)^2| \ + \ |u|^2 \ dv \ = \ -\int_M 2a(x)\,|u_t|^2(x)\ dv \ \leq \ 0\,. \tag{5.6.7}$$

This yields the uniform a priori estimate,

$$\exists C, \ \ \forall t \geq 0, \quad \|u(t)\|_{H^1} \ + \ \|u_t(t)\|_{L^2} \ \leq \ C\,\big(\|u(0)\|_{H^1} \ + \ \|u_t(0)\|_{L^2}\big)\,.$$

If u satisfies $Lu = 0$, then $v = (1 - \Delta + a(x))^{s/2}u$ satisfies $Lv = 0$.[3] The elliptic regularity theorem with elliptic operator $(1-\Delta+a(x))^{s/2}$ implies that $\|v_t\|_{L^2}$ and $\|v\|_{H^1}$ are norms equivalent to $\|u_t\|_{H^s}$ and $\|u\|_{H^{s+1}}$, respectively. Applying the a priori estimate proves that

$$\forall s, \ \exists C_s, \ \forall t \geq 0, \quad \|u(t)\|_{H^{s+1}} + \|u_t(t)\|_{H^s} \leq C_s\,\big(\|u(0)\|_{H^{s+1}} + \|u_t(0)\|_{H^s}\big)\,.$$

The strategy for proving existence by replacing ∂_x by symmetric difference quotients suitably globalized using partitions of unity, yields the following.

Theorem 5.6.1. *For any $s \in \mathbb{R}$, $f \in H^s(M)$, $g \in H^{s-1}(M)$ there is a unique*

$$u \ \in \ \bigcap_j C^j\Big([0,\infty[\ ; \ H^{s-j}(M)\Big)$$

satisfying

$$L\,u = 0 \quad \text{on} \quad \mathbb{R}\times M\,, \qquad u|_{t=0} = f\,, \quad u_t|_{t=0} = g\,.$$

Since the coefficients of L do not depend on t, it follows that if $u(t,x)$ is a solution, then so is $u(t+s,x)$ for any s. If $j \in C_0^\infty(\mathbb{R})$ with $\int j(t)\,dt = 1$, then

$$u^\delta(t,x) \ := \ \int u(t+\delta s, x)\ j(s)\ ds \ = \ \frac{1}{\delta}\int u(t+s)\ j(s/\delta)\ ds$$

satisfies $L\,u^\delta = 0$ since the Riemann sums that converge to the integral are solutions. In addition, $u^\delta \to u$ in $C^j([-T,T]\,;\,H^{s-j})$ as $\delta \to 0$.

Exercise 5.6.1. *Prove the last assertion by showing that the Cauchy data of u^δ converge to those of u in H^s and H^{s-1}.*

Proposition 5.6.2. *For each $\delta > 0$, $u^\delta \in C^\infty(\mathbb{R}\times M)$.*

---

[3]Readers may assume that s is an even integer to avoid fractional powers.

Proof. From the definition of u^δ and the continuity of u in time, it follows that $u^\delta \in C^\infty(\mathbb{R}\,;\, H^s)$ with

$$\partial_t^k u^\delta(t) \;=\; \int u^\delta(t+\delta s)\ \frac{d^k j(s)}{dt^k}\ ds \;\in\; C(\mathbb{R}\,;\, H^s)\,.$$

This gives a Sobolev regularity for $\partial_t^k u^\delta$ independent of k since denoting with a subscript comp the elements of compact support,

$$\begin{aligned} C(\mathbb{R}\,;\, H^s) \;\subset\; C_{\text{comp}}(\mathbb{R}\,;\, H^{-s})' \;\subset\; H^1_{\text{comp}}(\mathbb{R}\,;\, H^{-s})' \\ \subset\; H^{-s+1}_{\text{comp}}(\mathbb{R}\times M)' \;=\; H^{s-1}_{\text{loc}}(\mathbb{R}\times M)\,. \end{aligned}$$

Since $\text{Char}\, L \subset \tau \neq 0$, the operator ∂_t^k is microlocally elliptic on $\text{Char}\, L$. The microlocal elliptic regularity theorem applied to the equation $\partial_t^k u \in H^{s-1}_{\text{loc}}(\mathbb{R}\times M)$ implies that u is microlocally H^{k+s-1} on the characteristic variety of L.

The microlocal elliptic regularity theorem applied to $Lu = 0$ shows that u also has this regularity away from the characteristic variety of L. Therefore u is everywhere microlocally H^{k+s-1}. Therefore $u \in H^{k+s-1}_{\text{loc}}(\mathbb{R}\times M)$.

Since k is arbitrary, Sobolev's lemma completes the proof. □

Denote by $S(t)$ the operator

$$S(t)(f,g) \;:=\; \big(u(t), u_t(t)\big)\,,$$

where u is the solution with Cauchy data equal to (f,g). For any s, $t \mapsto S(t)$ is a C^0 one parameter group of operators on $H^s \times H^{s-1}$. It is a contraction semigroup on $H^1 \times H^0$ for $t \geq 0$. It is uniformly bounded on $H^s \times H^{s-1}$ for $t \geq 0$.[4]

If a is not identically equal to zero, then all solutions tend to zero as $t \to \infty$. The argument in the proof leading to the conservative solution v is called *Lasalle's invariance principal* in the dynamical systems community.

Iwasaki's Theorem 5.6.3 (Iwasaki, 1969). *If a is not identically equal to zero and the Cauchy data belong to $H^1 \times L^2$, then the solution of Theorem 5.6.1 satisfies*

$$\lim_{t\to\infty} \int_M |u_t|^2 \;+\; |d_x u|^2 \;+\; |u|^2\ dx \;=\; 0\,. \tag{5.6.8}$$

Proof. Since the integral on the left in (5.6.8) is a nonincreasing function of time, one has

$$\int_M |u_t|^2 \;+\; |du(x)^2| \;+\; |u|^2\ dv \;\to\; E_\infty \;\geq\; 0 \tag{5.6.9}$$

[4]It is contractive in a norm $\big(\|(1-\Delta+a)^{s/2}u_t\|^2_{L^2} + \|(1-\Delta+a)^{s/2}u\|^2_{H^1}\big)^2$ equivalent to the standard norm.

monotonically as $t \to \infty$.

In the regularization operator choose $j \in C_0^\infty(]0,1[)$ and define u^δ as above. The uniform in time a priori estimates imply that for $j = 0, 1$,

$$\lim_{\delta \to 0} \ \sup_{t \geq 0} \ \|\partial_t^j u^\delta(t) \ - \ \partial_t^j u(t)\|_{H^{1-j}(M)} \ = \ 0 \,.$$

Thus, it suffices to prove the result for u^δ. Therefore, it suffices to consider solutions satisfying

$$\forall j, s \quad \partial_t^j u \ \in \ L^\infty\big([0,\infty[\ ;\ H^s(M)\big)\,.$$

Let $v^n(t) := u(t+n)$. Ascoli's theorem implies that there is a subsequence so that $v^{n_k} \ \to \ v$ uniformly with all derivatives on the compact subsets of $[0,\infty[\ \times M$. Equation (5.6.9) implies that for all $t \geq 0$,

$$\int_M |v_t|^2 \ + \ |dv(x)^2| \ + \ |v|^2 \, dv(x) \ = \ E_\infty \ \geq \ 0\,. \tag{5.6.10}$$

Therefore (5.6.7) implies that for all $t \geq 0$,

$$\int a(x)\, |v_t|^2 \, dx \ = \ 0\,. \tag{5.6.11}$$

It follows that $a(x)\, v_t(t,x) = 0$. Therefore for $t \geq 0$, v satisfies the undamped wave equation $v_{tt} - \Delta\, v + v = 0$. The idea of the next argument dates at least to Carleman in the 1920s.

Introduce an orthonormal basis of eigenfunctions, ϕ_m, of Δ,

$$(1-\Delta)\,\phi_m = \lambda_m^2\,\phi_m\,, \qquad \lambda_m < \lambda_{m+1}, \qquad \|\phi_m\|_{L^2(M)} \ = \ 1\,. \tag{5.6.12}$$

Then

$$v \ = \ \sum_{m,\pm} a_{m,\pm} \ e^{\pm it\lambda_m} \ \phi_m(x)\,,$$

with rapidly decreasing Fourier coefficients $a_{m,\pm}$.

Since $v_t = 0$ on Ω, one has for $x \in \Omega$,

$$0 \ = \ \lim_{T\to\infty} \ \frac{1}{T} \int_0^T e^{\mp it\lambda_m} \ v(t,x)\, dt \ = \ a_{m,\pm}\phi_m(x)\,.$$

If one had $E_\infty > 0$, there would be a nonvanishing Fourier coefficient $a_{m,\pm}$. For that coefficient it would follow that $\phi_m = 0$ on Ω.

The manifold M is connected. Apply the unique continuation principle that asserts if a solution of a smooth homogeneous linear scalar second order elliptic equations vanishes on an open set, then that solution must vanish identically. The equation is $(1 - \lambda_m^2 - \Delta)\phi_m = 0$, and the conclusion is $\phi_m = 0$. This contradicts $\|\phi_m\|_{L^2} = 1$. Thus one cannot have $E_\infty > 0$, and (5.6.8) is proved. ∎

Our application of the propagation of singularities theorem is to find necessary and sufficient conditions on $a(x)$ so that solutions decay exponentially in time. The next result shows the equivalence of several notions of decay.

Proposition 5.6.4. *The following are equivalent.*

1. *For each* $(f,g) \in H^1 \times H^0$, *there are constants* $m(f,g) > 0$ *and* $\gamma(f,g) > 0$ *so that*

$$\forall t \geq 0, \quad \big\|S(t)(f,g)\big\|_{H^1\times H^0} \leq m\, e^{-\gamma t}.$$

2. *There are constants* $M > 0$ *and* $\Gamma > 0$ *so that*

$$\forall t \geq 0,\ \forall (f,g) \in H^1 \times H^0, \quad \big\|S(t)(f,g)\big\|_{H^1\times H^0} \leq M\, e^{-\Gamma t}\, \big\|(f,g)\big\|_{H^1\times H^0}.$$

3. *There is a* $T > 0$ *so that* $\|S(T)\| < 1$.

4. *For each* $t > 0$, *the spectral radius of* $S(t)$ *is strictly smaller than 1.*

Proof. 1 ⇒ 2. Define

$$\Omega_n := \Big\{(f,g) \in H^1 \times H^0 \ : \ \forall t \geq 0, \quad \big\|S(t)(f,g)\big\|_{H^1\times H^0} \leq n\, e^{-t/n}\, \big\|(f,g)\big\|_{H^1\times H^0}\Big\}.$$

Then Ω_n is closed. By hypothesis, $\bigcup_n \Omega_n = H^1 \times H^0$. The Baire category theorem implies that there is an n so that Ω_n contains a ball $B_r(U)$ with positive radius.

Therefore $S(t)$ has uniform exponential decay on $U + B_r(0)$. By linearity there is uniform exponential decay on $B_r(0)$. That is the desired conclusion.

2 ⇒ 1. Immediate.

2 ⇔ 3. Immediate.

3 ⇔ 4. Follows from the formula for the spectral radius, ρ,

$$\rho = \lim_{n\to\infty} \big\|S(t)^n\big\|^{1/n} = \lim_{n\to\infty} \big\|S(nt)\big\|^{1/n}. \qquad \square$$

Recall that the principal symbol of L is denoted h. The bicharacteristics are integral curves of the Hamilton field of h along which $h = 0$. Vanishing h is the condition that the curve lies in $\operatorname{Char} L$. The projections of these curves on $\mathbb{R} \times M$ are exactly the geodesics of M traversed at constant speed. Over a point (t, x, ξ) there pass two bicharacteristics, one for each of the two roots $\tau = \pm|\xi|$. The projections of the bicharacteristics on space-time are the same geodesic traversed in opposite directions.

The next result gives a necessary and sufficient condition for the equivalent conditions of Proposition 5.6.2 to hold.

Theorem 5.6.5 (Rauch and Taylor, 1974). *The following two properties are equivalent:*

1. $\|S(T)\|^2_{H^1\times H^0} < 1$.

2. *Each geodesic of length T passes through the set $\{a > 0\}$.*

In particular, solutions decay exponentially in the sense of Proposition 5.6.4 if and only if there is a $T > 0$ satisfying **2**.

Proof. **1** $\Rightarrow$ **2** (First proved by Ralston in 1969). We show that if **2** is violated, then $\|S(T)\|_{H^1\times H^0} = 1$.

When **2** is violated, there is a unit speed geodesic so that $\gamma([0,T])$ does not intersect $\{a > 0\}$. We construct smooth solutions on $[0,T]\times M$ that are concentrated on a small neighborhood of γ and therefore have little decay.

For each t, define $\xi(t) \in T^*_{\gamma(t)}(M)$ to be the dual vector satisfying

$$\xi(t)(w) \;=\; \langle \gamma'(t)\,,\, w\rangle\,, \qquad \forall w \in T_{\gamma(t)}(M)\,.$$

Then for one of the choices of sign, $\Gamma_\pm := (t, \gamma(t), \pm|\xi|, \xi(t))$ is a bicharacteristic of L that lies over γ. Call that bicharacteristic Γ.

Lemma 5.6.6. *There is a solution of $Lu = 0$ whose wavefront set coincides with the single bicharacteristic Γ.*

Proof of Lemma 5.6.6. Choose $f \in \mathcal{D}'(M)$ such that $WF(f) = (\gamma(0), \mathbb{R}_+\xi(0))$. Consider the solution of $Lv = 0$ with $v(0) = f$, $v_t(0) = 0$.

Over the point $(0, \gamma(0), \xi(0))$ there are two points in the characteristic variety, $(0, \gamma(0), \pm|\xi|, \xi(0))$. Denote by $\Gamma_\pm(t)$ the corresponding bicharacteristics. They pass over the geodesic γ traversed in opposite directions.

The global propagation of singularities theorem expresses

$$v \;=\; v_+ \;+\; v_- \;+\; C^\infty\,, \qquad L\,v_\pm \in C^\infty\,, \qquad WF\,v_\pm \subset \Gamma_\pm\,.$$

Suppose that Γ_+ is the bicharacteristic passing over the geodesic γ. The opposite case is similar. Define $w \in C^\infty$ to be the solution of

$$L\,w \;=\; -L\,v_+\,, \qquad w|_{t=0} = 0\,, \quad w_t|_{t=0} = 0\,.$$

Then $u := v_+ + w$ satisfies the conditions of the lemma. □

The next lemma completes the proof that **1** $\Rightarrow$ **2**.

Lemma 5.6.7. *If* **2** *is violated, then for any $\epsilon > 0$, there is a smooth solution with initial energy equal to one and with*

$$\int_0^T \int_M a(x)\,|u_t|^2(t,x)\;dx\,dt \;<\; \epsilon\,. \tag{5.6.13}$$

Proof of Lemma 5.6.7. Since the solution of Lemma 5.6.6 is not smooth, using the fact that the points of Char L are noncharacteristic for ∂_t^j as in the proof of Proposition 5.6.2, it follows that there is a $j \geq 0$ so that $\partial_t^j u \notin H^1(]0,T[\, \times M)$. Thus replacing u by $\partial_t^j u$, we may suppose that

$$u \notin H^1(]0,T[\, \times M)\,. \tag{5.6.14}$$

Define

$$v^\delta := \frac{u^\delta}{\int |u_t^\delta(0,x)|^2 + |d_x u^\delta(0,x)|^2 + |u^\delta(0,x)|^2 \, dv(x)}\,.$$

Then v^δ is a solution with initial energy equal to one. Therefore energy ≤ 1 for $t \geq 0$. Integrating over points where $a \leq \epsilon/2T$ yields

$$\int_0^T \int_{\{a \leq \epsilon/2T\}} a(x)\, |v_t^\delta|^2 \, dv(x)\, dt \leq \epsilon/2\,. \tag{5.6.15}$$

Thanks to (5.6.14), it follows that

$$\lim_{\delta \to 0} \|u^\delta\|_{H^1(]0,T[\, \times M)} = \infty\,.$$

The basic energy estimate shows that

$$\|u^\delta\|_{H^1(]0,T[\, \times M)} \leq C \int |u_t^\delta(0,x)|^2 + |d_x u^\delta(0,x)|^2 + |u^\delta(0,x)|^2 \, dv(x)\,,$$

so

$$\lim_{\delta \to 0} \int |u_t^\delta(0,x)|^2 + |d_x u^\delta(0,x)|^2 + |u^\delta(0,x)|^2 \, dv(x) = \infty\,. \tag{5.6.16}$$

The solution u is smooth on the complement of the geodesic which is the projection, $\pi(\Gamma)$ of Γ on space-time. It follows from (5.6.16) that for all α, $\partial_{t,x}^\alpha v^\delta$ converges uniformly to zero on compact subsets of $(\mathbb{R} \times M) \setminus \pi(\Gamma)$. Therefore, for δ small,

$$\int_{[0,T] \times \{a \geq \epsilon/2T\}} a(x)\, |v_t^\delta|^2 \, dv(x)\, dt \leq \epsilon/2\,. \tag{5.6.17}$$

Combining (5.6.15) and (5.6.17) shows that v^δ satisfies the conditions of the lemma for δ sufficiently small. □

This completes the proof that **1** ⇒ **2**. To prove that **2** ⇒ **1**, the key step is the following lemma relying on the propagation of singularities theorem.

Lemma 5.6.8. *Suppose that every geodesic of length T passes through the set $\{a > 0\}$. Then if $u \in L^2_{\rm loc}(\mathbb{R} \times M)$ satisfies $Lu = 0$ and*

$$u_t \in L^2_{\rm loc}(]0,T[\, \times \{a > 0\}), \tag{5.6.18}$$

then $u \in H^1$ on a neighborhood of $\{t = T/2\} \times M$.

Proof of Lemma 5.6.8. It suffices to show that for all $(T/2, x, \tau, \xi)$ that u belongs to H^1 microlocally at $(T/2, x, \tau, \xi)$.

Since $Lu = 0$, the microlocal elliptic regularity Theorem 4.6.1 implies that for all s, u is microlocally H^s on the complement of $\operatorname{Char} L$. Thus it suffices to consider $(T/2, x, \pm|\xi|, \xi) \in \operatorname{Char} L$.

The the global propagation of singularities Theorem 5.5.6 expresses

$$u = u_+ + u_- + C^\infty, \qquad WF\, u_\pm \subset \{\tau = \pm|\xi|\}.$$

It suffices to show that for all $x, \xi, \pm$,

$$u_\pm \in H^1(T/2, x, \pm|\xi|, \xi). \tag{5.6.19}$$

We treat the case $+$, the other being analogous.

Fix x, ξ and denote by $t \mapsto \Gamma(t)$ the bicharacteristic with $\Gamma(T/2) = (T/2, x, |\xi|, \xi)$. Hypothesis 2 of Theorem 5.6.5 implies that there is a $\underline{t} \in]0, T[$ so that $\Gamma(\underline{t}) = (\underline{t}, \underline{x}, |\underline{\xi}|, \underline{\xi})$ and $a(\underline{x}) > 0$.

Since $\int_{[0,T]\times M} a(x)\, |u_t|^2\, dt\, dv(x) < \infty$, we know that $\partial_t u$ is square integrable on a neighborhood of $(\underline{t}, \underline{x})$. Since ∂_t is elliptic at $(\underline{t}, \underline{x}, |\underline{\xi}|, \underline{\xi})$, the microlocal elliptic regularity theorem implies that $u \in H^1(\underline{t}, \underline{x}, |\underline{\xi}|, \underline{\xi})$. Since $(\underline{t}, \underline{x}, |\underline{\xi}|, \underline{\xi}) \notin WF(u_-)$, it follows that $u_+ \in H^1(\underline{t}, \underline{x}, |\underline{\xi}|, \underline{\xi})$.

Theorem 5.5.7.ii. implies that $WF_1(u_+)$ is invariant under the Hamilton flow with hamiltonian $\tau - |\xi|$. The image of the point $(\underline{t}, \underline{x}, |\underline{\xi}|, \underline{\xi})$ at time $T/2$ is the point $(T/2, x, |\xi|, \xi)$. Therefore, $u_+ \in H^1(T/2, x, |\xi|, \xi)$. This completes the proof of Lemma 5.6.8. □

Lemma 5.6.9. *If $Lu = 0$ and there is a time $\underline{t}$ so that u is H^1 on a neighborhood of $\{t = \underline{t}\} \times M$, then for $j = 0, 1$, $u \in C^j\big(\mathbb{R}\,;\, H^{1-j}\big)$.*

Proof of Lemma 5.6.9. It is given that there is a $\beta > 0$ so that $u \in H^1([\underline{t} - \beta, \underline{t} + \beta] \times M)$. The energy decay law shows that for all $t > \underline{t} + \beta$, $s \in]\underline{t} - \beta, \underline{t} + \beta[$, and smooth solutions v to $Lv = 0$,

$$\int_M |v_t(t)|^2 + |d_x v(t)|^2 + |v(t)|^2\, dv(x) \le \int_M |v_t(s)|^2 + |d_x v(s)|^2 + |v(s)|^2\, dv(x).$$

Therefore

$$\begin{aligned}
\sup_{t \ge \underline{t}+\beta} \int_M & |v_t(t)|^2 + |d_x v(t)|^2 + |v(t)|^2\, dv(x) \\
&\le \frac{1}{4\beta} \int_{\underline{t}-\beta/2}^{\underline{t}+\beta/2} \int_M |v_t(s)|^2 + |d_x v(s)|^2 + |v(s)|^2\, dv(x)\, ds \\
&\le C\, \big\|v\big\|_{H^1([\underline{t}-\beta/2, \underline{t}+\beta/2]\times M)},
\end{aligned}$$

with C independent of v.

Apply this inequality to the time regularizations u^δ with $\delta < \beta$ to find that u^δ is a Cauchy sequence in $C^j([\underline{t}+\delta,\infty[\,;\, H^{1-j}(M))$ for $j=0,1$. Since the limit of the u^δ in the sense of distributions is u, it follows that $u \in C^j([\underline{t}+\delta,\infty[\,;\, H^{1-j}(M))$ for $j=0,1$. □

This lemma strengthens the conclusion of Lemma 5.6.8 to

$$u \in C^j(\mathbb{R}\,;\, H^{1-j}(M))$$

for $j = 0, 1$.

Denote by $\mathcal{B}$ the set of $u \in L^2(]0,T[\times M)$ satisfying $Lu = 0$ and (5.6.18). It is a Hilbert space with norm squared equal to

$$\int_0^T \int_M |u|^2 \;+\; a(x)\,|u_t|^2\; dv(x)\,dt\,.$$

The strengthening of Lemma 5.6.8 shows that if T satisfies hypothesis 2 of Theorem 5.6.5, then

$$\mathcal{B} \quad\subset\quad \bigcap_{j=0,1} C^j([0,T]\,;\, H^{1-j}(M))\,.$$

The graph of the inclusion map is closed. Therefore, the closed graph theorem implies that the inclusion is continuous. Thus, there is a constant C so that $\forall\, u \in \mathcal{B}$,

$$\text{(5.6.20)}\qquad \int_M |u_t(0)|^2 \;+\; |d_x u(0)|^2 \;+\; |u(0)|^2\; dv(x) \;\le\; C \int_0^T \int_M |u|^2 \;+\; a(x)\,|u_t|^2\; dv(x)\,dt\,.$$

Lemma 5.6.10. *Suppose that hypothesis 2 of Theorem 5.6.5 is satisfied. Define*

$$K(T) \;:=\; \Big\{u \in L^2([0,T]\times M) \;:\; Lu = 0 \quad \text{and} \quad a(x)\,u_t = 0\Big\}\,.$$

Then K is finite dimensional subspace of finite energy solutions. $K(T) = \{0\}$ if and only if there is a constant C so that

$$\text{(5.6.21)}\qquad \forall\, u \in \mathcal{B},\quad \int_M |u_t(0)|^2 \;+\; |d_x u(0)|^2 \;+\; |u(0)|^2\; dv(x) \;\le\; C \int_{[0,T]\times M} a(x)\,|u_t|^2\; dt\, dv(x)\,.$$

Proof. On the space K, $au_t = 0$ so (5.6.20) implies the inequality

$$\int_M |u_t(0)|^2 \;+\; |d_x u(0)|^2 \;+\; |u(0)|^2\; dv(x) \;\le\; C \int_0^T \int_M |u|^2\; dv(x)\,dt\,.$$

Thus the unit ball in K is precompact proving that K is finite dimensional. At the same time one sees that the solutions have finite energy.

If K is not trivial, then (5.6.20) cannot hold for the nonzero elements of K, no matter how large C is chosen.

On the other hand, if (5.6.21) does not hold for any, C then there is a sequence u^n with

$$\int_0^T \int_M |u^n|^2 \; + \; a(x)\,|u_t^n|^2\, dv(x)\, dt \; = \; 1$$

$$\text{and} \quad \int_0^T \int_M a(x)\,|u_t^n|^2\, dv(x)\, dt \; \to \; 0\,.$$

By (5.6.20) the sequence is compact in $L^2([0,T] \times M)$. Passing to a convergent subsequence, we may assume that $u^n \to u$ in $L^2([0,T] \times M)$. The limit satisfies

$$\|u\|_{L^2([0,T]\times M)} \; = \; 1\,, \qquad Lu \; = \; 0\,, \qquad a(x)u_t \; = \; 0\,.$$

That is, u is a nontrivial element of K. □

The next step in the proof that **2** $\Rightarrow$ **1** in Theorem 5.6.5 is to show that $K(T) = \{0\}$ as soon as T satisfies the geometric condition.

Lemmas 5.6.8 and 5.6.9 show that the elements of K have finite energy. It then follows from the definition of K that $v \in K \Rightarrow \partial_t v \in K$. Since $\partial_t \in \mathrm{Hom}\,(K(T))$ and $K(T)$ is a finite dimensional space, it follows that if $K(T) \neq \{0\}$, then there is a $v \in K \setminus \{0\}$ and $\lambda \in \mathbb{C}$ so that $\partial_t v = \lambda v$.

Then $v = e^{\lambda t}\phi(x)$ with $\phi \neq 0$, satisfies $Lv = 0$ and $a(x)v_t = 0$. Since $av_t = 0$ the law of energy dissipation is for v a law of energy conservation. Thus v has constant nonzero energy violating Theorem 5.6.3. This contradiction shows that $K(T) = \{0\}$.

Therefore **2** implies inequality (5.6.21). That inequality implies

$$\|S(T)\|^2_{H^1\times H^0} \; \leq \; \frac{1}{1+C} \; < \; 1\,,$$

proving **1**. □

Exercise 5.6.2. *Show that when* **2** *of Theorem 5.6.5 is violated, then for any N and ϵ there is a linear space V of finite energy Cauchy data so that* $\dim V \geq N$ *and for all $(f,g) \in V$,*

$$\|S(T)(f,g)\|_{H^1\times H^0} \; \geq \; (1-\epsilon)\|f,g\|\,.$$

Hint. Find N nearby but distinct geodesics that nearly miss $\{a > 0\}$.
Discussion. This shows that $\{|z| = 1\}$ meets the essential spectrum of $S(T)$.

Appendix 5.I. Hamilton–Jacobi theory for the eikonal equation

5.I.1. Introduction. This section solves scalar real nonlinear first order partial differential equations,

$$F(y, d\phi) = 0, \tag{5.I.1}$$

by reducing to the solution of ordinary differential equations. The function $F(y, \eta)$ is assumed to be a smooth real valued function of its arguments on an open subset of $\mathbb{R}^n_y \times \mathbb{R}^n_\eta$. Written out, the equation takes the form

$$F\Big(y_1, \dots, y_n, \frac{\partial \phi(y)}{\partial y_1}, \dots, \frac{\partial \phi(y)}{\partial y_n}\Big) = 0\,. \tag{5.I.2}$$

In applications, the function ϕ is usually either a phase function or a function one of whose level sets represents a wavefront.

Examples 5.I.1. Three classical examples from optics are the equations

$$|\nabla_x \psi|^2 = 1\,, \qquad F(x, \xi) = |\xi|^2 - 1\,, \tag{5.I.3}$$

$$\phi_t^2 - |\nabla_x \phi|^2 = 0\,, \qquad F(t, x, \tau, \xi) = \tau^2 - |\xi|^2, \tag{5.I.4}$$

and

$$\phi_t^2 - c(x)^2\, |\nabla_x \phi|^2 = 0\,, \quad F(t, x, \tau, \xi) = \tau^2 - c(t, x)^2\, |\xi|^2\,, \quad c > 0\,. \tag{5.I.5}$$

The first describes solutions of the second that have the special form $\phi(t, x) = t \pm \psi(x)$. For equation (5.I.5) the rays bend or refract in a medium of variable speed of propagation $c(t, x)$. When $c(t, x)$ is independent of t, Solutions of (5.I.5) of the form $t \pm \psi(x)$ lead to a generalization of (5.I.3),

$$|\nabla_x \psi|^2 = 1/c(x)^2 := n^2(x)\,, \qquad F(x, \xi) = |\xi|^2 - n^2(x)\,. \tag{5.I.6}$$

The next three examples exhibit explicit solutions.

Example 5.I.2. Seek solutions of (5.I.4) as linear functions of the coordinates $\phi(t, x) = \tau t + \xi x$. It is a solution if and only if $\tau = \pm|\xi|$. Thus, for any linear initial function $g(x) = \xi x$ with $\xi \neq 0$, this yields two solutions of (5.I.4) with $\phi(0, x) = g$. The solutions come from two determinations of ϕ_t from ϕ_x. There is one solution for each of

$$\phi_t = \pm\, |\nabla_x \phi|\,. \tag{5.I.7}$$

Example 5.I.3. The linear solutions of (5.I.3) are precisely the functions ξx with $|\xi| = 1$. If one imposes initial data $\psi(0, x_2, \dots, x_n) = \xi_2 x_2 + \cdots + \xi_n x_n$, then if $|\xi_2, \dots, \xi_n| < 1$, there are two linear solutions given by $\xi_1 = \pm(1 - |\xi_2, \dots, \xi_n|^2)^{1/2}$. If $|\xi_2, \dots, \xi_n| > 1$, there are no solutions, since equation (5.I.1) cannot be solved even at a single point of the initial surface.

Example 5.I.4. The solutions of (5.I.3) that depend only on $r := |x|$ are of the form $\rho \pm r$ with constant ρ. These functions measure signed distance to the sphere $|x| = \rho$. More generally, if M is a piece of hypersurface in $\mathbb{R}^n$ and $\psi(x)$ is the signed distance to M, then ψ is well defined locally and solves (5.I.3). The case where M is a sphere shows that the solution need not exist globally as there is a singularity at $x = 0$.

Example 5.I.5. The spherically symmetric solutions $\phi(t, r)$ of (5.I.4) are functions $f(t \pm |x|)$ defined for $x \neq 0$. The level surfaces of ϕ are either outgoing or incoming spheres. The incoming (resp., outgoing) solutions degenerate to a point in finite positive (resp., negative) time.

Example 5.I.6. If $c = c(r)$ the spherically symmetric solutions $\phi(t, r)$ of (5.I.5) are the solutions of the equations

$$\big(\,\partial_t \pm c(r)\,\partial_r\big)\phi = 0\,.$$

The level surfaces of ϕ are spheres moving outward for the plus sign and inward with the minus sign. The speed $c(r)$ depends on the position. The integral curves of the vector fields $\partial_t \pm c(r)\partial_r$ describe refractive effects.

5.I.2. Determining the germ of ϕ at the initial manifold. Consider the initial value problem defined by equation (1) with initial data

$$\phi|_M = g\,. \tag{5.I.8}$$

Here M is a hypersurface in $\mathbb{R}^n$ and g is smooth on M. Differentiating (5.I.8) tangent to M determines $n - 1$ components of $d\phi$. Equivalently, knowing $\phi|_M$ determines the restriction of $d\phi$ to the tangent space of M,

$$d\phi\big|_{TM} = dg\,. \tag{5.I.9}$$

The differential equation (5.I.1) must be used to determine the remaining component of $d\phi$. The next example is a generalisation of (5.I.7).

Example 5.I.7. Consider equation (5.I.4) with the initial condition

$$\phi\big|_{t=0} = g(x)\,, \qquad dg \neq 0\,. \tag{5.I.10}$$

In this case $M = \{t = 0\}$. At $t = 0$, $\partial\phi/\partial x_j = \partial g/\partial x_j$ are known functions of x. The time derivative must be found by solving (5.I.4) for ϕ_t yielding

$$\frac{\partial \phi}{\partial t} = \pm\,|\nabla_x \phi| = \pm\,|\nabla_x g|\,. \tag{5.I.11}$$

If $dg = 0$, then the equation $F(t, x, \phi_t, \phi_x) = 0$ need not be smoothly solvable for ϕ_t as a function of the other variables. This is the case for example near $x = 0 \in \mathbb{R}^n$ with $g(x) = |x|^2$.

More generally, consider equation (5.I.1) and $M = \{y_1 = 0\} \subset \mathbb{R}^n$. The initial data (5.I.8) determine $\partial\phi/\partial y_2, \ldots, \partial\phi/\partial y_n$ along M. The missing derivative $\partial\phi/\partial y_1$ must be determined by solving equation (5.I.1). The preceding examples show that there may be multiple solutions or no solution at all. In favorable cases, picking a solution $\partial\phi/\partial y_1$ at one point $\underline{x} \in M$ uniquely determines $d\phi$ locally on M.

Infinitesimal Determination Lemma 5.I.1. *Suppose that $g \in C^\infty(M)$, $\underline{y} \in M$, and $\underline{\eta} \in T^*_{\underline{x}}(\mathbb{R}^n)$ satisfy*

$$F(\underline{y}, \underline{\eta}) = 0 \qquad \text{and} \qquad \underline{\eta}|_{T_{\underline{y}}M} = dg(\underline{y})\,. \tag{5.I.12}$$

Suppose in addition that $\nabla_\eta F(\underline{y}, \underline{\eta})$ is not tangent to M at $\underline{y}$. Then on a neighborhood $\omega \subset M$ of $\underline{y}$ in M there is one and only one smooth function $\omega \ni y \mapsto \eta(y)$ satisfying

$$F(y, \eta(y)) = 0 \quad \text{on} \quad M\,, \qquad \eta|_{TM} = dg\,, \qquad \text{and} \qquad \eta(\underline{y}) = \underline{\eta}\,. \tag{5.I.13}$$

In particular, a C^1 solution of the initial value problem (5.I.1), (5.I.8) satisfying $d\phi(\underline{y}) = \underline{\eta}$ must satisfy $d\phi(y) = \eta(y)$ for all $y \in \omega$.

Proof. Introduce coordinates near $\underline{y}$ so that M is locally given by $y_1 = 0$. Then

$$\underline{\eta} = \Big(\underline{\eta}_1, \frac{\partial g(\underline{y})}{\partial y_2}, \ldots, \frac{\partial g(\underline{y})}{\partial y_n}\Big)\,.$$

Denote $y' := (y_2, \ \ldots \ , y_n)$. Then

$$\eta(y') \ = \ \Big(\eta_1(y'), \frac{\partial g(y')}{\partial y_2}, \ldots, \frac{\partial g(y')}{\partial y_n}\Big)$$

must satisfy

$$F\Big(0, y', \eta_1, \frac{\partial g(y')}{\partial y_2}, \ldots, \frac{\partial g(y')}{\partial y_n}\Big) = 0\,, \qquad \eta_1(\underline{x}) = \underline{\eta}_1\,. \tag{5.I.14}$$

The Implicit Function Theorem shows that this uniquely determines $\eta_1(y')$ on a neighborhood of $\underline{y}$ provided that $0 \neq \partial F/\partial \eta_1$. This is equivalent to $\nabla_\eta F$ not being tangent to M. □

Remark. If M is connected and $\nabla_\eta F$ is nowhere tangent to M, then the implicit function theorem argument determines η on any compact subset of M.

Remark. The linearization of the equation (5.I.1) at a solution ϕ is the partial differential operator

$$\sum_\mu \frac{\partial F}{\partial \eta_\mu}\big(x, d\phi(x)\big)\, \frac{\partial}{\partial y_\mu}\,. \tag{5.I.15}$$

When $\nabla_\eta F$ is not tangent to M, the surface M is noncharacteristic for the linearized operator. That is equivalent to the surface M being noncharacteristic along the solution ϕ of (5.I.1) (see [Rauch, 1991, §1.5]). At a noncharacteristic surface, not only is $d\phi$ determined, but so are all the higher partial derivatives of ϕ.

5.I.3. Propagation laws for $\phi, d\phi$. To analyse equation (5.I.1), differentiate with respect to y_ν. This is an example of a general strategy whereby differentiating a fully nonlinear equation yields a quasilinear equation for the derivatives of a solution. In the case of a first order real scalar equation, this simple idea solves the problem. Differentiating (5.I.1) with respect to y_ν for $1 \le \nu \le n$ yields the n equations,

$$\frac{\partial F}{\partial y_\nu}(y, d\phi(y)) + \sum_\mu \frac{\partial F}{\partial \eta_\mu}(y, d\phi(y))\, \frac{\partial^2 \phi(y)}{\partial y_\nu \partial y_\mu} = 0\,, \qquad 1 \le \nu \le n\,. \tag{5.I.16}$$

Identity (5.I.16) shows that the derivative of $d\phi$ in the direction $\nabla_\eta F$ is equal to $-\nabla_y F$. These are the y and η parts of the Hamilton vector field

$$V_F := \sum_\mu \frac{\partial F(y,\eta)}{\partial \eta_\mu}\frac{\partial}{\partial y_\mu} - \frac{\partial F(y,\eta)}{\partial y_\mu}\frac{\partial}{\partial \eta_\mu} \tag{5.I.17}$$

associated to the hamiltonian F.

Propagation Lemma 5.I.2. *Suppose that ϕ is a solution of (5.I.1) and that $\big(y(s), \eta(s)\big)$ is an integral curve of the vector field V_F, that is*

$$\frac{dy(s)}{ds} = \frac{\partial F}{\partial \eta}(y(s), \eta(s))\,, \qquad \frac{d\eta(s)}{ds} = -\frac{\partial F}{\partial y}(y(s), \eta(s))\,. \tag{5.I.18}$$

If $\eta(0) = d\phi(y(0))$, then $\eta(s) = d\phi(y(s))$ so long as $y([0, s])$ belongs to the domain on which ϕ satisfies (5.I.1). If the system (5.I.18) is expanded to include

$$\frac{d\rho(s)}{ds} = \sum_\mu \eta_\mu(s)\, \frac{\partial F}{\partial \eta_\mu}(y(s), \eta(s))\,, \qquad \rho(0) = \phi(y(0))\,, \tag{5.I.19}$$

then $\rho(s) = \phi(y(s))$ on the same interval of s.

Proof. Define $Y(s)$ as the solution of $dY/ds = -\nabla_\eta F(Y(s), d\phi(Y(s))$ with $Y(0) = y(0)$. Define $\Xi(s) := d\phi(Y(s))$. Then equation (5.I.16) is equivalent to $d\Xi/ds = -\nabla_y F(Y(s), \Xi(s))$. Thus $(Y(s), \Xi(s))$ solves the same initial value problem as $(y(s), \eta(s))$. By uniqueness, $(y(s), \eta(s)) = (Y(s), \Xi(s))$.

Similarly,

$$\frac{d\phi(y(s))}{ds} = \sum_\mu \frac{\partial \phi(y(s))}{\partial y_\mu}\, \frac{d y_\mu(s)}{ds} = \sum_\mu \eta_\mu(s)\, \frac{\partial F}{\partial \eta_\mu}(y(s), \eta(s))\,,$$

so $(y(s), \eta(s), \phi(y(s))$ solves the y, η, ρ system of ordinary differential equations and the last assertion of the proposition also follows from uniqueness. □

Example 5.I.8. The integral curves of V in the case of equation (5.I.5) are solutions of the system of ordinary differential equations

$$\text{(5.I.20)} \qquad \frac{dt}{ds} = 2\tau\,, \qquad \frac{dx}{ds} = -2c^2\xi\,, \qquad \frac{d\tau}{ds} = 0\,, \qquad \frac{d\xi}{ds} = 2\xi^2 c\,\frac{\partial c(x)}{\partial x}\,.$$

The velocity is

$$\frac{dx}{dt} = \frac{dx/ds}{dt/ds} = \frac{-2c^2\xi}{2\,\tau} = \frac{-2c^2\xi}{\pm c|\xi|} = \mp c\,|\xi|.$$

This is equal to the group velocity associated to the root $\tau = \pm c|\xi|$. This gives a derivation of the group velocity complementary to the earlier ones.

If you know the values of $d\phi(0, x)$, then the proposition tells you that $d\phi(t(s), x(s)) = (\tau(s), \xi(s))$ where $(t(s), x(s), \tau(s), \xi(s))$ is a solution of (5.I.20). This allows you to compute $d\phi$ in $t > 0$ from its values at $t = 0$. The curves $(t(s), x(s), \tau(s), \xi(s))$ are called *bicharacteristic strips* or simply *bicharacteristics*, and their projections on (t, x) space are called *rays*. The bicharacteristics describe how the values of $d\phi$ are propagated along rays. The speed of the rays are equal to the local group velocity.

There are two intuitive ways to think of the integral curves $(y(s), \eta(s))$. The first is to note that $\eta(s) = d\phi(y(s))$ determines the tangent plane to the graph of ϕ at $y(s)$. One has a curve of tangent planes to the solution surface. Think of the level surfaces to ϕ as surfaces of constant phase. Then $d\phi(y(s))$ is conormal to the surfaces of constant phase. It can be viewed as giving an infinitesimal element of an oscillatory solution. Then the proposition can viewed as a propagation law for infinitesimal oscillations. The example $F = \tau^2 - \xi_1^2 - 4\xi_2^2$ shows that the direction of propagation dx/dt need not be parallel to ξ (see Exercise 2.4.9).

Covering Lemma 5.I.3. *Suppose that the restriction of $d\phi$ to M is known and that $\nabla_\eta F(\underline{y}, d\phi(\underline{y}))$ is not tangent to M at $\underline{y}$. For $q \in M$ denote by $(y(s, q), \eta(s, q))$ the solution of (5.I.18) with $y(0) = q$, $\eta(0) = d\phi(q)$. Then the map $s, q \mapsto y(s, q)$ is a diffeomorphism of a neighborhood of $(0, \underline{y})$ in $\mathbb{R} \times M$ to a neighborhood of $\underline{y}$ in $\mathbb{R}^n$. Equivalently, the family of rays $y(\cdot, q)$ parametrized by $q \in M$ simply covers a neighborhood of $\underline{y}$.*

Proof. Introduce local coordinates on M and consider the jacobian matrix of the mapping y at $(s, q) = (0, \underline{y})$. The last $n - 1$ columns of J span the tangent space of M. The first column is parallel to $dy/ds|_{(0,q)} = \nabla_\eta F(\underline{y}, \underline{\eta})$, which is not tangent to M. Thus the columns span $\mathbb{R}^n$, and the result follows from the Inverse Function Theorem. □

Main Theorem 5.I.4. *Suppose that data are given satisfying the following conditions.*

(i) *M is a hypersurface in $\mathbb{R}^n$.*

(ii) *$g \in C^\infty(M)$.*

(iii) *$(\underline{y}, \underline{\eta})$ satisfies $F(\underline{y}, \underline{\eta}) = 0$ and $(\sum_\mu \underline{\eta}_\mu\, dy_\mu)|_{T_{\underline{y}}(M)} = dg(\underline{y})$.*

(iv) *$\nabla_\eta F(\underline{y}, \underline{\eta})$ is not tangent to M at $\underline{y}$.*

Then, there is a smooth solution ϕ of (5.I.1) satisfying (5.I.17) on a neighborhood of $\underline{y}$ in M and $d\phi(\underline{y}) = \underline{\eta}$. Any two such solutions must coincide on a neighborhood of $\underline{y}$.

Proof. The previous results combine to prove uniqueness as follows. The Infinitesimal Determination Lemma determines $d\phi$ along M. Let $y(s,q), \eta(s,q), \rho(s,q)$ be the solutions of the system (5.I.18), (5.I.19). They are parametrized by $q \in M$. The Propagation Lemma implies that

$$\phi\big(y(s,q)\big) = \rho(s,q)\,. \tag{5.I.21}$$

The Covering Lemma shows that (5.I.21) uniquely determines ϕ on a neighborhood of $\underline{y}$.

To prove existence, we show that the function defined by (5.I.21) on a neighborhood of $\underline{y}$ in $\mathbb{R}^n$ furnishes a solution.

The initial conditions (5.I.19) for ρ show that ϕ defined by (5.I.21) satisfies $\phi|_M = g$.

The initial values $y(0,q), \eta(0,q)$ determined in the Infinitesimal Determination Lemma, satisfy $F(y,\eta) = 0$. The classic computation of conservation of energy shows that F is constant along solution curves of (5.I.18). This proves that $F\big(y(s,q), \eta(s,q)\big) = 0$. To prove that ϕ solves (5.I.1) it suffices to prove that

$$\frac{\partial \phi}{\partial y_\mu}(y(s,q)) = \eta_\mu(s,q)\,. \tag{5.I.22}$$

Define $\zeta(y) = \big(\zeta_1(y), \dots, \zeta_n(y)\big)$ on a neighborhood of $\underline{y}$ by

$$\zeta_\mu(x(s,q)) = \eta_\mu(s,q)\,. \tag{5.I.23}$$

The desired relation (5.I.22) is

$$\frac{\partial \phi}{\partial y_\mu} = \zeta_\mu\,, \qquad 1 \le \mu \le n\,. \tag{5.I.24}$$

Equation (5.I.24) asserts that the predicted value of the differential is the same as what one obtains by differentiating the predicted value of the function. Define a vector field on a neighborhood of $\underline{x}$ by

$$W := \sum_\mu \frac{\partial F}{\partial \eta_\mu}(y, \zeta(y))\, \frac{\partial}{\partial y_\mu}\,.$$

The defining relations for ϕ and ζ yield

$$W\,\phi \;=\; \sum_\mu \zeta_\mu(y)\,\frac{\partial F}{\partial \eta_\mu}\big(y,\zeta(y)\big)\,, \tag{5.I.25}$$

$$W\,\zeta_\mu \;=\; -\frac{\partial F}{\partial y_\mu}\big(y,\zeta(y)\big)\,, \qquad 1\le \mu\le n\,. \tag{5.I.26}$$

Differentiating (5.I.25) yields

$$W\,\frac{\partial\phi}{\partial y_\mu} + \Big(\sum_{\mu,\nu}\frac{\partial^2 F}{\partial\eta_\mu\partial y_\nu} + \frac{\partial^2 F}{\partial\eta_\mu\partial\eta_\nu}\,\frac{\partial\zeta_\nu}{\partial y_\mu}\Big)\,\frac{\partial\phi}{\partial y_\mu}$$
$$= \sum_\mu\left(\frac{\partial\zeta_\mu}{\partial y_\nu}\,\frac{\partial F}{\partial\eta_\mu} + \zeta_\mu\sum_\nu\Big(\frac{\partial^2 F}{\partial\eta_\mu\partial y_\nu} + \frac{\partial^2 F}{\partial\eta_\mu\partial\eta_\nu}\,\frac{\partial\zeta_\nu}{\partial y_\mu}\Big)\right).$$

Formulas (5.I.23) and (5.I.26) show that the first sum on the right is equal to $W\zeta_\mu$. Therefore

$$W\Big(\frac{\partial\phi}{\partial y_\mu} - \zeta_\mu\Big) + \Big(\sum_{\mu,\nu}\frac{\partial^2 F}{\partial\eta_\mu\partial y_\nu} + \frac{\partial^2 F}{\partial\eta_\mu\partial\eta_\nu}\,\frac{\partial\zeta_\nu}{\partial y_\mu}\Big)\Big(\frac{\partial\phi}{\partial y_\mu} - \zeta_\mu\Big) = 0\,. \tag{5.I.27}$$

This homogeneous linear ordinary differential equation shows that $\partial\phi/\partial y_\mu - \zeta_\mu$ vanishes on a ray as soon as it vanishes at the foot of that ray on M. Thus it suffices to show that (5.I.24) is satisfied on M.

To verify (5.I.24) along M it suffices to find n linearly independent vectors v so that $v.d\phi = v.\zeta$. The Infinitesimal Determination Lemma gives $\zeta(x) = \eta(x)$ on M so, in particular for v tangent to M, $v.\zeta = v.\eta = v.dg = v.d\phi$. For v equal to the field W, equation (5.I.25) yields $v.d\phi = \zeta.\nabla_\eta F(y,\zeta) = \zeta.v$. Hypothesis (iv) shows that W is not tangent to M near $\underline{y}$, so the fields just checked span and the proof is complete. □

Remark. The reality of F and the hypothesis that $\nabla_\eta F$ is not tangent to M, shows that (5.I.1) is a strictly hyperbolic partial differential equation at $\underline{y}$ with time-like direction given by the conormal to M. One generalization of the Main Theorem is a local solvability result for strictly hyperbolic nonlinear initial value problems. From that perspective, the proof just presented is the *method of characteristics* of §1.1 applied to fully nonlinear scalar equations.

5.I.4. The symplectic approach. The appearance of the Hamilton field V_F shows that the construction has a link with symplectic geometry. The connection offers an alternative way to prove the Main Theorem. Begin by reinterpreting the Propagation Lemma. Introduce the graph of the differential $d\phi$

$$\Lambda := \big\{\,(y,\eta)\;:\;\eta = d\phi(y)\,\big\}\,.$$

This is a smooth n dimensional surface in the $2n$ dimensional space of (y, η). It is defined by the n equations

$$\eta_\mu \;=\; \frac{\partial \phi(y)}{\partial y_\mu}\,, \qquad 1 \le \mu \le n\,. \tag{5.I.28}$$

Instead of looking for ϕ we look for Λ. Identify the space of (x, η) as the cotangent bundle of $\mathbb{R}^n$ with its symplectic form

$$\sigma \;:=\; \sum_{\mu=1}^{n} d\eta_\mu \wedge dy_\mu\,.$$

The equality of mixed partials shows that Λ is *Lagrangian* in the sense that it is an n-manifold such that $\sigma(v, w) = 0$ whenever v and w are tangent to Λ. Equation (5.I.1) is thus turned into the search for a Lagrangian manifold that lies in the set $\{(y, \eta) \,:\, F(y, \eta) = 0\,\}$.

Tangency Lemma 5.I.5. *If ϕ is a solution of (5.I.1), then the vector field V_F is tangent to the surface Λ. Equivalently, an integral curve of V_F that has one point in Λ lies in Λ.*

Proof. Since the surface Λ lies in the level set $\{F = 0\}$, one has $\langle dF, v\rangle = 0$ for all tangent vectors v to Λ. The definition of Hamilton field then shows that $\sigma(V_F, v) = \langle dF, v\rangle = 0$ for all such v. Thus V_F belongs to the σ annihilator of the tangent space of Λ. Since Λ is Lagrangian, this annihilator is the tangent space to Λ so V_F is tangent to Λ. □

An $n - 1$ dimensional piece Λ_0 of Λ is given in the Infinitesimal Determination Lemma. Precisely,

$$\Lambda_0 := \big\{\, (y, d\phi(y)) \,:\, y \in M \,\big\} \tag{5.I.29}$$

is known from the initial data. Then one takes the union of the bicharacteristics through Λ_0 to define Λ. The next result follows from the Inverse Function Theorem.

Flow Out Lemma 5.I.6. *If Λ_0 is a smooth embedded $n - 1$ dimensional surface in $2n$ dimensional (y, η) space that is nowhere tangent to the vector field V_F, then locally the union of integral curves of V_F starting in Λ_0 defines a smooth n dimensional manifold.*

The next question to resolve is whether Λ so defined is a graph; that is, are there smooth functions ζ_μ so that $\Lambda = \{(y, \zeta_1(y), \dots, \zeta_n(y))\}$. Denote by $\pi(x, \xi) := x$ the natural projection from (x, ξ) space to x space.

Clean Projection Lemma 5.I.7. *Suppose that Λ_0 is defined as in (5.I.29), $\underline{y} \in M$ and that Λ is constructed by the Flow Out Lemma. Then Λ is a graph on a neighborhood of $(\underline{y}, d\phi(\underline{y}))$ if and only if $\nabla_\eta F(\underline{y}, d\phi(\underline{y}))$ is not tangent to M at $\underline{y}$.*

Proof. The variety Λ is a smooth graph if and only if π is a diffeomorphism from a neighborhood of $(\underline{y}, d\phi(\underline{y}))$ in Λ to a neighborhood of $\underline{y}$ in $\mathbb{R}^n$. The Inverse Function Theorem implies that the necessary and sufficient condition is that the differential of π is an invertible map of tangent spaces. At a point $(y,\eta) \in \Lambda_0$ the tangent space is equal to $T_{y,\eta}(\Lambda_0) \oplus V_F(y,\eta)$. This implies that

$$d\pi\big(T_{(\underline{y},d\phi(\underline{y}))}\Lambda\big) \;=\; T_{\underline{y}}M \oplus \mathbb{R}\frac{\partial F}{\partial \eta}(\underline{y}, d\phi(\underline{y}))\,. \tag{5.I.30}$$

The right-hand side of (30) is all of $\mathbb{R}^n$ if and only if $\partial F/\partial\eta$ is not tangent to M at $\underline{y}$, proving the lemma. □

Alternate proof of the Main Theorem. To prove existence, it suffices to find a function ϕ defined on an $\mathbb{R}^n$ neighborhood of $\underline{y}$ so that

$$\frac{\partial \phi}{\partial y_\mu}(y) = \zeta_\mu(y)\,, \quad 1 \le \mu \le n\,, \qquad \phi(\underline{y}) = g(\underline{y})\,. \tag{5.I.31}$$

That such a function satisfies (5.I.8) follows from $d\phi|_{TM} = dg$. A necessary and sufficient condition for the existence of such a ϕ is the equality of mixed partials,

$$\frac{\partial \zeta_\mu}{\partial y_\nu} \;=\; \frac{\partial \zeta_\nu}{\partial y_\mu} \qquad \forall\, \mu \ne \nu\,. \tag{5.I.32}$$

Exercise 5.I.1. *Prove that (5.I.32) is equivalent to* $\sigma|_{T\Lambda} = 0$.

To complete the proof, it suffices to show that $\sigma|_{T\Lambda} = 0$. Since Λ is the flowout of Λ_0 and the flow by a hamiltonian vector field preserves the two form σ, it suffices to verify that $\sigma(v,w) = 0$ whenever v and w are tangent to Λ over a point of M.

The tangent space at such a point is the direct sum of the tangent space to Λ_0 and $\mathbb{R}\,V_F$. Thus it suffices to consider

$$v = v_0 + a\,V_F\,, \qquad w = w_0 + b\,V_F\,,$$

with v_0 and w_0 tangent to Λ_0 and real a, b. Use bilinearity to express $\sigma(v,w)$ as a sum of four terms. The term $\sigma(v_0,w_0)$ vanishes since this is the symplectic form of $T^*(M)$ evaluated at a pair of tangent vectors to the Lagrangian submanifold $\{\mathrm{graph}\, dg\}$. The term $ab\,\sigma(V_F,V_F)$ vanishes since σ is antisymmetric. Finally, the cross terms are evaluated using the definition of hamiltonian vector fields,

$$\sigma(v_0, V_F) = \langle dF, v_0\rangle\,, \qquad \sigma(V_F, w_0) = -\langle dF, w_0\rangle\,. \tag{5.I.33}$$

Since Λ_0 lies in the set $\{F = 0\}$, the tangent vectors v_0 and w_0 are annihilated by dF, so the terms (5.I.33) vanish. This completes the proof. □

Chapter 6

The Nonlinear Cauchy Problem

6.1. Introduction

Nonlinear equations are classified according to the strength of the nonlinearity. The key criterion is what order terms in the equation are nonlinear. A secondary condition is the growth of the nonlinear terms at infinity. Jeffrey Rauch ¡rauch@umich.edu¿ Among the nonlinear equations in applications, two sorts are most common. *Semilinear equations* are linear in their principal part. First order semilinear symmetric hyperbolic systems take the form

$$L(y, \partial_y)\, u + F(y, u) = f(y)\,, \qquad F(y, 0) \;=\; 0\,, \tag{6.1.1}$$

where L is a symmetric hyperbolic operator, and the nonlinear function is a smooth map from $\mathbb{R}^{1+d} \times \mathbb{C}^N \to \mathbb{C}^N$ whose partial derivatives of all orders are uniformly bounded on sets of the form $\mathbb{R}^{1+d} \times K$, with compact $K \subset \mathbb{C}^N$. The derivatives are standard partial derivatives and not derivatives in the sense of complex analysis. A translation invariant semilinear equation with principal part equal to the D'Alembertian is of the form

$$\Box u + F(u, u_t, \nabla_x u) \;=\; 0, \qquad F(0,0,0) \;=\; 0\,.$$

More strongly nonlinear, and typical of compressible inviscid fluid dynamics, are the quasilinear systems,

$$L(y, u, \partial_y)\, u \;=\; f(y)\,, \tag{6.1.2}$$

where the coefficients of

$$L(y,u,\partial_y) = \sum_{j=0}^{d} A_j(y,u)\,\partial_j \tag{6.1.3}$$

are smooth hermitian symmetric matrix valued functions with derivatives bounded on $\mathbb{R}^{1+d} \times K$ as above. $A_0(y,u)$ is assumed uniformly positive on such sets. The principal part is a linear operator applied to u where the linear operator depends nonlinearly on u.

Fully nonlinear equations involve nonlinear expressions in ∂u. With the exception of the first order scalar case, we will not discuss fully nonlinear equations, though a local existence theorem is not that difficult.

For semilinear equations there is a natural local existence theorem requiring data in $H^s(\mathbb{R}^d)$ for some $s > d/2$. The theorem gives solutions that are continuous functions of time with values in $H^s(\mathbb{R}^d)$. This shows that the spaces $H^s(\mathbb{R}^d)$ with $s > d/2$ are natural configuration spaces for the dynamics. Once a solution belongs to such a space, it is bounded and continuous so that $F(y,u)$ is well defined, bounded, and continuous. Nonlinear ordinary differential equations are a special case, so for general problems one expects at most a local existence theorem.

For quasilinear equations, the local existence theorem requires an extra derivative, that is initial data in $H^s(\mathbb{R}^d)$ with $s > 1+d/2$. Again the solution is a continuous function of time with values in $H^s(\mathbb{R}^d)$. The classic example is Burgers' equation

$$u_t + u\,u_x \;=\; 0\,.$$

We treat first the semilinear case. The quasilinear case is treated in §6.6. The key step in the proof uses Schauder's lemma showing that $u \mapsto F(y,u)$ takes $H^s(\mathbb{R}^d)$ to itself.

6.2. Schauder's lemma and Sobolev embedding

The fact that H^s is invariant under nonlinear maps is closely connected to the Sobolev embedding theorems $H^s \subset L^p$ for appropriate $p(s,d)$. The simplest such L^p estimate for Sobolev spaces is the following.

Theorem 6.2.1 (Sobolev). *If $s > d/2$, $H^s(\mathbb{R}^d) \subset L^\infty(\mathbb{R}^d)$ and*

$$\|w\|_{L^\infty(\mathbb{R}^d)} \;\le\; C(s,d)\,\|w\|_{H^s(\mathbb{R}^d)}\,. \tag{6.2.1}$$

Proof. Inequality (6.2.1) for elements of the Schwartz space $\mathcal{S}(\mathbb{R}^d)$ is an immediate consequence of the Fourier inversion formula,

$$w(x) = (2\pi)^{-d/2} \int_{\mathbb{R}^d} e^{-ix\xi}\, \hat{w}(\xi)\, d\xi = (2\pi)^{-d/2} \int_{\mathbb{R}^d} \frac{e^{-ix\xi}}{\langle\xi\rangle^s}\, \langle\xi\rangle^s \hat{w}(\xi)\, d\xi\,.$$

The Schwarz inequality yields

$$|w(x)| \;\le\; \Big\|\frac{1}{\langle\xi\rangle^s}\Big\|_{L^2(\mathbb{R}^d)}\, \|w\|_{H^s(\mathbb{R}^d)}.$$

The first factor on the right is finite if and only if $s > d/2$.

For $w \in H^s$, choose $w^n \in \mathcal{S}$ with

$$w^n \to w \quad \text{in} \quad H^s, \qquad \|w_n\|_{H^s(\mathbb{R}^d)} \;\le\; \|w\|_{H^s(\mathbb{R}^d)}\,.$$

Inequality (6.2.1) yields $\|w^n - w^m\|_{L^\infty(\mathbb{R}^d)} \le C\,\|w^n - w^m\|_{H^s(\mathbb{R}^d)}$. Therefore the w^n converge uniformly on $\mathbb{R}^d$ to a continuous limit γ. Therefore $w^n \to \gamma$ in $\mathcal{D}'(\mathbb{R}^d)$ with $\|\gamma\|_{L^\infty} \le C\,\|w\|_{H^s}$

However, $w^n \to w$ in H^s and therefore in $\mathcal{D}'$, so $w = \gamma$. This proves the continuity of w and the estimate (6.2.1). □

Consider the proof of a simple case of Schauder's lemma that $u \in H^2(\mathbb{R}^2)$ implies $u^2 \in H^2(\mathbb{R}^2)$. One must show that u^2, $\partial(u^2)$, and $\partial^2(u^2)$ are square integrable. The function u^2 and its first derivative $\partial(u^2) = 2u\partial u$, are both the product of a bounded function and a square integrable function, and so are in L^2. Compute the second derivative

$$\partial^2(u^2) \;=\; u\partial^2 u + 2(\partial u)^2\,.$$

The first is a product $L^\infty \times L^2$ so is L^2. For the second, one needs to know that $\partial u \in L^4$. The fact that $H^2(\mathbb{R}^2)$ is invariant is implied by the Sobolev embedding $H^1(\mathbb{R}^2) \subset L^4(\mathbb{R}^2)$.

Theorem 6.2.2 (Schauder's lemma). *Suppose that*

$$G(x,u) \in C^\infty(\mathbb{R}^d \times \mathbb{C}^N\,;\,\mathbb{C}^N)$$

such that $G(x,0) = 0$, *and for all* $|\alpha| \le s+1$ *and compact* $K \subset \mathbb{C}^N$, $\partial^\alpha_{x,u} G \in L^\infty(\mathbb{R}^d \times K)$. *Then the map* $w \mapsto G(x,w)$ *sends* $H^s(\mathbb{R}^d)$ *to itself provided* $s > d/2$. *The map is uniformly lipschitzean on bounded subsets of* $H^s(\mathbb{R}^d)$.

Proof of Schauder's lemma for integer s. Consider $G = G(w)$. The case of G depending on x is uglier but requires no additional ideas. The key step is to estimate the H^s norm of $G(w)$ assuming that $w \in \mathcal{S}$. We prove that

$$\forall\, R,\;\; \exists\, C = C(R),\;\; \forall\, w \in \mathcal{S}(\mathbb{R}^d),$$
$$\|w\|_{H^s(\mathbb{R}^d)} \le R \quad \Longrightarrow \quad \|G(w)\|_{H^s(\mathbb{R}^d)} \le C(R)\,.$$

This suffices to prove the first assertion of the theorem since for $w \in H^s$, choose $w^n \in \mathcal{S}$ with

$$w^n \to w \quad \text{in} \quad H^s, \qquad \|w^n\|_{H^s(\mathbb{R}^d)} \;\le\; \|w\|_{H^s(\mathbb{R}^d)}\,.$$

Then Sobolev's theorem implies that w^n converges uniformly on $\mathbb{R}^d$ to w, so $G(w^n)$ converges uniformly to $G(w)$. In particular, $G(w^n) \to G(w)$ in $\mathcal{D}'(\mathbb{R}^d)$.

However, $G(w^n)$ is bounded in H^s so, passing to a subsequence, we may suppose that $G(w^n) \to v$ weakly in H^s. Therefore $G(w^n) \to v$ in $\mathcal{D}'(\mathbb{R}^d)$. Equating the $\mathcal{D}'$ limits proves that $G(w) \in H^s$.

For $\|w\|_{H^s} \le R$ there is a constant so that $\|w\|_{L^\infty} \le C$. Choose $\Gamma > 0$ so that

$$\|w_j\|_{L^\infty} \le C \quad \Longrightarrow \quad |G(w_1) - G(w_2)| \le \Gamma\, |w_1 - w_2|\,.$$

Apply with $w_2 = G(w_2) = 0$ to show that $\|G(w_1)\|_{L^2} \le C$.

It remains to estimate the derivatives of $G(w)$. For $w \in \mathcal{S}$, Leibniz's rule implies that $\partial_x^\beta G(w(x))$ with $|\beta| \le s$ is a finite sum of terms of the form

$$G^{(\gamma)}(w) \prod_{j=1}^{J} \partial_x^{\alpha_j} w\,, \qquad |\gamma| = J \le |\beta|\,, \qquad \alpha_1 + \cdots + \alpha_J = \beta\,. \tag{6.2.2}$$

This is proved by induction on $|\beta|$. Increasing the order by one, the additional derivative either falls on the G term yielding an expression of the desired form with $|\gamma| = J$ increased by one, or on one of the factors in $\prod \partial_x^{\alpha_j} w$ yielding an expression of the desired form with the same value of γ.

Sobolev's theorem implies that

$$\begin{aligned}\|G^{(\gamma)}(w) \prod_{j=1}^{J} \partial_x^{\alpha_j} w\|_{L^2} &\le \|G^{(\gamma)}(w)\|_{L^\infty} \| \prod_{j=1}^{J} \partial_x^{\alpha_j} w\|_{L^2} \\ &\le C(R)\, \| \prod_{j=1}^{J} \partial_x^{\alpha_j} w\|_{L^2}.\end{aligned} \tag{6.2.3}$$

Following [Rauch, 1983], we use the Fourier transform to prove the key estimate.

Lemma 6.2.3. *If $s > d/2$, there is a constant $C = C(s,d)$ so that for all $w_j \in \mathcal{S}(\mathbb{R}^d)$ and all multi-indices α_j with $s' := \sum |\alpha_j| \le s$,*

$$\| \prod_{j=1}^{J} \partial_x^{\alpha_j} w_j \|_{L^2(\mathbb{R}^d)} \le C \prod_{j=1}^{J} \|w_j\|_{H^s(\mathbb{R}^d)}\,.$$

Example 6.2.1. For $u \in H^2(\mathbb{R}^2)$, $(\partial u\, \partial u) \in L^2$ so $\partial u \in L^4$.

Proof. By Plancherel's theorem, it suffices to estimate the L^2 norm of the Fourier transform of the product. Set

$$g_i \;:=\; \langle\xi\rangle^{s-|\alpha_i|}\mathcal{F}\big(\partial_x^{\alpha_i} w_i\big)\,,\qquad \langle\xi\rangle \;:=\; (1+|\xi|^2)^{1/2}\,,$$

$$\text{so}\qquad \|g_i\|_{L^2} \le c\|w_i\|_{H^s}\,.$$

Ignoring the factors of 2π, compute

(6.2.4)

$$\mathcal{F}\Big(\prod_{j=1}^{J}\partial_x^{\alpha_j} w_{l_j}\Big)(\xi_1) \;=\; \frac{g_1}{\langle\xi\rangle^{s-|\alpha_1|}} * \frac{g_2}{\langle\xi\rangle^{s-|\alpha_2|}} * \cdots * \frac{g_J}{\langle\xi\rangle^{s-|\alpha_J|}}(\xi_1)$$

$$= \int_{\mathbb{R}^{d(J-1)}} \frac{g_1(\xi_1-\xi_2)}{\langle\xi_1-\xi_2\rangle^{s-|\alpha_1|}}\,\frac{g_2(\xi_2-\xi_3)}{\langle\xi_2-\xi_3\rangle^{s-|\alpha_2|}}\cdots\frac{g_J(\xi_J)}{\langle\xi_j\rangle^{s-|\alpha_J|}}\;d\xi_2\ldots d\xi_J\,.$$

For each ξ, at least one of the J numbers $\langle\xi_1-\xi_2\rangle,\ldots,\langle\xi_{J-1}-\xi_J\rangle,\langle\xi_J\rangle$ is maximal. Suppose it's the bth number $\langle\xi_b-\xi_{b+1}\rangle$ with the convention that $\xi_{J+1}\equiv 0$. Then since $\sum|\alpha_i|\le s$,

$$\langle\xi_b-\xi_{b+1}\rangle^{s-|\alpha_b|} \;\ge\; \langle\xi_b-\xi_{b+1}\rangle^{\sum_{j\neq b}|\alpha_j|} \;\ge\; \prod_{j\neq b}\langle\,\xi_j-\xi_{j+1}\rangle^{|\alpha_j|}\,.$$

Therefore

$$\prod_{j=1}^{J}\langle\xi_j-\xi_{j+1}\rangle^{s-|\alpha_j|}$$

$$=\;\langle\xi_b-\xi_{b+1}\rangle^{s-|\alpha_b|}\prod_{j\neq b}\langle\xi_j-\xi_{j+1}\rangle^{s-|\alpha_j|} \;\ge\; \prod_{j\neq b}\langle\xi_j-\xi_{j+1}\rangle^{s}\,.$$

Thus the integrand on the right-hand side of (6.2.4) is dominated by

$$\Big|\,g_b(\xi_b-\xi_{b+1})\prod_{j\neq b}\frac{g_j(\xi_j-\xi_{j+1})}{\langle\xi_j-\xi_{j+1}\rangle^{s}}\,\Big|\,. \tag{6.2.5}$$

Thus for any $\xi_1\in\mathbb{R}^d$, the integrand in (6.2.4) is dominated by the sum over b of the terms (6.2.5). Hence

$$\|\mathcal{F}\Big(\prod_{j=1}^{J}\partial_x^{\alpha_j} w_{l_j}\Big)\|_{L^2(\mathbb{R}^d)}$$

$$\le\;\Big\|\sum_{b=1}^{J}\frac{|g_1|}{\langle\xi\rangle^s}*\cdots*|g_b|*\frac{|g_{b+1}|}{\langle\xi\rangle^s}*\cdots*\frac{|g_J|}{\langle\xi\rangle^s}\Big\|_{L^2}$$

$$\le\;\sum_{b=1}^{J}\Big\|\frac{g_1}{\langle\xi\rangle^s}\Big\|_{L^1}\cdots\Big\|\frac{g_{b-1}}{\langle\xi\rangle^s}\Big\|_{L^1}\|g_b\|_{L^2}\Big\|\frac{g_{b+1}}{\langle\xi\rangle^s}\Big\|_{L^1}\cdots\Big\|\frac{g_J}{\langle\xi\rangle^s}\Big\|_{L^1},$$

where the last step uses Young's inequality.

As in Sobolev's theorem, $s > d/2 \Rightarrow \langle\xi\rangle^{-s} \in L^2(\mathbb{R}^d)$ and the Schwarz inequality yields

$$\Big\| \frac{g_j}{\langle\xi\rangle^s} \Big\|_{L^1} \ \le\ C_1 \,\|g_j\|_{L^2} \ \le\ C_2 \,\|w_j\|_{H^s}.$$

Using this in the previous estimate proves the lemma. □

To prove the lipschitz continuity asserted in Schauder's lemma it suffices to show that for all R there is a constant $C(R)$ so that

$$w_j \in \mathcal{S}(\mathbb{R}^d) \quad \text{for } j = 1, 2 \qquad \text{and} \qquad \|w_j\|_{H^s(\mathbb{R}^d)} \le R$$

imply

$$\|G(w_1) - G(w_2)\|_{H^s(\mathbb{R}^d)} \ \le\ C \,\|w_1 - w_2\|_{H^s(\mathbb{R}^d)}\,.$$

Taylor's theorem expresses

$$G(w_1) - G(w_2) = \int_0^1 G'(w_2 + \theta(w_1 - w_2))\, d\theta \ (w_1 - w_2)\,.$$

The estimates of the first part show that the family of functions $G'(w_2 + \theta(w_1 - w_2))$ parametrized by θ is bounded in $H^s(\mathbb{R}^d)$. Thus

$$\Big\| \int_0^1 G'(w_2 + \theta(w_1 - w_2))\, d\theta \Big\|_{H^s(\mathbb{R}^d)} \ \le\ C(R)\,.$$

Applying the lemma to the expression for $G(w_1) - G(w_2)$ as a product of two terms completes the proof. □

The standard proof of Schauder's lemma for integer s uses the L^p version of the Sobolev Embedding Theorem.

Sobolev Embedding Theorem 6.2.4. *If $1 \le s \in \mathbb{R}$ and $\alpha \in \mathbb{N}^d$ is a multi-index with $0 < s - |\alpha| < d/2$, there is a constant $C = C(\alpha, s, d)$ independent of $u \in H^s(\mathbb{R}^d)$ so that*

$$\| \,\partial_y^\alpha u\, \|_{L^{p(\alpha)}} \ \le\ C \,\|\, |\xi|^s \,\hat{u}(\xi)\, \|_{L^2(\mathbb{R}^d)}\,, \tag{6.2.6}$$

where

$$p(\alpha) \ :=\ \frac{2d}{d - 2s + 2|\alpha|}\,. \tag{6.2.7}$$

For $s - |\alpha| > d/2$, $\partial_y^\alpha u$ is bounded and continuous and

$$\|\partial_y^\alpha u\|_{L^\infty} \ \le\ C\|u\|_{H^s(\mathbb{R}^d)}\,.$$

For $s - |\alpha| = d/2$, one has

$$\|\partial_y^\alpha u\|_{L^p(\mathbb{R}^d)} \ \le\ C(p, s, \alpha)\, \|u\|_{H^s(\mathbb{R}^d)}$$

for all $2 \le p < \infty$.

Proofs can be found in [Hörmander, 1983] and [Taylor, 1997]. When $p(\alpha)$ is an integer, the estimate can be proved using Lemma 6.2.3. The formula for $p(\alpha)$ is forced by dimensional analysis. For a fixed nonzero $\psi \in C_0^\infty$, consider $u_\lambda(x) := \psi(\lambda x)$. The left-hand side of (6.2.6) then is of the form $c\lambda^a$ for some a. Similarly, the right-hand side is of the form $c'\lambda^b$ for some b. In order for the inequality to hold, one must have $\lambda^a \le c''\lambda^b$ for all $\lambda \in]0,\infty[$, so it is necessary that $a = b$.

Exercise 6.2.1. *Show that $a = b$ if and only if p is given by (6.2.7).*

Another way to look at the scaling argument is that for dimensionless u the left-hand side of (6.2.6) has dimensions length$^{(d-p|\alpha|)/p}$ while the right-hand side has dimensions length$^{(d-2s)/2}$. The formula for p results from equating these two expressions.

Standard proof of Schauder's lemma. The usual proof for integer s uses the Sobolev estimates together with Hölder's inequality. Hölder's inequality yields

$$\sum_{k=1}^{J} \frac{1}{p_k} = \frac{1}{2} \qquad \Longrightarrow \qquad \big\| \, \partial_x^{\alpha_1} w_{l_1} \cdots \partial_x^{\alpha_J} w_{l_J} \, \big\|_{L^2} \le \prod_{k=1}^{J} \|\partial_x^{\alpha_k} w_{l_k}\|_{L^{p_k}} \,.$$

Since each factor $\partial_x^{\alpha_k} w_{l_k}$ belongs to L^2, it suffices to find q_k so that

$$\partial_x^{\alpha_k} w_{l_k} \in L^{q_k} \quad \text{and} \quad \sum \frac{1}{q_k} \le \frac{1}{2}\,.$$

Let $\mathcal{B}$ denote the set of $k \in \{1,\dots,J\}$ so that $s - |\alpha_k| > d/2$. For these indices the factor in our product is bounded, and so for $k \in \mathcal{B}$ set $q_k := \infty$.

Let $\mathcal{A} \subset \{1,\dots,J\}$ denote those indices i for which $s - |\alpha_i| < \frac{d}{2}$. For $k \in \mathcal{A}$, q_k is chosen as in Sobolev's theorem,

$$q_k := \frac{2d}{d - 2s + 2|\alpha_k|}\,.$$

If $s - |\alpha_k| = \frac{d}{2}$, the factor in the product belongs to L^p for all $2 \le p < \infty$, and the choice of q_k in this range is postponed.

With these choices, the Sobolev embedding theorem estimates

$$\|\partial_x^{\alpha_k} w_{l_k}\|_{L^{q_k}} \le C\|w\|_{H^s(\mathbb{R}^d)}\,.$$

Then since $\sum |\alpha_i| \le s$, and $s > d/2$,

$$\sum_{i\in\mathcal{A}\cup\mathcal{B}} \frac{1}{q_i} = \sum_{i\in\mathcal{A}} \frac{1}{q_i} = \sum_{i\in\mathcal{A}} \frac{d - 2s + 2|\alpha_i|}{2d} \le \frac{Jd - 2Js + 2s}{2d}$$

$$= \frac{Jd - 2(J-1)s}{2d} < \frac{Jd - (J-1)d}{2d} = \frac{1}{2}\,.$$

This shows there is room to pick large q_k corresponding to the case $s-|\alpha_k| = d/2$ so that $\sum 1/q_k < 1/2$, and the proof is complete. □

Another nice proof of Schauder's lemma can be found in [Beals, 1989, pp. 11–12]. Other arguments can be built on the Littlewood–Paley decomposition of $G(w)$ as in [Bony, 1981] and [Meyer, 1981] and presented in [Alinhac and Gerard, 2007] and [Taylor, 1997], or on the representation

$$G(u) \;=\; \int \hat{G}(\xi)\,(e^{iu\xi}-1)\;d\xi\,.$$

The latter requires that one prove a bound on the norm of $e^{iu\xi}-1$ (see [Rauch and Reed, 1982]) growing at most polynomially in ξ. The last two arguments have the advantage of working when s is not an integer.

6.3. Basic existence theorem

The basic local existence theorem follows from Schauder's lemma and the linear existence theorem. Schauder (1935) proved a quasilinear second order scalar version, but his argument, which is recalled in [Courant and Hilbert, 1953, §VI.10], works without essential modification once the linear energy inequalities of Friedrichs are added. The following existence proof is inspired by Picard's argument for ordinary differential equations. As in §1.1, Picard's bounds (6.3.8) replace the standard and less precise contraction argument.

Theorem 6.3.1. *If $s > d/2$ and $f \in L^1_{\rm loc}\big([0,\infty[\,;\,H^s(\mathbb{R}^d)\big)$, then there is a $T \in\,]0,1]$ and a unique solution $u \in C([0,T]\,;\,H^s(\mathbb{R}^d))$ to the semilinear initial value problem defined by the partial differential equation (6.1.1) together with the initial condition*

$$u(0,x) \;=\; g(x) \in H^s(\mathbb{R}^d). \tag{6.3.1}$$

The time T can be chosen uniformly for f and g from bounded subsets of $L^1([0,1]\,;\,H^s(\mathbb{R}^d))$ and $H^s(\mathbb{R}^d)$, respectively. Consequently, there is a $T^* \in]0,\infty]$ and a maximal solution $u \in C\big([0,T^*[\,;\,H^s(\mathbb{R})^d)\big)$. If $T^* < \infty$, then

$$\lim_{t\to T^*} \|u(t)\|_{H^s(\mathbb{R}^d)} \;=\; \infty\,. \tag{6.3.2}$$

Proof. The solution is constructed as the limit of Picard iterates. The first approximation is not really important. Set

$$\forall\, t,x\,, \qquad u^1(t,x) \;:=\; g(x)\,.$$

For $\nu > 1$, the basic linear existence theorem implies that the Picard iterates defined as solutions of the initial value problems

$$L(y,\partial_y)\,u^{\nu+1} + F(y,u^\nu) \;=\; f(y)\,, \qquad u^{\nu+1}(0) = g,$$

are well defined elements of $C([0,\infty[\,;\,H^s(\mathbb{R}^d)\,)\,.$

Let C denote the constant in the linear energy estimate (2.2.2). Choose a real number

$$R \;>\; 2C\,\|g\|_{H^s(\mathbb{R}^d)}\,. \tag{6.3.3}$$

Schauder's lemma implies that there is a constant $B(R) > 0$ so that

$$\|w(t,\cdot)\|_{H^s(\mathbb{R}^d)} \le R \qquad \Longrightarrow \qquad \|F(t,\cdot,w(\cdot))\|_{H^s(\mathbb{R}^d)} \;\le\; B\,.$$

Thanks to (6.3.3) one can choose $T > 0$ so that

$$C\left(e^{CT}\|g\|_{H^s(\mathbb{R}^d)} + \int_0^T e^{C(T-\sigma)}\big(B + \|f(\sigma)\|_{H^s(\mathbb{R}^d)}\big)d\sigma\right) \le R\,. \tag{6.3.4}$$

Using (2.2.2) shows that for all $\nu \ge 1$ and all $0 \le t \le T$,

$$\|u^\nu(t)\|_{H^s(\mathbb{R}^d)} \;\le\; R\,. \tag{6.3.5}$$

Schauder's lemma implies that there is a constant Λ so that for all $t \in [0,T]$ and $w_j \in H^s(\mathbb{R}^2)$ with $\|w_j\|_{H^s(\mathbb{R}^d)} \le R$,

$$\|F(t,x,w_1(x)) - F(t,x,w_2(x))\|_{H^s(\mathbb{R}^d_x)} \le \Lambda\|w_1 - w_2\|_{H^s(\mathbb{R}^d_x)}\,. \tag{6.3.6}$$

Then for $\nu \ge 2$, (2.2.2) applied to the difference $u^{\nu+1} - u^\nu$ implies that (6.3.7)

$$\|u^{\nu+1}(t) - u^\nu(t)\|_{H^s(\mathbb{R}^d)} \;\le\; C\,\Lambda \int_0^t e^{C(t-\sigma)}\,\|u^\nu(\sigma) - u^{\nu-1}(\sigma)\|_{H^s(\mathbb{R}^d)}\,d\sigma\,.$$

Define

$$M_1 \;:=\; \sup_{0\le t\le T}\, \|u^1(t) - u^2(t)\|_{H^s(\mathbb{R}^d)} \qquad \text{and} \quad M_2 \;:=\; C\,\Lambda\, e^{CT}.$$

An induction on ν using (6.3.7) shows that for all $\nu \ge 2$

$$\|u^{\nu+1}(t) - u^\nu(t)\|_{H^s(\mathbb{R}^d)} \;\le\; M_1\,\frac{(M_2 t)^{\nu-1}}{(\nu-1)!}\,. \tag{6.3.8}$$

Exercise 6.3.1. *Prove (6.3.8).*

Estimate (6.3.8) shows that $\{u^\nu\}$ is a Cauchy sequence in $C([0,T]\,;\,H^s(\mathbb{R}^d))$. Denote by u the limit. Passing to the limit in the initial value problem defining $u^{\nu+1}$ shows that u satisfies the initial value problem (6.1.1), (6.3.1). This completes the proof of existence.

Uniqueness is a consequence of the inequality

$$\|u_1(t) - u_2(t)\|_{H^s(\mathbb{R}^d)} \;\le\; C_1 \int_0^t e^{C(t-\sigma)}\,\|u_1(\sigma) - u_2(\sigma)\|_{H^s(\mathbb{R}^d)}\,d\sigma\,, \tag{6.3.9}$$

which is proved exactly as (6.3.7). Gronwall's inequality implies that $\|u_1 - u_2\| \equiv 0$. □

Remarks. Similar estimates show that there is continuous dependence of the solutions when the data f and g converge in $L^1_{\rm loc}(\mathbb{R}\,;\, H^s(\mathbb{R}^d))$ and $H^s(\mathbb{R}^d)$, respectively.

Exercise 6.3.2. *Prove this.*

Exercise 6.3.3. *Show that if the source term f satisfies*

$$\partial_t^k f \in L^1_{\rm loc}([0,T^*[\,;\, H^{s-k}(\mathbb{R}^d))$$

for $k = 1, 2, \dots, m$ as in Theorem 2.2.2, then $u \in \bigcap_k C^k\big([0,T^*[\,;\, H^{s-k}(\mathbb{R}^d)\big)$ for the same k.

Finite speed of propagation for nonlinear equations is usually proved by writing a linear equation for the difference of two solutions. When $Lu_j + G(y, u_j) = 0$, denote by

$$w \;:=\; u_1 - u_2, \qquad B(y) \;:=\; \int_0^1 G'(y, u_2 + \theta(u_2(y) - u_1(y)))\; d\theta\,.$$

Taylor's theorem implies that

$$G(y, u_2) - G(y, u_1) = B(y)(u_2 - u_1), \qquad \text{so} \qquad L\,w + B(y)\,w \;=\; 0\,.$$

This is a linear equation with coefficient $B \in C(H^s)$ that need not be smooth. For the L^2 estimates that are used to prove finite speed, it is sufficient to know that $B \in L^\infty$.

The finite speed of propagation is determined entirely by the linear operator $L(y,\partial)$. Sharp estimates were proved in §2.5.

6.4. Moser's inequality and the nature of the breakdown

The breakdown (6.3.2) could, in principle, occur in a variety of ways. For example, the function might stay bounded and become more and more rapidly oscillatory. In fact this does not occur. Where the domain of existence ends, the maximal amplitude of the solution must diverge to infinity. To prove this requires more refined inequalities than those of Sobolev and Schauder.

The proofs of Schauder's lemma show that

$$\|G(y,w)\|_{H^s(\mathbb{R}^d_y)} \;\le\; h(\|w\|_{H^s(\mathbb{R}^d_y)})\,,$$

with a nonlinear function h that depends on G. There is a sharper estimate growing linearly in $\|w\|_{H^s}$ when one has L^∞ bounds.

Theorem 6.4.1 (Moser's Inequality). *With the same hypotheses as Schauder's lemma, there is a smooth function $h : [0,\infty[\, \to [0,\infty[$ so that for all $w \in H^s(\mathbb{R}^d)$ and t,*

$$\|G(t,x,w)\|_{H^s(\mathbb{R}^d_x)} \le h(\|w\|_{L^\infty(\mathbb{R}^d_x)})\, \|w\|_{H^s(\mathbb{R}^d_x)}\,. \tag{6.4.1}$$

This is proved by using Leibniz's rule and Hölder's inequality as in the standard proof of Schauder's lemma. However in place of the Sobolev inequalities one uses the Gagliardo–Nirenberg interpolation inequalities.

Theorem 6.4.2 (Gagliardo–Nirenberg Inequalities). *If $w \in H^s(\mathbb{R}^d) \cap L^\infty(\mathbb{R}^d)$ and $0 < |\alpha| < s$, then*

$$\partial_x^\alpha \, w \in L^{2s/|\alpha|}(\mathbb{R}^d).$$

In addition, there is a constant $C = C(|\alpha|, s, d)$ so that

$$\|\partial^\alpha w\|_{L^{2s/|\alpha|}(\mathbb{R}^d)} \leq C \, \|w\|_{L^\infty(\mathbb{R}^d)}^{1-|\alpha|/s} \Big(\sum_{|\beta|=s} \|\partial^\beta w\|_{L^2(\mathbb{R}^d)} \Big)^{|\alpha|/s}. \tag{6.4.2}$$

Remarks. 1. The second factor on the right in (6.4.2) is equivalent to the L^2 norm of the operator $|\partial_x|^s$ applied to u where $|\partial_x|^s$ is defined to be the Fourier multiplier by $|\xi|^s$. This gives the correct extension to noninteger s.

2. The indices in (6.4.2) are nearly forced. Consider which inequalities

$$\|\partial^\alpha w\|_{L^p(\mathbb{R}^d)} \leq C \, \|w\|_{L^\infty(\mathbb{R}^d)}^{1-\theta} \Big(\sum_{|\beta|=s} \|\partial^\beta w\|_{L^2(\mathbb{R}^d)} \Big)^\theta$$

homogeneous of degree one in w might be true. The test functions $w = e^{ix\xi/\epsilon}\psi(x)$ with $\epsilon \to 0$ show that a necessary condition is $|\alpha| \leq s\theta$. The idea is to use the L^∞ norm as much as possible and the s-norm as little as possible, which yields $|\alpha| = s\theta$. Considering $w = \psi(\epsilon x)$ or, equivalently, comparing the dimensions of the two sides, forces $p = 2s/\alpha$.

3. The works [Evans, 1998], [Hörmander, 1997], [Chemin et al., 2006], and [Taylor, 1997] are convenient references.

Proof of Moser's inequality. For $w \in \mathcal{S}(\mathbb{R}^d)$, G is independent of (t, x), and $\sigma := |\alpha| \leq s$, the quantity $\partial_x^\alpha(G(w))$ is a sum of terms of the form

$$G^{(\gamma)}(w) \prod_{j=1}^{J} \partial_x^{\alpha_j} w, \qquad |\gamma| = J, \qquad \alpha_1 + \cdots + \alpha_J = \alpha\,. \tag{6.4.3}$$

The first factor in (6.4.3) is bounded with L^∞ norm bounded by a nonlinear function of the L^∞ norm of w.

For the second factor, the Hölder inequality yields

$$\| \, \partial_x^{\alpha_1} w \cdots \partial_x^{\alpha_J} w \, \|_{L^2} \leq \prod_{k=1}^{J} \|\partial_x^{\alpha_k} w\|_{L^{2/\lambda_k}},$$

provided the nonnegative λ_k satisfy $\sum \lambda_k = 1$.

The Gagliardo–Nirenberg inequalities yield

$$\|\partial_x^{\alpha_k} w\|_{L^{2\sigma/|\alpha_k|}} \leq C \, \|w\|_{L^\infty}^{(\sigma-|\alpha_k|)/\sigma} \, \||\partial|^\sigma w\|_{L^2}^{|\alpha_k|/\sigma}.$$

With these choices

$$\sum \lambda_k = \sum \frac{|\alpha_k|}{\sigma} = 1\,,$$

and one has

$$\| \, \partial_x^{\alpha_1} w \cdots \partial_x^{\alpha_J} w \, \|_{L^2} \ \leq \ C \, \|w\|_{H^s}\,. \qquad \square$$

Exercise 6.4.1. *Carry out the proof when G depends on x.*

Theorem 6.4.3. *If $T^* < \infty$ in Theorem 6.3.1, then*

$$\limsup_{t \nearrow T^*} \ \|u(t)\|_{L^\infty} \ = \ \infty\,. \tag{6.4.4}$$

Proof. It suffices to show that it is impossible to have $T^* < \infty$ and $|u| \leq R < \infty$ on $[0, T^*[\, \times \mathbb{R}^d$. The strategy is to show that if $|u(t,x)| \leq R < \infty$ on $[0, T^*[\, \times \mathbb{R}^d$, then (6.3.2) is violated.

Use the linear inequality for $0 \leq t < T$,

$$\|u(t)\|_{H^s(\mathbb{R}^d)} \ \leq \ C \Big(\, \|u(0)\|_{H^s(\mathbb{R}^d)} + \int_0^t \ \|(Lu)(\sigma)\|_{H^s(\mathbb{R}^d)} \ d\sigma \Big)\,. \tag{6.4.5}$$

Then use Moser's inequality to give

$$\begin{aligned} \|(Lu)(\sigma)\|_{H^s(\mathbb{R}^d)} &= \|F(\sigma, x, u(\sigma,x)) - f(\sigma,x)\|_{H^s(\mathbb{R}^d)} \\ &\leq \ C(R) \Big(\, \|u(\sigma)\|_{H^s(\mathbb{R}^d)} + 1 \Big)\,. \end{aligned} \tag{6.4.6}$$

Insert (6.4.5) in (6.4.6) to find

$$\|u(t)\|_{H^s(\mathbb{R}^d)} \leq C \Big(\, \|u(0)\|_{H^s(\mathbb{R}^d)} + \int_0^t \ \big(\, \|u(\sigma)\|_{H^s(\mathbb{R}^d)} + 1 \, \big) \ d\sigma \Big)\,. \tag{6.4.7}$$

Gronwall's inequality shows that there is a constant $C'' < \infty$ so that for $t \in [0, T^*[$,

$$\|u(t)\|_{H^s(\mathbb{R}^d)} \leq C''\,. \tag{6.4.8}$$

This violates (6.3.2), and the proof is complete. $\square$

A mild sharpening of this argument (following [Yudovich, 1963]) shows that weaker norms than L^∞, for example the BMO norm, must also blow up at T^*.

Corollary 6.4.4. *If the data f and g in Theorem 6.3.1 belong to $L^1_{\text{loc}}\big([0,\infty[\,;\, H^s(\mathbb{R}^d)\big)$ and $H^s(\mathbb{R}^d)$, respectively, then they belong for all $d/2 < \tilde{s} \leq s$. The blowup time $T^*(\tilde{s})$ is independent of $\tilde{s}$. In particular, if the data belong to $H^s(\mathbb{R}^d)$ for all s, then the solution belongs to $C\big([0,T^*[\,;\, H^s(\mathbb{R}^d)\big)$ for all s.*

Proof. For $s \geq \tilde{s} > d/2$, denote by $u_{\tilde{s}}(t,x)$ the corresponding maximal solution. Since u_s is a $C(H^{\tilde{s}})$ solution, it follows that if u_s is defined on $[0,T]$, then $u_s = u_{\tilde{s}}$ on this this interval so $T^*(\tilde{s}) \geq T$. Therefore, $T^*(\tilde{s}) \geq T^*(s)$.

On the other hand if $T^*(\tilde{s}) > T^*(s)$, it follows that

$$u_{\tilde{s}} \in L^\infty([0,T^*(s)] \times \mathbb{R}^d).$$

By uniqueness of $H^{\tilde{s}}$ valued solutions one has

$$u_s = u_{\tilde{s}} \qquad \text{for} \qquad 0 \leq t < T^*(s)\,.$$

So

$$\|u_{\tilde{s}}\|_{L^\infty([0,T^*(s)[\times\mathbb{R}^d)} \;=\; \|u_{\tilde{s}}\|_{L^\infty([0,T^*(s)[\times\mathbb{R}^d)} \;<\; \infty\,,$$

violating the blowup criterion of Theorem 6.4.1. □

6.5. Perturbation theory and smooth dependence

In this section the dependence of solutions on data is investigated. The first result yields two versions of Lipschitz dependence.

Theorem 6.5.1. i. *If u and v are two solutions in $C([0,T]\,;\,H^s(\mathbb{R}^d))$, then there is a constant C depending only on $\sup_{[0,T]} \max\{\|u(t)\|_s, \|v(t)\|_s\}$ so that*

$$\forall\, 0 \leq t \leq T, \qquad \|u(t) - v(t)\|_s \;\leq\; C\,\|u(0) - v(0)\|_s\,. \tag{6.5.1}$$

ii. *If $u \in C([0,T]\,;\,H^s(\mathbb{R}^d))$ is a solution, then there are constants $C, \delta > 0$ so that if $\|u(0) - h\|_s < \delta$, then the solution v with $v(0) = h$ belongs to $C([0,T]\,;\,H^s(\mathbb{R}^d))$ and $\sup_{0\leq t\leq T} \|v(t) - u(t)\|_s \leq C\,\|h\|_s$.*

Proof. i. Choose Λ so that for w_1 and w_2 in H^s with

$$\|w_j\|_s \;\leq\; \sup_{[0,T]} \max\{\|u(t)\|_s, \|v(t)\|_s\}\,,$$

and $0 \leq t \leq T$,

$$\|F(t,x,w_1(x)) - F(t,x,w_2(x))\|_{H^s(\mathbb{R}^d_x)} \leq \Lambda \|w_1 - w_2\|_{H^s(\mathbb{R}^d_x)}\,.$$

Then subtracting the equations for u and v yields

$$\|u(t)-v(t)\|_{H^s(\mathbb{R}^d)} \;\leq\; \|u(0)-v(0)\|_s + \Lambda \int_0^t e^{C(t-\sigma)}\, \|u(\sigma)-v(\sigma)\|_{H^s(\mathbb{R}^d)}\, d\sigma\,.$$

Gronwall's inequality completes the proof of **i**.

To prove **ii**, it suffices to consider $\delta < 1$. Thanks to **i**, it suffices to prove the existence of v with $\sup_{[0,T]} \|v\|_s \leq 2 + \sup_{[0,T]} \|u\|_s$. Write $v = u + w$. The initial value problem defining v is equivalent to

$$L\,w + F(u+w) - F(u) = 0\,, \qquad w(0) = h\,.$$

So long as

$$\sup_{[0,t]} \|w(s)\| \leq 2,$$

one estimates $\|F(w+u) - F(u)\|_s \leq K\|w\|_s$ to find

$$\|w(t)\|_s \leq \|h\|_s + \int_0^t K\, \|w(\sigma)\|_s \, d\sigma .$$

Gronwall implies that

$$\|w(t)\|_s \leq \|h\|_s \, e^{Kt} .$$

Choose $C := e^{KT}$ and $0 < \delta < 1$ so small that $\delta C < 2$. Then a solution $w \in C([0,\underline{t}]\, , H^s)$ with $\underline{t} \leq T$ satisfies $\sup_{[0,\underline{t}]} \|w(t)\|_s < 2$. This existence of v together with the desired bound follows. □

Given a solution u, we compute a perturbation expansion for the solution with initial data $u(0) + g$ with small g. To simplify the notation, consider the semilinear equation

$$L(y, \partial)\, u + F(u) = 0, \qquad F(0) = 0, \quad F'(0) = 0 .$$

Denote by $\mathcal{N} : H^s(\mathbb{R}^d) \to C\big([0,T]\, ; H^s(\mathbb{R}^d)\big)$ the map $u(0) \mapsto u$. We will show that $\mathcal{N}$ is infinitely differentiable. For the moment we compute the Taylor expansion, assuming that it exists. Assuming smoothness, the solution with data $u(0) + g$ has the expansion

$$\mathcal{N}(u(0)+g) \sim u + M_1(g) + M_2(g) + \cdots \sim \sum_{j=1}^{\infty} M_j(g) , \tag{6.5.2}$$

where the M_j are continuous symmetric j-linear operators from $H^s(\mathbb{R}^d)$ to $C([0,T]\, ; H^s(\mathbb{R}^d))$.

To compute them, fix g and consider the initial data equal to $u(0) + \delta g$ with real δ. The resulting solution has an expansion in δ

$$\mathcal{N}(u+\delta g) \sim u_0 + \delta u_1 + \delta^2 u_2 + \cdots . \tag{6.5.3}$$

However,

$$\mathcal{N}(u+\delta g) \sim u + M_1(\delta g) + M_2(\delta g) + \cdots \sim u + \sum_{j=1}^{\infty} \delta^j \, M_j(g) .$$

Comparing with (6.5.2) yields

$$\forall\, j \geq 1, \qquad u_j = M_j(g, g, \ldots, g), \qquad j \text{ copies of } g .$$

To compute u_j, set $t = 0$ in the expansion to find

$$u_0(0) = g, \qquad \text{and for all } j \geq 1, \quad u_j(0) = 0 . \tag{6.5.4}$$

Plug the expansion (6.5.2) into the equation

$$L(y,\partial)\Big(u_0+\sum_{j\ge1}\delta^j u_j\Big) \;+\; F\Big(u_0+\sum_{j\ge1}\delta^j u_j\Big) \;\sim\; 0\,.$$

Expanding the left-hand side in powers of δ, the terms u_j are determined by setting the coefficients of the successive powers of δ equal to zero. Introduce the compact notation for the Taylor expansion

$$F(v+h)\;\sim\; F(v)+F_1(v;h)+F_2(v;h,h)+\cdots\,,$$

where for $v\in\mathbb{C}^N$, $F_j(v\,;\,\cdot)$ is a symmetric j linear map from $(\mathbb{C}^N)^j\to\mathbb{C}^N$. The simplest is $F_1(v;h)=F'(u_0)h$.

Setting the coefficients of δ^j equal to zero for $j=0,1,2,3$ yields the initial value problems

(6.5.5) $$L\,u_0 \;=\; 0\,,\qquad u_0(0,x)\;=\;u(0,x)\,,$$

(6.5.6) $$L\,u_1+F'(u_0)\,u_1\;=\;0\,,\quad u_1(0,x)=g\,.$$

(6.5.7) $$L\,u_2\;+\;F'(u_0)\,u_2\;+\;F_2(u_0;u_1,u_1)\;=\;0\,,$$

$$u_2(0,x)\;=\;0\,.$$

$$L\,u_3\;+\;F'(u_0)\,u_3\;+\;F_3(u_0;u_1,u_1,u_1)\;+\;2\,F_2(u_0;u_1,u_2)\;=\;0\,,$$

$$u_2(0,x)\;=\;0\,.$$

From (6.5.4) one finds $u_0=u$. And the next determines u_1, then u_2, then u_3. The initial value problem determining u_j is of the form $Lu_j+F'(u_0)u_j$ equal to an expression depending polynomially on the previously determined coefficients $u_{<j}$. An induction using Schauder's lemma shows that $u_j\in C([0,T]\,;\,H^s(\mathbb{R}^d))$.

Exercise 6.5.1. *Suppose that the u_j are determined by solving these initial value problems. Then define $u_{\text{approx}}(\delta)$ using Borel's theorem so that*

$$u_{\text{approx}}(\delta)\;\sim\;\sum_{j\ge0}\delta^j\,u_j\quad in\quad C([0,T]\,;\,H^s(\mathbb{R}^d))\,.$$

Prove that for δ sufficiently small, the exact solution of the initial value problem exists on $[0,T]$ and $u_{\text{exact}}(\delta)-u_{\text{approx}}(\delta)\sim 0$ in $C([0,T]\,;\,H^s(\mathbb{R}^d))$. **Hint.** Compute a nonlinear equation for the error that has source terms $O(\delta^\infty)$. Use the method of Theorem 6.5.1.ii. **Discussion.** The key element is the stability argument at the end showing that a nonlinear problem with infinitely small sources has an infinitely small solution. In science texts it is routine to overlook the need for such stability arguments.

The next result is stronger than that of the exercise.

Theorem 6.5.2. *If $u\in C([0,T]\,;\,H^s(\mathbb{R}^d))$ is a solution, then the map $\mathcal{N}$ from initial data to solution is smooth from a neighborhood of $u(0)$ in $H^s(\mathbb{R}^d)$*

to $C([0,T]\,;\,H^s(\mathbb{R}^d))$. The derivative is given by $\mathcal{N}_1(u(0);g) = u_1$ from (6.5.5). Derivatives of each order are uniformly bounded on the neighborhood.

Proof. The preceding computations show that if $\mathcal{N}$ is differentiable, then $\mathcal{N}_1(u(0);g) = u_1$ from (6.5.5). It suffices to show that this is the derivative of $\mathcal{N}$ and that the map from $u(0)$ to $\mathcal{N}_1(u(0);\cdot)$ is locally bounded and smooth with values in the linear maps from $H^s(\mathbb{R}^d)$ to $C([0,T]\,;\,H^s(\mathbb{R}^d))$.

To prove differentiability, let $u := \mathcal{N}(u(0))$ be the base solution, and let $v := u + u_1$ be the first approximation. Then

$$L\,u + F(u) \;=\; 0\,, \qquad L\,v + F(v) \;=\; F(u+u_1) - F_1(u;u_1)\,.$$

Schauder's lemma together with Taylor's theorem shows that

$$\big\|F(u+u_1) - F_1(u,u_1)\big\|_{C([0,T]\,;\,H^s)} \;\le\; C\,\big\|u_1\big\|^2_{C([0,T]\,;\,H^s)}.$$

Since the initial values of u and v are equal, the basic linear energy estimate proves that

$$\big\|u - v\big\|_{C([0,T]\,;\,H^s)} \;\le\; C\,\big\|u_1\big\|^2_{C([0,T]\,;\,H^s)}\,.$$

This proves that $\mathcal{N}$ is differentiable and also the formula for the derivative. The formula for $\mathcal{N}_1$ implies that the derivative is locally bounded.

We sketch the higher order differentiability. To prove that $\mathcal{N}$ is twice differentiable, one must differentiate the map $u \mapsto \mathcal{N}_1(u;\cdot)$, where $u_1 = \mathcal{N}_1(u;h)$ is computed by solving (6.5.5). Since u is a differentiable function of $u(0)$ with locally bounded derivative, it follows by Schauder's lemma that $u \mapsto F_1(u;\cdot)$ is differentiable with locally bounded derivative. As in the proof of differentiability, it follows that $u \mapsto \mathcal{N}_1(u(0),\cdot)$ is a differentiable function with locally bounded derivative. Higher derivatives are similar. □

6.6. The Cauchy problem for quasilinear symmetric hyperbolic systems

The quasilinear local existence theorem is more difficult but similar to the semilinear case. The differentiable dependence is fundamentally different. The mapping $\mathcal{N}$ is not locally lipschitzean.

For ease of reading, we present only the case of real solutions of real equations. The equations have the form

$$L(u,\partial)u \;:=\; \sum_{\mu=0}^{d} A_\mu(u)\,\partial_\mu u \;=\; f\,, \tag{6.6.1}$$

where the coefficient matrices A_μ are smooth symmetric matrix valued functions of u defined on an open subset of $\mathbb{R}^d$. The leading coefficient, $A_0(u)$, is assumed to be strictly positive. One can almost as easily treat coefficients that are function of y and u.

The existence theorem is local in time, and for small times the values of u are close to values of the initial data. Thus for convenience we can modify the coefficients outside a neighborhood of the values taken by the initial data to arrive at a system with everywhere defined smooth matrix valued coefficients. Even more we may suppose that the coefficients take constant values outside a compact subset of u space.

In contrast to the linear case, one cannot reduce to the case $A_0 = I$. However, if one is interested only in solutions that take values near a constant value $\underline{u}$, by changing the variable to $v := A_0(\underline{u})^{1/2}u$, one can reduce to the case $A_0(\underline{u}) = I$. This is useful for quasilinear geometric optics.

6.6.1. Existence of solutions.

Local existence is analogous to Theorem 6.3.1, except that it is important that the coefficients $A_\mu(u(y))$ are Lipschitz continuous functions of y. For this reason we work in Sobolev spaces $H^s(\mathbb{R}^d)$ with $s > 1 + d/2$. The importance of the lipschitzean condition is that it is needed to control the growth of the $L^2(\mathbb{R}^d)$ norm. The basic energy law when $f = 0$ is

$$\frac{d}{dt}\Big(A_0(u)\,u(t)\,,\,u(t)\Big) = \Big(\big(\sum_\mu \partial_\mu(A_\mu(u))\big)\;u(t)\,,\,u(t)\Big)\,. \tag{6.6.2}$$

To control the growth of the L^2 norm, one needs $\partial_\mu\big(A_\mu(u(y))\big) \in L^\infty$. That in turn requires $\partial_y u \in L^\infty$. It is true and not obvious that the constraint on s required to bound the L^2 norm also suffices to control the growth of derivatives up to order s. The following theorem is essentially due to Schauder in the sense that once the class of symmetric hyperbolic systems was introduced by Friedrichs, the technique used by Schauder for scalar second order equations immediately applies.

Theorem 6.6.1. *Suppose that $\mathbb{N} \ni s > 1 + d/2$, $g \in H^s(\mathbb{R}^d)$, $f \in L^1_{\rm loc}([0,\infty[:\ H^s(\mathbb{R}^d))$, and $\partial_t^j f \in C^j\big([0,\infty[\,;\,H^{s-j-1}(\mathbb{R}^d)\big)$ for $1 \le j \le s-1$. Then there is a $T > 0$ and a unique solution*

$$u \in \bigcap_{0\le j\le s} C^j([0,T]\,;\,H^{s-j}(\mathbb{R}^d))$$

to the initial value problem

$$L(u,\partial)u = f\,, \qquad u(0,x) = g(x)\,. \tag{6.6.3}$$

The time T can be chosen uniformly for g, f bounded in $H^s(\mathbb{R}^d)\times L^1\big([0,1]\,;\,H^s(\mathbb{R}^d)\big)$. Therefore, there is a $T_* \in\,]0,\infty]$ and a maximal solution in $\bigcap_j C^j([0,T_*[\,;\,H^{s-j}(\mathbb{R}^d))$. If $T_*<\infty$, then

$$\limsup_{t\nearrow T_*}\ \|u(t), \nabla_y u(t)\|_{L^\infty(\mathbb{R}^d)} \;=\; \infty\,. \tag{6.6.4}$$

Remark. If we had not modified the coefficients to be everywhere defined and smooth, the blow up criterion would be that either (6.6.4) occurs or the values of u approach the boundary of the domain where the coefficients are defined. This is so since if one has a solution of the original system whose values are taken in a compact subset K of the domain of definition of the coefficients, one can modify the coefficients outside a neighborhood of K. Theorem 6.6.2 then implies that there is a solution on a larger time interval.

The standard proof of Theorem 6.6.1 proceeds by considering the sequence of approximate solutions defined by the quasilinear iteration

$$L(u^\nu,\partial)u^{\nu+1} \;=\; f\,, \qquad u^{\nu+1}\big|_{t=0} \;=\; g\,.$$

The linear equation satisfied by $u^{\nu+1}$ has coefficients $A_\mu(u^\nu)$ depending on u^ν for which one has only H^s control. The key to the proof is to derive a priori estimates for solutions of linear symmetric hyperbolic initial value problems with coefficient matrices having only H^s regularity (see [Métivier, 1986] and [Lax, 1963]).

Schauder's approach was to approximate the functions A_μ by polynomials in u and the data g, f by real analytic functions and to use the Cauchy–Kovalevskaya theorem (see [Courant and Hilbert, 1962]). A priori estimates are used to control the approximate solutions on a fixed, possibly small, time interval. We solve the equation by the method of finite differences. A disadvantage of this method is that it reproves the linear existence theorem. An advantage is that the basic a priori estimate for the difference scheme allows one to prove existence and the sharp blowup criterion at the same time. The proof is technical. We have tried to bring the essential ideas to the fore without writing out all the gory details.

Proof. For ease of reading, we present the case $f=0$. The approximate solution u^h is the unique local solution of the ordinary differential equation in $H^s(\mathbb{R}^d)$,

$$A_0(u^h)\partial_t u^h \;+\; \sum_j A_j(u^h)\,\delta_j^h u^h \;=\; 0\,, \qquad u^h(0,x) \;=\; g(x)\,. \tag{6.6.5}$$

For each fixed $h>0$, the map

$$w \;\mapsto\; A_0(w)^{-1}\sum_j A_j(w)\,\delta_j^h w$$

from $H^s(\mathbb{R}^d)$ to itself is uniformly lipschitzean on bounded subsets. It follows that there is a unique maximal solution

$$u^h \in C^1([0,T_*^h[\,;\, H^s(\mathbb{R}^d))\,, \qquad T_*^h \in \,]0,\infty]\,.$$

If $T_*^h < \infty$, then $\lim_{t\nearrow T_*^h} \|u^h(t)\|_{H^s(\mathbb{R}^d)} = \infty$.

The heart of the existence proof is uniform estimates for u^h on an h independent interval. The starting point is the $L^2(\mathbb{R}^d)$ identity,

$$\frac{d}{dt}\Big(A_0(u^h)\,u^h(t)\,,\,u^h(t)\Big) = \Big((\partial_t(A_0(u^h))\,u^h\,,\,u^h\Big) + \sum_j \Big((A_j\delta_j^h + (A_j\delta_j^h)^*)u^h\,,\,u^h\Big)\,.$$

Thanks to the symmetry of A_j and the antisymmetry of δ_j^h,

$$A_j\delta_j^h + (A_j\delta_j^h)^* = A_j\delta_j^h + (\delta_j^h)^* A_j^* = A_j\delta_j^h - \delta_j^h\,A_j = [A_j(u^h)\,,\,\delta_j^h]\,.$$

If A_j were constant, this would vanish. In general there is a constant, $C = C(A_\mu)$, so that

$$\|\partial_t A_0(u^h)\|_{\mathrm{Hom}(L^2(\mathbb{R}^d))} + \|A_j\delta_j^h + (A_j\delta_j^h)^*\|_{\mathrm{Hom}(L^2(\mathbb{R}^d))} \le C\,\|\nabla_y u^h(t)\|_{L^\infty(\mathbb{R}^d)}\,. \tag{6.6.6}$$

This controls the $L^2(\mathbb{R}^d)$ norm provided one has a bound on $\|\nabla_y u^h\|_{L^\infty}$.

Next we derive similar estimates for the L^2 norm of $\partial^\alpha u$ with with $|\alpha| \le s$ and $\partial = \partial_x$. Compute

(6.6.7)

$$\begin{aligned}\frac{d}{dt}\Big(A_0(u^h)\,\partial^\alpha u^h(t)\,,\,\partial^\alpha u^h(t)\Big) &= \Big(A_0(u^h)\partial^\alpha\partial_t u^h\,,\,\partial^\alpha u^h\Big)\\ &\quad + \Big(A_0(u^h)\partial^\alpha u^h\,,\,\partial^\alpha\partial_t u^h\Big) + \Big((\partial_t A_0(u^h))\partial^\alpha u^h\,,\,\partial^\alpha u^h\Big)\\ &:= \Big(A_0(u^h)\partial^\alpha\partial_t u^h\,,\,\partial^\alpha u^h\Big) + \Big(A_0(u^h)\partial^\alpha u^h\,,\,\partial^\alpha\partial_t u^h\Big) + E_1\,,\end{aligned}$$

where

$$E_1 := \Big((\partial_t A_0(u^h))\partial^\alpha u^h\,,\,\partial^\alpha u^h\Big)$$

beginning the collection of error terms E_j that we prove are not too large.

The first term on the right of (6.6.7) is equal to

$$\begin{aligned}\Big(\partial^\alpha A_0(u^h)\partial_t u^h\,,\,\partial^\alpha u^h\Big) &+ \Big([A_0(u^h),\partial^\alpha]\partial_t u^h, \partial^\alpha u^h\Big)\\ &:= \Big(\partial^\alpha A_0(u^h)\partial_t u^h\,,\,\partial^\alpha u^h\Big) + E_2\,.\end{aligned} \tag{6.6.8}$$

Analogously, the symmetry of A_0 shows that the second term in (6.6.7) is equal to

(6.6.9)
$$\begin{aligned}\Big(\partial^\alpha u^h\,,\,A_0(u^h)\partial^\alpha\partial_t u^h\Big)\\ &= \Big(\partial^\alpha u^h\,,\,\partial^\alpha A_0(u^h)\partial_t u^h\Big) \;+\; \Big(\partial^\alpha u^h, [A_0(u^h),\partial^\alpha]\partial_t u^h\Big)\\ &:= \Big(\partial^\alpha u^h\,,\,\partial^\alpha A_0(u^h)\partial_t u^h\Big) \;+\; E_3\,.\end{aligned}$$

Using the differential equation, the sum of the error terms in (6.6.8), (6.6.9) is equal to the sum on j of

(6.6.10)
$$\begin{aligned}&\Big(\partial^\alpha u^h\,,\,\partial^\alpha A_j(u^h)\,\delta_j^h u^h\Big) \;+\; \Big(\partial^\alpha A_j(u^h)\,\delta_j^h u^h\,,\,\partial^\alpha u^h\Big)\\ &\quad:= \Big(\partial^\alpha u^h\,,\,A_j(u^h)\,\delta_j^h\partial^\alpha u^h\Big) \;+\; \Big(A_j(u^h)\,\delta_j^h\partial^\alpha u^h\,,\,\partial^\alpha u^h\Big) \;+\; E_4\\ &\quad= \Big(\big(A_j(u^h)\,\delta_j^h + (A_j(u^h)\delta_j^h)^*\big)\partial^\alpha u^h\,,\,\partial^\alpha u^h\Big) \;+\; E_4\\ &\quad:= E_5 \;+\; E_4\,,\end{aligned}$$

where

$$E_4 \;:=\; \Big(\partial^\alpha u^h\,,\,[\partial^\alpha, A_j(u^h)]\,\delta_j^h u^h\Big) \;+\; \Big([\partial^\alpha, A_j(u^h)]\,\delta_j^h u^h\,,\,\partial^\alpha u^h\Big)\,.$$

Denote

$$\mathcal{E}(w) \;:=\; \sum_{|\alpha|\le s}\Big(A_0(w)\partial_x^\alpha w\,,\,\partial_x^\alpha w\Big)\,.$$

Since A_0 is strictly positive, there is a constant C independent of w so that

$$\frac{1}{C}\sum_{|\alpha|\le s}\|\partial_x^\alpha w(t)\|^2_{L^2(\mathbb{R}^d)} \;\le\; \mathcal{E}(w) \;\le\; C\sum_{|\alpha|\le s}\|\partial_x^\alpha w\|^2_{L^2(\mathbb{R}^d)}\,. \tag{6.6.11}$$

Summing over all $|\alpha|\le s$ yields

$$\frac{d\mathcal{E}(u^h(t))}{dt} \;=\; \sum_{j=1}^{5} E_j\,. \tag{6.6.12}$$

Lemma 6.6.2. *For all $R>0$, $1\le j\le 5$, and $0<h<1$, there is a constant $C(R)$ depending only on L and R so that*

$$\big\|u^h(t),\nabla_y u^h(t)\big\|_{L^\infty(\mathbb{R}^d)} \;\le\; R \quad\Longrightarrow\quad |E_j| \;\le\; C(R)\sum_{|\alpha|\le s}\|\partial_x^\alpha u^h(t)\|^2_{H^s(\mathbb{R}^d)}\,.$$

Proof of Lemma 6.6.2. The cases $j=1$ and $j=5$ follow from (6.6.6). The remaining three cases are similar, and we present only $j=3$, which is the hardest. To treat that case it suffices to show that

$$\| [\partial^\alpha, A_j(u^h)]\, \partial_t u^h \|^2_{L^2(\mathbb{R}^d)} \ \le\ C(R) \sum_{|\alpha|\le s} \| \partial_x^\alpha u^h(t) \|^2_{H^s(\mathbb{R}^d)} \,. \tag{6.6.13}$$

The ordinary differential equation satisfied by u^h shows that the quantity on the left of (6.6.13) is a linear combination of terms

$$\partial^\beta (A_j(u^h))\, \partial^\gamma \big((A_0^{-1} A_j)(u^h) \delta_j^h\, u^h \big)\,, \qquad \beta + \gamma = \alpha, \quad \beta \ne 0\,.$$

Since $\beta \ne 0$, this is equal to

$$\big(\partial^\beta (A_j(u^h) - A_j(0))\, \partial^\gamma \big((A_0^{-1} A_j(u^h) - A_0^{-1} A_j(0)) \delta_j^h\, u^h \big) \\ + \ \partial^\beta (A_j(u^h) - A_j(0))\, (A_0^{-1} A_j)(0) \partial^\gamma \delta_j^h\, u^h\,.$$

Estimate

$$\|A_j(u^h) - A_j(0)\|_{L^\infty} \ +\ \|(A_0^{-1}A_j)(u^h)\ -\ (A_0^{-1}A_j)(0)\|_{L^\infty} \ \le\ C(R)\,.$$

Then Moser's inequality yields

$$\|A_j(u^h) - A_j(0)\|_{H^s} \ +\ \|(A_0^{-1}A_j)(u^h)\ -\ (A_0^{-1}A_j)(0)\|_{H^s} \\ \le\ C(R)\, \|u^h\|_{H^s}\,.$$

The Gagliardo–Nirenberg estimates then imply (6.6.13). □

Exercise 6.6.1. *Give a detailed proof of this lemma for all j.*

The local solution is constructed so as to take values in the set

$$\mathcal{W} \ :=\ \Big\{ w \in H^s(\mathbb{R}^d)\ ;\ \mathcal{E}(w) \le \mathcal{E}(g) + 1 \Big\}\,.$$

Choose $R > 0$ so that

$$w \in \mathcal{W} \quad \Longrightarrow \quad \|w\|_{L^\infty} + \|\nabla_x w\|_{L^\infty} + \|\sum_j A_j(w) \delta_j^h w\|_{L^\infty} \ <\ R\,.$$

So long as $u^h(t)$ stays in $\mathcal{W}$, one has

$$\frac{d\mathcal{E}(u^h(t))}{dt} \ \le\ C(R)\, C\, \mathcal{E}(u^h(t)) \ \le\ C(R)\, C\, \Big(\mathcal{E}(g) + 1 \Big)$$

with C from (6.6.11). Therefore,

$$\mathcal{E}(u^h(t)) - \mathcal{E}(g) \ \le\ T\, C(R)\, C\, \Big(\mathcal{E}(g) + 1 \Big)\,.$$

Define $T > 0$ by

$$T\, C(R)\, C\, \Big(\mathcal{E}(g) + 1 \Big) \ =\ \frac{1}{2}\,.$$

If follows that for all h, u^h takes values in $\mathcal{W}$ for $0 \le t \le T$. This T is uniform on bounded sets of $g \in H^s(\mathbb{R}^d)$ as required in the theorem.

Since the u^h belong to $\mathcal{W}$, $\{u^h \ : \ 0 < h \le 1\}$ is bounded in $L^\infty([0,T]; H^s(\mathbb{R}^d))$. Using the equation shows that for $1 \le j \le s$, $\{\partial_t^j u^h\}$ is bounded in $L^\infty([0,T]; H^{s-j}(\mathbb{R}^d))$. The estimate $\mathcal{E}(u^h(t))' \le C\mathcal{E}$ shows that $\mathcal{E}(u^h(t))'$ is bounded in $L^\infty([0,T])$. The weak star compactness of bounded sets in duals, the Rellich compactness theorem and the Arzelà–Ascoli theorem imply that there is subsequence $h(k) \to 0$ so that as $k \to \infty$

$$\begin{aligned}
u^{h(k)} &\to u && \text{weak star in } L^\infty([0,T]\,;\, H^s(\mathbb{R}^d)),\\
\partial^h u^{h(k)} &\to \partial^h u && \text{weak star in } L^\infty([0,T]\,;\, H^{s-j}(\mathbb{R}^d)),\\
& && \qquad 1 \le J \le s,\\
u^{h(k)} &\to u && \text{strongly in } C^j([0,T]\,;\, H^{s-j-1}(\mathbb{R}^d))\\
& && \qquad \text{for } 1 \le j \le s-1\,, \text{and}\\
\mathcal{E}(u^{h(k)}(t)) &\to \mathcal{E}(u(t)) && \text{uniformly on } [0,T]\,.
\end{aligned}$$

It follows that the limit $u(t)$ is weakly continuous with values in $H^s(\mathbb{R}^d)$ and that $\mathcal{E}(u(t))$ is continuous. The continuity of $\mathcal{E}$ together with the uniform continuity of u as an $\mathbb{R}^N$ valued function on $[0,T] \times \mathbb{R}^d$ implies that $t \mapsto \|u(t)\|_{H^s(\mathbb{R}^d)}$ is continuous. With the weak star continuity of $t \mapsto u(t)$, this implies that $u \in C([0,T]\,;\, H^s(\mathbb{R}^d))$. Then $\partial_t^j u \in C([0,T]\,;\, H^{s-j}(\mathbb{R}^d))$ follows by using the differential equation to express these derivatives in terms of spatial derivatives as in the semilinear case.

Uniqueness is proved by deriving a linear equation for the difference $w := u - v$ of two solutions u and v. Toward that end compute

$$A_\mu(u)\partial_\mu u - A_\mu(v)\partial_\mu v \;=\; A_\mu(u)\partial_\mu(u-v) \;+\; (A_\mu(u) - A_\mu(v))\partial_\mu v\,.$$

Write $A_\mu(u) - A_\mu(v) \;=\; \mathcal{G}_\mu(u,v)\,(u-v)$, to find

$$A_\mu(u)\partial_\mu u - A_\mu(v)\partial_\mu v = \mathcal{A}_\mu \partial_\mu w + \mathcal{B}_\mu w,$$

with

$$\mathcal{A}_\mu(y) := A_\mu(u(y)), \qquad \mathcal{B}_\mu(y) := \mathcal{G}_\mu(u(y), v(y))\,\partial_\mu v(y).$$

Therefore,

$$\mathcal{L}(y,\partial)\,w \;=\; 0\,, \qquad \mathcal{L}(y,\partial_y) := \sum \Big(\mathcal{A}_\mu\partial_\mu + \mathcal{B}_\mu\Big)\,. \tag{6.6.14}$$

The energy method yields

$$\frac{d}{dt}\big(\mathcal{A}_0\, w(t), w(t)\big) \;\le\; C\,\big(\mathcal{A}_0\, w(t)\,,\, w(t)\big)\,, \tag{6.6.15}$$

Since $w|_{t=0} = 0$, it follows that $w = 0$ proving uniqueness.

All that remains is the proof of the precise blowup criterion (6.6.4). This is immediate since if the Lipschitz norm does not blow up, then Lemma 6.6.2 implies that the $H^s(\mathbb{R}^d)$ norm does not blow up. Then the uniformity of the domain of existence on bounded sets of data gives existence beyond T. This completes the proof of Theorem 6.6.1. □

Example 6.6.1. In addition to the numerous examples from mathematical physics, we point out Garabedian's elegant reduction of the Cauchy–Kovalevskaya theorem to the solution of quasilinear symmetric hyperbolic initial value problems (see [Garabedian, 1964] and [Taylor, 1997]) yielding a holomorphic extension of the solution to complex values of t, y.

6.6.2. Examples of breakdown. In this section we exhibit the mechanism for the breakdown of solutions with u bounded and $\nabla_x u$ diverging to infinity as $t \nearrow T^*$. The method of proof using the method of characteristics leads to two Liouville type theorems.

The classic example is Burgers' equation

$$u_t + u\, u_x \;=\; 0\,. \tag{6.6.16}$$

For a smooth solution on $[0,T]\times\mathbb{R}^d$, the equation shows that u is constant on the integral curves of $\partial_t + u\,\partial_x$. Therefore those integral curves are straight lines.

For the solution of the initial value problem with

$$u(0,x) \;=\; g(x) \in C_0^\infty(\mathbb{R})\,, \tag{6.6.17}$$

the value of u on the line $(t, x + g(x)t)$ must be equal to $g(x)$. This is an implicit equation,

$$u(t, x + tg(x)) \;=\; g(x)\,, \tag{6.6.18}$$

uniquely determining a smooth solution for t small.

However, if g is not monotone increasing, consider the lines starting from two points $x_1 < x_2$ where $g(x_1) > g(x_2)$. The lines intersect in $t > 0$ at which point the conditions that u take value $g(x_1)$ and $g(x_2)$ contradict. Thus the solution must break down before this time. While the solution is smooth, $u(t)$ is a rearrangement of $u(0)$, so the sup norm of u does not blow up. The existence theorem shows that the gradient must explode.

That the gradient explodes can be proved by differentiating the equation to show that $v := \partial_x u$ satisfies

$$v_t + u\,\partial_x v + v^2 \;=\; 0\,.$$

Burgers' equation shows that u is constant on the line $(t, x + g(x)t)$. The equation for v is exactly solvable since

$$\frac{d}{dt}\, v(t, x + g(x)t) \;=\; v_t + u\,\partial_x v \;=\; -\,v^2\,.$$

Therefore,

$$v(t, x + g(x)t) \;=\; \frac{g'(x)}{1 - g'(x)t}\,.$$

Proposition 6.6.3. *The maximal solution of the initial value problem (6.6.4)–(6.6.5) satisfies*

$$(6.6.19) \qquad T_* \;=\; \frac{1}{-\min g'(x)}\,.$$

Proof. Denote by $T := 1/(-\min g')$. The preceding computations show that $T_* \le T$. On the other hand, if $0 < \underline{t} < T$ and u is a solution on $[0,\underline{t}] \times \mathbb{R}$, the formulas for u and ∂u show that $u, \nabla u$ is bounded on $[0,\underline{t}] \times \mathbb{R}$ so the blowup time must be larger than $\underline{t}$. Therefore $T_* \ge \underline{t}$. □

The method of proof yields the following results of Liouville type.

Theorem 6.6.4. i. *The only global solutions $u \in C^1(\mathbb{R}^{1+d})$ of Burgers' equation 6.6.5 are the constants.*

ii. *The only global solutions $\psi(x) \in C^3(\mathbb{R}^d)$ of the eikonal equation $|\nabla_x \psi| = 1$ are affine functions.*

Proof. i. Denote $g(x) := u(0,x)$. If there is a point with $g'(\underline{x}) < 0$, the above proof shows that $u_x(t, \underline{x} + g(\underline{x})t)$ diverges as $t \nearrow T^*$. If there is a point with $g'(\underline{x}) > 0$, then an analogous argument shows that $u_x(t, \underline{x}+g(\underline{x})t)$ diverges as $t \searrow -1/g'(\underline{x})$. Therefore, g is constant and the result follows.

ii. Denote by

$$V \;:=\; 2 \sum \partial_j \psi \; \partial_j$$

a C^2 vector field. The equation $\partial_j \sum (\partial_k \psi)^2 = 0$ yields the equation $V \partial \psi = 0$. Differentiate again, and one has the pair of equations

$$(6.6.20) \quad V\,\partial\psi \;=\; 0\,, \qquad 0 \;=\; V\,\partial^2\psi + 2\sum_j (\partial_j \partial\psi)^2 \;\ge\; V\,\partial^2\psi + (\partial^2\psi)^2\,.$$

The first implies that $\nabla_x \psi$ is constant on the integral curves of V. Therefore the integral curves are stationary points or straight lines $\underline{x} + s\nabla_x\psi(\underline{x})$.

If ψ is not affine, there is a point $\underline{x}$ at which the matrix of second derivatives at $\underline{x}$ is not equal to zero. The same holds on a neighborhood of $\underline{x}$ so changing $\underline{x}$ if needed we have $\nabla_x \psi(\underline{x}) \ne 0$. A linear change of coordinates yields $\partial_1^2 \psi(\underline{x}) \ne 0$.

Then

$$h(s) \;:=\; \partial_1^2 \psi(\underline{x} + 2s\nabla_x \psi(\underline{x})) \quad \text{satisfies} \quad \frac{dh}{ds} \;\le\; -h(s)^2\,.$$

If $h(0) < 0$, then h diverges to $-\infty$ at a finite positive value of s. Similarly, if $h(0) > 0$, then h diverges to $+\infty$ at a finite negative value of s. Thus, ψ cannot be globally C^2. □

6.6.3. Dependence on initial data. Theorem 6.6.1 shows that the map from $u(0)$ to $u(t)$ maps $H^s(\mathbb{R}^d)$ to itself and takes bounded sets to bounded sets. In contrast with the case of semilinear equations, this mapping is not smooth. It is not even lipschitzean. It is lipschitzean as a mapping from $H^s(\mathbb{R}^d)$ to $H^{s-1}(\mathbb{R}^d)$.

Suppose that $v \in C([0,T]\,;\, H^s(\mathbb{R}^d))$ with $s > 1 + d/2$ solves (6.6.1). Denote by $\mathcal{N}$ the map $u(0) \mapsto u(\cdot)$ from initial data to solution. It is defined on a neighborhood $\mathcal{U}$ of $v(0)$ in $H^s(\mathbb{R}^d)$ to $\bigcap_{j\geq 0} C^j([0,T]\,;\, H^{s-j}(\mathbb{R}^d))$.

Theorem 6.6.5. *The map*

$$H^s(\mathbb{R}^d) \supset \mathcal{U} \ni u(0) \mapsto u(\cdot) \in C([0,T]\,;\, H^{s-1}(\mathbb{R}^d))$$

is uniformly lipschitzean on closed bounded subsets of $\mathcal{U}$.

Proof. The assertion follows from the linear equation (6.6.15) for the difference of two solutions. The coefficients $\mathcal{A}_\mu$ belong to $C^j([0,T] : H^{s-j}(\mathbb{R}^d))$ for $0 \leq j \leq s$. On the other hand, the coefficients $\mathcal{B}_\mu \in C^j([0,T] : H^{s-j-1}(\mathbb{R}^d))$ for $0 \leq j \leq s-1$ have one less derivative. For this linear equation, the change of variable $\tilde{w} = \mathcal{A}_0^{-1/2} w$ reduces to the case $\mathcal{A}_0 = I$.

The estimate is proved by computing

$$\frac{d}{dt} \sum_{|\alpha| \leq s-1} (\partial^\alpha \tilde{w}(t)\,,\, \partial^\alpha \tilde{w})\,.$$

The restriction to $s-1$ comes from the fact that $\mathcal{B}$ is only $s-1$ times differentiable.

Exercise 6.6.2. *Carry out this proof using the proof of Theorem 6.6.1 as a model.* □

We next prove differentiable dependence by the perturbation theory method of §6.5. Suppose that

$$L(v, \partial)\, v = 0\,,$$

and consider the perturbed problem

$$L(u, \partial) u = 0\,, \qquad u|_{t=0} = v(0) + g\,, \tag{6.6.21}$$

with g small. To compute the Taylor expansion, introduce the auxiliary problems

(6.6.22)
$$L(\tilde{u}, \partial)\tilde{u} = 0\,, \quad \tilde{u}|_{t=0} = v(0) + \delta g\,, \qquad \tilde{u} \sim u_0 + \delta u_1 + \delta^2 u_2 + \cdots\,.$$

Then $L(\tilde u, \partial)\tilde u$ has an expansion in powers of δ computed from the expression

$$0 = \sum_\mu \Big(A_\mu(u_0) + \delta A'_\mu(u_0)(u_1) + O(\delta^2)\Big) \\ \times \partial_\mu\Big(u_0 + \delta u_1 + \delta^2 u_2 + \cdots\Big).$$

The $O(\delta^0)$ term yields

$$L(u_0,\partial)u_0 \;=\; 0\,, \qquad u_0|_{t=0} \;=\; v(0)\,. \tag{6.6.23}$$

The unique solution of this quasilinear Cauchy problem is $u_0 = v$.

The $O(\delta)$ term yields

$$\sum_\mu A_\mu(v)\partial_\mu u_1 \;+\; \sum_\mu \big[A'_\mu(v)u_1\big]\partial_\mu v \;=\; 0\,, \qquad u_1|_{t=0} \;=\; g\,. \tag{6.6.24}$$

Introduce the linearization of L at the solution v by

$$\mathbf{L}\,w \;:=\; \sum_\mu A_\mu(v)\partial_\mu w \;+\; \sum_\mu \big[A'_\mu(v)(w)\big]\partial_\mu v\,. \tag{6.6.25}$$

The equation of first order perturbation theory becomes

$$\mathbf{L}\,u_1 \;=\; 0\,, \qquad u_1|_{t=0} \;=\; g\,. \tag{6.6.26}$$

In the zero order term of $\mathbf{L}$, the coefficient depends on ∂v so, in general, u_1 will be one derivative less regular than v.

The $O(\delta^2)$ terms yield

$$\mathbf{L}\,u_2 + \sum_\mu \big[A'_\mu(v)(u_1)\big]\partial_\mu u_1 + \sum_\mu \big[A''_\mu(v)(u_1,u_1)\big]\partial_\mu v = 0\,, \qquad u_2|_{t=0} \;=\; 0\,. \tag{6.6.27}$$

There is a source term depending on ∂u_1, so typically u_2 will be one derivative less regular than u_1 and therefore two derivatives less regular than v.

Continuing in this fashion yields initial value problems determining u_j as symmetric j-multilinear functionals of g provided that v is sufficiently smooth.

Theorem 6.6.6. *Suppose that $s > 1 + d/2$, and $v \in C([0,T] \,:\, H^s(\mathbb{R}^d))$ satisfies (6.6.1). Then the map $\mathcal{N}$, from initial data to solution, is a differentiable function from a neighborhood of $v(0)$ in $H^s(\mathbb{R}^d)$ to $C([0,T]; H^{s-1}(\mathbb{R}^d))$. The derivative is locally bounded. If $s - j > d/2$, then $\mathcal{N}$ is j times differentiable as a map with values in $C([0,T]\,;\, H^{s-j}(\mathbb{R}^d))$. The derivatives are locally bounded.*

Sketch of Proof. The linear equation determining u_1 has coefficients which involve the first derivative of v. As a result u_1 will in general be one derivative less regular than v. That is as bad as it gets. It is not difficult to show using an estimate as in Theorems 6.5.2 and 6.6.4 that

$$\Big\|\mathcal{N}(u(0)+g) - \big(\mathcal{N}(u(0)) + u_1\big)\Big\|_{C\left([0,T]\,;\,H^{s-1}(\mathbb{R}^d)\right)} \;\le\; C\,\|g\|^2_{H^s(\mathbb{R}^d)}\,.$$

This yields differentiability, the formula for the derivative, and local boundedness.

Similarly, the calculations before the theorem show that if $\mathcal{N}$ is twice differentiable, then one must have

$$\mathcal{N}_2(v(0), g, g) \;=\; u_2\,,$$

where u_2 is the solution of (6.6.28). It is straightforward to show that $\mathcal{N}_2$ so defined is a continuous quadratic map from $H^s \mapsto C\big([0,T]\,;\,H^{s-2}(\mathbb{R}^d)\big)$.

A calculation like that in Theorem 6.5.2 shows that

$$\Big\|\mathcal{N}(u(0)+g) - \big(\mathcal{N}(u(0)) + u_1 + u_2\big)\Big\|_{C\left([0,T]\,;\,H^{s-2}(\mathbb{R}^d)\right)} \;\le\; C\,\|g\|^3_{H^s(\mathbb{R}^d)}\,.$$

This is not enough to conclude that $\mathcal{N}$ is twice differentiable. What is needed is a formula for the variation of $\mathcal{N}_1(v(0), g)$ when $v(0)$ is varied. The derivative $\mathcal{N}_1(v(0), g) = u_1$ is determined by solving the linear Cauchy problem (6.6.27) which has the form

$$L(v, \partial)\, u_1 + B(v, \partial v)\, u_1 \;=\; 0\,, \qquad u_1(0) \;=\; g\,.$$

The map from $v(0)$ to the coefficients in (6.6.26) is differentiable and locally bounded from $H^s \to C([0,T]\,;\,H^{s-1})$. Provided that $s-1 > d/2+1$, it follows from a calculation like that used to show that $\mathcal{N}$ is differentiable, that the map from $v(0)$ to u_1 is differentiable from H^s to $C([0.T]\,;\,H^{s-2}(\mathbb{R}^d))$, that $\mathcal{N}$ is twice differentiable, and the second derivative is locally bounded. The straightforward but notationally challenging computations are left to the reader.

The inductive argument for higher derivatives is similarly passed to the reader. □

Exercise 6.6.3. *Carry out the details.*

We next show by example that the loss of one derivative expressed in Theorems 6.6.4 and 6.6.5 is sharp. Choose

$$0 \le \chi \in C_0^\infty, \qquad \chi = 1 \ \text{ on } \ \{|x| \le 1/2\},$$

and denote $x_+ \;=\; \max\{x, 0\}$. Consider Burgers' equation, $v_t + v\, v_x \;=\; 0$, with initial data

$$v(0,x) = (x_+)^{3/2+\delta}\chi(x) \qquad 0 < \delta < 1/2,$$

belonging to $H^2(\mathbb{R})$ but not $H^3(\mathbb{R})$. Choose $\underline{t} > 0$ so that the local solution v valued in H^2 exists for $0 \le t \le \underline{t}$. That solution is compared with the solution u with initial value equal to $v(0,x) + \epsilon\chi(x)$.

The solution u vanishes for $x \le 1 + \epsilon t$. So, The difference $(u - v)(\underline{t})$ is equal to v on an interval of length $\epsilon \underline{t}$ to the right of the origin. Therefore

$$\|(u-v)(\underline{t})\|^2_{H^2(\mathbb{R})} \ \ge \ \int_0^{\epsilon \underline{t}} (v_{xx})^2 dx \ \ge \ C \int_0^{\epsilon \underline{t}} x^{-1+2\delta}\, dx \ \ge \ C\,\epsilon^{2\delta} \ \ne \ O(\epsilon)\,.$$

Since the initial data differ by $O(\epsilon)$ in H^2, the example shows that the map from data to solution is not lipschitzean on H^2. If we had taken $\delta > 1/2$, then the data would be H^3 and, consistent with the theorems, the map would be lipschitzean with values in H^2. At the opposite extreme, by choosing $0 < \delta$ as small as we like shows that the map is not Hölder continuous from H^2 to itself with any positive exponent.

Exercise 6.6.4. *Prove that the map $\mathcal{N}$ is continuous at g with values in $C([0,T]\,;\,H^s(\mathbb{R}^d))$ for any $T < T_*(g)$.* **Hint.** If $g_n \to g$, show that for n large, $T_*(g_n) > T$ and that the corresponding solution u_n converges to u in $C([0,T] \times H^s(\mathbb{R}^d)$ by writing $u_n = u + v_n$ and estimating v_n.

6.7. Global small solutions for maximally dispersive nonlinear systems

Consider solutions of linear constant coefficient symmetric hyperbolic systems with no lower order terms and no hyperplanes in their characteristic variety. For Cauchy data in $L^1(\mathbb{R}^d)$, $\|u(t)\|_{L^\infty}$ tends to zero as $t \to \infty$. For $d > 1$, the maximally dispersive systems decay as fast as is possible, consistent with L^2 conservation. Solutions of the nonlinear system

$$L(\partial)\, u + G(u) \ = \ 0\,, \qquad G(0) = 0\,, \quad \nabla_u G(0) = 0,$$

that have small initial data, say $u|_{t=0} \ = \ \epsilon\, f$, are approximated by solutions of the linearized equation $L(\partial)u \ = \ 0$, with the same initial data. On bounded time intervals, the error is $O(\epsilon^2)$ since the nonlinear term is at least quadratic at the origin. When solutions of $L\,u = 0$ decay in L^∞, $G(u)$ decays even faster. There is a tendency to approach linear behavior for large times. For $G = O(|u|^p)$ at the origin, the higher p is, the stronger the tendency. The higher the dimension is, the more dispersion is possible and the stronger can be the effect. We prove that, for maximally dispersive systems in dimension $d \ge 4$ with $p \ge 3$, the Cauchy problem is globally solvable for small data. This line of investigation has been the subject of much research. The important special case of perturbations of the wave equation was the central object of a program of F. John in which the contributions of S. Klainerman were capital. I recommend the books of Sogge (1995),

Hörmander (1997), Shatah and Struwe (1998), and Strauss (1989) for more information. We present ideas predating the John–Klainerman revolution. A quasilinear version including refined estimates for scattering operators can be found in [Satoh, 2005] and [Kajitani and Satoh, 2004]. The sharper result in the spirit of John and Klainerman is that there is global existence of small solutions when $(d-1)(p-1)/2 > 1$. Estimates sufficient for the sharper result in the setting of maximally dispersive systems are proved in the article of Georgiev, Lucente, and Zillotti (2004).

The sharp condition can be understood as follows. The nonlinear equation is like a linear equation with potential $\sim u^{p-1} \sim t^{-(d-1)(p-1)/2}$. The Cook criterion (see [Reed and Simon, 1983]) suggests that there is scattering behavior when this is integrable in time, that is $(d-1)(p-1)/2 > 1$.

Example 6.7.1. The global existence result is in sharp contrast to the example

$$\frac{\partial u}{\partial t} + \frac{\partial u}{\partial x_1} - u^2 = 0, \qquad u(0,x) = \epsilon\,\phi(x), \quad 0 \le \phi \in C_0^\infty(\mathbb{R}^d) \setminus 0,$$

for which solutions blow up in time $O(\epsilon^{-1})$ independent of dimension. The associated linear problem is completely nondispersive.

Assumption 1. *The space dimension $d \ge 2$ and $L(\partial)$ is a homogeneous maximally dispersive symmetric hyperbolic system with constant coefficients as in §3.4.*

Assumption 2. *$G(u)$ is a smooth nonlinear function whose leading Taylor polynomial at the origin is homogeneous of degree $p \ge 3$.*

Theorem 6.7.1. *Suppose that $(d-1)/2 > 1$ and $(d-1)(p-2)/2 > 1$, and σ is an integer greater than $(d+1)/2$. For each $\delta_1 > 0$, there is a $\delta_0 > 0$ so that if*

(6.7.1)
$$\|f\|_{H^\sigma(\mathbb{R}^d)} + \|f\|_{W^{\sigma,1}(\mathbb{R}^d)} \le \delta_0\,, \qquad \Big(\|f\|_{W^{\sigma,1}(\mathbb{R}^d)} := \sum_{|\alpha|\le\sigma} \|\partial_x^\alpha f\|_{L^1(\mathbb{R}^d)}\Big)\,,$$

then the solution of the Cauchy problem

$$Lu + G(u) = 0, \qquad u\big|_{t=0} = f\,, \tag{6.7.2}$$

exists globally and satisfies for all $t \in \mathbb{R}$,

$$\|u(t)\|_{L^\infty(\mathbb{R}^d)} \le \langle t\rangle^{-(d-1)/2}\,\delta_1 \quad \text{and} \quad \|u(t)\|_{H^\sigma(\mathbb{R}^d)} \le \delta_1\,. \tag{6.7.3}$$

There is a $c > 0$, so that for δ_1 small, one can take $\delta_0 = c\,\delta_1$.

Proof. We treat the case of $t \ge 0$. For simplicity we treat only the case of G equal to a homogeneous polynomial. The general case is the subject of Exercise 6.7.1.

Decreasing δ_1 makes the task more difficult. If $\delta_1 \le 1$ is given, choosing δ_0 sufficiently small, the solution satisfies (6.7.3) on some maximal interval $[0,T[$, $T \in]0,\infty]$. The proof relies on a priori estimates for the solution on this maximal interval.

Denote by $S(t) := e^{-it\sum_j A_j\partial_j}$ the unitary operator on $H^s(\mathbb{R}^d)$ giving the time evolution for the linear equation $Lu = 0$,

$$\|S(t)\, f\|_{H^s(\mathbb{R}^d)} \;=\; \|f\|_{H^s(\mathbb{R}^d)}\,. \tag{6.7.4}$$

Theorem 3.4.5 yields

(6.7.5)

$$\begin{aligned}\|S(t)\, f\|_{L^\infty(\mathbb{R}^d)} \;&\le\; C_0\, \langle t\rangle^{-(d-1)/2} \Big(\|f\|_{H^s(\mathbb{R}^d)} + \sum_{j=-\infty}^{\infty} \big\|\, |D|^{(d+1)/2} f_j\big\|_{L^1}\Big)\\ &\le\; C_1 \langle t\rangle^{-(d-1)/2}\, \delta_0\,.\end{aligned}$$

Duhamel's formula is

$$u(t) \;=\; S(t)\, f \;+\; \int_0^t S(t-s)\; G(u(s))\; ds\,. \tag{6.7.6}$$

For the homogeneous polynomial G, Moser's inequality is

$$\|G(u)\|_{H^\sigma(\mathbb{R}^d)} \;\le\; C_2\, \|u\|^{p-1}_{L^\infty(\mathbb{R}^d)}\; \|u\|_{H^\sigma(\mathbb{R}^d)}\,. \tag{6.7.7}$$

Exercise 6.7.1. *Prove (6.7.7).*

Use this and (the sharp condition) $(d-1)(p-1)/2 > 1$ to estimate

$$\begin{aligned}\|u(t)\|_{H^\sigma} \;&\le\; \delta_0 \;+\; \int_0^t C_2\, \big(\langle t-s\rangle^{-(d-1)/2}\, \delta_1\big)^{p-1}\; \delta_1\; ds\\ &\le\; \delta_0 + C_3\, \delta_1^p\,, \qquad C_3 := C_2 \int_0^\infty \langle t\rangle^{-(p-1)(d-1)/2}\, dt\,.\end{aligned} \tag{6.7.8}$$

The L^∞ norm satisfies,

$$\|u(t)\|_{L^\infty} \;\le\; C_1\, \langle t\rangle^{-(d-1)/2}\, \delta_0 \;+\; \int_0^t \big\|S(t-s)\; G(u(s))\big\|_{L^\infty}\; ds\,. \tag{6.7.9}$$

The dispersive estimate (3.4.8)–(3.4.9) yields

$$\big\|S(t-s)\, G(u(s))\big\|_{L^\infty} \;\le\; C_6\, \langle t-s\rangle^{-(d-1)/2}\, \|G(u(s))\|_{W^{\sigma,1}}\,. \tag{6.7.10}$$

The weighted L^2 estimates of John and Klainerman give a qualitatively better estimate than (6.7.10).

Lemma 6.7.2. *There is a constant C so that for all u one has*

$$\|G(u)\|_{W^{\sigma,1}} \;\le\; C\, \|u\|^{p-2}_{L^\infty}\; \|u\|^2_{H^\sigma}\,. \tag{6.7.11}$$

Proof of Lemma 6.7.2. Leibniz's rule shows that it suffices to show that if $|\alpha_1 + \cdots + \alpha_p| = s \leq \sigma$, then

$$\big\|\partial^{\alpha_1} u\, \partial^{\alpha_2} u \ \cdots \ \partial^{\alpha_p} u\big\|_{L^1} \ \leq \ C\, \|u\|^{p-2}_{L^\infty}\, \|u\|_{L^2} \|\, |D|^s u\|_{L^2}\,.$$

Both sides have the dimensions ℓ^{d-s}.

Define $\theta_i := |\alpha_i|/s$ so $\sum \theta_i = 1$. The Gagliardo–Nirenberg estimate interpolating between $u \in L^\infty$ and $|D|^s u \in L^2$ is

$$\|\partial^{\alpha_i} u\|_{L^{p_i}} \ \leq \ C\, \|u\|^{1-\theta_i}_{L^\infty}\, \|\, |D|^s u\|^{\theta_i}_{L^2}\,, \qquad \frac{1}{p_i} = \frac{1-\theta_i}{\infty} + \frac{\theta_i}{2} = \frac{\theta_i}{2}\,.$$

Define $\theta := 1/(p-1)$ and interpolate between $\partial^{\alpha_i} u \in L^{p_i}$ and $\partial^{\alpha_i} u \in L^2$ to find

$$\|\partial^{\alpha_i} u\|_{L^{r_i}} \ \leq \ \|\partial^{\alpha_i} u\|^{1-\theta}_{L^{p_i}}\, \|\partial^{\alpha_i} u\|^{\theta}_{L^2}\,, \qquad \frac{1}{r_1} \ = \ \frac{1-\theta}{p_i} + \frac{\theta}{2}\,.$$

Therefore,

$$\|\partial^{\alpha_i} u\|_{L^{r_i}} \ \leq \ \|u\|^{(1-\theta_i)(1-\theta)}_{L^\infty}\, \|u\|^{(1-\theta_i)\theta}_{L^2}\, \|\, |D|^s u\|^{\theta_i}_{L^2}\,, \qquad 1 = \sum 1/r_i\,.$$

Hölder's inequality implies

$$\big\|\partial^{\alpha_1} u\, \partial^{\alpha_2} u \ \cdots \ \partial^{\alpha_p} u\big\|_{L^1} \ \leq \ \prod_{i=1}^{p} \|\partial^{\alpha_i} u\|_{L^{r_i}} \ \leq \ C\, \|u\|^{p-2}_{L^\infty}\, \|u\|_{L^2}\, \|\, |D|^s u\|_{L^2}\,.$$

□

Estimates (6.7.10)–(6.7.11) yield

$$\int_0^t \big\|S(t-s)\, G(u(s))\big\|_{L^\infty}\, ds \ \leq \ C_7 \int_0^t \langle t-s\rangle^{-(d-1)/2}\, \langle s\rangle^{-(p-2)(d-1)/2}\, \delta_1^2\, ds\,.$$

Consider $0 \leq s \leq 1$ to see that this integral cannot decay faster than $\langle t\rangle^{-(d-1)/2}$. On the other hand, on $s \geq t/2$ (resp., $s \leq t/2$), the first (resp., second) factor in the integral is bounded above by $C\langle t\rangle^{-(d-1)/2}$ and the other factor is uniformly integrable using (the not sharp hypotheses) $(d-1)/2 > 1$ and $(p-2)(d-1)/2 > 1$. Therefore,

$$\int_0^t \big\|S(t-s)\, G(u(s))\big\|_{L^\infty}\, ds \ \leq \ C_8\, \langle t\rangle^{-(d-1)/2}\, \delta_1^2\,. \tag{6.7.12}$$

Combining yields

$$\|u(t)\|_{L^\infty} \ \leq \ \Big(C_1\, \delta_0 + C_8\, \delta_1^2\Big)\, \langle t\rangle^{-(d-1)/2}\,.$$

Since $p > 1$, decreasing δ_1 if necessary, we may suppose that

$$C_3\, \delta_1^p \ < \ \frac{\delta_1}{2} \qquad \text{and} \qquad C_8\, \delta_1^2 \ < \ \frac{\delta_1}{2}\,.$$

Then, choose $\delta_0 > 0$ so that

$$\delta_0 + C_3\,\delta_1^2 < \frac{\delta_1}{2} \qquad \text{and} \qquad C_1\delta_0 + C_8\,\delta_1^2 < \frac{\delta_1}{2}.$$

With these choices, the estimates show that on the maximal interval of existence $[0,T[$, one has

$$(6.7.13) \qquad \|u(t)\|_{L^\infty(\mathbb{R}^d)} \le \langle t\rangle^{-(d-1)/2}\,\frac{\delta_1}{2} \quad \text{and} \quad \|u(t)\|_{H^s(\mathbb{R}^d)} \le \frac{\delta_1}{2}.$$

If T were finite, the second estimate would contradict the maximality of T. Therefore $T = \infty$. The estimate (6.7.13) on $[0,T[\, = [0,\infty[$ completes the proof. □

Exercise 6.7.2. *If G is not homogeneous, show that there are smooth functions H_α so that $G(u) = \sum_{|\alpha|=p} u^\alpha\, H_\alpha(u)$. Modify the Moser inequality arguments appropriately to prove the general result.*

6.8. The subcritical nonlinear Klein–Gordon equation in the energy space

6.8.1. Introductory remarks. The mass zero nonlinear Klein–Gordon equation is

$$(6.8.1) \qquad \Box_{1+d} u + F(u) = 0,$$

where

$$(6.8.2) \qquad F \in C^1(\mathbb{R}), \qquad F(0) = 0, \qquad F'(0) = 0.$$

The classic examples from quantum field theory are the equations with $F(u) = u^p$ with $p \ge 3$ an odd integer. For ease of reading we consider only real solutions.

The equation (6.8.1) is Lorentz invariant and if G denotes the primitive,

$$(6.8.3) \qquad G'(s) = F(s), \qquad G(0) = 0,$$

then the local energy density is defined as

$$(6.8.4) \qquad e(u) := \frac{u_t^2 + |\nabla_x u|^2}{2} + G(u).$$

Solutions $u \in H^2_{\rm loc}(\mathbb{R}^{1+d})$ satisfy the differential energy law,

$$(6.8.5) \qquad \partial_t e - \operatorname{div}\big(u_t\,\nabla_x u\big) = u_t\Big(\Box u + F(u)\Big) = 0.$$

The corresponding integral conservation law for solutions suitably small at infinity is

$$(6.8.6) \qquad \partial_t \int_{\mathbb{R}^d} \frac{u_t^2 + |\nabla_x u|^2}{2} + G(u)\,dx = 0,$$

is one of the fundamental estimates in this section. Solutions are stationary for the Lagrangian,

$$\int_0^T \int_{\mathbb{R}^d} \frac{u_t^2 - |\nabla_x u|^2}{2} \;-\; G(u)\; dt\, dx\,.$$

When F is smooth, the methods of §6.3–6.4 yield local smooth existence.

Theorem 6.8.1. *If $F \in C^\infty$, $s > d/2$, $f \in H^s(\mathbb{R}^d)$, and $g \in H^{s-1}(\mathbb{R}^d)$, then there is a unique maximal solution*

$$u \;\in\; C\big([0,T_*[\;;\; H^s(\mathbb{R}^d)\big) \;\cap\; C^1\big([0,T_*[\;;\; H^{s-1}(\mathbb{R}^d)\big)$$

satisfying

$$u(0,x) = f\,, \quad u_t(0,x) = g\,.$$

If $T_* < \infty$, then

$$\limsup_{t\to T_*} \;\|u(t)\|_{L^\infty(\mathbb{R}^d)} \;=\; \infty\,.$$

In favorable cases, the energy law (6.8.6) gives good control of the norm of $u, u_t \in H^1 \times L^2$. Controlling the norm of the difference of two solutions is, in contrast, a very difficult problem with many fundamental questions unresolved.

An easy first case is nonlinearities F that are uniformly lipschitzean. In this case, there is global existence in the energy space.

Theorem 6.8.2. *If F satisfies $F' \in L^\infty(\mathbb{R})$, then for all Cauchy data $f, g \in H^1(\mathbb{R}^d) \times L^2(\mathbb{R}^d)$ there is a unique solution*

$$u \in C\big(\mathbb{R}\;;\; H^1(\mathbb{R}^d)\big) \;\cap\; C^1\big(\mathbb{R}\;;\; L^2(\mathbb{R}^d)\big)\,.$$

For any finite T, the map from data to solution is uniformly lipschitzean from $H^1 \times L^2$ to $C([-T,T\,;\,H^1) \cap C^1([-T,T]\,;\,L^2)$. If $f, g \in H^2 \times H^1$, then

$$u \;\in\; L^\infty\big(\mathbb{R};\; H^2(\mathbb{R}^d)\big)\,, \qquad u_t \;\in\; L^\infty\big(\mathbb{R};\; H^1(\mathbb{R}^d)\big)\,.$$

If $f, g \in H^s \times H^{s-1}$ with $1 \le s < 2$, then

$$u \;\in\; C\big(\mathbb{R};\; H^s(\mathbb{R}^d)\big)\,, \qquad u_t \;\in\; C\big(\mathbb{R};\; H^{s-1}(\mathbb{R}^d)\big)\,.$$

Sketch of Proof. The key estimate is the following. If u and v are solutions, then

$$\square(u-v) \;=\; F(v) - F(u), \qquad |F(u) - F(v)| \;\le\; C|u-v|\,.$$

Multiplying by $u_t - v_t$ yields

$$\frac{d}{dt}\int (u_t - v_t)^2 + |\nabla_x(u-v)|^2 dx = 2\int (u_t - v_t)\,\big(F(v) - F(u)\big)\,dx$$
$$\le C\,\|u_t - v_t\|_{L^2}^2\,\|u-v\|_{L^2}^2\,.$$

It follows that for any T there is an a priori estimate

$$\sup_{|t|\le T} \Big(\|u(t)-v(t)\|_{H^1} \ + \ \|u_t - v_t\|_{L^2} \Big)$$
$$\le \ C(T)\Big(\|u(0)-v(0)\|_{H^1} \ + \ \|u_t(0)-v_t(0)\|_{L^2} \Big) .$$

This estimate exactly corresponds to the asserted Lipschitz continuity of the map from data to solutions.

Applying the estimate to $v = u(x+h)$ and taking the supremum over small vectors h, yields an a priori estimate

$$\sup_{|t|\le T} \Big(\|u(t)\|_{H^2} \ + \ \|u_t\|_{L^2} \Big) \ \le \ C(T)\Big(\|u(0)\|_{H^2} \ + \ \|u_t(0)\|_{H^1} \Big) ,$$

which is the estimate corresponding to the H^2 regularity. □

Higher regularity for dimensions $d \ge 10$ is an outstanding open problem. For example, for $d \ge 10$ and $C_0^\infty(\mathbb{R}^d)$ initial data, and $F \in C_0^\infty$ or $F = \sin u$, it is not known if the above global unique solutions are smooth. For $d \le 9$ the result is proved in Brenner and von Wahl (1981). Smoothness would follow if one could prove that $u \in L^\infty_{\rm loc}$. What is needed is to show that the solutions do not get large in the pointwise sense. Compared to the analogous open regularity problem for Navier–Stokes this problem has the advantage that solutions are known to be unique and depend continuously on the data.

6.8.2. The ordinary differential equation and non-lipshitzean F. Concerning global existence for functions $F(u)$ that may grow more rapidly than linearly as $u \to \infty$, the first considerations concern solutions that are independent of x and therefore satisfy the ordinary differential equation,

$$u_{tt} \ + \ F(u) \ = \ 0 . \tag{6.8.7}$$

Global solvability of the ordinary differential equation is analyzed using the energy conservation law

$$\Big(\frac{u_t^2}{2} \ + \ G(u) \Big)' \ = \ u_t \Big(u_{tt} + F(u) \Big) \ = \ 0 .$$

Think of the equation as modeling a nonlinear spring. The spring force is attractive, that is it pulls the spring toward the origin when

$$F(u) > 0 \quad \text{when} \quad u > 0 \qquad \text{and} \qquad F(u) < 0 \quad \text{when} \quad u < 0 .$$

In this case one has $G(u) > 0$ for all $u \ne 0$. Conservation of energy then gives a pointwise bound on u_t uniform in time

$$u_t^2(t) \ \le \ u_t^2(0) + 2G(u(0)) , \qquad |u_t(t)| \ \le \ \big(u_t^2(0) + 2G(u(0)) \big)^{1/2} .$$

Integrating yields the bound

$$|u(t)| \;\le\; |u(0)| + |t|\big(u_t^2(0) + 2G(u(0))\big)^{1/2}.$$

In particular the ordinary differential equation has global solutions.

In the extreme opposite case consider the repulsive spring force $F(u) = -u^2$ and $G(u) = -u^3/3$. The energy law asserts that $u_t^2/2 - u^3/3 := E$ is independent of time. Consider solutions with

$$u(0) > 0, \quad u_t(0) > 0, \qquad \text{so} \quad E > -\frac{u^3(0)}{3}.$$

For all $t > 0$,

$$|u_t| \;=\; \Big|\frac{u^3}{3} + E\Big|^{1/2}.$$

At $t = 0$, one has

$$u_t(0) \;=\; \Big(\frac{u^3(0)}{3} + E\Big)^{1/2} > 0\,.$$

Therefore u increases and $u^3/3 + E$ stays positive and one has for $t \ge 0$

$$u_t(t) \;=\; \Big(\frac{u^3(t)}{3} + E\Big)^{1/2} > 0\,.$$

Both u and u_t are strictly increasing.

Since

$$\frac{du}{\big(\frac{u^3}{3} + E\big)^{1/2}} \;=\; dt\,,$$

$u(t)$ approaches ∞ at time

$$T \;:=\; \int_{u(0)}^{\infty} \frac{du}{\big(\frac{u^3}{3} + E\big)^{1/2}}\,.$$

Exercise 6.8.1. *Show that if there is an $M > 0$ so that $G(s) < 0$ for $s \ge M$ and*

$$\int_M^\infty \frac{1}{\sqrt{|G(s)|}}\, ds \;<\; \infty,$$

then there are solutions of the ordinary differential equation that blow up in finite time.

Keller's Blowup Theorem 6.8.3 (1957). *If*

$$a, \delta > 0, \qquad d \le 3, \qquad E := \delta^2/2 - a^3/3, \qquad T \;:=\; \int_a^\infty \Big|\frac{u^3}{3} + E\Big|^{-1/2} du\,,$$

and $\phi, \psi \in C^\infty(\mathbb{R}^d)$ satisfy

$$\phi \ge a \quad \text{and} \quad \psi \ge \delta \qquad \text{for} \quad |x| \le T\,,$$

then the the smooth solution of

$$\Box_{1+d} u \;-\; u^2\,, \qquad u(0) = \phi, \quad u_t(0) = \psi$$

blows up on or before time T.

Proof. Denote by $\underline{u}$ the solution of the ordinary differential equation with initial data $\underline{u}(0) = a$, $\underline{u}_t(0) = \delta$.

If $u \in C^\infty\big([0, \underline{t}] \times \mathbb{R}^d\big)$, then finite speed of propagation and positivity of the fundamental solution of $\square_{1+d}$ imply that

$$u \ \geq \ \underline{u} \qquad \text{on} \quad \big\{|x| \leq T - \underline{t}\big\}\,.$$

Since $\underline{u}$ diverges as $t \to T$, it follows that $\underline{t} \leq T$ □

In the case of attractive forces where $G \geq 0$, one can hope that there is global smooth solvability for smooth initial data. This question has received much attention and is very far from being understood.

6.8.3. Subcritical nonlinearities. In the remainder of this section we study solvability in the energy space defined by $u, u_t \in H^1 \times L^2$. This regularity is suggested by the basic energy law. For uniformly lipschitzean nonlinearities the global solvability is given by Theorem 6.8.2. The interest is in attractive nonlinearities with superlinear growth at infinity.

A crucial role is played by the rate of growth of F at infinity. There is a critical growth rate so that for nonlinearities that are subcritical and critical there is a good theory based on Strichartz estimates. The analysis is valid in all dimensions.

To concentrate on essentials, we present the family of attractive (resp., repulsive) nonlinearities $F = u|u|^{p-1}$ (resp., $F = -u|u|^{p-1}$) with potential energies given by $\pm \int |u|^{p+1}/(p+1)dx$. The idea is to gauge the size of the nonlinear term. The natural comparison is $\|\nabla_{t,x} u(t)\|_{L^2(\mathbb{R}^d)}$, the energy of the linear problem. Start with four natural notions of subcriticality. They are in increasing order of strength. One could call p subcritical when

1. $H^1(\mathbb{R}^d) \subset L^p(\mathbb{R}^d)$ so the nonlinear term makes sense for elements of H^1.

2. $H^1(\mathbb{R}^d) \subset L^{p+1}(\mathbb{R}^d)$ so the nonlinear potential energy makes sense for elements of H^1.

3. $H^1(\mathbb{R}^d)$ is compact in $L^{p+1}_{\text{loc}}(\mathbb{R}^d)$ so the nonlinear potential energy is in a sense small compared to the linear potential energy.

4. $H^1(\mathbb{R}^d) \subset L^{2p}(\mathbb{R}^d)$ so the nonlinear term belongs to $L^2(\mathbb{R}^d)$. Recall that sources $L^1_{\text{loc}}(\mathbb{R} : L^2(\mathbb{R}^d))$ yields solutions of the linear problem that are continuous with values in H^1.

The Sobolev embedding is

$$H^1(\mathbb{R}^d) \ \subset \ L^q(\mathbb{R}^d)\,, \quad \text{for} \quad q \ = \ \frac{2d}{d-2}\,. \tag{6.8.8}$$

The above conditions then read (with the values for $d = 3$ given in parentheses),

1. $p \le 2d/(d-2) \qquad (p \le 6)$,
2. $p + 1 \le 2d/(d-2) \qquad \{\text{equiv. } \ p \le (d+2)/(d-2)\} \qquad (p \le 5)$,
3. $p < (d+2)/(d-2) \qquad (p < 5)$,
4. $p \le d/(d-2) \qquad (p \le 3)$.

The best answer is **3**. Much that follows can be extended to the critical case $p = (d+2)/(d-2)$. When **1** is satisfied but **2** is violated, the problem is supercritical. It is known that in the supercritical case, solutions are very sensitive to initial data. The dependence is not lipschitzean, and it is lipschitzean in the subcritical and critical cases. The books of Sogge (1995), Shatah and Struwe (1998), Tao (2006), and the original 1985 article of Ginibre and Velo are good references. The sensitive dependence is a result of Lebeau (2001 and 2005).

Notation. *Denote by* $L^q_tL^r_x([0,T])$ *the space* $L^q_tL^r_x([0,T]\times\mathbb{R}^d)$. *For the half open interval* $[0,T[$,

$$L^q_tL^r_x([0,T[) \ := \ \bigcup_{0<\underline{T}<T} L^q_tL^r_x([0,\underline{T}])\,.$$

Theorem 6.8.4. i. *If p is subcritical for H^1, that is $p < (d+2)/(d-2)$, then for any $f \in H^1(\mathbb{R}^d)$ and $g \in L^2(\mathbb{R}^d)$ there is $T_* > 0$ and a unique solution*

$$u \in C([0,T_*[\ H^1(\mathbb{R}^d)) \cap C^1([0,T_*[;\ L^2(\mathbb{R}^d)) \cap L^p_tL^{2p}_x([0,T_*[) \tag{6.8.9}$$

of the repulsive problem

$$\Box u \ - \ u|u|^{p-1} \ = \ 0\,, \qquad u(0) = f, \quad u_t(0) = g\,. \tag{6.8.10}$$

If $T_* < \infty$, then

$$\liminf_{t\nearrow T_*} \ \|\nabla_{t,x}u\|_{L^2(\mathbb{R}^d)} \ = \ \infty\,. \tag{6.8.11}$$

The energy conservation law (6.8.6) is satisfied.

ii. *For the attractive problem*

$$\Box u \ + \ u|u|^{p-1} \ = \ 0\,, \qquad u(0) = f, \quad u_t(0) = g\,, \tag{6.8.12}$$

one has the same result with $T_* = \infty$ and with $u \in L^p_tL^{2p}_x([0,\infty[)$. For any $T > 0$, the map from Cauchy data to solution is uniformly lipschitzean on bounded sets as a mapping

$$H^1 \times L^2 \ \ \to \ \ C([0,T]\,;\, H^1) \cap C^1([0,T]\,;\, L^2) \cap L^p_tL^{2p}_x([0,T]))\,.$$

In the proof of this result and all that follows, a central role is played by the linear wave equation. Recall the basic energy estimate

$$\|\nabla_{t,x} u(t)\|_{L^2(\mathbb{R}^d)} \ \le\ \|\nabla_{t,x} u(0)\|_{L^2(\mathbb{R}^d)} \ +\ \int_0^t \|\Box u(t)\|_{L^2(\mathbb{R}^d)}\ dt\,.$$

This is completed by the L^2 estimate

$$\|u(t)\|_{L^2(\mathbb{R}^d)} \ \le\ \int_0^t \|u_t(t)\|_{L^2(\mathbb{R}^d)}\ dt\,.$$

Notation. *Denote by $\Box^{-1}h$ the solution u of the initial value problem*

$$\Box u \ =\ h\,, \qquad u(0)\ =\ 0\,, \quad u_t(0)\ =\ 0\,.$$

For $h \in L^1_{\text{loc}}\big(\mathbb{R}\,;\, L^2(\mathbb{R}^d)\big)$, one has

$$\Box^{-1}h\ \in\ C\big(\mathbb{R}\,;\, H^1(\mathbb{R}^d)\big)\ \cap\ C^1\big(\mathbb{R}\,;\, L^2(\mathbb{R}^d)\big)\,.$$

In order to take advantage of this, we seek solutions so that

$$F_p(u)\ :=\ \pm\, u|u|^{p-1}\ \in\ L^1_t L^2_x\,.$$

Compute

$$\|F_p(u)\|_{L^1_t L^2_x}\ =\ \int_0^T \Big(\int |u^p|^2\ dx\Big)^{1/2}\ dt$$

using

$$\Big(\int |u|^{2p}\ dx\Big)^{1/2}\ =\ \Big[\Big(\int |u|^{2p}\Big)^{1/2p}\Big]^p\ =\ \|u\|^p_{L^{2p}(\mathbb{R}^d)}\,,$$

to find

$$\|F_p(u)\|_{L^1_t L^2_x}\ =\ \int_0^T \|u(t)\|^p_{L^{2p}(\mathbb{R}^d_x)}\ dt\ =\ \|u\|^p_{L^p_t L^{2p}_x}\,. \tag{6.8.13}$$

The above calculation proves the first part of the next lemma.

Lemma 6.8.5. *The mapping $u \mapsto F_p(u)$ takes $L^p_t L^{2p}_x([0,T])$ to $L^1_t L^2_x([0,T])$. It is uniformly lipschitzean on bounded subsets.*

Proof. Write

$$F_p(v)\ -\ F_p(w)\ =\ G(v,w)\big(v-w\big)\,, \qquad |G(v,w)|\ \le\ C\Big(|v|^{p-1}\ +\ |w|^{p-1}\Big)\,.$$

Write

$$\Big\|G(v,w)(v-w)\Big\|^2_{L^2_x}\ =\ \int |G|^2\ |v-w|^2\ dx\,.$$

Use Hölder's inequality for $L^{p/(p-1)}_x \times L^p_x$ to estimate by

$$\le\ \Big(\int |G(v,w)|^{2p/(p-1)}dx\Big)^{\frac{p-1}{p}}\ \Big(\int |v-w|^{2p}dx\Big)^{\frac{1}{p}}\,.$$

Then

$$\|F_p(v) - F_p(w)\|_{L^2} \le C\,\|v,w\|^{p-1}_{L^{2p}_x}\,\|v-w\|_{L^{2p}_x}\,.$$

Finally estimate the integral in time using Hölder's inequality for $L_t^{p/(p-1)} \times L_t^p$. □

It is natural to seek solutions $u \in L_t^p L_x^{2p}([0,T])$. With that as a goal, we ask when it is true that

$$\Box^{-1}(L_t^1 L_x^2) \subset L_t^p L_x^{2p}\,.$$

This is one of the questions answered by the Strichartz inequalities. The next lemma gives the form of those inequalities adapted to the present situation.

Lemma 6.8.6. *If*

$$q > 2 \qquad \text{and} \qquad \frac{1}{q} + \frac{d}{r} = \frac{d}{2} - 1\,, \tag{6.8.14}$$

then there is a constant $C > 0$ so that for all $T > 0$, $\{h, f, g\} \in L_t^1(L_x^2) \times H^1 \times L^2$ the solution of

$$\Box u = h\,, \qquad u(0) = f\,, \qquad u_t(0) = g\,,$$

satisfies

$$\|u\|_{L_t^q L_x^r([0,T])} \le C\Big(\|h\|_{L_t^1 L_x^2([0,T])} + \|\nabla_x f\|_{L^2(\mathbb{R}^d)} + \|g\|_{L^2(\mathbb{R}^d)}\Big)\,. \tag{6.8.15}$$

Proof. 1. Rewrite the wave equation as a maximally dispersive symmetric hyperbolic pseudodifferential system. Motivated by D'Alembert's solution of the $1-d$ wave equation factor,

$$\partial_t^2 - \Delta = (\partial_t + i|D|)\,(\partial_t - i|D|) = (\partial_t + i|D|)\,(\partial_t - i|D|)\,,$$

introduce

$$v_\pm := (\partial_t \mp i|D|)u\,, \qquad V := (v_+, v_-)\,,$$

so

$$V_t + \begin{pmatrix} 1 & 0 \\ 0 & -1 \end{pmatrix} i|D|\,V = \begin{pmatrix} h \\ h \end{pmatrix}.$$

The proof of Lemma 3.4.8 implies that for $\sigma = d-1$, $q > 2$, (q,r) σ-admissible, and h, f, g with spectrum in $\{R_1 \le |\xi| \le R_2\}$, one has

$$\begin{aligned}\|u\|_{L_t^q L_x^r} &\le C\,\|\nabla_{t,x} u\|_{L_t^q L_x^r}\\ &\le C\,\|V\|_{L_t^q L_x^r} \le C\Big(\|h\|_{L_t^1 L_x^2} + \||D|f\|_{L^2} + \|g\|_{L^2}\Big)\,.\end{aligned}$$

2. Denote by ℓ the dimensions of t and x. With dimensionless u, the terms on right of this inequality have dimension $\ell^{d/2-1}$. The dimension of the left is equal to $\big(\ell^{dq/r}\,\ell\big)^{1/q} = \ell^{(d/r)+1/q}$. The two sides have the same dimensions if and only if

$$\frac{d}{r} + \frac{1}{q} = \frac{d}{2} - 1\,. \tag{6.8.16}$$

When (6.8.16) holds it follows that the same inequality holds, with the same constant C for data with spectrum in $\lambda R_1 \le |\xi| \le \lambda R_2$.

Comparing (6.8.16) with σ-admissibility that is equivalent to

$$\frac{d}{r} + \frac{1}{q} \le \frac{d}{2} - \frac{1}{2} - \frac{1}{r}$$

shows that (6.8.16) implies admissibility since $r \ge 2$.

3. Lemma 6.8.6 follows using the Littlewood–Paley theory as at the end of §3.4.3. □

We now answer the question of when $\square^{-1}$ maps $L^1_t L^2_x$ to $L^p_t L^{2p}_x$. This is the crucial calculation. In Lemma 6.8.6, take $r = 2p$ to find

$$\frac{1}{q} + \frac{d}{2p} = \frac{d-2}{2}\,,$$

so

$$\frac{1}{q} = \frac{d-2}{2} - \frac{d}{2p} = \frac{p(d-2)-d}{2p}\,, \qquad q = p\Big(\frac{2}{p(d-2)-d}\Big)\,.$$

We want $q \ge p$, that is

$$\frac{2}{p(d-2)-d} \ge 1 \iff p(d-2)-d \le 2 \iff p \le \frac{d+2}{d-2}\,.$$

The critical case is that of equality. We treat the subcritical case defined by strict inequality. For $d = 3$ the critical power is $p = 5$ and for $d = 4$ it is $p = 3$. In the subcritical case Hölder's inequality implies that $\square^{-1}$ with zero Cauchy data has small norm for $T \ll 1$.

Exercise 6.8.2. *Prove this.*

The strategy of the proof is to write the solution u as a perturbation of the solution of the linear problem. Define u_0 to be the solution of

$$\square u_0 = 0\,, \qquad u_0(0) = f, \qquad \frac{\partial u_0}{\partial t}(0) = g\,. \tag{6.8.17}$$

Write

$$u = u_0 + v \tag{6.8.18}$$

with the hope that v will be small for t small.

Lemma 6.8.7. *If $u = u_0 + v$ with $v \in L^p_t L^{2p}_x\big([0,T]\big)$ satisfying*

$$v \;=\; \pm\Box^{-1}F_p(u_0+v)\,, \tag{6.8.19}$$

then

$$u \;\in\; C([0,T]\,;\,H^1(\mathbb{R}^d)) \;\cap\; C^1([0,T]\,;\,L^2(\mathbb{R}^d)) \;\cap\; L^p_tL^{2p}_x([0,T]) \tag{6.8.20}$$

satisfies

$$\Box u \;\pm\; F_p(u) \;=\; 0\,, \qquad u(0)=f, \quad u_t(0)=g\,. \tag{6.8.21}$$

Conversely, if u satisfies (6.8.21)–(6.8.22), then $v := u - u_0 \in L^p_tL^{2p}_x([0,T])$ and satisfies (6.8.20)

Proof. The Strichartz inequality implies that $u_0 \in L^p_tL^{2p}_x$ and by hypothesis the same is true of v. Therefore u_0+v belongs to $L^p_tL^{2p}_x$ so $F_p(u_0+v) \in L^1_tL^2_x$.

Therefore $v = \pm\Box^{-1}F_p$ is $C(H^1)\cap C^1(L^2)$. The differential equation and initial condition for v are immediate.

The converse is similar, not used below, and left to the reader. □

Exercise 6.8.3. *Prove the converse.*

Proof of Theorem 6.8.4. For $K>0$ arbitrary but fixed, we prove unique solvability with continuous dependence for $0\le t\le T$ with T uniform for all data f,g with

$$\|f\|_{H^1} \;+\; \|g\|_{L^2} \;\le\; K\,.$$

Choose $R=R(K)$ so that for such data,

$$\|u_0\|_{L^p_tL^{2p}_x([0,1])} \;\le\; \frac{R}{2}\,.$$

Define

$$B \;=\; B(T) \;:=\; \Big\{v\in L^p_tL^{2p}_x([0,T]) \;:\; \|v\|_{L^p_tL^{2p}_x([0,T])} \;\le R\Big\}\,.$$

It follows that for $T=T(K)$ sufficiently small, the map $v\mapsto \Box^{-1}F_p(v)$ is a contraction from B to itself. This is a consequence of three facts that have already been established.

1. Lemma 6.8.5 shows that F_p is uniformly lipschitzean from B to $L^1_tL^2_x([0,T])$ uniformly for $0<T\le 1$.

2. Lemma 6.8.6 together with subcriticality shows that there is an $r>p$ so that $\Box^{-1}$ is uniformly lipschitzean from $L^1_tL^2_x$ to $L^r_tL^{2p}_x$ uniformly for $0<T<1$.

3. The injection $L^r_tL^{2p}_x \mapsto L^p_tL^{2p}_x$ has norm which tends to zero as $T\to 0$. This is enough to carry out the existence parts of Theorem 6.8.4.

For uniqueness, reason as follows. If there are two solutions u, v with the same initial data, compute

$$\Box(u-v) \;=\; G(u,v)(u-v)\,.$$

Lemma 6.8.6 together with subcriticality shows that with r slightly larger than p,

$$\|u-v\|_{L^r_tL^{2p}_x} \;\le\; C\,\|G(u,v)(u-v)\|_{L^1_tL^2_x} \;\le\; C\,\|u-v\|_{L^p_tL^{2p}_x}\,.$$

Use this estimate for $0 \le t \le T \ll 1$ noting that Hölder's inequality shows that for $T \to 0$,

$$\|u-v\|_{L^p_tL^{2p}_x} \;\le\; C\,T^\rho\,\|u-v\|_{L^r_tL^{2p}_x} \;\le\; C\,T^\rho\|u-v\|_{L^p_tL^{2p}_x}\,, \qquad \rho>0\,,$$

to show that the two solutions agree for small times. Thus the set of times where the solutions agree is open and closed, proving uniqueness.

To prove the energy law, note that $F_p(u) \in L^1_tL^2_x$ so the linear energy law shows that

$$\int \frac{|u_t|^2+|\nabla_x u|^2}{2}\,dx\Big|_{t=0}^{t} \;=\; \mp\int_0^t\int u_t\;F_p(u)\;dx\,dt\,. \tag{6.8.22}$$

Now

$$u_t \;\in\; L^\infty_tL^2_x\,, \qquad \text{and} \qquad F_p(u) \in L^1_tL^2_x\,.$$

Hölder's inequality shows that

$$\int |u_t\;F_p(u)|dx \;\le\; \|u_t(t)\|_{L^2_x}\;\|F_p(u(t)\|_{L^2_x}\,.$$

The latter is the product of a bounded and an integrable function so

$$\forall T, \quad u_t\,F_p(u) \;\in\; L^1([0,T]\times\mathbb{R}^d)\,.$$

Let

$$w \;:=\; \frac{|u|^{p+1}}{p+1}\,.$$

Since p is subcritical, one has for some $0<\epsilon$,

$$\|w(t)\|_{L^1_x} \;\le\; C\|u(t)\|_{H^{1-\epsilon}(\mathbb{R}^d)} \;\in\; L^\infty([0,T])\,.$$

In particular $w \in L^1([0,T]\times\mathbb{R}^d)$ and the family $\{w(t)\}_{t\in[0,T]}$ is precompact in L^1_{loc}.

Formally differentiating yields

$$w_t \;=\; u_tF_p(u) \;\in\; L^1([0,T]\times\mathbb{R}^d)\,. \tag{6.8.23}$$

Using the above estimates, it is not hard to justify (6.8.23).

Exercise 6.8.4. *Verify.*

It follows that $w \in C([0,T]\,;\, L^1(\mathbb{R}^d))$ and

$$\int w(t,x)\, dx \Big|_{t=0}^{t=T} = \int_0^T \int u_t\, F_p(u) dx\, dt\,.$$

Together with (6.8.23) this yields the energy identity (6.8.6).

Once the energy law is known, one concludes global solvability in the attractive case since the blowup criterion (6.8.11) is ruled out by energy conservation. □

Exercise 6.8.5. *Prove that the solutions of the theorem satisfy the local energy law* $\partial_t\big(\nabla_{t,x} u|^2/2 + |u|^{p+1}/(p+1)\big) = \operatorname{div}(u_t \nabla_x u)$ *in the sense of distributions.*

Chapter 7

One Phase Nonlinear Geometric Optics

In this chapter we construct asymptotic expansions that are nonlinear analogs of the Lax construction. There are two important nonlinear effects that must be understood in order to arrive at the appropriate *ansatz*, rectification, and the generation of harmonics. The *ansatz* involves a trigonometric generating function for the amplitudes of the harmonics, and the equations that determine it involve an averaging operator. Solvability is as in the linear case, but is only local because of nonlinearity. The end result is the construction of an approximate solution with infinitely small residual. Chapter 8 contains the proof that that there are infinitely close exact solutions.

7.1. Amplitudes and harmonics

For linear equations, any solution may be multiplied by a constant to yield another solution. This is not the case for nonlinear equations. If one studies short wavelength oscillatory solutions, the propagation and interactions depend crucially on the amplitudes. The easiest case to understand, and therefore a natural starting point, is small oscillations. For that we perform a (regular) perturbation analysis.

Consider the semilinear equation

$$L(y,\partial)\, u \;+\; F(u) \;=\; 0 \tag{7.1.1}$$

with nonlinear function satisfying

$$F(0) = 0\,, \quad F'(0) = 0\,. \tag{7.1.2}$$

Suppose that $v(\epsilon, y) = a(\epsilon, y)\, e^{i\phi(y)/\epsilon}$ is a family Lax solution (as in §5.4) of the linear problem $Lv \sim 0$. Suppose that a has compact support for each t. Consider the semilinear initial value problem with the initial data

$$g(\epsilon, x) = \epsilon^m\, a(\epsilon, 0, x)\, e^{i\phi(0,x)/\epsilon}\,. \tag{7.1.3}$$

The power m scales the amplitude as a function of the wavelength. The larger m is, the smaller the data. The initial data is bounded in $H^s(\mathbb{R}^d)$ if and only if $s \le m$. If this is satisfied for some $s > d/2$, then Theorem 6.3.1 proves existence for the semilinear problem on a domain independent of ϵ. For $m \le d/2$, the theorem guarantees existence only on a domain that shrinks with ϵ because the H^s norm of the data grows to ∞ for all $s > d/2$.

If $m > d/2$, then the data converges to zero in $H^s(\mathbb{R}^d)$ for all $s \in]d/2, m[$. The perturbation theory of §6.5 proves that solutions exist on an ϵ-independent neighborhood and are given by a Taylor series,

$$u(\epsilon, y) \;\sim\; \sum_{j=1}^{\infty} M_j(g(\epsilon, x)) \;:=\; \sum_j u_j(\epsilon, x),$$

with u_j a j-linear function of g, hence $\|u_j\|_{H^s(\mathbb{R}^d)} = O(\|g\|^j_{H^s(\mathbb{R}^d)})$. The leading u_j are determined by equations (6.5.4) through (6.5.6).

We will see that for $m \ge 0$, there is existence on an ϵ-independent domain. The simple explicitly solvable example

$$\partial_t u(\epsilon, y) = u(\epsilon, y)^2\,, \qquad u(\epsilon, 0, x) = \epsilon^m\, e^{ix.\xi/\epsilon}$$

shows that the domain may shrink to zero for $m < 0$.

Exercise 7.1.1. *Verify.*

Equations (6.5.4)–(6.5.5) show that the two leading terms in perturbation theory are determined by

$$L\, u_0 \;=\; 0\,, \qquad L\, u_1 \;+\; F_2\big(0;\, u_1, u_1\big) \;=\; 0\,, \tag{7.1.4}$$

with initial conditions

$$u_1(\epsilon, 0, x) \;=\; \epsilon^m\, a(\epsilon, 0, x)\, e^{i\phi(0,x)/\epsilon}\,, \qquad u_2(0, x) \;=\; 0\,. \tag{7.1.5}$$

Equations (7.1.4) and (7.1.5) show that as $\epsilon \to 0$, u_1 is given asymptotically by the Lax solution $\epsilon^m\, a(\epsilon, y)\, e^{i\phi(y)/\epsilon}$. Once u_1 is known, the next term u_2 can be found. And so on.

To see the form of u_2, consider the source term. It is a quadratic expression in u_1. The term u_1 oscillates with phase $\phi(y)/\epsilon$. Squaring such a term yields a source oscillating with phase $2\phi(y)/\epsilon$. The square of the complex conjugate, which is a second example of a smooth quadratic expression, yields a phase $-2\phi(y)/\epsilon$. Finally an expression in the product of u with its conjugate yields a nonoscillatory source. The source term has the form

$$\epsilon^{2m}\left(c_{-2}(y)\, e^{-i2\phi(y)/\epsilon} + c_0(y) + c_{+2}(y) e^{i2\phi(y)/\epsilon}\right). \tag{7.1.6}$$

From Lax's Theorem 5.3.5 with source oscillating with phase satisfying the eikonal equation, the oscillatory parts of this source yields terms of the form

$$\sum_{\pm} \epsilon^{2m}\left(a_{\pm 2}(y) + O(\epsilon)\right) e^{\pm i2\phi(y)/\epsilon}$$

in the solution $u_2(\epsilon, y)$.

The key observation is that the Taylor expansion begins with an $O(\epsilon^m)$ term that is linear in the initial data and is equal to the Lax solution. The next term, quadratic in the initial data, is of order ϵ^{2m} and has terms oscillating with the new phases $\pm 2\phi(y)/\epsilon$. It may also have a nonoscillating term of order ϵ^{2m}. The cubic and higher order terms in the Taylor expansion are of order ϵ^{jm} for integer j and will have terms oscillating with phases including higher integer multiples of $\pm\phi(y)/\epsilon$.

This generation and interaction of harmonics is one of the key signatures of nonlinear problems. Note that the wavelength of the jth harmonic is $1/j$ times the original wavelength. Thus the interaction is also an interaction between different length scales. A classic experiment in 1961 by Franken, Weinreich and Hall at the University of Michigan passed monochromatic laser light through a carefully selected medium and observed the second harmonic. This was the birth of *nonlinear optics.* At the energies of that experiment, a small data perturbation theory like that just sketched is appropriate. Such an analysis can be found in the classic text of Nobel laureate, Bloembergen (1964).

Remark. In the setting of dispersive geometric optics, the underlying operator is $L_1(y,\partial) + \epsilon^{-1}L_0(y)$ and the natural eikonal equation is $\det\big(L_1(y,d\phi) + L_0(y)\big) = 0$. For such equations the phases $\pm 2\phi$ usually do not satisfy the eikonal equation. The response to the source terms is given by an elliptic resolution, so it is local and not propagating. The cases where one or both of $\pm\phi$ is eikonal are rare in the dispersive setting. They are important when they occur.

The computation so far is justified for $m > d/2$; it is an interesting indication that something better is true. Formally, the expansion seems to work, provided only that $m > 0$, in which case the supposedly higher order corrections are indeed higher order in ϵ. In fact, using local existence results tailored to oscillatory data, as in §8, the expansions can be justified for $m > 0$ on an arbitrary but fixed interval of time. On the other hand, for $m < 0$, we know by example that the domain of existence can shrink to zero.

A fundamental lesson to be learned is that for $m > 0$ linear phenomena are accompanied by *creation of harmonics at higher order in* ϵ. This leads to correction terms in the solution that have amplitudes with higher powers of ϵ and phases that are integer multiples of $\phi(y)/\epsilon$. The higher m is, the smaller the initial data and the greater the gap between the amplitudes of the principal term and the harmonics. Equivalently, the smaller m is, the larger the data and the more important the nonlinear effects.

There is another important lesson. The leading nonlinear term is of order ϵ^{2m}, while the Lax solution enters at order ϵ^m. As $m \to 0$, these orders approach each other. This leads the courageous to suspect that there may be something interesting occurring when $m = 0$, in which case the harmonics should appear in the principal term. This, in fact, is the case. For $m = 0$, oscillations can be described on an ϵ-independent domain, and the leading term in the expansion involves a nonlinear interaction among oscillations with phases $j\phi(y)/\epsilon$. This critical scaling of the amplitudes is called *nonlinear geometric optics* or *weakly nonlinear geometric optics*, depending on the author. For this scaling, the nonlinear terms are not negligible at leading order. For this reason we say that the time of nonlinear interaction is ~ 1. For this scaling to leading order, the nonlinear terms can be neglected only for times $o(1)$ as $\epsilon \to 0$.

Nonlinear geometric optics described here are more complicated than, but descendant from, earlier work on pulses of width ϵ and height one in spatial dimension 1. A description of the pulses and the relation to wave trains can be found in [Hunter, Majda, Rosales, 1986] and in the survey article of [Majda, 1986]. Wave trains are blessed with interesting nonlinear interactions that go under the name of *resonance*. Generation of harmonics is the simplest case. Resonance for the $m = 0$ scaling of geometric optics are also described in the articles just cited. Resonance phenomena are studied in Chapters 9–11.

7.2. Elementary examples of generation of harmonics

Here are three ordinary differential equation calculations that introduce the creation of harmonics.

Exercise 7.2.1. *Consider the solution $x(\epsilon, t)$ of the nonlinear initial value problem*

$$\frac{d^2x}{dt^2} + \omega^2 x + x^2 = 0\,, \qquad x|_{t=0} = \epsilon\,, \quad \frac{dx}{dt}\Big|_{t=0} = 0\,.$$

Then x is an analytic function of ϵ, t on its domain of existence. Compute the first three terms in the Taylor expansion

$$x(\epsilon, t) = a_0(t) + \epsilon a_1(t) + \epsilon^2 a_2(t) + \cdots .$$

Note the presence of harmonics when they appear, and the amplitude of the harmonics.

In Exercise 7.2.1, the harmonics appeared in a regular perturbation expansion of small solutions to a nonlinear equation. An entirely equivalent problem is the expansion of solutions of fixed amplitude with a weak nonlinearity.

Exercise 7.2.2. *Consider the solution $x(\epsilon, t)$ of the weakly nonlinear initial value problem*

$$\frac{d^2x}{dt^2} + \omega^2 x + \epsilon x^2 = 0\,, \qquad x|_{t=0} = 1\,, \quad \frac{dx}{dt}\Big|_{t=0} = 0\,.$$

Then x is an analytic function of ϵ, t on its domain of existence. Compute the first three terms in the Taylor expansion

$$x(\epsilon, t) = a_0(t) + \epsilon a_1(t) + \epsilon^2 a_2(t) + \cdots .$$

Note the presence of harmonics when they appear, and the amplitude of the harmonics.

Finally, here is an example of the generation of harmonics for forced oscillations.

Exercise 7.2.3. *Consider the solution $x(\epsilon, t)$ of the nonlinear initial value problem*

$$\frac{d^2x}{dt^2} + x + x^2 = \epsilon \cos \beta t\,, \qquad x|_{t=0} = 0\,, \quad \frac{dx}{dt}\Big|_{t=0} = 0\,, \qquad \beta \neq 0, \pm 1\,.$$

Then x is an analytic function of ϵ, t on its domain of existence. Compute the first three terms in the Taylor expansion

$$x(\epsilon, t) = a_0(t) + \epsilon a_1(t) + \epsilon^2 a_2(t) + \cdots .$$

Note the presence of harmonics when they appear, and the amplitude of the harmonics.

7.3. Formulating the ansatz

The results of §7.1 lead us to consider semilinear initial value problems with initial data of the form $a(\epsilon, 0, x)\, e^{i\phi(0,x)/\epsilon}$ that are initial data of a Lax solution in the linear case. The key fact is that the amplitude is of order ϵ^0. For this amplitude one expects harmonics to be present in the leading ϵ^0 term and for these harmonics to interact. We describe these phenomena.

The computations suggest that the solution will have oscillations with all the phases $n\phi(y)/\epsilon$. Thus the principal term is expected to be at least as complicated as the sum of leading terms, one for each harmonic. The amplitude of the nth harmonic is denoted

$$a_0(n,\epsilon,y) \sim a_0(n,y) + \epsilon\, a_1(n,y) + \cdots . \tag{7.3.1}$$

It seems that the natural thing to do is to derive dynamic equations for the infinite set of amplitudes $a_0(n,y)$. One has the linear hyperbolic propagation properties given by rays and transport equations for each $a_0(n,y)$ and also nonlinear interaction terms that express at least the idea that if one starts with $a_0(1,y) \neq 0$ and all others vanishing, then the other modes will tend to be illuminated.

There is a very effective method for managing this infinity of unknowns. The expected form for the leading terms is

$$\sum_{n\in\mathbb{Z}} a_0(n,y)\, e^{in\phi(y)/\epsilon} .$$

The sum on n suggests the Fourier series

$$U_0(y,\theta) := \sum_{n=-\infty}^{\infty} a_0(n,y)\, e^{in\theta} . \tag{7.3.2}$$

The leading terms take the elegant form $U_0(y,\phi(y)/\epsilon)$. The nonoscillatory terms are present from the $n=0$ term. The function U_0 is periodic in θ, and the amplitudes $a_0(n,y)$ are the Fourier coefficients of U. Knowing U is equivalent to knowing the $a_0(n,y)$ for all $n\in\mathbb{Z}$.

Adding correctors, we *seek asymptotic solutions of first order semilinear symmetric hyperbolic systems in the form*

$$u(\epsilon,y) = U(\epsilon,y,\phi(y)/\epsilon), \tag{7.3.3}$$

where $U(\epsilon,y,\theta)$ is periodic in θ and has asymptotic expansion

$$U(\epsilon,y,\theta) \sim \sum_{j=0}^{\infty} \epsilon^j\, U_j(y,\theta) . \tag{7.3.4}$$

We pause to review just how much thought has gone into developing this *ansatz*. First there is the treatment of linear problems starting with Euler to arrive at Lax. Then there are the considerations of perturbation theory to understand the generation of harmonics and the critical $m=0$ amplitude scaling. Finally the generating function U is introduced.

The leading term, $U_0(y,\phi(y)/\epsilon)$, presents two scales. If $U_0(y,\theta)$ and $\phi(y)$ vary on the length scale 1 then the leading term varies on the scale 1, and the scale ϵ. The expansion (7.3.3) is called a two scale or multiscale expansion. In the case of ordinary differential equations, where there is only one independent variable, often called time, such expansions are often called two timing, after the presence of two time scales.

Our approach can be viewed as originating in the work of [Choquet-Bruhat, 1969], [Hunter, Majda, Rosales, 1986] and [Keller, 1957], where a variety of very smart approximate solutions are proposed. The latter authors explored resonance, including the multiphase case. We draw attention to the important note of [Joly, 1983], which includes a proof of accuracy in the very rich but special case of one dimensional constant coefficient linear principal part and linear phases. The remainder of this book is devoted to describing developments flowing from these sources.

In one phase problems, the only resonances are the harmonics. In this case correctors as in (7.3.2) can be constructed. The multiphase theory must often content itself with leading order asymptotics only as in the paper of [Joly, 1983].

7.4. Equations for the profiles

Once the *ansatz* (7.3.3)–(7.3.4) is formulated, the key question is whether it is possible to find profiles $U_j(y,\theta)$ so that

$$L(y,\partial_y)\,U(\epsilon,y,\phi(y)/\epsilon) + F(y,U(\epsilon,y,\phi(y)/\epsilon)\,) \;\sim\; 0\,. \tag{7.4.1}$$

Since

$$\begin{aligned}\partial_j\Big(U(\epsilon,y,\phi(y)/\epsilon)\Big) &= \frac{\partial U}{\partial y}\Big|_{\epsilon,y,\phi(y)/\epsilon} + \frac{\partial_j\phi}{\epsilon}\,\frac{\partial U}{\partial\theta}\Big|_{\epsilon,y,\phi(y)/\epsilon}\\ &= \Big(\frac{\partial}{\partial y} + \frac{\partial_j\phi}{\epsilon}\,\frac{\partial}{\partial\theta}\Big)U\big|_{\theta=\phi(y)/\epsilon},\end{aligned}$$

one has

$$L(y,\partial_y)\,U(\epsilon,y,\phi(y)/\epsilon) \;=\; \Big[L\Big(y\,,\,\partial_y + \frac{d\phi(y)}{\epsilon}\,\frac{\partial}{\partial\theta}\Big)\,U(\epsilon,y,\theta)\Big]_{\theta=\phi(y)/\epsilon}\,.$$

Define the θ periodic profile

$$\begin{aligned}W(\epsilon,y,\theta) &:= L\Big(y\,,\,\partial_y + \frac{d\phi(y)}{\epsilon}\,\frac{\partial}{\partial\theta}\Big)\,U(\epsilon,y,\theta) + F\big(U(\epsilon,y,\theta)\big)\\ &= \frac{1}{\epsilon}\,L(y,d\phi(y))\,\frac{\partial}{\partial\theta}\,U(\epsilon,y,\theta) + L(y,\partial_y)U(\epsilon,y,\theta) + F\big(U(\epsilon,y,\theta)\big)\,.\end{aligned}$$

Then the left-hand side of (7.4.1) is equal to

$$W(\epsilon,y,\phi(y)/\epsilon) \;=\; W(\epsilon,y,\theta)\Big|_{\theta=\phi(y)/\epsilon}\,.$$

The middle term in the second formula for $W(\epsilon,y,\theta)$ has asymptotic expansion,

$$L(y,\partial_y)\,U(\epsilon,y,\theta) \;\sim\; \sum \epsilon^j\,L(y,\partial_y)\,U_j(y,\theta)\,.$$

Taylor expansion about U_0 yields

$$F(y, U_0 + \epsilon U_1 + \cdots) \sim F(y, U_0) + \epsilon\, F_u(y, U_0)\, U_1 + \text{h.o.t.} \tag{7.4.2}$$

The linear terms in U_1 are real linear and not necessarily complex linear, since F is assumed to be smooth but not necessarily holomorphic. These two expansions show that

(7.4.3)
$$W(\epsilon, y, \theta) \sim \sum_{j=-1}^{\infty} \epsilon^j\, W_j(y, \theta) = \epsilon^{-1} W_{-1}(y, \theta) + W_0(y, \theta) + \epsilon^1\, W_1(y, \theta) + \cdots .$$

The first terms are given by

$$W_{-1}(y, \theta) \;=\; L_1(y, d\phi(y))\, \partial_\theta U_0\,, \tag{7.4.4}$$

$$W_0(y, \theta) \;=\; L_1(y, d\phi(y))\, \partial_\theta\, U_1 + L(y, \partial_y)\, U_0 + F(y, U_0(y, \theta))\,, \tag{7.4.5}$$

$$W_1(y, \theta) = L_1(y, d\phi(y))\, \partial_\theta\, U_2 + L(y, \partial_y)\, U_1 + F_u(y, U_0)\, U_1\,, \tag{7.4.6}$$

(7.4.7)
$$W_2 = L_1(y, d\phi(y))\, \partial_\theta\, U_3 + L(y, \partial_y)\, U_2 + F_u(y, U_0)\, U_2 + F_{uu}(U_0)(U_1, U_1)\,.$$

The term W_{-1} is $O(\epsilon^{-1})$ and comes from the terms in (7.4.1) where the y derivatives fall on the $\phi(y)/\epsilon$ part.

The general case is of the form

(7.4.8)
$$W_j = L_1(y, d\phi(y))\, \partial_\theta\, U_{j-1} + L(y, \partial_y)\, U_j + F_u(y, U_0)\, U_j + G_j(U_0, \ldots, U_{j-1})\,,$$

where $G_j(U_0, \ldots, U_{j-1})$ is a nonlinear function of the preceding profiles. The sum $F_u U_j + G_j$ is the $O(\epsilon^j)$ term in the Taylor expansion (7.4.2)

We construct the U_j so that all the W_j vanish identically.

In order for there to be nontrivial oscillations, one must have $\partial_\theta U_0 \neq 0$, so the first constraint we place on the expansion is that the matrix $L_1(y, d\phi(y))$ have nontrivial kernel. Equivalently, ϕ must satisfy the familiar eikonal equation

$$\det\, L_1(y, d\phi(y)) = 0\,. \tag{7.4.9}$$

Setting $W_{-1} = 0$ then yields the equation

$$U_0 \;\in\; \ker \Big(L_1(y, d\phi(y))\, \frac{\partial}{\partial \theta} \Big)\,. \tag{7.4.10}$$

Setting $W_0 = 0$ yields an equation which mixes U_0 and U_1. As in the linear case, information about U_0 is contained in the assertion

$$L(y, \partial_y)\, U_0(y, \theta) + F(y, U_0(y, \theta)) \;\in\; \text{range} \Big(L_1(y, d\phi(y))\, \frac{\partial}{\partial \theta} \Big)\,. \tag{7.4.11}$$

Equations (7.4.10) and (7.4.11) are our first form of the profile equations of nonlinear geometric optics. Written this way, it is not at all clear that they determine U_0 from its initial data. They are an invitation to study the action of the operator $L_1(y, d\phi(y))\partial_\theta$ on periodic functions of θ.

Suppose that (7.4.9) is satisfied, and in addition that the constant rank hypothesis from §5.4 is satisfied on $\overline{\Omega}$ where Ω is open in $\mathbb{R}^{1+d}$. Denote by $\pi(y)$ the orthogonal projection of $\mathbb{C}^N$ onto the $\ker L(y, d\phi(y))$ and by $Q(y)$ the partial inverse. They are smooth thanks to Theorem 3.I.1.

The operator $L_1(y, d\phi(y))\partial_\theta$ maps $\mathcal{D}'(\Omega \times S^1)$ to itself with the subspace $C^\infty(\Omega \times S^1)$ also mapped to itself. The kernel and image can be computed by expanding in Fourier series in θ,

$$V(y,\theta) \;=\; \sum_{n=-\infty}^{\infty} V_n(y)\, e^{in\theta}\,. \tag{7.4.12}$$

When V is a distribution, the coefficient $V_n \in \mathcal{D}'(\Omega)$ is defined by

$$\big\langle V_n, \psi(y)\big\rangle \;:=\; \frac{1}{2\pi}\big\langle V, e^{-in\theta}\big\rangle\,, \qquad \psi \,\in\, C_0^\infty(\Omega)\,.$$

$V_0(y)$ is the nonoscillating contribution, and $(V - V_0)(y, \phi/\epsilon)$ is the oscillating part. One has

$$L_1(y, d\phi(y))\,\partial_\theta V \;=\; \sum L_1(y, d\phi(y))\, V_n(y) in\, e^{in\theta}\,. \tag{7.4.13}$$

Thus the kernel consists of functions such that for $n \neq 0$, V_n takes values in the kernel of $L_1(y, d\phi(y))$. Equivalently,

$$V \subset \ker L_1(y, d\phi)\partial_\theta \quad\Longleftrightarrow\quad \forall n \geq 1,\;\; \pi(y)\, V_n(y) \;=\; V_n(y)\,. \tag{7.4.14}$$

Formula (7.4.13) shows that the image of $L_1(y, d\phi(y))\partial_\theta$ consists of those Fourier series whose constant term vanishes, and whose other coefficients lie in the image of $L_1(y, d\phi(y))$. Equivalently,

(7.4.15)

$$V \subset \operatorname{range} L_1(y, d\phi)\,\partial_\theta \Longleftrightarrow V_0 = 0 \;\text{ and, }\; \forall n \geq 1,\; (I - \pi(y))\, V_n(y) = V_n(y)\,.$$

Define a projection operator $\mathbf{E}$ on Fourier series by

$$\mathbf{E} \sum_{n=-\infty}^{\infty} V_n(y)\, e^{in\theta} \;:=\; V_0 \,+\, \pi(y) \sum_{n\neq 0} V_n(y)\, e^{in\theta}\,. \tag{7.4.16}$$

Then

$$\mathbf{E}\, V \;=\; V_0 + \pi(y)\big(V - V_0\big) \;=\; V_0 + \pi(y)\Big(V \;-\; \frac{1}{2\pi}\int_0^{2\pi} V(y,\theta)\; d\theta\Big)\,.$$

For each y, $\mathbf{E}$ acts as an orthogonal projection in $L^2(S^1)$. It follows that $\mathbf{E}$ is an orthogonal projection on $L^2(B \times S^1)$ for B a subset of $\{t = \text{const}\}$ or of $\mathbb{R}^{1+d}$.

Formulas (7.4.15) and (7.4.16) show that acting on either distributions, smooth periodic functions, or formal trigonometric series, one has

$$V \subset \ker L_1(y, d\phi)\, \partial_\theta \quad \Longleftrightarrow \quad \mathbf{E}\, V \;=\; V \tag{7.4.17}$$

and

$$V \subset \operatorname{range} L_1(y, d\phi)\, \partial_\theta \quad \Longleftrightarrow \quad \mathbf{E} V \;=\; 0\,. \tag{7.4.18}$$

Thus $\mathbf{E}$ projects onto the kernel of $L_1(y, d\phi)\partial_\theta$ along its range. The operators $L_1(y, d\phi(y))\partial_\theta$ and $\mathbf{E}$ satisfy

$$\Big(L_1(y, d\phi(y))\partial_\theta\Big)\, \mathbf{E} \;=\; \mathbf{E}\, \Big(L_1(y, d\phi(y))\partial_\theta\Big) \;=\; 0 \tag{7.4.19}$$

and

(7.4.20)

$$\Big(L_1(y, d\phi(y))\partial_\theta\Big)\, (I - \mathbf{E}) \;=\; (I - \mathbf{E})\, \Big(L_1(y, d\phi(y))\partial_\theta\Big) \;=\; L_1(y, d\phi(y))\partial_\theta\,.$$

Define the partial inverse $\mathbf{Q}$ of the operator $L_1(y, d\phi(y))\partial_\theta$ by

$$\mathbf{Q}\Big(\sum V_n(y)\, e^{in\theta}\Big) \;:=\; Q(y) \sum_{n \neq 0} \frac{1}{in}\, V_n(y)\, e^{in\theta}\,. \tag{7.4.21}$$

Then

(7.4.22)

$$\mathbf{E}\,\mathbf{Q} = \mathbf{Q}\mathbf{E} = 0 \quad \text{and} \quad \mathbf{Q}\, \Big(L_1(y, d\phi(y))\partial_\theta\Big) = \Big(L_1(y, d\phi(y))\partial_\theta\Big)\, \mathbf{Q} = I - \mathbf{E}\,.$$

Equation (7.4.10) is equivalent to

$$\mathbf{E}\, U_0 \;=\; U_0\,. \tag{7.4.23}$$

This equation shows that the oscillating part of U_0 satisfies the familiar polarization from Chapter 5.

Equation (7.4.11) is equivalent to

$$\mathbf{E}\, \Big(\, L(y, \partial_y)\, U_0(y, \theta) + F(y, U_0(y, \theta))\, \Big) \;=\; 0\,. \tag{7.4.24}$$

The pair of equations (7.4.23)–(7.4.24) is analogous in form to the pair of equations (5.3.10) and (5.3.11) that determine a_0. Equations (7.4.23)–(7.4.24) hold if and only if

$$W_{-1} = 0 \qquad \text{and} \qquad \mathbf{E}\, W_0 = 0\,. \tag{7.4.25}$$

A note about our strategy here. Each equation $W_j = 0$ is equivalent to a pair of equations

$$W_j \;=\; 0 \qquad \Longleftrightarrow \qquad \mathbf{E}\, W_j \;=\; 0 \quad \text{and} \quad (I - \mathbf{E})\, W_j \;=\; 0\,.$$

The second equation is often transformed using

$$(I - \mathbf{E})\, W_j \;=\; 0 \qquad \Longleftrightarrow \qquad \mathbf{Q}\, W_j = 0\,.$$

The equations for the profiles U_j are found by induction. Suppose that $j \geq 1$ and that $U_0, \dots, U_{j-1}$ have been determined so that

$$W_{-1} = \cdots = W_{j-1} = 0 \qquad \text{and} \qquad \mathbf{E} W_j = 0\,. \tag{7.4.26}$$

The equations determining U_j from suitable initial data are equivalent to

$$(I - \mathbf{E})\, W_{j-1} = 0 \qquad \text{and} \qquad \mathbf{E}\, W_j = 0\,. \tag{7.4.27}$$

To illustrate the procedure, we find the equations for U_1. Equation (7.4.5) shows that $(I - \mathbf{E})W_0 = 0$ if and only if

$$\begin{aligned} (I - \mathbf{E})\, L_1(y, d\phi(y))\, \partial_\theta U_1 \;&=\; -(I - \mathbf{E}) \Big(L(y, \partial_y)\, U_0 + F(u, U_0) \Big) \\ &:=\; \Phi_0(y, U_0)\,, \end{aligned} \tag{7.4.28}$$

where the right-hand side, denoted Φ_0, is a function of the profile U_0 and its derivatives that are assumed known. The dependence on the derivatives is not indicated in the notation, since for the sequel it is not important just how many derivatives occur in the terms Φ_j. This equation determines U_1 modulo the kernel of the operator $(I - \mathbf{E})\, L_1(y, d\phi(y))\partial_\theta$. Equation (7.4.20) shows that the kernel is is equal to $\ker\big(L_1(y, d\phi(y))\partial_\theta\big)$.

Multiplying by $\mathbf{Q}$ shows that equation (7.4.28) is equivalent to

$$(I - \mathbf{E})\, U_1 = -\mathbf{Q}\, \Phi_0(y, U_0)\,. \tag{7.4.29}$$

The determination of $\mathbf{E}\, U_1$ comes from setting $\mathbf{E}\, W_1 = 0$. Multiplying (7.4.6) by $\mathbf{E}$ eliminates the $\partial_\theta U_2$ term and yields the second equation for the profile U_1,

$$\mathbf{E} \Big(L(y, \partial_y)\, U_1(y, \theta) + F_u(y, U_0)\, U_1(y, \theta) \Big) = 0\,. \tag{7.4.30}$$

The pattern is now established. Setting $(I - \mathbf{E})W_1 = 0$ yields

$$(I - \mathbf{E})\, U_2 \;=\; \Phi_1(y, U_0, U_1)\,, \tag{7.4.31}$$

and the equation $\mathbf{E}W_2 = 0$ yields

$$\mathbf{E} \Big(L(y, \partial_y)\, U_2(y, \theta) + F_u(y, U_0)\, U_2(y, \theta) + F_{uu}(y, U_0)(U_1, U_1) \Big) = 0\,. \tag{7.4.32}$$

Here F_{uu} is the order two term in (7.4.2) so is a symmetric quadratic form in in U_1.

Continuing in this fashion yields for all $j \geq 1$ a pair of equations

$$(I - \mathbf{E})\, U_j = \Phi_{j-1}(y, U_0, U_1, \dots, U_{j-1}) \tag{7.4.33}$$

and

$$\mathbf{E} \Big(L(y, \partial_y)\, U_j(y, \theta) + F_u(y, U_0)\, U_j(y, \theta) + G_j(y, U_0, \dots, U_{j-1}) \Big) = 0 \tag{7.4.34}$$

that are equivalent to (7.4.27).

Remarks. In (7.4.34) the sum $F_u U_j + G_j$ is the $O(\epsilon^j)$ term in (7.4.2). In contrast, the right-hand side of (7.4.33) is a shorthand hiding the fact that it also depends on derivatives of the previously determined profiles $U_0, \ldots, U_{j-1}$.

These computations are summarized as a sufficient condition for infinitely small residual.

Theorem 7.4.1. *Suppose that $\phi \in C^\infty(\Omega)$ satisfies the eikonal equation (7.4.9) with nowhere vanishing differential and* dim ker $L(y, d\phi(y))$ *independent of y. In addition suppose $U_j \in C^\infty(\Omega \times \mathbb{T}^1)$ are profiles such that the principal profile U_0 satisfies (7.3.23)–(7.3.24) and the for $j \geq 1$ the profiles satisfy (7.4.33)–(7.4.34). If $U(\epsilon, y, \theta) \sim \sum_{j=0}^\infty \epsilon^j\, U_j(y,\theta)$ in $C^\infty(\Omega \times \mathbb{T}^1)$ and $u^\epsilon(y) := U(\epsilon, y, \phi(y)/\epsilon)$, then*

$$L(y, \partial_y)\, u^\epsilon + F(u^\epsilon) \;\sim\; 0 \qquad in \quad C^\infty(\Omega)\,.$$

Proof. The equations for the profiles are equivalent to solving $W_j = 0$ for all j. Thus if the profiles U_j satisfy the profile equations and $U(\epsilon, y, \theta) \sim \sum_{j=0}^\infty \epsilon^j\, U_j(y,\theta)$, then

$$L(y, \partial_y)\, U(\epsilon, y, \phi(y)/\epsilon) + F(y, U((\epsilon, y, \phi(y)/\epsilon))\,) \;\sim\; 0\,. \qquad \square$$

Remarks. 1. The equation for the principal profile U_0 is nonlinear in U_0, whereas the equations for the higher profiles U_j with $j \geq 1$, are $\mathbb{R}$ linear in U_j.

2. It is not at all obvious that the profile equations have solutions. One must prove analogs of Theorem 5.3.5 for the more complicated profile equations.

7.5. Solving the profile equations

This subsection shows that the equations derived above determine the profiles U_j from suitable initial data. The corresponding asymptotic expansion yields a family of approximate solutions with infinitely small residual in the limit $\epsilon \to 0$. In Chapter 8 it is proved that the approximate solution is asymptotic to the exact solution that has the same initial data.

To see that U_0 is determined from its initial data, start with the fact that (7.4.23) and (7.4.24) together imply that

$$\mathbf{E}\, L(y, \partial_y)\, \mathbf{E} U_0 + \mathbf{E}\, F(y, \mathbf{E} U_0(y,\theta)) = 0\,. \tag{7.5.1}$$

Applying $(I - \mathbf{E})\, L(y, \partial_y)$ to (7.4.23) yields

$$(I - \mathbf{E})\, L(y, \partial_y)\, (I - \mathbf{E})\, U_0 = 0\,. \tag{7.5.2}$$

Adding these two equations yields

(7.5.3)
$$(I - \mathbf{E})\, L(y, \partial_y)\, (I - \mathbf{E})\, U_0 \;+\; \mathbf{E}\, L(y, \partial_y)\, \mathbf{E}\, U_0 \;+\; \mathbf{E}\, F(y, \mathbf{E}\, U_0(y, \theta)) \;=\; 0\,,$$

an analogue of (5.3.16).

Define the linear operator

$$\mathbf{L} \;:=\; \sum_{\mu=0}^{d} \mathbf{A}_\mu\, \partial_\mu \;+\; \mathbf{B}\,, \tag{7.5.4}$$

where the coefficients are the operators

$$\begin{aligned} \mathbf{A}_\mu \;&:=\; (I - \mathbf{E})\, A_\mu(y)\, (I - \mathbf{E}) + \mathbf{E}\, A_\mu(y)\, \mathbf{E}\,, \\ \mathbf{B} \;&:=\; (I - \mathbf{E})\, B(y)\, (I - \mathbf{E}) + \mathbf{E}\, B(y)\, \mathbf{E}\,. \end{aligned} \tag{7.5.5}$$

The unknown U_0 is a $\mathbb{C}^N$ valued function of t, x, θ with $\theta \in S^1$. The notation is chosen so that $\mathbf{L}$ looks like a differential operator. Some care must be taken since the coefficient operators are not simple matrix multiplications. However, the idea behind the basic energy estimate for symmetric hyperbolic operators extends nearly immediately to $\mathbf{L}$.

First of all the operators $\mathbf{A}_\mu$ are selfadjoint in $L^2(\omega \times S^1)$ because A_μ and $\mathbf{E}$ are. Since $\mathbf{E}$ commutes with differentiation, one has

$$[\mathbf{A}_\mu, \partial] \;=\; (I - \mathbf{E})\, (\partial A_\mu(y))\, (I - \mathbf{E}) + \mathbf{E}\, (\partial A_\mu(y))\, \mathbf{E}\,.$$

This is bounded on any $L^2(\omega \times S^1)$ with norm bounded independent of ω.

The operator $\mathbf{A}_0$ is strictly positive. If ω is arbitrary and $(\,\cdot\,,\cdot\,)$ denotes the $L^2(\omega \times S^1)$ scalar product, one has

$$\begin{aligned} (V, \mathbf{A}_0 V) \;&=\; \big(V\,,\, (I - \mathbf{E})(A_0 (I - \mathbf{E}) V + \mathbf{E} A_0 \mathbf{E} V\big) \\ &=\; \big((I - \mathbf{E}) V\,,\, A_0 (I - \mathbf{E}) V\big) + \big(\mathbf{E} V\,,\, A_0 \mathbf{E} V\big). \end{aligned}$$

If C is a lower bound for A_0, this is

$$\geq\; C\Big(\|(I - \mathbf{E}) V\|_{L^2}^2 + \|\mathbf{E} V\|_{L^2}^2\Big) \;=\; C\, \|V\|_{L^2}^2\,,$$

proving strict positivity with a bound independent of ω. The same argument proves

$$\sum \eta_\mu A_\mu \geq 0 \qquad \Longrightarrow \qquad \sum \eta_\mu \Big((I - \mathbf{E}) A_0 (I - \mathbf{E}) + \mathbf{E} A_0 \mathbf{E}\Big) \;\geq\; 0\,.$$

In order to treat phases that need not be globally defined, we work on domains of the form $\Omega \times S^1$, where Ω denotes a domain of determinacy for $L(y, \partial)$ as in Corollary 2.3.7 or more generally as in §2.6. In either case the outward conormals η to the lateral boundaries satisfy $\sum \eta_\mu A_\mu \geq 0$, asserting that the flux corresponding to the energy density (A_0, u, u) is outward. This is reasonable for a domain that is not influenced by what goes on outside.

Next derive an L^2 energy estimate for the operator $\mathbf{L}$ in the set $\Omega \times S^1$. Recall that $\Omega_T := \Omega \cap \{0 \le t \le T\}$, while $\Omega(\underline{t}) := \Omega \cap \{t = \underline{t}\}$ is the section at time $\underline{t}$. By definition,

$$\begin{aligned}(\mathbf{L} V\,,\, V)_{L^2(\Omega_t \times S^1)} &= \big(\mathbf{E} L \mathbf{E} V\,,\, V\big) + \big((I - \mathbf{E}) L (I - \mathbf{E}) V\,,\, V\big)\\ &= \big(L \mathbf{E} V\,,\, \mathbf{E} V\big) + \big(L (I - \mathbf{E}) V\,,\, (I - \mathbf{E}) V\big)\,.\end{aligned}$$

In the last expressions θ plays the role of a parameter, and one has expressions $(L\,\cdot\,,\,\cdot)$ applied to $\mathbf{E}V$ and to $(I - \mathbf{E})V$.

Since the energy flux is outward at the boundary, the energy balance identity (2.3.1), implies that for $W \in C^1(\Omega_t \times S^1)$,

$$(7.5.6)\quad 2\mathrm{Re}\,\big(L\,W\,,\,W\big)_{L^2(\Omega_t \times S^1)} \ge \big(A_0\,W\,,\,W\big)_{L^2(\Omega(t)\times S^1)}\Big|_0^t - \big(ZW\,,\,W\big)_{L^2(\Omega_t\times S^1)}\,.$$

Applying this with $W = \mathbf{E}\,V$ and $(I - \mathbf{E})W$ and adding the results shows that for $U \in C^1(\Omega_t \times S^1)$,

$$(7.5.7)\quad \big(U(t), \mathbf{A}_0\,U(t)\big)_{L^2(\Omega(t)\times S^1)}\Big|_0^t \le 2\mathrm{Re}\,\big(U(t), \mathbf{L}\,U(t)\big)_{L^2(\Omega_t\times S^1)} + C\,\big(U(t)\,,\,U(t)\,\big)_{L^2(\Omega_t\times S^1)}\,.$$

Since $\mathbf{A}_0$ is a strictly positive operator, (7.5.7) implies

$$(7.5.8)\quad \|U(t)\|_{L^2(\Omega(t)\times S^1)} \le C(L,\phi)\,\Big(\,\|U(0)\|_{L^2(\Omega(0)\times S^1)} + \int_0^t \|(\mathbf{L}\,U(\sigma))\|_{L^2(\Omega(\sigma)\times S^1)}\,d\sigma\Big)\,.$$

A commutation argument like that in §2.1 yields the more general estimate for derivatives $\partial^\alpha_{x,\theta}U$ with $|\alpha| \le s \in \mathbb{N}$,

(7.5.9)

$$\|U(t)\|_{H^s(\Omega(t)\times S^1)} \le C(s,L,\phi)\,\Big(\,\|U(0)\|_{H^s(\Omega(0)\times S^1)} + \int_0^t \|(\mathbf{L}\,U(\sigma))\|_{H^s(\Omega(\sigma)\times S^1)}\,d\sigma\Big)\,.$$

The existence theorem corresponding to this estimate is the following.

Theorem 7.5.1. *If $s \in \mathbb{N}$, $g \in H^s(\Omega(0) \times S^1)$, and $f \in L^1(\Omega \times S^1)$ satisfy*

$$(7.5.10)\qquad \int_0^T \Big(\int_{\Omega(t)\times S^1} \sum_{|\alpha|\le s} |\partial^\alpha_{x,\theta} f(t,x,\theta)|^2\,dx d\theta\Big)^{1/2} dt < \infty\,,$$

then there is a unique $U \in L^2(\Omega \times S^1)$ satisfying

$$(7.5.11)\qquad \mathbf{L}U = f \qquad \text{and} \qquad U|_{t=0} = g\,.$$

The solution satisfies the estimate (7.5.10). If $g \in C^\infty(\Omega(0) \times S^1)$ and $f \in C^\infty(\Omega \times S^1)$, then $U \in C^\infty(\Omega \times S^1)$.

Sketch of Proof. The strategy is to replace derivatives ∂_j by centered differences δ_j^h, derive estimates uniform in h, and pass to the limit $h \to 0$. There is a difficulty. Though the solution of the differential equation is determined in Ω_T by the data in Ω_T, the same is not true of the difference scheme. To overcome this, one extends the data and the differential operator $\mathbf{L}$ to all of $[0,T] \times \mathbb{R}^d$.

The coefficients of the differential operator are everywhere defined by the projection $\pi(y)$ and therefore $\mathbf{E}$ is only defined where ϕ is defined. Choose two extensions $\pi_j(y)$ as hermitian symmetric matrix valued functions on $[0,T] \times \mathbb{R}^d$ whose partial derivatives of all orders are bounded on $[0,T] \times \mathbb{R}^d$. The extensions are chosen so that there is a constant $c > 0$ so that on $[0,T] \times \mathbb{R}^d$ one has

$$\pi_1^* \pi_1 + (I - \pi_2)^*(I - \pi_2) \le cI.$$

In so doing, it is good to choose π_1 large and π_2 small. There is no constraint that the extensions be projectors or a constant rank. One can choose π_1 equal to the identity and π_2 equal to zero outside a small neighborhood or Ω_T.

Then define operators $\mathbf{E}_j$ using the projectors π_j and extend $\mathbf{L}$ to $[0,T] \times S^1 \times \mathbb{R}^d$ as

$$\mathbf{E}_1 L(y, \partial) \mathbf{E}_1 + (I - \mathbf{E}_2) L(y, \partial)(I - \mathbf{E}_2),$$

for which estimates analogous to (7.5.9) hold. Choose extensions of f and g, still denoted by the same letter, and solve

$$(\mathbf{E}_1 L(y, \partial) \mathbf{E}_1 + (I - \mathbf{E}_2) L(y, \partial)(I - \mathbf{E}_3))\, U = f, \quad U(0, x, \theta) = g(x, \theta),$$

by the finite difference method.

In the original domain $\Omega_T \times S^1$ one has a solution of (7.5.11). Estimate (7.5.9) shows that the solution in this domain is determined by the data in that domain. □

To treat nonlinear problems as in Chapter 6, note that for $s > (d+1)/2$, Schauder's lemma implies that the map

$$U(t) \quad \mapsto \quad \mathbf{E}\, F(y, \mathbf{E}\, U(t))$$

is a locally lipschitzean map of $H^s(\Omega(t) \times S^1)$ to itself, uniformly for $0 \le t \le T$.

Standard Picard iteration,

$$\mathbf{L}\, U^{\nu+1} + \mathbf{E}\, F(y, \mathbf{E}\, U^{\nu}) = 0\,, \qquad U^{\nu+1}|_{t=0} = g, \tag{7.5.12}$$

as in Chapter 6 leads to the basic nonlinear local existence theorem. Existence is proved on $\Omega_T \times S^1$.

Theorem 7.5.2 (Local Solvability of the Principal Profile Equation). *If $(d+1)/2 < s \in \mathbb{N}$ and $g_0 \in H^s(\Omega(0) \times S^1)$, then there is a $0 < T$ and unique $U_0 \in C(\Omega_T \times S^1)$ satisfying (7.5.3) together with the initial condition $U_0(0, \cdot) = g$. If $g_0 \in C^\infty(\Omega(0) \times S^1)$, then $U_0 \in C^\infty(\Omega_T \times S^1)$.*

It is important to note that what is solved here is equation (7.5.3) that follows from the desired equations (7.4.23)–(7.5.24). Thus we have shown that the latter equations determine uniquely U_0 but we have not yet shown that there exists a U_0 satisfying (7.4.23)–(7.4.24). If the initial data g do not satisfy (7.4.23), then there is no chance for that equation.

Lemma 7.5.3. *If in addition to the hypotheses of the Theorem 7.5.2, $\mathbf{E}\, g = g$ in $\Omega(0) \times S^1$, then the resulting solution U_0 satisfies (7.4.23)–(7.4.24).*

Proof. As in the analysis of §5.4 an important first step is to observe that the left-hand side of (7.5.3) is the sum of two orthogonal parts so that equation (7.5.3) implies that both vanish. Equivalently, multiplying (7.5.3) by $\mathbf{E}$ shows that U_0 satisfies the pair of equations

(7.5.13)

$$(I - \mathbf{E})\, L(y, \partial_y)\, (I - \mathbf{E})\, U_0 = 0 \quad \text{and} \quad \mathbf{E}\, L(y, \partial_y)\, \mathbf{E}\, U_0 + \mathbf{E}\, f(y, \mathbf{E}\, U_0(y, \theta)) = 0\,.$$

It follows that $\mathbf{E}\, U_0$ also satisfies both of these equations and, therefore, equation (7.5.3). Since $\mathbf{E}\, U_0$ has the same initial data as U_0, it follows by uniqueness of solutions of the initial value problem for (7.5.3) that $\mathbf{E}\, U_0 = U_0$, which is equation (7.4.23).

Finally, (7.5.3) and the second equation of (7.5.14) imply (7.4.24). □

Fix $0 < T$ as in Theorem 7.5.2. Then the higher order profiles can be found on $\Omega_T \times S^1$ so as to satisfy (7.4.33)–(7.4.34). The argument is as follows. In (7.4.34) write using (7.4.33),

$$U_j \;=\; \mathbf{E}\, U_j + (I - \mathbf{E})\, U_j \;=\; \mathbf{E}\, U_j + \Phi_{j-1}\,.$$

This yields an equation of the form

$$\mathbf{E}\, L(y, \partial_y)\, \mathbf{E}\, U_j + \text{linear in } \; \mathbf{E}\, U_j = \text{known}\,.$$

To this equation add $(I - \mathbf{E})\, L(y, \partial_y)$ applied to (7.4.33) to find an equation for

(7.5.14)
$$C = \mathbf{E} U_j$$

of the form

(7.5.15)
$$\mathbf{L}\, C + \text{linear in } C = \text{known}\,.$$

This linear equation determines C from its initial data. Imitating arguments, which by now should be familiar, one shows that if the solution C satisfies

$\mathbf{E}\,C = C$ at $t = 0$, then it does so throughout $\Omega_T \times S^1$ and that $U_j := C + \mathbf{Q}\,\Phi_{j-1}$ satisfies the two profile equations (7.4.33)–(7.4.34).

Exercise 7.5.1. *Flesh out the details of this argument.*

Theorem 7.5.4 (Joly and Rauch)**.** *Suppose that $g_j = \mathbf{E}\,g_j \in C^\infty(\Omega(0) \times S^1)$, and that $T > 0$ is chosen as Theorem 7.5.2. Then there are uniquely determined profiles $U_j(y,\theta) \in C^\infty(\Omega_T \times S^1)$ with*

$$\mathbf{E}\,U_j\big|_{t=0} = g_j \quad \text{on} \quad \Omega(0) \times S^1 \tag{7.5.16}$$

and satisfying the profile equations (7.4.23)–(7.4.24) and (7.4.33)–(7.4.34). If

$$U(\epsilon, y, \theta) \;\sim\; \sum_{j=0}^{\infty} \epsilon^j\, U_j(y,\theta) \quad \text{in } C^\infty(\Omega_T \times S^1) \tag{7.5.17}$$

$$\text{and} \quad u^\epsilon(y) := U(\epsilon, y, \phi(y)/\epsilon)\,,$$

then

$$L(y, \partial_y)\, u^\epsilon + f(y, u^\epsilon) \;\sim\; 0 \qquad \text{in} \qquad C^\infty(\Omega_T)\,. \tag{7.5.18}$$

This completes the construction of an infinitely accurate family of approximate solutions u^ϵ. One point of view toward this, and that expressed in most science texts, is that the partial differential equations involve parameters that are only known approximately, so an infinitely accurate approximation is, for all practical purposes, as good as an exact solution.

Hadamard offered a deeper appreciation of this remark. He observes that since there are uncertainties in the equations and data, in order for the equations to lead to well defined predictions, it is crucial that the predictions be unchanged or only very slightly changed when the equations and data are changed within the limits of the uncertainties. This led to his notion of *well posed problems.*

In our case, the point of view of Hadamard leads to the problem of showing that a pair of infinitely accurate approximate solutions with infinitely close initial data are in fact close. This does not follow from the basic existence theorem of Chapter 6, because the approximate solutions tend to infinity in the configuration space H^s with $s > d/2$, and the sensitivity of the equation to perturbations grows for large data. One approach to circumventing this is to find a different configuration space in which a good existence theory is available and in which the approximate solutions do not grow. In the case $d = 1$, L^∞ does the trick. For higher dimensions, the space of bounded stratified solutions introduced by [Rauch and Reed, 1989] works and is the heart of the proof in [Joly and Rauch, 1992]. In the next chapter we give a different proof using ideas from [Guès, 1992] and [Donnat, 1994].

Chapter 8

Stability for One Phase Nonlinear Geometric Optics

In Chapter 7, the profiles $U_j(y, \theta)$, periodic in θ, were constructed so that if

$$(8.0.1)\qquad U(\epsilon, y, \theta) \sim \sum_{j=0}^{\infty} \epsilon^j\, U_j(y, \theta)\,,$$

then

$$(8.0.2)\qquad u^\epsilon(y) := U(\epsilon, y, \phi(y)/\epsilon)$$

satisfies

$$(8.0.3)\qquad L(y, \partial)\, u^\epsilon + F(y, u^\epsilon) \sim 0\,.$$

Denote by $v^\epsilon(y)$ the exact solution of

$$(8.0.4)\qquad L(y, \partial)\, v^\epsilon + F(y, v^\epsilon) = 0\,, \qquad v^\epsilon\big|_{t=0} = u^\epsilon\big|_{t=0}\,.$$

To show that the asymptotic expansion is correct amounts to showing that

$$(8.0.5)\qquad u^\epsilon(y) \sim v^\epsilon(y).$$

The difference between the equations defining the exact and approximate solutions is an infinitely small source term on the right-hand side of (8.0.3). The task is to show that this small source can only lead to small changes in the solution. This is a stability problem. The technical challenge is that the stability is needed near a family of solutions u^ϵ that is bounded in L^∞ but is unbounded in H^s for each $s > 0$. The nonlinear evolution is well behaved in H^s with $s > d/2$. The initial data of our family diverges to infinity in this space.

An important part of the proof is that the exact solution v^ϵ exists on an ϵ-independent time interval. Since the $H^s(\mathbb{R}^d)$ norm of the initial data grows infinitely large, this is not obvious. It is nearly as hard as proving (8.0.5).

8.1. The $H^s_\epsilon(\mathbb{R}^d)$ norms

A key to the analysis is the introduction of ϵ-dependent Sobolev norms. The asymptotic solution has the form (8.0.2). The derivatives grow as ϵ decreases, but the operator $\epsilon\partial$ applied to the asymptotic solution is bounded independent of ϵ. This suggests that one estimates $(\epsilon\partial)^\alpha\{u^\epsilon, v^\epsilon\}$. This strategy was introduced by Guès (1993, 1992) to study the quasilinear version of the one phase theorems. It is also important in the semiclassical limit in quantum mechanics where operators in $\hbar\partial$ take center stage.

Definition. *For $s \in \mathbb{Z}$, $0 < \epsilon \le 1$, and $w \in H^s(\mathbb{R}^d)$, define the $H^s_\epsilon(\mathbb{R}^d)$ norm by*

$$\|w\|^2_{H^s_\epsilon(\mathbb{R}^d)} := \sum_{|\alpha|\le s} \| (\epsilon\partial_x)^\alpha w \|^2_{L^2(\mathbb{R}^d)} . \tag{8.1.1}$$

A family w^ϵ is bounded in $H^s_\epsilon(\mathbb{R}^d)$ when

$$\sup_{0<\epsilon\le 1} \|w^\epsilon\|_{H^s_\epsilon(\mathbb{R}^d)} < \infty .$$

Example 8.1.1. For t fixed, the family $u^\epsilon(y)$ defined in (8.0.2) is bounded in $H^s_\epsilon(\mathbb{R}^d)$ provided that the support of $U(\epsilon, t, x, \theta)$ is bounded in x and contained in the domain of definition of ϕ.

The norm in $H^s_\epsilon(\mathbb{R}^d)$ is equivalent to the norm whose square is equal to

$$\int_{\mathbb{R}^d} (1 + |\epsilon\,\xi|^2)^s\, |\hat{u}(\xi)|^2\, d\xi . \tag{8.1.2}$$

This gives the generalization to noninteger s.

The $H^s_\epsilon(\mathbb{R}^d)$ Sobolev inequalities are immediate consequences of the scaling identity,

$$v(x) := w(\epsilon x) \implies \partial_x v = (\epsilon\partial_x w)(\epsilon x) . \tag{8.1.3}$$

Thus,

$$\partial_x^\alpha v = \big((\epsilon\partial_x)^\alpha w\big)\,(\epsilon x) , \tag{8.1.4}$$

so

$$\|\partial_x^\alpha v\|^2_{L^2(\mathbb{R}^d)} = \int \big|(\epsilon\partial_x^\alpha) w\,(\epsilon x)\big|^2\, dx .$$

The change of variable $X := \epsilon x$ shows that this is equal to

$$\int |\,(\epsilon\partial_x)^\alpha w(X)\,|^2\ \epsilon^{-d}\ dX \;=\; \epsilon^{-d}\,\|\,(\epsilon\partial_x)^\alpha\, w\,\|^2_{L^2(\mathbb{R}^d)}\,.$$

Summing shows that

$$\epsilon^{d/2}\,\|v\|_{H^s(\mathbb{R}^d)} \;=\; \|w\|_{H^s_\epsilon(\mathbb{R}^d)}\,. \tag{8.1.5}$$

For $s > d/2$,

$$\|w\|_{L^\infty(\mathbb{R}^d)} = \|v\|_{L^\infty(\mathbb{R}^d)} \le C(s,d)\,\|v\|_{H^s(\mathbb{R}^d)} = \epsilon^{-d/2}\,C(s,d)\,\|w\|_{H^s_\epsilon(\mathbb{R}^d)}\,. \tag{8.1.6}$$

Similarly for smooth $F(w)$ with $F(0) = 0$, one has an $H^s_\epsilon(\mathbb{R}^d)$ version of Schauder's lemma for $s > d/2$. A first attempt is the estimate $\|\,F(w)\,\|_{H^s(\mathbb{R}^d)} \le H(\|v\|_{H^s(\mathbb{R}^d)})$ with a nonlinear function H. Then,

$$\begin{aligned}\|\,F(w)\,\|_{H^s_\epsilon(\mathbb{R}^d)} &= \epsilon^{d/2}\,\|\,F(v)\,\|_{H^s(\mathbb{R}^d)} \\ &\le \epsilon^{d/2}\,H(\|v\|_{H^s(\mathbb{R}^d)}) \;=\; \epsilon^{d/2}\,H(\epsilon^{-d/2}\,\|w\|_{H^s_\epsilon(\mathbb{R}^d)}\,)\,.\end{aligned} \tag{8.1.7}$$

The negative power of ϵ in the argument of H is undesirable. Moser's inequality is much better behaved,

$$\begin{aligned}\|\,F(w)\,\|_{H^s_\epsilon(\mathbb{R}^d)} &= \epsilon^{d/2}\,\|\,F(v)\,\|_{H^s(\mathbb{R}^d)} \\ &\le \epsilon^{d/2}\,G(\|v\|_{L^\infty})\,\|v\|_{H^s(\mathbb{R}^d)} \;=\; G(\|w\|_{L^\infty})\,\|w\|_{H^s_\epsilon(\mathbb{R}^d)}\,.\end{aligned} \tag{8.1.8}$$

The cancellation of powers of $\epsilon^{\pm d/2}$ in the last inequality shows that *the Moser inequality for $H^s_\epsilon(\mathbb{R}^d)$ is independent of ϵ.*

In the justification of the asymptotic expansion, it is crucial to estimate the difference $F(u) - F(v)$ when u and v are close. In our application, the function u is our approximate solution and v is the exact solution. In this case, more is known of u than of v. There is sup norm control of the operators $\epsilon\partial$ applied to u which will be used to get L^2 control on these operators applied to v.

Lemma 8.1.1. *For any $R > 0$ and s, there is a constant $C = C(F, R, s)$ so that if u satisfies*

$$\|(\epsilon\partial)^\alpha u\|_{L^\infty(\mathbb{R}^d)} \le R \qquad \text{for all} \quad |\alpha| \le s$$

and w satisfies the weaker estimate

$$\|w\|_{L^\infty(\mathbb{R}^d)} \le R\,,$$

then for $0 < \epsilon$,

$$\|F(y, u + w) - F(y, u)\|_{H^s_\epsilon(\mathbb{R}^d)} \;\le\; C\,\|w\|_{H^s_\epsilon(\mathbb{R}^d)}\,.$$

Proof. To simplify the exposition, suppose that F does not depend on y. It suffices to prove the assertion with $\epsilon = 1$ since both sides of the inequality scale as $\epsilon^{d/2}$.

To prove the assertion for $\epsilon = 1$, write

$$F(u+w) - F(w) \;=\; \Big(\int_0^1 F'(u+\sigma w)\, d\sigma\Big) w \;:=\; \mathcal{G}(u,w)\, w\,. \tag{8.1.9}$$

Expanding $\partial^\nu(\mathcal{G}(u,w)\,w)$ using Leibniz's rule yields a finite number of terms of the form

$$H_{\alpha,\beta}(u,w)\ (\partial^{\alpha_1}u)\cdots(\partial^{\alpha_m}u)\ (\partial^{\beta_1}w)\cdots(\partial^{\beta_n}w)$$

with $\sum \alpha_k + \sum \beta_l = \nu$. The product of the first $m+1$ factors has sup norm bounded by $C(R)$. The proof of Moser's inequality shows that the product of the last n has $L^2(\mathbb{R}^d)$ norm bounded by $C(R)\|w\|_{H^s(\mathbb{R}^d)}$. This completes the poof. □

Exercise 8.1.1. *Prove the lemma for F that depend on y.*

Since phases may be only defined locally, we need localized versions. As in Chapter 7, suppose that ϕ is defined on a domain of determinacy, Ω, and denote

$$\Omega_T \;:=\; [0,T]\cap\Omega \qquad \text{and} \qquad \Omega(t) := \{x \,:\, (t,x)\in\Omega\}\,.$$

The usual reflection operators construct linear extension operators $v \to E(t)v$ from $H^s(\Omega(t))$ to $H^s(\mathbb{R}^d)$ so that $Ev = v$ on $\Omega(t)$. To study H^s_ϵ by scaling, one needs extensions from $\Omega(t)/\epsilon$ to $\mathbb{R}^d$ for $0<\epsilon<1$. These are domains with boundaries becoming less oscillatory. The standard constructions yield extension operators with norms bounded independently of $0<\epsilon\le 1$, $0\le t\le T$ under very mild regularity assumptions on Ω. The next exercise recalls the construction for balls. An analogous construction works for half spaces and then via coordinate charts for regular Ω.

Exercise 8.1.2. *For $s=0$, extending v to vanish outside B works. For $s=1$ and $|x|>1$, denote by $R(x) := x/|x|^2$ the reflected point in the unit sphere. Choose a smooth function χ that is equal to 1 on a neighborhood of 1 and which vanishes outside $]1/2,3/2[$. Show that setting $Ev(x) := \chi(|x|)\, v(R(x))$ for $|x|>1$ works for $s=1$. For larger s construct an appropriate extension operator by setting*

$$Ev \;:=\; \sum_{j=1}^{s} c_j\, v(R(x_j))\,, \qquad x_j \;:=\; \Big(1+2^j(|x|-1)\Big)\frac{x}{|x|}\,.$$

The key is the choice of the constants c_j so that $s-1$ derivatives match at the boundary of the ball. This elegant idea is called the Lions reflection after J. L. Lions.

Assumption. *Assume that uniformly bounded extension operators from $H^s_\epsilon(\Omega(t))$ to $H^s_\epsilon(\mathbb{R}^d)$ exist for $0 < \epsilon \le 1$ and $0 \le t \le T$.*

Lemma 8.1.2. *When the assumption holds, Lemma 8.1.1 holds with $\mathbb{R}^d$ replaced by $\Omega(t)$ with constant independent of $0 < \epsilon \le 1$ and $0 \le t \le T$.*

Exercise 8.1.3. *Write out the details of the proof using the extension operators.*

8.2. H^s_ϵ estimates for linear symmetric hyperbolic systems

In addition to the estimates of the last section, the analysis relies on the fact that linear hyperbolic systems propagate the $H^s_\epsilon(\mathbb{R}^d)$ norms. This follows from the basic linear estimate and commutation identities between the operator L and the operators $(\epsilon\partial)^\alpha$. The argument is entirely analogous to the commutation arguments in sections 1.1, 2.1, and 2.2. Square brackets are used to denote the commutator.

Introducing the new variable $\underline{u} := A_0^{-1/2} u$ and multiplying the resulting system for $\underline{u}$ by $A_0^{-1/2}$ yields a semilinear equation of the same form as before with new coefficient matrices $\underline{A}_\mu := A_0^{-1/2} A_\mu A_0^{-1/2}$. In particular, the coefficient of the time derivative is equal to the identity matrix. Thus, without loss of generality, we suppose that $A_0 = I$.

Lemma 8.2.1. *If $A_0 = I$, then for any $\alpha \in \mathbb{N}^d$ there are matrix valued functions $C_{\alpha\beta}(\epsilon, y)$ with uniformly bounded derivatives on $]0,1] \times \mathbb{R}^{1+d}$ so that*

$$[L(y, \partial_y)\,,\,(\epsilon\partial_x)^\alpha\,] = \sum_{|\beta|\le|\alpha|} C_{\alpha\beta}(\epsilon, y)\,(\epsilon\partial_x)^\beta\,.$$

Remark. If A_0 depended on time, there would be time derivatives in the commutators. It is to avoid these that we transform to the case $A_0 = I$.

Proof. The proof is by induction on $|\alpha|$. For $|\alpha| = 1$ compute

$$[L(y, \partial_y)\,,\,\epsilon\partial_j] = -\sum_k (\partial_j A_k)\,\epsilon\partial_k + \epsilon(\partial_j B).$$

Suppose next that $m \ge 1$, and the result is true for derivatives of length less than or equal to m. A differentiation of length $m+1$ is of the form $\epsilon\partial_j(\epsilon\partial_x)^\alpha$ with $|\alpha| = m$. Then

$$\begin{aligned} L\,\epsilon\partial_j\,(\epsilon\partial_x)^\alpha - \epsilon\partial_j\,(\epsilon\partial_x)^\alpha\,L &= [L, \epsilon\partial_j]\,(\epsilon\partial_x)^\alpha + \epsilon\partial_j\,L\,(\epsilon\partial_x)^\alpha - \epsilon\partial_j\,(\epsilon\partial_x)^\alpha\,L \\ &= [L, \epsilon\partial_j]\,(\epsilon\partial_x)^\alpha + \epsilon\partial_j\,[L, (\epsilon\partial_x)^\alpha]\,. \end{aligned}$$

Using the inductive hypothesis to express the commutators, the result follows. □

Theorem 8.2.2. *If $A_0 = I$, then for any $s \in \mathbb{N}$ and $T \in]0, \infty[$ there is a constant $C = C(s, T, L)$ so that for all $0 \le \underline{t} \le T$, and $u \in C([0, \underline{t}]\,;\, H^s(\mathbb{R}^d))$ with $Lu \in L^1([0, \underline{t}]\,;\, H^s(\mathbb{R}^d))$,*

$$\|u(\underline{t})\|_{H^s_\epsilon(\mathbb{R}^d)} \le C \left(\|u(0)\|_{H^s_\epsilon(\mathbb{R}^d)} + \int_0^{\underline{t}} \|(Lu)(\sigma)\|_{H^s_\epsilon(\mathbb{R}^d)}\, d\sigma \right). \tag{8.2.1}$$

Proof. For $|\alpha| \le s$ use the commutation lemma to write

$$L\,(\epsilon\partial_x)^\alpha u = (\epsilon\partial_x)^\alpha Lu + \sum C_{\alpha\beta}(\epsilon, y)\,(\epsilon\partial_x)^\beta u\,. \tag{8.2.2}$$

The basic linear estimate (2.1.18) then implies that for any $0 \le t \le \underline{t}$,

$$\begin{aligned}\|(\epsilon\partial_x)^\alpha u(t)\|_{L^2(\mathbb{R}^d)} \le C \Big(&\|(\epsilon\partial_x)^\alpha u(0)\|_{L^2(\mathbb{R}^d)} \\ &+ \int_0^t \big\{\|(\epsilon\partial_x^\alpha)(Lu)(\sigma)\|_{L^2(\mathbb{R}^d)} + \|u(\sigma)\|_{H^s_\epsilon(\mathbb{R}^d)}\big\}\, d\sigma \Big).\end{aligned} \tag{8.2.3}$$

Summing over all $|\alpha| \le s$ yields with a new constant

$$\|u(t)\|_{H^s_\epsilon(\mathbb{R}^d)} \le C \left(\|u(0)\|_{H^s_\epsilon(\mathbb{R}^d)} + \int_0^t \big\{\|(Lu)(\sigma)\|_{H^s_\epsilon(\mathbb{R}^d)} + \|u(\sigma)\|_{H^s_\epsilon(\mathbb{R}^d)}\big\}\, d\sigma \right). \tag{8.2.4}$$

Gronwall's inequality completes the proof. □

The following local estimate is sufficient for our needs. The proof is like the proof of the estimate in $H^s(\mathbb{R}^d)$ except that there are boundary terms that have a favorable sign because Ω is a domain of determinacy.

Theorem 8.2.3. *If Ω_T is a domain of determinacy and $s \in \mathbb{N}$, there is a constant $C = C(s, L, \Omega)$ so that for all $u \in C^\infty(\Omega_T)$ and all $t \in [0, T]$,*

$$\|u(t)\|_{H^s_\epsilon(\Omega(t))} \le C \left(\|u(0)\|_{H^s_\epsilon(\Omega(0))} + \int_0^t \|Lu(\sigma)\|_{H^s_\epsilon(\Omega(\sigma))}\, d\sigma \right). \tag{8.2.5}$$

8.3. Justification of the asymptotic expansion

Theorem 8.3.1 (Joly and Rauch, 1992). *Suppose that the phase ϕ and smooth profiles $U_j(y, \theta)$ satisfy the profile equations on the domain of determinacy Ω_T and the approximate solution u^ϵ is defined by (8.0.2) with $U(\epsilon, y, \theta) \sim \sum \epsilon^j\, U_j(y, \theta)$ in $C^\infty(\Omega_T \times S^1)$. Then for ϵ small the exact solution v^ϵ defined in (8.0.4) exists and is smooth on Ω_T and*

$$v^\epsilon \sim u^\epsilon \qquad \text{in} \quad C^\infty(\Omega_T)\,.$$

This result has nothing to do with the form of the profile equations and the algorithm to construct the approximate solutions. It is a special case of a stability result about families of approximate solutions with bounded $\epsilon\partial$ derivatives.

Theorem 8.3.2 (Guès, 1993, and Donnat, 1994). *Suppose that $u^\epsilon \in C^\infty(\Omega_T)$ is $\epsilon\partial$ bounded in the sense that for all $\alpha \in \mathbb{N}^{d+1}$*

$$\sup_{0<\epsilon<1} \left\| \big(\epsilon\,\partial\big)^\alpha u^\epsilon \right\|_{L^\infty(\Omega_T)} < \infty.$$

Suppose that it is an infinitely accurate family of approximate solutions in the sense that

$$L(u^\epsilon) + F(u^\epsilon) \;\sim\; 0 \quad \text{in} \quad C^\infty(\Omega_T)\,.$$

Then for ϵ small the exact solution v^ϵ defined in (8.0.4) exists and is smooth on Ω_T and

$$v^\epsilon \;\sim\; u^\epsilon \qquad \text{in} \quad C^\infty(\Omega_T)\,. \tag{8.3.1}$$

Proof. Fix $d/2 < s \in \mathbb{N}$. The local existence theorem implies either the existence of a smooth solution v^ϵ on Ω_T or the existence of a $T^*(\epsilon) \le T$ so that v^ϵ is smooth on

$$\Omega_* \;:=\; \Omega \cap \{0 \le t < T^*\} \tag{8.3.2}$$

and v^ϵ blows up at $T^*(\epsilon)$,

$$\lim_{t\to T^*(\epsilon)} \|v^\epsilon(t)\|_{H^s_\epsilon(\Omega(t))} \;=\; \infty\,. \tag{8.3.3}$$

We show that for ϵ sufficiently small, the second alternative does not occur and that (8.3.1) holds.

The $\epsilon\partial$ boundedness of u^ϵ implies that there is an $R > 0$ so that for $0 < \epsilon \le 1$ and $|\alpha| \le s$

$$\sup_{0\le t\le T} \Big(\|(\epsilon\partial_x)^\alpha u^\epsilon\|_{L^\infty(\Omega(t))} \Big) \;\le\; R/2\,. \tag{8.3.4}$$

Denote by $r^\epsilon(y)$ the residual in the equation for the approximate solution

$$L(y,\partial_y)\, u^\epsilon + F(y,u^\epsilon) \;:=\; r^\epsilon\,. \tag{8.3.5}$$

By hypothesis,

$$r^\epsilon(y) \;\sim\; 0 \qquad \text{in} \quad C^\infty(\Omega_T)\,. \tag{8.3.6}$$

Introduce the error

$$w^\epsilon \;:=\; v^\epsilon - u^\epsilon\,. \tag{8.3.7}$$

An initial value problem for the error is derived by subtracting (8.3.5) from (8.0.4). Suppressing the y dependence of F, this yields

$$L\,w^\epsilon + F(u^\epsilon + w^\epsilon) - F(u^\epsilon) \;=\; -r^\epsilon\,, \quad \text{on} \quad \Omega_*\,, \tag{8.3.8}$$

$$w^\epsilon\big|_{t=0} \;=\; 0\,. \tag{8.3.9}$$

Estimate (8.2.5) gives a C independent of ϵ and t so that for $0 \le t < T^*(\epsilon)$,

$$\|w^\epsilon(t)\|_{H^s_\epsilon(\Omega(t))} \le C\Big(\int_0^t \|F(u^\epsilon + w^\epsilon) - F(u^\epsilon)\|_{H^s_\epsilon(\Omega(\sigma))}\, d\sigma + \int_0^t \|r^\epsilon\|_{H^s_\epsilon(\Omega(\sigma))}\, d\sigma \Big).$$

So long as

$$\sup_{0\le\sigma\le t} \Big(\|w^\epsilon\|_{L^\infty(\Omega(\sigma))} \Big) \le R/2\,, \tag{8.3.10}$$

Lemma 8.1.1 yields with new C,

$$\|w^\epsilon(t)\|_{H^s_\epsilon(\Omega(t))} \le C\Big(\int_0^t \|w^\epsilon\|_{H^s_\epsilon(\Omega(\sigma))}\, d\sigma + \int_0^t \|r^\epsilon\|_{H^s_\epsilon(\Omega(\sigma))}\, d\sigma \Big). \tag{8.3.11}$$

The first application of this estimate is to show that $T^*(\epsilon) = T$ for ϵ small. If not, since $w^\epsilon(0) = 0$ and $\|w^\epsilon\|_{H^s(\Omega(t))} \to \infty$ as $t \nearrow T^*(\epsilon)$, there is a smallest $\underline{t} \in]0, T[$ so that

$$\|w^\epsilon\|_{H^s_\epsilon(\Omega(\underline{t}))} + \|w^\epsilon\|_{L^\infty(\Omega(\underline{t}))} = R/2. \tag{8.3.12}$$

Then from the definition of $\underline{t}$, (8.3.11) holds for $0 \le t \le \underline{t}$ and Gronwall's inequality implies that

$$\sup_{0\le t<\underline{t}} \|w^\epsilon(t)\|_{H^s_\epsilon(\Omega(t)} \le C' \int_0^T \|r^\epsilon\|_{H^s_\epsilon(\Omega(\sigma))}\, d\sigma \le C_{n,s}\, \epsilon^n\,. \tag{8.3.13}$$

The H^s_ϵ Sobolev inequality (8.1.6) implies that for $0 \le t \le \underline{t}$,

$$\sup_{0\le t<\underline{t}} \|w^\epsilon\|_{L^\infty(\Omega(t))} \le C_{n,s}\, \epsilon^{n-d/2}\,. \tag{8.3.14}$$

Equations (8.3.12)–(8.3.13) imply that there is an $\epsilon_0 > 0$ so that for $\epsilon \le \epsilon_0$,

$$\sup_{0\le t\le\underline{t}} \|w^\epsilon\|_{H^s_\epsilon(\Omega(t))} + \|w^\epsilon\|_{L^\infty(\Omega(t))} \le R/4\,.$$

For $t = \underline{t}$, this contradicts (8.3.12). It follows that v^ϵ is smooth on Ω_T and (8.3.10) holds with $t = T$.

Therefore, (8.3.13)–(8.3.14) holds with $\underline{t} = T$. Since $n > s > d/2$ are arbitrary, it follows that

$$\forall \epsilon \le \epsilon_0 \ \ \forall n, \ \ \forall 0 \le t \le T, \ \ \exists C, \quad \|w^\epsilon\|_{H^s_\epsilon(\Omega(t))} + \|w^\epsilon\|_{L^\infty(\Omega(t))} \le C\,\epsilon^n\,. \tag{8.3.15}$$

Estimate (8.3.15) is nearly equivalent to $w^\epsilon \sim 0$. What is missing is an analogous estimate for the time derivatives.

Express

$$\partial_t w = -\sum A_j \partial_j w + \mathcal{G}(u^\epsilon, w^\epsilon)\, w^\epsilon - r^\epsilon\,.$$

The H^s_ϵ Moser inequality shows that $\mathcal{G}(u^\epsilon, w^\epsilon)$ is bounded in $H^s_\epsilon(\Omega(t))$ since both u^ϵ and w^ϵ are uniformly bounded. Therefore (8.3.15) implies the case $j = 1$ of

$$\forall j, \ \ \forall s, \ \ \forall n, \ \ \forall \epsilon < \epsilon_0, \qquad \sup_{0\le t\le T} \ \|\partial_t^j w^\epsilon(t)\|_{H^s_\epsilon(\Omega(t))} \ \le \ C_{n,s,j}\, \epsilon^n \,.$$

The proof for arbitrary j is by induction. Write

$$\partial_t^{j+1} w \ = \ \partial_t^j\Big(-\sum A_j \partial_j w + \mathcal{G}(u^\epsilon, w^\epsilon)\, w^\epsilon - r^\epsilon \Big)\,. \tag{8.3.16}$$

The inductive hypothesis shows that for $k \le j$ and s arbitrary,

$$\sup_{0\le t\le T} \ \|\partial_t^k \mathcal{G}(u^\epsilon, w^\epsilon)\|_{H^s_\epsilon(\Omega(t))} \ = \ O(1)$$

$$\text{and} \qquad \sup_{0\le t\le T} \ \|\partial_t^k w^\epsilon\|_{H^s_\epsilon(\Omega(t))} \ = \ O(\epsilon^\infty)\,.$$

Therefore the $H^s_\epsilon(\Omega(t))$ norm of the right-hand side of (8.3.16) is $O(\epsilon^\infty)$ uniformly on $[0, T]$, completing the induction. □

8.4. Rays and nonlinear transport

In the linear case, the equations for the leading amplitudes simplify to transport equations when the smooth variety hypothesis is satisfied. With suitable hypotheses on F and initial data, one has a similar simplification in the nonlinear case.

The leading profile U_0 is determined from its initial data as the solution of (7.4.23)–(7.4.24) that we repeat here suppressing the y dependence of F,

$$\mathbf{E}\big(L\,U_0 + F(U_0)\big) \ = \ 0\,, \qquad \mathbf{E}U_0 \ = \ U_0. \tag{8.4.1}$$

Equations for periodic functions in θ are split into their oscillating and nonoscillating parts. Denote with an underline the average value of a periodic function of θ

$$\underline{g}(\theta) \ := \ \frac{1}{2\pi}\int_0^{2\pi} g(\theta)\, d\theta\,.$$

The oscillatory part is denoted with an asterisk,

$$g^*(\theta) \ := \ g - \underline{g}\,.$$

Splitting the equations for $U_0 = \underline{U} + U^*$ into their oscillating and nonoscillating parts yields the equivalent pair of equations

$$L(y, \partial_y)\, \underline{U} + \underline{F(y, \underline{U} + U^*)} = 0\,, \tag{8.4.2}$$

$$\pi(y)\Big(L\,U^* \ + \ F(\underline{U} + U^*)^*\Big) \ = \ 0\,, \qquad \pi(y)\,U^* \ = \ U^*\,. \tag{8.4.3}$$

Neither the mean $\underline{U}$ nor the oscillatory part U^* can be found by itself. They interact.

The equations for the principal profile are an integro-differential system that is essentially a hyperbolic problem with one more space variable, namely θ. The equation does not have θ derivatives. To find the principal profile is a little harder than solving a single hyperbolic Cauchy problem. The payoff is a family of approximate solution problems parametrized by a short wavelength ϵ. As pointed out in the introduction, these small structures make such a family particularly difficult to solve by numerical methods. If $\operatorname{rank} \pi(y) = k$, then the unknown function U_0 takes values in a k dimensional space. The number of unknown functions is reduced from N to k. The equation for the profile is usually simpler than solving a single initial value problem for the original problem.

The pair of equations becomes significantly simpler when one can guarantee that $\underline{U} = 0$. The next result gives two such situations.

Proposition 8.4.1. **i.** *If the nonlinear map $U \mapsto F(U)$ is odd, that is $F(-U) = -F(U)$ and the initial value $U\big|_{t=0}$ is odd in θ, then the solution U is odd in θ.*

ii. *If $F(U)$ is a linear combination of polynomials of odd degree in U and its complex conjugate $\overline{U}$ and $U\big|_{t=0}$ has spectrum contained in the odd integers, then the solution U has spectrum contained in the odd integers.*

In both cases $\underline{U} = 0$.

Proof. **i.** The assumptions imply that the function $-U(y, -\theta)$ is a solution with the same initial data. By uniqueness, $U(y, \theta) = -U(y, -\theta)$.

ii. Denote the initial data $g(x, \theta) = U\big|_{t=0}$. The Picard iterates converging to the solution are defined by $U^1(t, x, \theta) = g(x, \theta)$ and

$$\mathbf{E}\big(L\,U^{\nu+1} + F(U^\nu)\big) = 0\,, \qquad \mathbf{E}U^{\nu+1} = U^{\nu+1}, \qquad U^{\nu+1}\big|_{t=0} = g\,.$$

If U^ν has odd spectrum, then the same is true of $F(U^\nu)$. Expand $F(U^\nu)$ and g in a Fourier series in θ. The solutions of

$$\mathbf{E}(LW - f(y)e^{in\theta}) = 0, \qquad \mathbf{E}W = W, \qquad W|_{t=0} = \gamma(x)e^{in\theta} = \mathbf{E}(\gamma e^{in\theta})$$

are equal to $V(y)e^{in\theta}$ with V determined as the solution of

$$\pi(y, d\phi)(L(V) - f) = 0, \qquad \pi(y, d\phi)V = V, \qquad V|_{t=0} = \gamma\,.$$

An induction shows the U^ν have spectrum contained in the odd integers. □

When $\underline{U} = 0$, the profile equation becomes

$$\pi\, L\, \pi\, U + \pi(y)\, F(U) = 0\,, \qquad \pi(y)\, U = U\,.$$

Next *suppose that the smooth characteristic variety hypothesis is satisfied at $(y, d\phi(y))$ with $y \in \Omega_T$.* In this case using the equation displayed before

(5.4.4), the profile equation simplifies to the nonlinear transport equation

$$\big(\partial_t + \mathbf{v}.\partial_x + \gamma\big)\, U + \pi(y)\, F(U) = 0\,, \qquad \pi(y)\, U = U\,.$$

For each fixed θ this is a semilinear ordinary differential equation for U_0 along the integral curves of $\partial_t + \mathbf{v}\partial_x$. Solving such a family of equations is radically simpler than solving a multidimensional hyperbolic system. *When the smooth variety hypothesis is satisfied as well as* $\underline{U} = 0$ *(e.g., as in Proposition 8.4.1), the construction of the approximate solutions reduces to solving nonlinear ordinary differential equations along the rays.*

In special cases there are explicit solutions that give insight into the underlying dynamics defined by the nonlinear hyperbolic equation. It is in this way that the subject is often used in the scientific community. The reader is encouraged to browse the references given in the bibliography to find interesting applications, both mathematical and physical. In the applied literature the method often goes under the name *slowly varying envelope approximation.*

Example 8.4.1. A striking example is the analysis of *self-phase modulation* when a laser beam passes through glass. We will not introduce the appropriate nonlinear Maxwell equations but content ourselves with a cartoon that shares the key features. Consider the semilinear system

$$\frac{\partial u}{\partial t} + \begin{pmatrix} 1 & 0 \\ 0 & -1 \end{pmatrix} \frac{\partial u}{\partial x_1} + \begin{pmatrix} 0 & 1 \\ 1 & 0 \end{pmatrix} \frac{\partial u}{\partial x_2} + F(u) \;=\; 0\,, \qquad F \;=\; (F_1, F_2)\,.$$

The characteristic equation is $\tau^2 - |\xi|^2 = 0$. Consider the phase $\phi(t,x) = t - x_1$ with group velocity equal to $(1,0)$. The associated spectral projection and polarization are given by

$$\pi \;=\; \begin{pmatrix} 1 & 0 \\ 0 & 0 \end{pmatrix}, \qquad U_0(t,x,\theta) \;=\; (a_0(t,x,\theta), 0)\,.$$

When the hypotheses of Proposition 8.4.1 are satisfied, the principal profile equation is

$$(8.4.4) \quad \frac{\partial a_0}{\partial t} + \frac{\partial a_0}{\partial x_1} + F_1(a_0, 0) \;=\; 0\,, \qquad a_0(0, x_1, x_2, \theta) \;=\; g(x_1, x_2, \theta).$$

For the special case where $F = i|u|^2 u$ and a_0 with spectrum in the odd integers, Proposition 8.4.1.ii applies and the profile equation is

$$(8.4.5) \quad \frac{\partial a_0}{\partial t} + \frac{\partial a_0}{\partial x_1} + i|a_0|^2 a_0 \;=\; 0\,, \qquad a_0(0, x_1, x_2, \theta) \;=\; g(x_1, x_2, \theta).$$

Exercise 8.4.1. *Prove that if* $a_0 \in C^1(\mathbb{R}\,;\, H^1(\mathbb{R}^d \times S^1))$ *satisfies (8.4.5), then* $\int |a_0|^2\, dx\, d\theta$ *is independent of* t. **Hint.** Differentiate the quantity with respect to time. Alternatively, multiply by $\overline{a_0}$, take real part, and integrate.

Corresponding to this conservation, one has the ray-by-ray conservation law proved by considering small tubes of rays (see §5.4.3). Precisely, solutions of (8.4.2) satisfy

$$\big(\partial_t + \mathbf{v}.\partial_x\big)|a_0|^2 \;=\; 0\,.$$

Exercise 8.4.2. *Prove this.* **Hint.** Take the real part of the product of (8.4.2) with $\overline{a_0}$ as suggested in Exercise 8.4.1.

Thus $|a_0|$ is constant on rays and equation (8.4.5) becomes a linear equation exactly solved by

$$a_0(t,x_1,x_2,\theta) \;=\; e^{-i|g(x_1-t,x_2,\theta)|^2 t}\, g(x_1-t,x_2,\theta)\,. \tag{8.4.6}$$

The leading term in the approximation solution is

$$u_{\text{approx}} \;=\; e^{-i|g(x_1-t,x_2,(t-x_1)/\epsilon)|^2 t}\; g(x_1-t,x_2,(t-x_1)/\epsilon)\,.$$

A particularly simple case is when g is monochromatic, $g \;=\; g(x)\,e^{i\theta}$, in which case the solution simplifies to

$$u_{\text{approx}} \;=\; e^{i(t-x_1)/\epsilon}\, e^{-i|g(x_1-t,x_2,(t-x_1)/\epsilon)|^2 t}\, b(x_1-t,x_2)\,. \tag{8.4.7}$$

The approximate solution in the linear case, $F=0$, would be

$$u_{\text{approx}} \;=\; e^{i(t-x_1)/\epsilon}\; b(x_1-t,x_2)\,.$$

Compared to the linear case, what has happened is that the phase has been modified. Along rays there is a phase lag which grows linearly in time and is proportional to the square of the amplitude. In optics, this is called *self-phase modulation.*

If the nonlinearity were multiplied by -1, the phase lag would be converted to a phase advance. The linear solution is moving with speed exactly equal to one. Such phase advance should not be confused with movement faster than light. Such confusion is common in the science literature. No information moves faster than one, as proved in sections 2.3 and 6.3.

Example 8.4.2. We describe a second nonlinear optical phenomenon revealed by the nonlinear transport equation. For Maxwell's equations, the projectors π have rank two and the smooth variety hypothesis is satisfied. Therefore the nonlinear transport equation governs the dynamics of a function with values in a two dimensional space. In the important case of the commonly occurring cubic Kerr nonlinearity, the equations are explicitly solvable almost as in the preceding example. The exact solutions have $U_0(y,\theta)$ real with spectrum contained in $\pm n$ with $n \neq 0$. The leading term has electric field of the form

$$\sin((t-x_1)/\epsilon)\;\mathbf{e}_1(t,x) \;+\; \cos((t-x_1)/\epsilon)\;\mathbf{e}_2(t,x)\,,$$
$$\mathbf{e}_1 \perp (1,0,0), \quad \mathbf{e}_2 \perp (1,0,0)\,.$$

The pair $\mathbf{e}_1, \mathbf{e}_2$ is determined by a coupled cubic nonlinear transport equation (see [Donnat, 1994] or [Métivier, 2008]). On scales small compared to one and large compared to ϵ, solutions look like the exponential plane wave solutions of Maxwell's equations described in §2.4. There are solutions with $\mathbf{e}_j$ collinear that are linearly polarized. The transport equation shows that the solutions stay polarized and the axis of polarization is constant along rays. In the generic case the waves are elliptically polarized. *For the Kerr nonlinearity, solutions of the transport equation with elliptically polarized data remain elliptically polarized with constant eccentricity and the axis of polarization rotates at a constant speed in the plane perpendicular to* $(1, 0, 0)$. This explanation of an observed physical phenomenon is a second striking success of the nonlinear geometric optics approximation. The predictions were made long before the approximations were proved to be accurate in the 1990s.

Chapter 9

Resonant Interaction and Quasilinear Systems

This chapter describes two extensions. The first is the construction of asymptotic solutions with distinct phases. This is *multiphase nonlinear geometric optics.* In the linear case one simply sums solutions for the corresponding phases. In the nonlinear case the waves with different phases can interact. In addition they can launch new waves with phases not present at the start. These phenomena are called *resonance.* The second extension is from the semilinear to the quasilinear case. That is needed to study interactions for compressible inviscid fluid dynamics in Chapter 11.

9.1. Introduction to resonance

Even at the level of formal asymptotic expansions, resonance poses a challenge. Majda and Rosales (1986) got it right. The approach presented in this chapter is that of [Joly, Métivier, and Rauch, Duke, 1993]. The essence of the phenomenon is illustrated by the simple explicitly solvable example.

Example 9.1.1. Consider the semilinear Cauchy problem

$$(9.1.1)\qquad \begin{aligned} (\partial_t+\partial_x)u_1 &= 0\,, & u_1\big|_{t=0} &= a_1(x)\,e^{ix/\epsilon}\,,\\ \partial_t u_2 &= u_1 u_3\,, & u_2\big|_{t=0} &= 0\,,\\ (\partial_t-\partial_x)u_3 &= 0\,, & u_3\big|_{t=0} &= a_3(x)\,e^{ix/\epsilon}\,. \end{aligned}$$

with initial amplitudes $a_j \in C_0^\infty(\mathbb{R})$. The solution is given by

$$u_1(t,x) = a_1(x-t)\,e^{i(x-t)/\epsilon}, \qquad u_3(t,x) = a_3(x+t)\,e^{i(x+t)/\epsilon},$$

$$u_2 = \int_0^t u_1(t,x)\,u_3(t,x)\,dt\,.$$

The phases, $(x \pm t)/\epsilon$, that appear in the integrand for u_2 sum to $2ix/\epsilon$. Taking this outside the integral yields

$$u_2 = e^{i2x/\epsilon} \int_0^t a_1(x-t)\,a_3(x+t)\,dt\,. \tag{9.1.2}$$

The terms u_1 *and* u_3 *are wave trains with phases* $\phi_1 := (x-t)/\epsilon$, *and* $\phi_2 := (x+t)/\epsilon$, *respectively. They interact to generate the wave train* u_2 *with phase* $\phi_3 := 2x/\epsilon$. The phases satisfy the resonance relation

$$\phi_1 \;+\; \phi_2 \;=\; \phi_3\,.$$

The amplitude of the new wave is of the same order, ϵ^0, as the waves from which it is formed.

The underlying linear operator,

$$L(\partial_t, \partial_x) \;:=\; \begin{pmatrix} 1 & 0 & 0 \\ 0 & 1 & 0 \\ 0 & 0 & 1 \end{pmatrix} \partial_t \;+\; \begin{pmatrix} 1 & 0 & 0 \\ 0 & 0 & 0 \\ 0 & 0 & -1 \end{pmatrix} \partial_x\,,$$

has principal symbol

$$L(i\tau, i\xi) \;:=\; i \begin{pmatrix} \tau+\xi & 0 & 0 \\ 0 & \tau & 0 \\ 0 & 0 & \tau-\xi \end{pmatrix}.$$

The characteristic variety of L has equation $0 = \tau(\tau+\xi)(\tau-\xi)$. The eikonal equation is

$$\phi_t(\phi_t + \phi_x)(\phi_t - \phi_x) = 0\,.$$

Each of the ϕ_j satisfies this equation. A function ϕ defined on a connected set and satisfying $\nabla_{t,x}\phi \neq 0$ is a solution of this eikonal equation if and only if it is a solution of exactly one of the equations

$$\phi_t = 0, \qquad \phi_t + \phi_x = 0, \qquad \text{or} \qquad \phi_t - \phi_x = 0\,.$$

For example, if $\phi_t + \phi_x$ vanishes at one point $\underline{y}$, one proves that it is everywhere vanishing as follows. The set of points where it vanishes is closed, so we need only show that it is open. At $\underline{y}$,

$$|\phi_t| \;=\; |(\phi_t + \phi_x) - \phi_x| \;=\; |\phi_x| \;=\; |\nabla_{t,x}\phi|/\sqrt{2} \;\neq\; 0\,.$$

By continuity ϕ_t is nonzero on a neighborhood of $\underline{y}$. A similar argument shows that $\phi_t - \phi_x$ is nonvanishing on a neighborhood. The eikonal equation shows that $\phi_t + \phi_x = 0$ on a neighborhood of y. Therefore, it is everywhere zero and the preceding argument shows that the other two expressions vanish nowhere.

Variants of the preceding example illustrate two properties of resonance.

Example 9.1.2. If the initial condition for u_3 is changed to $u_3(0,x) = a_3(x)e^{i\psi(x)}$, then $u_3 = a_3(x+t)\, e^{i\psi(x+t)/\epsilon}$. The integral defining u_2 is an oscillatory integral in time with phase $(x - t + \psi(x+t))/\epsilon$. If $d\psi(x)/dx$ is nowhere equal to 1, then the time derivative of the phase is $O(1/\epsilon)$. The method of nonstationary phase shows that $u_2 = O(\epsilon)$. The resonant interaction is destroyed.

Exercise 9.1.1. *Prove that, more generally, if* $\{x : \psi'(x) = 1\}$ *has Lebesgue measure zero, then* $u_2 = o(1)$ *as* $\epsilon \to 0$. **Hint.** Write the integral as a sum of a nonstationary integral and an integral over a small set. **Discussion.** The offspring wave is smaller than the parents, but not as small as in Example 9.1.2.

For those who know about Young measures, it is interesting to note that the Young measures of the initial data are independent of the function ψ so long as $\psi' \neq 0$. Thus, *data with the same Young measures yield solutions with different Young measures.*

Introduce the symmetric form $\sum \phi_j = 0$ for resonance relations. If ψ_k satisfy $\sum n_k\, \psi_k = 0$, then the phases $\phi_k := n_k \psi_k$ satisfy the symmetric form. The symmetric form is often easier to manipulate.

Example 9.1.3. Find all triples of resonant linear eikonal phases with pairwise independent differentials for $L = \partial_t + \text{diag}\,(\lambda_1, \lambda_2, \lambda_3)\partial_x$ with λ_j distinct real numbers. Seek such ϕ_j satisfying the resonance relation $\sum \phi_j = 0$. The independent differentials together with the eikonal relation force (up to permutation),

$$\phi_j(t,x) \;=\; \alpha_j(x - \lambda_j\, t), \qquad \alpha_j \neq 0.$$

The resonance relation is equivalent to the pair of equations

$$\sum_j \alpha_j \;=\; 0 \qquad \text{and} \qquad \sum_j \alpha_j\, \lambda_j \;=\; 0\,.$$

These determine α up to multiplication by a scalar.

Exercise 9.1.2. *For* $(\lambda_1, \lambda_2, \lambda_3) = (1, 0, -1)$, *show that, if* $f, g, h \in C^\infty(\mathbb{R})$, *each has nonvanishing derivative at the origin and at least one of them has nonvanishing second derivative, then the three phases*

$$f(t), \qquad g(t-x), \qquad \text{and} \qquad h(t+x)$$

cannot be resonant on a neighborhood of the origin. **Discussion.** For this constant coefficient strictly hyperbolic operators on $\mathbb{R}^{1+1}$, linear phases are the only possibilities for resonant triples that have pairwise independent differentials.

The example and exercises show that the *phenomenon of resonance is both rare and sensitive when viewed from the perspective of perturbing the phases.* On the other hand, *wave trains with resonant phases interact much more strongly, increasing their importance.*

9.2. The three wave interaction partial differential equation

One can understand a rich variety of resonance phenomena by studying the following special example. It illustrates many important principles that are part of a general theory discussed later. Its analysis is needed to understand the general situation.

Begin with a real valued three wave interaction system

$$
\begin{aligned}
(\partial_t + \partial_x)u_1 &= c_1\, u_3\, u_2\,, \\
\partial_t u_2 &= c_2\, u_1 u_3\,, \\
(\partial_t - \partial_x)u_3 &= c_3\, u_1\, u_2\,,
\end{aligned}
\tag{9.2.1}
$$

with $0 \neq c_j \in \mathbb{R}$. This equation maximizes the intermode interaction. The absence of a term in u_j^2 in the jth equation has a consequence that harmonics are not generated by self-interaction.

Multiplying the first equation by $a_1\, u_1$, the second by $a_2\, u_2$, and the third by $a_3\, u_3$, shows that if a_1, a_2, and a_3 are real numbers so that $\sum a_j c_j = 0$, then for solutions one has the differential conservation law

$$
\frac{\partial}{\partial t}\left(a_1\, u_1^2 + a_2\, u_2^2 + a_3\, u_3^2\right) + \frac{\partial}{\partial x}\left(a_1\, u_1^2\, -\, a_3\, u_3^2\right) \;=\; \Big(2 \sum_j\, a_j\, c_j\Big) u_1 u_2 u_3 \;=\; 0\,.
$$

Integrating dx yields the integral conservation laws for solutions sufficiently smooth and sufficiently small at ∞,

$$
\frac{d}{dt}\int a_1\, u_1^2 + a_2\, u_2^2 + a_3\, u_3^2\ dx \;=\; 0\,.
$$

This is a two dimensional space of conservation laws parametrized by the a.

In order to take advantage of the complex exponential function, we are interested in complex solutions. For complex solutions, conservation laws involving $|u_j|^2$ can yield L^2 bounds while those involving u_j^2 cannot. The complex analog of (9.2.1) with such strong conservation laws is

$$
\begin{aligned}
(\partial_t + \partial_x)u_1 &= c_1\, u_2\, u_3^*, \\
\partial_t u_2 &= c_2\, u_1 u_3, \\
(\partial_t - \partial_x)u_3 &= c_3\, u_1^*\, u_2\,,
\end{aligned}
\tag{9.2.2}
$$

with $0 \neq c_j \in \mathbb{R}$.

Multiplying the jth equation by $a_j u_j^*$ and taking the real part shows that if $a_j \in \mathbb{R}$ satisfy $\sum a_j\, c_j = 0$, then solutions satisfy

$$\partial_t\Big(\sum_j a_j |u_j|^2\Big) \;+\; \partial_x\Big(a_1\,|u_1|^2 \;-\; a_3\,|u_3|^2\Big)$$
$$= \;2\,\Big(\sum_j a_j\,c_j\Big)\;\mathrm{Re}\Big(u_1^*\,u_2\,u_3^*\Big) \;=\; 0\,.$$

If the c_j do not all have the same sign, then there are conservation laws of this type with all the $a_j > 0$. This yields an L^2 bound on solutions. On the other hand, if the c_j are all positive, then initial data with $u_j(0,x)$ real and positive yield real solutions such that for all j, u_j is nondecreasing along j characteristics. For sufficiently positive data there is finite time blowup. The next result is similar to Theorem 6.8.3.

Theorem 9.2.1. *Suppose that $c_j \ge c > 0$ and the real valued initial data satisfy*

$$\forall j,\;\; \forall |x| \le R, \qquad (C^\infty \cap L^\infty)(\mathbb{R}) \;\ni\; u_j(0,x) \;\ge A \;>\; 0.$$

Denote by $u(t,x) \in C^\infty([0,t_*[\;\times\;\mathbb{R})$ the maximal solution and by $y = A/(1-cAt)$ the solution of $y' = c\,y^2$, $y(0) = A$.

i. *Then $u_j(t,x) \ge y(t)$ for $0 \le t < t_*$ and $|x| \le R - t$.*

ii. *If $T_* := (cA)^{-1}$ is the blowup time for y and $R > T_*$, then u blows up on or before time T_*. That is $t_* \le T_*$.*

Proof. The second assertion follows from the first. To prove the first, observe that since the speed of propagation is no larger than 1, the values of u in $|x| \le R - t$ are unaffected by the values of the Cauchy data for $|x| > R$. Therefore, it suffices to prove that $u_j \ge y(t)$ when the data satisfy $u_j(0,x) \ge A$ for all $x \in \mathbb{R}$.

Define

$$m(t) \;:=\; \min_{x\in\mathbb{R}\,,\,j}\; u_j(t,x)\,.$$

Since the u_j are nondecreasing on j-characteristics, it follows that $m(t)$ is nondecreasing. And, $m(0) \ge A > 0$. In addition one has the lower bound obtained by integration along j characteristics,

$$u_j(t,x) \;\ge\; m(0) + c\int_0^t m(t)^2\; dt\,.$$

Taking the infimum on x yields

$$m(t) \;\ge\; m(0) + c\int_0^t m(t)^2\; d \;\ge\; A + c\int_0^t m(t)^2\, dt\,.$$

The function y is characterized as the solution of

$$y(t) \;=\; A + c\int_0^t y(t)^2\,dt\,.$$

For $\epsilon > 0$ small, let y^ϵ be the solution of $(y^\epsilon)' = c(y^\epsilon)^2$ with $y^\epsilon(0) = A - \epsilon$, so

$$y^\epsilon(t) \;=\; A - \epsilon + c\int_0^t y^\epsilon(t)^2\,dt\,.$$

It follows that $m(t) > y^\epsilon(t)$ for all $0 \le t < t_*$. If this were not so, there would be a smallest $\underline{t} \in\,]0, t_*[$ where $m(\underline{t}) = y^\epsilon(\underline{t})$. Then

$$y^\epsilon(\underline{t}) \;=\; m(\underline{t}) \;\ge\; A + c\int_0^{\underline{t}} m(t)^2\,dt \;>\; A - \epsilon + c\int_0^t (y^\epsilon(t))^2\,dt \;=\; y^\epsilon(\underline{t})\,.$$

This contradiction establishes $m > y^\epsilon$. Passing to the limit $\epsilon \to 0$ proves $m \ge y$. This is the desired conclusion. □

Only the signs of the c_j play a role in the qualitative behavior of the equation (9.2.2).

Proposition 9.2.2. *There is exactly one positive diagonal linear transformation*

$$u \;:=\; (d_1\,v_1\,,\, d_2\,v_2\,,\, d_3\,v_3)\,, \qquad d_j > 0\,,$$

that transforms the the system to the analogous system with interaction coefficients $\{c_1, c_2, c_3\}$ *replaced by*

$$\Big\{\frac{c_1}{|c_1|}\,,\,\frac{c_2}{|c_2|}\,,\,\frac{c_3}{|c_3|}\Big\}\,.$$

Proof. The change of variables yields an analogous system for v with the interaction coefficients replaced by

$$\Big\{\frac{d_2\,d_3}{d_1}\,c_1\,,\,\frac{d_3\,d_1}{d_2}\,c_2\,,\,\frac{d_1\,d_2}{d_3}\,c_3\Big\}\,.$$

The d_j must satisfy

$$\frac{d_2\,d_3}{d_1}\,c_1 = \frac{c_1}{|c_1|}\,, \qquad \frac{d_3\,d_1}{d_2}\,c_2 = \frac{c_2}{|c_2|}\,, \qquad \frac{d_1\,d_2}{d_3}\,c_3 = \frac{c_3}{|c_3|}\,.$$

Multiplying the jth equation by d_j^2 yields the equivalent system,

$$\frac{d_1^2}{|c_1|} \;=\; \frac{d_2^2}{|c_2|} \;=\; \frac{d_3^2}{|c_3|} \;=\; d_1 d_2 d_3\,.$$

The first two equalities hold if and only if

$$(d_1^2, d_2^2, d_3^2) = a(|c_1|, |c_2|, |c_3|) \qquad \text{with} \qquad a > 0\,.$$

This uniquely determines d up to the positive scalar multiple a. If this is satisfied, then the third equation holds if and only if

$$a \;=\; a^3\,|c_1c_2c_3|\,.$$

This uniquely determines a, and therefore d. □

Remark. For general $d_j \neq 0$, the three quantities d_1d_2/d_3, d_2d_3/d_1, d_3d_1/d_2 have the same sign. Using $d_j \neq 0$ allows us to multiply the three interaction coefficients by -1 if desired. Thus every system is transformed to one with interaction coefficients all equal to $+1$ or two equal to $+1$. There are four equivalence classes, the last three depending on the location of the coefficient -1.

Theorem 9.2.3. i. *If the real interaction coefficients $c_j \neq 0$ do not all have the same sign, then the Cauchy problem for (9.2.1) has a unique global solution $u \in \bigcap_s C^s([0,\infty[\,;\, H^s(\mathbb{R}))$ for arbitrary Cauchy data in $\bigcap_s H^s(\mathbb{R})$.*

ii. *If the c_j have the same sign, there are smooth compactly supported data so that the solution of the Cauchy problem blows up in finite time.*

Proof. For real data, this equation reduces to the previous one and **ii** follows from Theorem 9.2.1.ii.

To prove **i**, the results of §6.4 show that it suffices to prove for every $T > 0$, an a priori bound for the $L^\infty([0,T]\times\mathbb{R})$ norm.

From the conservation law, one has

$$\sup_{t\in[0,T]} \int \sum_j |u_j|^2\, dx \;\le\; K \;<\; \infty\,.$$

The equation for u_2 yields

$$|u_2(t,\underline{x})| \;\le\; |u_2(0,\underline{x})\;| + \int_0^t |u_1\, u_3(t,\underline{x})|\; dt\,. \tag{9.2.3}$$

The key idea is to estimate the integral on the right-hand side using energy estimates for u_1 and u_3.

For any $\underline{x} \in \mathbb{R}$ integrate the identity

$$(\partial_t + \partial_x)|u_1|^2 \;=\; 2\,\mathrm{Re}\Big(u_1^*(\partial_t+\partial_x)u_1\Big) \;=\; 2\,\mathrm{Re}\, c_1 u_1^* u_2 u_3$$

over the strip $[0,t]\times\,]-\infty,\underline{x}]$ to find that

$$\int_0^t |u_1(t,\underline{x})|^2\, dt \;\le\; 2K + 2|c_1| \int_{[0,t]\times\mathbb{R}} |u_1u_2u_3|\; dt\, dx\,.$$

Estimate the integral dx on the right using the $L^\infty\times L^2\times L^2$ Hölder inequality to find

$$\int_0^t |u_1(t,\underline{x})|^2\, dt \;\le\; 2K \;+\; 2|c_1| \int_0^t K\,\|u_2(t)\|_{L^\infty(\mathbb{R})}\; dt\,.$$

By symmetry,

$$\int_0^t |u_3(t,\underline{x})|^2\, dt \;\le\; 2K \;+\; 2|c_3| \int_0^t K\, \|u_2(t)\|_{L^\infty(\mathbb{R})}\, dt\,.$$

The Cauchy–Schwarz inequality implies that
(9.2.4)

$$\int_0^t |u_1(t,\underline{x})\, u_3(t,\underline{x})|\, dt \;\le\; 2K \;+\; 2\max\{|c_1|,|c_3|\} \int_0^t \|u_2(t)\|_{L^\infty(\mathbb{R})}\, dt\,.$$

Estimate the integral on the right in (9.2.3) using (9.2.4) to find

$$|u_2(t,\underline{x})| \;\le\; C \;+\; \int_0^t C\, \|u_2(t)\|_{L^\infty(\mathbb{R})}\, dt\,,$$

with C independent of $(t,\underline{x}) \in [0,T]\times\mathbb{R}$. Taking the supremum of the left-hand side over $\underline{x}$ yields

$$\|u_2(t)\|_{L^\infty(\mathbb{R})} \;\le\; C \;+\; \int_0^t C\, \|u_2(t)\|_{L^\infty(\mathbb{R})}\, dt\,, \qquad 0\le t\le T\,.$$

Gronwall's inequality bounds the sup norm of u_2 over bounded time intervals.

To estimate u_1 one needs L^2 estimates for u_2 and u_3 on the speed one characteristics $x = \underline{x} + t$. These are obtained by integrating $\partial_t|u_2|^2$ and $(\partial_t - \partial_x)|u_3|^3$ over $\{(s,x) : 0\le s\le t,\ \text{and}\ x \ge \underline{x}+s\}$.

A similar argument works for u_3. □

Exercise 9.2.1. *Prove the $L^\infty([0,T]\times\mathbb{R})$ estimate for u_3.*

9.3. The three wave interaction ordinary differential equation

For the three wave partial differential equation (9.2.2) and phases equal to the resonant triplet, waves of each pair of families influence, by resonant interaction, the wave of the third. The simplest examples showing this are solutions of the special form
(9.3.1)

$$u_1 = A_1(t)\, e^{i(t-x)/\epsilon}\,, \qquad u_2 = A_2(t)\, e^{-i2x/\epsilon}\,, \qquad u_3 := A_3(t) e^{-i(t+x)/\epsilon}\,,$$

with amplitudes A_j independent of x. The oscillatory structure evolves in time, but is uniform in space. Equation (9.2.2) is satisfied if and only if the amplitudes A_j satisfy the *three wave interaction ordinary differential equation*

$$A_1' = c_1\, A_2\, A_3^*\,, \qquad A_2' = c_2\, A_1\, A_3\,, \qquad A_3' = c_3\, A_1^*\, A_2\,. \tag{9.3.2}$$

This is a nonlinear system of ordinary differential equations for three complex quantities A_j. The phase space is $\mathbb{C}^3$, hence it is six dimensional as a real vector space. It is the same equation that one would find if one sought solutions of the three wave interaction partial differential equation that are independent of x.

The equilibria are the points where (at least) two of the three $\{A_j\}$ vanish. There are three linear subspaces of equilibria, each with real dimension equal to 2,

$$\big\{A_2 = A_3 = 0\big\}\,, \qquad \big\{A_3 = A_1 = 0\big\}\,, \qquad \text{and} \qquad \big\{A_1 = A_2 = 0\big\}\,.$$

Each pair of planes intersect at the origin. The system (9.3.2) is highly symmetric.

Proposition 9.3.1. **i.** *The quantity* $\mathrm{Im}\big(A_1(t)\,A_2^*(t)\,A_3(t)\big)$ *is constant on orbits of (9.3.2).*

ii. *If* a_j *are real numbers so that* $\sum a_j\,c_j = 0$, *then the quantity* $\sum_j a_j\,|A_j(t)|^2$ *is constant on orbits of (9.3.2).*

iii. *If* A *is a solution and* $\theta \in \mathbb{R}$, *then* $\widetilde{A}$ *obtained by each of the three gauge transformations*

$$\widetilde{A} := \big(e^{i\theta}\,A_1\,,\, A_2\,,\, e^{-i\theta}\,A_3\big), \quad \widetilde{A} := \big(A_1\,,\, e^{i\theta}\,A_2\,,\, e^{i\theta}\,A_3\big),$$
$$\widetilde{A} := \big(e^{i\theta}\,A_1\,,\; e^{i\theta}\,A_2\,,\, A_3\big),$$

is also a solution. The conserved quantities in **i** *and* **ii** *are invariant under the gauge transformations.*

iv. *If* A *is a solution and* $\sigma \in \mathbb{R}\setminus 0$, *then* $\widetilde{A}$ *obtained by the scaling*

$$\widetilde{A}_j(t) \;=\; \sigma A_j(\sigma t)$$

is also a solution.

Proof. **i.** Compute

$$\begin{aligned}(A_1A_2^*A_3)_t &= (A_1)_tA_2^*A_3 + A_1(A_2^*)_tA_3 + A_1A_2^*(A_3)_t\\ &= c_1A_2A_3^*A_2^*A_3 + c_2A_1A_1^*A_3^*A_3 + c_3A_1A_2^*A_1^*A_2\\ &= c_1|A_2A_3|^2 \;+\; c_1|A_1A_3|^2 \;+\; c_1|A_2A_1|^2 \;\in\; \mathbb{R}\,.\end{aligned}$$

ii. Compute

$$\begin{aligned}\frac{d}{dt}|A_1|^2 &= 2\mathrm{Re}\,A_1^*\frac{d}{dt}A_1 &&= 2c_1\mathrm{Re}(A_1^*A_2A_3^*),\\ \frac{d}{dt}|A_2|^2 &= 2\mathrm{Re}\,A_1^*\frac{d}{dt}A_1 &&= 2c_2\mathrm{Re}(A_2^*A_1A_3),\\ \frac{d}{dt}|A_3|^2 &= 2\mathrm{Re}\,A_1^*\frac{d}{dt}A_1 &&= 2c_3\mathrm{Re}(A_1^*A_2A_3^*).\end{aligned}$$

The real parts are of $A_1 A_2^* A_3$ or its complex conjugate, so are equal. Therefore one has

$$\frac{d}{dt}\Big(\sum a_j\,|A_j(t)|^2\Big) \;=\; \Big(2\sum_j a_j\,c_j\Big)\mathrm{Re}(A_2^* A_1 A_3) \;=\; 0\,.$$

The assertions **iii** and **iv** are immediate. □

Remarks. 1. When the $c_j \neq 0$ do not all have the same sign, one can choose the $a_j > 0$. In this case, the three wave interaction system is globally solvable.

2. When the three c_j have the same sign, there exist solutions that blow up in finite time. This is proved by comparison with an explosive Ricatti equation, as for the three wave interaction partial differential equation.

Exercise 9.3.1. *Suppose that the c_j have the same sign and that $A(t)$ is a solution defined for $0 \leq t < T_*$ so that $\limsup_{t\to T_*} \|A(t)\| = \infty$. Prove that for all j, $\limsup_{t\to T_*} \|A_j(t)\| = \infty$.* **Hint.** Use quadratic conservation laws.

3. The gauge transformations commute. The third gauge transformation is the product of the preceding two. The abelian group of gauge transformations is a two dimensional torus of mappings

$$A \quad\mapsto\quad \big(e^{i\theta_1}\,A_1\,,\, e^{i\theta_2}\,A_2\,,\, e^{i\theta_2}\,e^{-i\theta_1}\,A_3\big)\,.$$

Theorem 9.3.2. i. *The equilibrium $(0,0,0)$ is unstable if and only if the three c_j have the same sign.*

ii. For i, j, k distinct, the equilibrium $A_i = A_j = 0\,,\ \underline{A}_k \neq 0$ of (9.3.2), is unstable if the interaction coefficients c_i and c_j have the same sign. In this case, the stable, unstable, and center manifolds of the linearization have real dimension equal to 2.

iii. For the same equilibrium, if c_i and c_j have opposite signs, orbits of the linearized equation are bounded. For initial data starting close to the equilibrium, the solutions of the nonlinear system exist for all time and $A_i(t), A_j(t)$, and $|A_k(t)|$ stay close to their initial values uniformly in time. The equilibrium is unstable. If $\underline{A}_k$ is real, then the equilibrium is stable for the restriction of the dynamics to $A \in \mathbb{R}^3$.

Proof. i. The stability of the origin when the c_j do not have the same sign follows from the conservation of $\sum a_j |A_j|^2$ with positive a_j. The instability is proved using explosive positive (resp., negative) solutions when the c_j are positive (resp., negative).

ii. For ease of reading, consider the equilibrium $(0, 0, \underline{A}_3) \neq 0$. The linearized equation at this equilibrium is

$$B' = \begin{pmatrix} 0 & c_1 \underline{A}_3^* & 0 \\ c_2 \underline{A}_3 & 0 & 0 \\ 0 & 0 & 0 \end{pmatrix} B .$$

The eigenvalues of the coefficient matrix are the solutions λ of

$$\lambda(\lambda^2 - c_1 c_2 |\underline{A}_3|^2) = 0 .$$

If c_1 and c_2 have the same sign, then the roots are $0, \pm|c_1 c_2|^{1/2}|\underline{A}_3|$. The positive eigenvalue implies that the equilibrium is unstable. The stable, unstable, and center manifolds of the linearization have complex dimension equal to 1 and therefore real dimension equal to 2.

iii. If c_1, c_2 have opposite signs, then $|c_2||B_1|^2 + |c_1||B_2|^2$ and B_3 are constant on orbits of the linearized equation. It follows that each orbit is uniformly bounded in time.

In the case of opposite signs, the functional $|c_2|^2|A_1|^2 + |c_1||A_2|^2$ is constant on orbits of (9.3.2). For initial data that start near $(0, 0, \underline{A}_3)$, the components $A_1(t), A_2(t)$ stay close to zero for all time.

A conserved quantity $a_1|A_1|^2 + a_2|A_2|^2 + |A_3|^2$ together with the control of $A_1(t), A_2(t)$ implies that $|A_3(t)|$ stays close to $|A_3(0)|$. In particular, the orbit exists for all time. For real solutions this implies stability since in that case the sign of $A_3(t)$ does not change and

$$|A_3(t) - A_3(0)| = \Big| |A_3(t)| - |A_3(0)| \Big| .$$

For complex solutions write $A_3 = \rho\, e^{i\theta}$, $\rho := |A_3|$, and compute[1]

$$A_3' = \rho' e^{i\theta} + i\rho e^{i\theta}\theta' = \rho e^{i\theta}\Big(\rho'/\rho + i\theta'\Big),$$

$$\theta' = \operatorname{Im}\left(\frac{A_3'}{A_3}\right) = \operatorname{Im}\left(\frac{c_1 A_1^* A_2}{A_3}\right) = \frac{\operatorname{Im} c_1 A_1^* A_2 A_3^*}{|A_3|^2} .$$

To prove instability, choose $\delta = |\underline{A}_3| > 0$. For $\epsilon > 0$, choose complex initial data $(A_1(0), A_2(0), \underline{A}_3)$ with $|A_1(0), A_2(0)| < \epsilon$ and $\operatorname{Im} A_1 A_2^* A_3 \neq 0$. In the expression for θ', the numerator is a nonzero constant, and the denominator is always $\approx |\underline{A}_3|$, so the angle θ has derivative bounded below. There is a $t > 0$ so that $\theta(t) - \theta(0) = \pi$. Therefore $|A_3(t) - A_3(0)| > |A_3(0)| > \delta$, proving instability. □

For any triple of interaction coefficients, there exists $i \neq j$ so that c_i and c_j have the same sign. Then the equilibria defined by $A_i = A_j = 0$

[1]Thanks to G. Métivier for this short proof. See [Alber et al., 1998] for complementary information on this system.

is unstable. The unstable equilibrium exists even in the globally solvable case where the c_j do not all have the same sign. For example, if c_1 and c_2 have the same sign and c_3 the opposite, then there is a conserved euclidean norm $\sum a_j|A_j|^2$. On the other hand, most orbits starting near $A_1 = 0, A_2 = 0, A_3 = 1$ stray far from this state. This situation is described as saying oscillations on the third mode generate *frequency conversion* to modes 1 and 2. The solution cannot grow, but it can wander far from its initial state. The energy originally localized nearly entirely on mode 3, moves substantially away. An appreciable portion of the energy passes to modes 1 and 2.

The analysis of the interactions in the highly oscillatory family (9.2.2) reduces to the analysis of a system of nonlinear ordinary differential equations. This is a special case of a general phenomenon for *homogeneous oscillations*, that is oscillations that are the same at all positions of space. We return to the construction of high frequency asymptotic solutions, this time with several phases and in the quasilinear case.

9.4. Formal asymptotic solutions for resonant quasilinear geometric optics

We give a self-contained but rapid derivation of the equations of quasilinear geometric optics. Consider the quasilinear system of partial differential operators,

$$L(u,\partial)u \;:=\; \sum_{\mu=0}^{d} A_\mu(u)\,\partial_\mu u\,.$$

Suppose that the system is symmetric in the sense of the first paragraph of §6.6. Consider solutions whose values are close to a constant state $\underline{u}$. The change of independent variable $u \mapsto u - \underline{u}$ reduces to the case $\underline{u} = 0$. Without loss of generality we study solutions close to 0.

As in the last paragraph of §6.6, the change dependent variable, $u := A_0(0)^{1/2}v$, yields the equivalent symmetric hyperbolic system

$$\sum_{\mu=0}^{d} \widetilde{A}_\mu(v)\,\partial_\mu v \;=\; 0, \qquad \widetilde{A}_\mu(v) \;:=\; A_0(0)^{-1/2}\,A_\mu(A_0(0)^{1/2}v)\,A_0(0)^{-1/2}\,,$$

with

$$\widetilde{A}_\mu \;=\; \widetilde{A}_\mu^* \qquad \text{and} \qquad \widetilde{A}_0(0) \;=\; I\,.$$

We suppose that such a change has been performed and suppress the tildes.

For $u \approx 0$, use the approximation

$$A_\mu(u) \;\approx\; A_\mu(0) \;+\; A_\mu'(0)u$$

to show that the nonlinear terms are equal to

$$\big(A'_\mu(0)u\big)\,\partial_\mu u \;+\; \text{higher order terms.}$$

We assess the time of nonlinear interaction for solutions built from oscillatory wave trains $\epsilon^p\, e^{i\phi(y)/\epsilon}$. The power p is chosen so that this time is ~ 1.

For Burgers' equation $u_t + uu_x = 0$, with compactly supported initial data with $\|\partial_x u(0,x)\|_{L^\infty} \sim 1$, solutions break down at times $t \sim 1$. Thus for initial data $\epsilon^p a(x)\, e^{i\phi(0,x)/\epsilon}$ the lifetime is $O(1)$ when $p = 1$ and is much longer (resp., shorter) when $p > 1$ (resp., $p < 1$). This shows that nonlinear effects are important for $t \sim 1$ for the critical power $p = 1$.

A second estimate yielding the same critical power is the following. Assume that $A'_\mu(0) \neq 0$ for some μ, so that the leading nonlinear terms are quadratic. For the important examples from inviscid fluid dynamics, the hypothesis of quadratic nonlinearity is usually verified. Consider solutions built from wave trains $\epsilon^p a(y)\, e^{i\phi(y)/\epsilon}$ whose amplitudes are $O(\epsilon^p)$ and whose derivatives are $O(\epsilon^{p-1})$. The nonlinear terms then have amplitude $O(\epsilon^{2p-1})$. Phases satisfying the eikonal equation terms of this size yield a response that is $O(\epsilon^{2p-1})$ for $t \sim O(1)$. Seek amplitudes so that the time of nonlinear interaction is $O(1)$ so $\epsilon^{2p-1} \sim \epsilon^p$. This yields the critical power $p = 1$. This heuristic is confirmed by the theorems that show that when $p = 1$, nonlinear effects usually affect the leading order asymptotics for times $t \sim 1$.

Exercise 9.4.1. i. *Perform a perturbation computation as in §6.5, to show that the solution of*

$$\partial_t u + u\partial_x u \;=\; 0, \qquad u(0,x) \;=\; 1 + \delta g(x)$$

is given by

$$u \;\sim\; u_0 \;+\; \delta u_1 \;+\; \delta^2 u_2 \;+\; \cdots,$$

where $u_0(t,x) = 1$ is the unperturbed state, $u_1(t,x) = g(x-t)$ solves the linearized equation, and the leading nonlinear term u_2 is determined by

$$(\partial_t + \partial_x)u_2 + g(x-t)\, g'(x-t) \;=\; 0, \qquad u_2(0,x) = 0\,.$$

ii. *As in §7.1 take $g = \epsilon^p\, a(x)\, e^{ix/\epsilon}$ and $\delta = 1$ to see that for $t \sim 1$ the leading nonlinear term is small compared to the leading term when $p > 1$ and become comparable as $p \to 1$.* **Discussion.** This is a third motivation for the critical exponent $p = 1$.

Consider the interaction of waves with linear phases $\phi_j(y)$ that by nonlinear interaction yield possible phases $\sum n_j\phi_j$ with $n_j \in \mathbb{Z}$. Each of these candidate phases is a linear function αy with $\alpha \in \mathbb{R}^{1+d}$. Thus the expression

$$\epsilon\, U(y, y/\epsilon) \qquad \text{with} \qquad U(y,Y) \sim \sum_{\alpha\in\mathbb{R}^d} U_\alpha(y)\, e^{i\alpha Y}$$

is of the critical amplitude and includes all the anticipated terms. In a formal trigonometric sum over α it is understood that there are at most a countable number of nonvanishing coefficients U_α.

For the first computations, consider $U(y, Y)$ as a formal trigonometric series in Y with coefficients that are smooth functions of y. To solve profile equations and prove accuracy will require important and not obvious supplementary hypotheses. They do not play a role in the formal derivation of the profile equations.

Pose the *ansatz*

$$u^\epsilon = \epsilon\, U(\epsilon, y, y/\epsilon)\,, \tag{9.4.1}$$

$$U(\epsilon, y, Y) \;\sim\; \sum_{j=0}^{\infty} \epsilon^j\, U_j(y, Y) \;\sim\; U_0(y, Y) + \epsilon\, U_1(y, Y) + \cdots\,, \tag{9.4.2}$$

$$U_j(y, Y) \;\sim\; \sum_{\alpha\in\mathbb{R}^d} U_{j,\alpha}(y)\, e^{i\alpha Y}\,. \tag{9.4.3}$$

Write

$$L(u^\epsilon, \partial)u^\epsilon \;=\; L(\epsilon U,\, \partial)\Big(\epsilon\, U(\epsilon, y, y/\epsilon)\Big)\,. \tag{9.4.4}$$

Expand in a Taylor series at $\epsilon = 0$,

$$A_\mu\big(\epsilon U(\epsilon, y, Y)\big) \;\sim\; A(0) \;+\; \epsilon A'_\mu(0) U_0 \;+\; \cdots\,, \tag{9.4.5}$$

to find

$$L(\epsilon U(\epsilon, y, Y)\,,\, \partial_y) \;\sim\; L_0 \;+\; \epsilon L_1 \;+\; \cdots \;=\; \sum_{j=0}^{\infty} \epsilon^j\, L_j(y, Y, \partial_y)\,.$$

The L_j are operators whose coefficients are functions of y, Y involving the derivatives $\partial_u^\beta A_\mu(0)$ and the profiles U_k with $k \le j-1$. The most important come from the leading terms in (9.4.5),

$$L_0 = L(0, \partial_y) \qquad \text{and} \qquad L_1 = \sum_\mu A'_\mu(0) U_0(y, Y)\, \partial_\mu\,.$$

The chain rule shows that

$$\frac{\partial}{\partial y_\mu} U(\epsilon, y, y/\epsilon) \;=\; \Big(\frac{\partial}{\partial y_\mu} + \frac{1}{\epsilon}\frac{\partial}{\partial Y_\mu}\Big) U(\epsilon, y, Y)^\epsilon\Big|_{Y=y/\epsilon}\,. \tag{9.4.6}$$

So

$$L(u^\epsilon, \partial)u^\epsilon \;=\; W(\epsilon, y, Y)\Big|_{Y=y/\epsilon}\,, \tag{9.4.7}$$

where

$$W(\epsilon, y, Y) \;=\; L\Big(\epsilon U(\epsilon, y, Y)\,,\; \frac{\partial}{\partial y_\mu} + \frac{1}{\epsilon}\frac{\partial}{\partial Y_\mu}\Big)\, \epsilon\, U(\epsilon, y, Y)\,.$$

Expand to find

$$
\begin{aligned}
&W(\epsilon, y, Y) \\
&\quad \sim \Big(\sum_{j=0}^{\infty} \epsilon^j L_j\Big(y, Y, \frac{\partial}{\partial y_\mu} + \frac{1}{\epsilon}\frac{\partial}{\partial Y_\mu}\Big)\Big)\Big(\epsilon \sum_{k=0}^{\infty} \epsilon^k U_k(y, Y)\Big) \\
&\quad \sim \Big(\sum_{j=0}^{\infty} \Big[\epsilon^j L_j\Big(y, Y, \frac{\partial}{\partial y_\mu}\Big) + \epsilon^{j-1} L_j\Big(y, Y, \frac{\partial}{\partial Y_\mu}\Big)\Big]\Big)\Big(\epsilon \sum_{k=0}^{\infty} \epsilon^k U_k(y, Y)\Big) \\
&\quad \sim \sum_{j=0}^{\infty} \epsilon^j \, W_j(y, Y) \,.
\end{aligned}
$$

The two leading terms are

$$
W_0(y, Y) = L(0, \partial_Y)\, U_0(y, Y) \tag{9.4.8}
$$

and

$$
\begin{aligned}
W_1(y, Y) &= L(0, \partial_y)\, U_0 + L_1(y, Y, \partial_Y) U_0 + L(0, \partial_Y) U_1 \\
&= L(0, \partial_y)\, U_0 + \sum_\mu A'_\mu(0) U_0\, \partial_{Y_\mu} U_0 + L(0, \partial_Y)\, U_1 \,.
\end{aligned} \tag{9.4.9}
$$

As is typical for multiscale methods, the formula for W_1 involves both U_0 and U_1. The quadratically quasilinear terms $A'_\mu(0) U_0\, \partial_Y U_0$ involve derivatives in the fast variables Y. For $j \geq 2$ the formula for W_j is

$$
\begin{aligned}
W_j \;&=\; \sum_{k+\ell=j} \Big(L_k(y, Y, \partial_y) + L_{k+1}(y, Y, \partial_Y)\Big) U_\ell \\
&=\; L(0, \partial_y)\, U_j + L_1(y, Y, \partial_Y) U_j + L(0, \partial_Y)\, U_{j+1} \\
&\qquad + \text{terms in } U_0, U_1, \ldots, U_{j-1} \,.
\end{aligned}
$$

The strategy is to choose profiles U_j so that $W_j(y, Y) = 0$ for all y, Y, not just on the $d+2$ dimensional subset $\{Y = y/\epsilon\}$ parametrized by $(\epsilon, y) = (\epsilon, t, x)$.

Setting $W_0 = 0$ in (9.4.8) shows that W_0 must lie in the kernel of $L(0, \partial_Y)$. In order for it to be possible to choose a U_1 so that (9.4.9) holds requires the second of the equations,

$$
\begin{aligned}
&U_0 \in \operatorname{kernel} L(0, \partial_Y) \\
&\qquad\qquad \text{and} \quad L(0, \partial_y)\, U_0 + \sum_\mu A'_\mu(0) U_0\, \partial_{Y_\mu} U_0 \in \operatorname{range} L(0, \partial_Y).
\end{aligned}
$$

To understand (9.4.10) requires a study of the operator $L(0,\partial_Y)$. This is straightforward using the Fourier representation,
(9.4.10)
$$L(0,\partial_Y)\,U \;=\; L(0,\partial_Y)\sum_\alpha U_\alpha(y)\,e^{i\alpha Y} \;=\; i\sum_\alpha L(0,\alpha)\,U_\alpha(y)\,e^{i\alpha Y}\,.$$

As an operator acting on formal trigonometric series, $L(0,\partial_Y)$ has kernel consisting of those series whose αth coefficient belongs to the kernel of $L(0,\alpha)$. Recall the definition of $\pi(\alpha)$ as the projection onto the kernel of $L(0,\alpha)$ along its range. The kernel of $L(0,\partial_Y)$ is then the set of trigonometric series such that $\pi(\alpha)\,U_\alpha = U_\alpha$. The range is the set of series with U_α belonging to the range of $L(0,\alpha)$. Equivalently, $\pi(\alpha)\,U_\alpha = 0$.

Define an operator $\mathbf{E}$ from formal trigonometric series to themselves by

$$\mathbf{E}\sum_\alpha U_\alpha(y)\,e^{i\alpha Y} := \sum_\alpha \pi(\alpha)\,U_\alpha(y)\,e^{i\alpha Y}\,. \tag{9.4.11}$$

The previous remarks show that on formal trigonometric series, the operator $\mathbf{E}$ projects along the range of $L(0,\partial_Y)$ onto its kernel. Therefore, the two conditions in (9.4.10) are equivalent to the pair of equations

$$\mathbf{E}\,U_0 \;=\; U_0 \tag{9.4.12}$$

and

$$\mathbf{E}\Big(L(0,\partial_y)\,U_0 + \sum_\mu A'_\mu(0)U_0\,\partial_{Y_\mu}U_0\Big) \;=\; 0\,. \tag{9.4.13}$$

These are the fundamental equations of resonant quasilinear geometric optics. They are analogs of (7.4.23) and (7.4.24).

Since $A(0) = I$, equation (9.4.13) is equivalent to

$$\partial_t U_0 + \mathbf{E}\Big(\sum_j A_j(0)\,\partial_j U_0 + \sum_\mu A'_\mu(0)U_0\,\partial_{Y_\mu}U_0\Big) = 0\,. \tag{9.4.14}$$

Written this way, the equation looks like an evolution equation for U_0. Since the operator $\mathbf{E}$ does not depend on t, multiplying (9.4.15) by $I-\mathbf{E}$ yields, at least formally,

$$\partial_t\,(I-\mathbf{E})U_0 = 0\,.$$

Thus, the constraint (9.4.12) is satisfied as soon as it is satisfied at $t = 0$. It is reasonable to expect that U_0 can be determined from its initial data required to satisfy $\mathbf{E}U_0(0,x,Y) = U_0(0,x,Y)$.

The equation $W_1 = 0$ is equivalent to the pair of equations $\mathbf{E}\,W_1 = 0$ and $(I-\mathbf{E})\,W_1 = 0$. The first, $\mathbf{E}\,W_1 = 0$ implies (9.4.13).

Define $Q(\alpha)$ to be the partial inverse of $L(0,\alpha)$ that is,

$$Q(\alpha)\,\pi(\alpha) \;=\; 0, \qquad Q(\alpha)\,L(0,\alpha) \;=\; I-\pi(\alpha)\,.$$

Introduce the operator $\mathbf{Q}$ on a trigonometric series by

$$\mathbf{Q} \sum_\alpha U_\alpha(y)\, e^{i\alpha Y} := \sum_\alpha Q(\alpha)\, U_\alpha(y)\, e^{i\alpha Y}\,, \tag{9.4.15}$$

where $\mathbf{Q}$ is a partial inverse to $L(0, \partial_Y)$. It is determined by

$$\mathbf{Q}\,\mathbf{E} = 0, \qquad \mathbf{Q}\, L(0, \partial_Y) = I - \mathbf{E}\,. \tag{9.4.16}$$

Since $Q(\alpha)$ commutes with $L(0,\alpha)$, it follows that $\mathbf{Q}$ commutes with $L(0, \partial_Y)$.

The second part, $(I - \mathbf{E})W_1 = 0$, of the equation $W_1 = 0$ is equivalent to $\mathbf{Q}W_1 = 0$. Multiplying (9.4.9) by $\mathbf{Q}$ shows this is equivalent to

$$(I - \mathbf{E})U_1 = -\mathbf{Q}\Big(L(0, \partial_y)\, U_0 + \sum_\mu A'_\mu(0) U_0\, \partial_{Y_\mu} U_0 \Big)\,. \tag{9.4.17}$$

Once U_0 is determined, this determines $(I - \mathbf{E})U_1$.

Multiplying $W_{j-1} = 0$ by $\mathbf{Q}$ and $W_j = 0$ by $\mathbf{E}$ shows that the equations $(I-\mathbf{E})W_{j-1} = 0$ together with $\mathbf{E}W_j = 0$ are equivalent to a pair of equations

$$(I - \mathbf{E})\, U_j = \mathbf{Q}\, (\text{terms in } U_0, U_1, \dots, U_{j-1}) \tag{9.4.18}$$

and

(9.4.19)

$$\mathbf{E}\Big(L(0, \partial_y)\, U_j + \sum_\mu A'_\mu(0) U_0\, \partial_{Y_\mu} U_j \Big) = \mathbf{E}\, (\text{terms in } U_0, U_1, \dots, U_{j-1})\,.$$

Under suitable hypotheses, once $U_0, \dots, U_{j-1}$ are determined, equations (9.4.18) and (9.4.19) serve to determine U_j from initial values $\mathbf{E}U_j(0, x, Y)$.

9.5. Existence for quasiperiodic principal profiles

An essential step is to pass from formal trigonometric series in Y to a more manageable class. The class of profiles $U_0(y, Y)$ that are periodic in Y will serve us well. Though this suffices for the most interesting examples we construct, the general theory should and does go further. To see the necessity, consider the one dimensional problem with leading part $\partial_t + \operatorname{diag}(\lambda_1, \lambda_2, \lambda_3)\partial_x$ from §9.1. One wants to treat functions oscillating with the resonant trio of phases $\alpha_j(x - \lambda_j t)$. For the phases ny/ϵ that appear for periodic profiles, the ratio of the coefficients of t and x are rational. Thus one could only treat the case of λ parallel to an element of $\mathbb{Q}^3$. Quasiperiodic functions as in the next definition are sufficient to treat a wide variety of problems including general λ.

Definitions. *Suppose that the real linear functions $\{\phi_j(Y)\}_{j=1}^m$ are linearly independent over the rationals. To a function $\mathcal{U}(y, \theta_1, \dots, \theta_m)$ smooth and 2π multiply periodic in θ, associate the quasiperiodic profile $U(y, Y) :=$*

$\mathcal{U}\big(y, \phi_1(Y), \dots, \phi_m(Y)\big)$. *An induced operator* $\mathcal{E}$ *mapping periodic functions to themselves is defined by*

$$\mathcal{E}\Big(\sum_{n\in\mathbb{Z}^m} \mathcal{U}_n(y) e^{in\theta}\Big) := \sum_{n\in\mathbb{Z}^m} \pi\big(\sum_k n_k d\phi_k\big)\, \mathcal{U}_n(y) e^{in\theta},$$

so that $(\mathcal{E}\mathcal{U})(y, \phi(Y)) := \mathbf{E}\, U(y, Y)$. *Similarly, define the partial inverse,*

$$\mathcal{Q}\mathcal{U} := \sum_{n\in\mathbb{Z}^m} Q\big(\sum_k n_k d\phi_k\big)\, \mathcal{U}_n(y) e^{in\theta}.$$

Introduce the shorthand, $nd\phi := \sum_k n_k d\phi_k$.

To write (9.4.12)–(9.4.13) as an equation for $\mathcal{U}$, note that

$$\frac{\partial}{\partial Y_\mu} \mathcal{U}(y, \phi_1(Y), \dots, \phi_m(Y)) = \sum_{k=1}^m \frac{\partial \phi_k}{\partial Y_\mu} \frac{\partial \mathcal{U}_0}{\partial \theta_k}.$$

The profile equation for U_0 is equivalent to

$$\text{(9.5.1)} \qquad \mathcal{E}\,\mathcal{U}_0 = \mathcal{U}_0\,, \quad \mathcal{E}\Big(L(0, \partial_y)\,\mathcal{U}_0 + \sum_\mu A'_\mu(0)\mathcal{U}_0 \sum_k \frac{\partial \phi_k}{\partial Y_\mu} \frac{\partial \mathcal{U}_0}{\partial \theta_k} \Big) = 0\,,$$

for $\mathcal{U}_0$. This equation has the form

$$\partial_t \mathcal{U}_0 + G(\mathcal{U}_0, \partial_{y,\theta})\,\mathcal{U}_0 = 0\,,$$

where

$$\begin{aligned} G(\mathcal{U}, \partial_{y,\theta}) &:= \mathcal{E}\Big(\sum_{j=0}^d A_j(0)\, \partial_{y_j} + \sum_\mu A'_\mu(0)\mathcal{U} \sum_k \frac{\partial \phi_k}{\partial Y_\mu} \frac{\partial}{\partial \theta_k} \Big)\mathcal{U} \\ &:= \mathcal{E}\, K(\mathcal{U}, \partial_{y,\theta})\mathcal{U}\,. \end{aligned}$$

The notation suggests a quasilinear hyperbolic system. But, the operator $\mathcal{E}$ is nonlocal in θ. The operator $\mathcal{E}$ is an orthogonal projection operator in $L^2(\mathbb{R}^d_x \times \mathbb{T}^m_\theta)$ and commutes with $\partial_{y,\theta}$.

Theorem 9.5.1 (Joly, Métivier, and Rauch, Duke, 1994). *Suppose that* $H_0(x, \theta) \in \bigcap_s (H^s(\mathbb{R}^d \times \mathbb{T}^m)$ *satisfies the constraint* $\mathcal{E} H_0 = H_0$. *Then there is* $T_* > 0$ *and a unique maximal solution*

$$\mathcal{U}_0 \in \bigcap_s C^s\big([0, T_*[\,;\ H^s(\mathbb{R}^d \times \mathbb{T}^m)\big)$$

satisfying (9.5.1) together with the initial condition $\mathcal{U}\big|_{t=0} = H_0$. *If* $T_* < \infty$, *then*

$$\lim_{t \nearrow T_*} \big\| \mathcal{U}_0\,,\ \nabla_{x,\theta}\mathcal{U}_0 \big\|_{L^\infty([0,t]\times\mathbb{R}^d\times\mathbb{T}^m)} = \infty\,.$$

Sketch of Proof. The key idea is to derive a priori estimates as in the case of quasilinear hyperbolic systems. One applies $\partial^\beta_{x,\theta}$, and takes the real part of the $L^2(\mathbb{R}^d \times \mathbb{T}^m)$ scalar product with $\partial^\beta_{x,\theta}\mathcal{U}_0$ (suppressing the subscript 0) to find

$$\frac{d}{dt}\Big(\frac{1}{2}\|\partial^\beta_{x,\theta}\mathcal{U}\|^2_{L^2(\mathbb{R}^d\times\mathbb{T}^m)}\Big) \;=\; \mathrm{Re}\,\Big(\partial^\beta_{x,\theta}\mathcal{U}\,,\,\partial^\beta_{x,\theta}\mathcal{E}K(\mathcal{U},\partial_{y,\theta})\mathcal{U}\Big)_{L^2(\mathbb{R}^d\times\mathbb{T}^m)}\,.$$

Using the commutation and symmetry properties of $\mathcal{E}$ yields

$$\begin{aligned}
&\big(\partial^\beta_{x,\theta}\mathcal{U}\,,\,\partial^\beta_{x,\theta}\mathcal{E}K(\mathcal{U},\partial_{y,\theta})\mathcal{U}\big)_{L^2(\mathbb{R}^d\times\mathbb{T}^m)}\\
&\qquad = \big(\mathcal{E}\,\partial^\beta_{x,\theta}\mathcal{U}\,,\,\partial^\beta_{x,\theta}K(\mathcal{U},\partial_{y,\theta})\mathcal{U}\big)_{L^2(\mathbb{R}^d\times\mathbb{T}^m)}\\
&\qquad = \big(\partial^\beta_{x,\theta}\mathcal{U}\,,\,\partial^\beta_{x,\theta}K(\mathcal{U},\partial_{y,\theta})\mathcal{U}\big)_{L^2(\mathbb{R}^d\times\mathbb{T}^m)}\,,
\end{aligned}$$

the last equality using $\mathcal{E}\mathcal{U} = \mathcal{U}$. The last *is* a quasilinear hyperbolic expression. Using Gagliardo–Nirenberg estimates as in the treatment of the quasilinear Cauchy problem yields

$$\mathrm{Re}\,\big(\partial^\beta_{x,\theta}\mathcal{U}\,,\,\partial^\beta_{x,\theta}K(\mathcal{U},\partial_{y,\theta})\mathcal{U}\big)_{L^2(\mathbb{R}^d\times\mathbb{T}^m)} \;\le\; C(\|\mathcal{U}\|_{\mathrm{Lip}(\mathbb{R}^d\times\mathbb{T}^m)})\,\|\mathcal{U}\|^2_{H^{|\beta|}(\mathbb{R}^d\times\mathbb{T}^m)}\,.$$

Summing on $|\beta| \le s \in \mathbb{N}$ yields

$$\frac{d}{dt}\|\mathcal{U}(t)\|^2_{H^s(\mathbb{R}^d\times\mathbb{T}^m)} \;\le\; C(\|\mathcal{U}\|_{\mathrm{Lip}(\mathbb{R}^d\times\mathbb{T}^m)})\,\|\mathcal{U}\|^2_{H^s(\mathbb{R}^d\times\mathbb{T}^m)}\,.$$

Local well-posedness in H^s for $\mathbb{N} \ni s > 1 + (d+m)/2$ so that $H^s \subset \mathrm{Lip}$ is then proved as for quasilinear hyperbolic systems.

In particular, the principal profile is constructed as the limit of solutions $\mathcal{U}^h(t,x,\theta)$ of difference approximations,

$$\partial_t\mathcal{U}^h \;+\; \mathcal{E}\Big(L(y,(0,\delta^h_x)\mathcal{U}^h \;+\; \sum_\mu\sum_k A'_\mu(0)\mathcal{U}^h\,\frac{\partial\phi_k}{\partial Y_\mu}\,\delta^h_{\theta_k}\mathcal{U}^h\Big) \;=\; 0,$$

$$\mathcal{U}^h\big|_{t=0} = g\,.$$

This equation is an ordinary differential equation in $H^s(\mathbb{R}^d \times \mathbb{T}^m)$. □

One has the following upper bound on the spectrum of $\mathcal{U}$ as a function of θ.

Definition. *If $V(\theta_1,\dots,\theta_m)$ is a periodic distribution, then the* **spectrum** *of V is the set of $n \in \mathbb{Z}^m$ so that the nth Fourier coefficient of V is not equal to zero. The spectrum is denoted* spec V.

Theorem 9.5.2. *Suppose that $\mathcal{U}$ is as in Theorem 9.5.1. Denote by $\mathbb{S}$ the smallest $\mathbb{Z}$-module containing the spectrum of $\mathcal{U}(0,x,\theta)$ for all $x \in \mathbb{R}^d$ and closed under complex conjugation. Then*

$$\forall\ (t,x) \in [0,T_*[\times\mathbb{R}^{d+m}, \qquad \mathrm{spec}\,\mathcal{U}(t,x,\theta) \;\subset\; \mathbb{S}\cap\mathrm{Char}\,L(0,\partial)\,.$$

Sketch of proof. It suffices to show that the spectrum of $\mathcal{U}^h(t) \subset \mathbb{S} \cap \operatorname{Char} L$. The set of functions in H^s with spectrum contained in $\mathbb{S}$ is a Hilbert space. For $s > (d+m)/2$ it is an algebra. For those s, the map

$$\mathcal{V} \quad \mapsto \quad \mathcal{E}\Big(L(y,(0,\delta_x^h)\mathcal{V} + \sum_\mu \sum_k A'_\mu(0)\mathcal{V}\frac{\partial \phi_k}{\partial Y_\mu}\delta_{\theta_k}^h \mathcal{V}\Big)$$

is locally lipschitzean from functions with spectrum in $\mathbb{S} \cap \operatorname{Char} L$ to itself. Thus Picard's existence theorem proves local existence of an H^s solution $\widetilde{\mathcal{U}}^h$ with spectrum in $\mathbb{S} \cap \operatorname{Char} L$. By uniqueness of the $H^s(\mathbb{R}^d \times \mathbb{T}^m)$ valued solution, one has $\widetilde{\mathcal{U}}^h = \mathcal{U}^h$. This applies to the local solution and therefore to the maximal solution. Therefore the spectrum of $\mathcal{U}^h$ is contained in $\mathbb{S} \cap \operatorname{Char} L$. □

Examples 9.5.1. i. If the initial data has spectrum contained in $\mathbb{Z}\alpha$ with $\alpha \in \mathbb{Z}^m$, then $\operatorname{spec}\mathcal{U} \subset \mathbb{Z}\alpha$ and one finds the expansions of one phase geometric optics.

ii. In the extreme opposite case, the $\mathbb{Z}$-span of the spectrum of the initial data can meet the characteristic variety of L in a set much larger than the initial spectrum. This suggests the possibility of the creation of many new oscillatory modes. In Chapter 11, we show that an amazingly rich set of such new oscillations can be generated for the compressible inviscid Euler equations in dimensions $d \geq 2$.

9.6. Small divisors and correctors

This section is concerned with a surprising fact. The nonlinear problem determining the principal profile $\mathcal{U}_0$ has just been solved. The surprise is that the *linear* equations determining the correctors $\mathcal{U}_j$ for $j \geq 1$ are not solvable without making additional hypotheses.

The equations for the correctors $\mathcal{U}_j$ with $j \geq 1$, involve the operator $\mathcal{Q}$, for example (9.4.18). *Without further hypotheses, $\mathcal{Q}$ may be very ill behaved.* The matrices $Q(\alpha)$ may grow very rapidly as α grows. This has two consequence. First, $\mathbf{Q}$ may not even map smooth profiles in θ to distributions in θ. In that case the equations for the higher profiles do not make sense. Second, there are examples with weaker supplementary hypotheses where the error of approximation by the leading term is $o(\epsilon^p)$ but is not $O(\epsilon^{p+\delta})$ for any $\delta > 0$ [Joly, Métivier, and Rauch, 1992].

What is needed in order to get a reasonably well behaved operator $\mathcal{Q}$ is that the matrices $Q(nd\phi)$ grow no faster than polynomially in $|n|$. The trouble spots for $\mathcal{Q}$ are eigenvalues of $L(0, nd\phi)$ that, though not equal to zero, are very close to zero.

The proof in §8.3 is short and relatively simple in part because the approximate solution constructed is infinitely accurate. For example, the loss of $d/2$ powers of epsilon in passing from (8.3.13) to (8.3.14) is absorbed by the small size of the residual. While the equations for the principal profile derived in §9.4 is robust, there are serious problems with the equations for the correctors. We need the correctors to use a stability argument as in Chapter 8. This section presents a simple example illustrating the nature of the problem.

Begin by considering a very special case that captures the essence of the difficulty posed by small divisors. Suppose that $\phi = \alpha y$ is a linear phase that satisfies the eikonal equation for $L(0, \partial)$. Consider a nonlinear problem with initial data that oscillate with the same phase but with different and incommensurate frequencies. For example one can take the initial data corresponding to the Lax solution

$$a(\epsilon, y)\, e^{i\phi(y)/\epsilon} \;+\; a(\epsilon, y)\, e^{i\rho\,\phi(y)/\epsilon}\,,$$

where ρ is irrational. The two phases $\phi_1 = \phi$ and $\phi_2 = \rho\,\phi$ are linearly independent over the rationals. By resonant interaction one expects the solution to involve at least the phases $n_1\phi_1 + n_2\phi_2$ with $n_j \in \mathbb{Z}$. The leading profile is expected to be at least as complicated as

$$U(y, \theta) \;=\; \sum_{n\in\mathbb{Z}^2} a_n(y)\; e^{i(n_1+\rho\, n_2)\theta}\,.$$

The set of numbers $n_1 + \rho\, n_2$ is countable but dense in $\mathbb{R}$. Even if the a_n are smooth and rapidly decreasing, we have an almost periodic function with a dense set of frequencies.

The operator $\mathbf{Q}$ is given by

$$\mathbf{Q}\,U \;:=\; \sum_n Q((n_1 + \rho\, n_2)\phi)\, a_n(y)\; e^{i(n_1+\rho\, n_2)\theta}\,.$$

This is perfectly well defined on formal trigonometric series. However since Q is the partial inverse of $L(0\,,\,(n_1 + \rho\, n_2)\phi) = (n_1 + \rho\, n_2)L(0\,,\,\phi)$,

$$Q((n_1 + \rho n_2)\phi) \;=\; \frac{1}{n_1 + \rho\, n_2}\, Q(\phi)\,.$$

When $n_1 + \rho n_2$ is small, this matrix is large. There are divisors $n_1 + \rho n_2$ arbitrarily close to 0. The operator Q is not bounded on L^2. The divisor is small when $\rho \approx -n_1/n_2$. The mapping properties depend on how well the irrational number ρ is approximated by rational numbers.

If ρ is algebraic, that is, the solution of an irreducible integer polynomial of degree $D \geq 2$, then Liouville's theorem asserts that there is a constant

$c > 0$ depending on ρ so that for all integers p, q,

$$\Big|\rho - \frac{p}{q}\Big| \ \geq\ \frac{c}{q^D}\,.$$

In this case,

$$\big|n_1 + \rho\, n_2\big| \ =\ \Big|n_2\Big(\frac{n_1}{n_2} + \rho\Big)\Big| \ \geq\ \frac{c\,|n_2|}{|n_2|^D}\,.$$

Therefore $Q\big((n_1 + \rho\, n_2)\phi\big)$ can grow no faster than polynomially in n. This implies that $\mathbf{Q}$ is bounded $H^s \to L^2$ for s sufficiently large.

If ρ were exceptionally well approximable by rationals (for example for the Liouville number $\rho = \sum_{j=1}^{\infty} 10^{-(j!)}$), then $\mathbf{Q}$ would not have this desirable property and the construction of correctors hits a serious snag.

The next hypothesis describes a class of small divisors that can be tolerated.

Small divisor hypothesis. *There is a $C > 0$ and an integer N so that for all $n \in \mathbb{Z}^m \setminus 0$, if $\lambda \neq 0$ is an eigenvalue of $L(0, \sum_k n_k d\phi_k)$, then*

$$|\lambda| \ \geq\ \frac{C}{|n|^N}\,. \tag{9.6.1}$$

Examples 9.6.1. In the example of §9.5 the small divisor hypothesis is satisfied for any irrational number ρ that is not exceptionally well approximated by rationals, that is, when

$$\exists c > 0,\ \ \exists N,\ \ \forall p, q \in \mathbb{Z}, \qquad \Big|\rho - \frac{p}{q}\Big| \ \geq\ \frac{c}{|q|^N}\,.$$

For example, if ρ is the an algebraic number of degree $d \geq 2$.

On the other hand if ρ is too well approximable by rationals, for example, the Liouville number, then the hypothesis is violated.

Examples 9.6.2. If $\varphi_j = \alpha_j y$ with $\alpha \in \mathbb{R}^m$, then there is a measure zero set $\mathcal{R} \subset \mathbb{R}^d$ so that the hypothesis is satisfied for all $\alpha \notin \mathcal{R}$ (see [Joly, Métivier, and Rauch, Duke Math. J., 1993]).

Proposition 9.6.1. *If the small divisor hypothesis is satisfied, then there is a constant $C > 0$ and an integer M so that for all $n \in \mathbb{Z}^m$,*

$$\|Q(nd\phi)\|_{\mathrm{Hom}\,(\mathbb{C}^N)} \ \leq\ C\,\langle n\rangle^M\,.$$

Proof. The small divisor hypothesis implies that the nonzero eigenvalues of $Q(nd\phi)$ lie in an annulus $2c/\langle n\rangle^N \leq |z| \leq C\langle n\rangle^N/2$. Define a larger annulus containing the eigenvalues strictly in its interior by

$$D(n) \ :=\ \big\{z \,:\, c/\langle n\rangle^N \leq |z| \leq C\langle n\rangle^N\big\}\,.$$

Then

$$Q(nd\phi) \ =\ \frac{1}{2\pi i}\oint_{\partial D(n)} \frac{1}{z}\,\big(zI - L(0, nd\phi)\big)^{-1}\,dz\,.$$

For $z \in \partial D(n)$, $\|zI - L(0, nd\phi)\| \le C\langle n\rangle^N$. The nearest eigenvalue is no closer than $C\langle n\rangle^{-N}$. Therefore $\|(zI - L(0, nd\phi))^{-1}\| \le C\langle n\rangle^{N'}$, and the proposition follows. □

The proposition implies that when the small divisor hypothesis is satisfied, $\mathbf{Q}$ maps $\bigcap_s H^s(\mathbb{R}^d \times \mathbb{T}^m)$ continuously to itself. The next theorem is linear and easier than Theorem 9.5.1.

Theorem 9.6.2. *Suppose that the small divisor hypothesis is satisfied and that $\mathcal{U}_0$ is as in Theorem 9.5.1 and for $j \ge 1$ initial profiles $H_j(x,\theta) \in \bigcap_s (H^s(\mathbb{R}^d \times \mathbb{T}^m)$ satisfy $\mathcal{E} H_j = H_j$. Then higher order profiles*

$$\mathcal{U}_j \in \bigcap_s C^s([0, T_*[\,;\, H^s(\mathbb{R}^d \times \mathbb{T}^m)) \text{ for } j \ge 1$$

are uniquely determined by the initial conditions $\mathcal{E}\,\mathcal{U}_j = H_j$ and the transcriptions of (9.4.18) and (9.4.19) to the reduced profiles.

Suppose that the profiles $\mathcal{U}_j$ of all orders are determined as in (9.5.1) and (9.6.2). Borel's theorem constructs

(9.6.2)
$$C^\infty([0,1]_\epsilon \times [0, T_*[_t : \bigcap_s H^s(\mathbb{R}^d_x \times \mathbb{T}^m_\theta)) \ni \mathcal{U}(\epsilon, y, \theta) \sim \sum_j \epsilon^j\, \mathcal{U}_j(y, \theta)\,.$$

Define approximate solutions

(9.6.3)
$$u^\epsilon(t,x) := \epsilon\, \mathcal{U}(\epsilon, t, x, \phi_1(t,x)/\epsilon, \dots, \phi_m(t,x)/\epsilon) \in \bigcap_s C^s([0, T_*[\,;\, H^s(\mathbb{R}^d)).$$

Theorem 9.6.3. *With the above definitions, the residual*

$$r^\epsilon := L(u^\epsilon, \partial)\, u^\epsilon \tag{9.6.4}$$

is infinitely small in the sense that

(9.6.5) $\forall T \in [0, T_*[\,,\ \gamma \in \mathbb{N}^{d+1}\,,\ N \in \mathbb{N}\,,\ \exists c > 0\,,\ \forall \epsilon \in \,]0,1]$,
$$\|\partial_y^\gamma r^\epsilon\|_{L^2([0,T]\times\mathbb{R}^d)} \le c\,\epsilon^N\,.$$

The small divisor hypothesis is needed to construct correctors. Without it, the leading profile U_0 still exists. There are cases where an approximation with relative error $o(1)$ as $\epsilon \to 0$ can be justified without the small divisor hypothesis [Joly, Métiver, and Rauch, Ann. Inst. Fourier, 1994].

9.7. Stability and accuracy of the approximate solutions

The approximate solutions are of size $O(\epsilon)$. Derivatives are larger by a factor ϵ^{-1}. Thus $(\epsilon\partial_y)^\gamma$ applied to the approximate solutions yields a term that is $O(\epsilon)$. The next theorem implies that the approximate solutions are infinitely close to the exact solutions with the same initial values. The result differs from Theorem 8.3.1 in two ways. First it is on $\mathbb{R}^d$ rather than local in Ω_T. More important it is quasilinear instead of semilinear and that requires some changes in the proof. The reader is referred to the original papers, for example [Joly, Métiver, and Rauch, Duke, 1993] for details. We present statements only. Examples are discussed in the next two chapters.

Stability Theorem 9.7.1. *Suppose that $T > 0$ and that u^ϵ is a family of smooth approximate solutions to $L(u,\partial)\,u = 0$ which are $O(\epsilon)$ in the sense that*

(9.7.1)
$$\forall \gamma \in \mathbb{N}^{d+1},\ \exists c(\gamma),\ \forall \epsilon \in\,]0,1], \qquad \|\,(\epsilon\partial_y)^\gamma u^\epsilon\,\|_{(L^\infty \cap L^2)([0,T]\times\mathbb{R}^d)} \ \le\ c(\gamma)\,\epsilon\,.$$

Suppose that the residuals $r^\epsilon := L(u^\epsilon,\partial)\,u^\epsilon$ are infinitely small in the sense that

$$\text{(9.7.2)}\quad \forall \gamma \in \mathbb{N}^{d+1}\,,\ N\in\mathbb{N}\,,\ \exists c(N)>0\,,\ \forall \epsilon\in\,]0,1],$$
$$\|\partial_y^\gamma r^\epsilon\|_{L^2([0,T]\times\mathbb{R}^d)} \le c(N)\,\epsilon^N\,.$$

Define $v^\epsilon \in C^\infty([0,T_*(\epsilon)[\times\mathbb{R}^d)$ to be the maximal solution of the initial value problem

$$\text{(9.7.3)}\qquad L(v^\epsilon,\partial)\,v^\epsilon \ =\ 0\,,\qquad v^\epsilon(0,x)\ =\ u^\epsilon(0,x)\,.$$

Then there is an $\epsilon_0 > 0$ so that for $\epsilon < \epsilon_0$, $T_*(\epsilon) > T$, and the approximate solution u^ϵ is infinitely close to the exact solution v^ϵ in the sense that for all integers s and N

$$\|u^\epsilon - v^\epsilon\|_{H^s([0,T]\times\mathbb{R}^d)} \ \le\ c(s,N)\,\epsilon^N\,.$$

9.8. Semilinear resonant nonlinear geometric optics

The simplest examples, like those in §9.2, are semilinear. The first examples in Chapter 10 are semilinear. In this section we state the form of the *ansatz* and profile equations in the semilinear case. The precise theorem statements and proofs closely resemble the quasilinear case and can be found in the bibliography.

For a semilinear system

$$L(\partial)\,u\ +\ f(u)\ =\ 0\,,\qquad L(\partial)\ :=\ \sum_{\mu=0}^{d} A_\mu\,\partial_\mu\,,$$

recall that $\pi(\alpha)$ is orthogonal projection on the kernel of $L(\alpha)$ and $\mathbf{E}$ is the operator on formal trigonometric series $\mathbf{E}\sum a_\alpha(y)\,e^{i\alpha.\theta} := \sum \pi(\alpha)\,a_\alpha(y)\,e^{i\alpha.\theta}$.

The critical size for semilinear problems is amplitudes $O(\epsilon^p)$ with $p = 0$. The approximate solutions have the form

$$u^\epsilon \sim U_0^\epsilon(y, y/\epsilon)\,, \tag{9.8.1}$$

$$U_0(y, Y) \sim \sum_{\alpha \in \mathbb{R}^d} U_{0,\alpha}(y)\, e^{i\alpha Y}\,. \tag{9.8.2}$$

The amplitudes are $O(1)$ as $\epsilon \to 0$ in contrast to the quasilinear case where the amplitudes were $O(\epsilon)$. The size agrees with the one phase semilinear theory in Chapters 7 and 8.

The profile equations for U_0 are

$$\mathbf{E}\, U_0 = U_0\,, \tag{9.8.3}$$

$$\mathbf{E}\Big(L(\partial_y)\, U_0(y,\theta) + f\big(U_0(y,\theta)\big)\Big) = 0\,. \tag{9.8.4}$$

Solutions of the profile equation of the quasiperiodic form

$$\begin{aligned} U(\epsilon, y, Y) &= \mathcal{U}(\epsilon, y, \phi_1(Y), \dots, \phi_m(Y)) \\ &\in\ C^\infty\big([0, \epsilon_0]\,;\, \bigcap_s H^s([0,T] \times \mathbb{R}^d \times \mathbb{T}^m)\big)\,, \end{aligned}$$

with

$$\mathcal{U}(\epsilon, y, \theta_1, \dots, \theta_m) \sim \sum_{j=0}^{\infty} \epsilon^j\, \mathcal{U}_j(y, \theta)\,,$$

in the sense of Taylor series, exist provided the small divisor hypothesis of the preceding section holds with $L(nd\phi)$ in place of $L(0, nd\phi)$. This yields approximate solutions with infinitely small residual. The accuracy of these solutions follows from the stability Theorem 8.3.2.

The equations for $\mathcal{U}$ involve the nonlinear term $f(\mathcal{U})$. To prove the analog of Theorem 9.5.2 in the semilinear context requires the following lemma.

Lemma 9.8.1. *If $V \in L^\infty(\mathbb{T}^m\,;\,\mathbb{C}^N)$ and $F \in C(\mathbb{C}^N\,;\,\mathbb{C}^N)$, denote by $\mathbb{S}$ the smallest $\mathbb{Z}$ module containing* spec V *and invariant under complex conjugation. Then,*

$$\operatorname{spec} F(V) \subset \mathbb{S}\,. \tag{9.8.5}$$

Proof. The Weierstrass approximation theorem allows us to choose polynomials P^ν in $U, \overline{U}$ so that $P^\nu(W) \to F(W)$ uniformly on $\{|W\| \le \|V\|_{L^\infty}\}$. Then $P^\nu(V) \to F(V)$ uniformly.

Since $\operatorname{spec}(UV) \subset \operatorname{spec} U + \operatorname{spec} V$, it follows that

$$\operatorname{spec} P^\nu(V) \subset \mathbb{S}\,.$$

Passing to the limit $\nu \to \infty$ proves (9.7.5). □

Chapter 10

Examples of Resonance in One Dimensional Space

10.1. Resonance relations

The examples in this chapter share a common spectral structure. The semilinear examples have

$$A_0 = I = \begin{pmatrix} 1 & 0 & 0 \\ 0 & 1 & 0 \\ 0 & 0 & 1 \end{pmatrix}, \qquad A_1 = \operatorname{diag}\{1, 0, -1\} = \begin{pmatrix} 1 & 0 & 0 \\ 0 & 0 & 0 \\ 0 & 0 & -1 \end{pmatrix}.$$

The quasilinear examples have $A_0 = I$ and $A_1'(0) = \operatorname{diag}\{1, 0, -1\}$. The operator

$$L := \partial_t + \begin{pmatrix} 1 & 0 & 0 \\ 0 & 0 & 0 \\ 0 & 0 & -1 \end{pmatrix} \partial_x \tag{10.1.1}$$

is equal to $L(\partial)$ in the first case and to $L(0, \partial)$ in the second.

The examples have profiles that are 2π periodic in Y,

$$U(y, Y) = \sum_{n \in \mathbb{Z}^2} a_n \, e^{inY}, \tag{10.1.2}$$

so the formal trigonometric series from §9.4 are Fourier series. In the language of reduced profiles $\mathcal{U}$ from §9.5, this corresponds to taking $m = 2$ and phases $\phi_\mu(y) := y_\mu$, $\mu = 0, 1$. For the more general operator, $\partial_t +$

diag $(\lambda_1, \lambda_2, \lambda_3)\partial_x$, the quasiperiodic setting is required in order to capture the triad of resonant phases computed in Example 9.1.3 after Exercise 9.1.1.

Proposition 10.1.1. *For the operator (10.1.1) and phases $\phi_\mu = y_\mu$, the small divisor hypothesis is satisfied.*

Proof. The matrix

$$L(nd\phi) = L(n_0, n_1) = \begin{pmatrix} n_0 + n_1 & 0 & 0 \\ 0 & n_0 & 0 \\ 0 & 0 & n_0 - n_1 \end{pmatrix}$$

has eigenvalues $n_0 + n_1$, n_0, $n_0 - n_1$. For $n \in \mathbb{Z}^2$, the eigenvalues are integers.

When an eigenvalue is nonzero, it is bounded below by 1 in modulus. This proves small divisor hypothesis (9.6.1) with $N = 0$ and $C = 1$. □

Denote the standard basis elements of $\mathbb{C}^3$ by

$$r_1 := (1,0,0)\,, \qquad r_2 := (0,1,0)\,, \qquad r_3 := (0,0,1)\,. \tag{10.1.3}$$

The matrix $L(n)$ is diagonal in this basis, and when $n_1 \neq 0$ the eigenvalues are distinct and the corresponding eigenprojectors are

$$\pi_1 := \begin{pmatrix} 1 & 0 & 0 \\ 0 & 0 & 0 \\ 0 & 0 & 0 \end{pmatrix}, \qquad \pi_2 := \begin{pmatrix} 0 & 0 & 0 \\ 0 & 1 & 0 \\ 0 & 0 & 0 \end{pmatrix}, \qquad \pi_3 := \begin{pmatrix} 0 & 0 & 0 \\ 0 & 0 & 0 \\ 0 & 0 & 1 \end{pmatrix}.$$

For $n_1 \neq 0$, $\pi(n_0, n_1)$ is nonzero in exactly three circumstances:

$$\begin{aligned} n_0 + n_1 = 0\,, &\quad \text{in which case} \quad \pi(n) = \pi_1\,, \\ n_0 = 0\,, &\quad \text{in which case} \quad \pi(n) = \pi_2\,, \\ n_0 - n_1 = 0\,, &\quad \text{in which case} \quad \pi(n) = \pi_3\,. \end{aligned} \tag{10.1.4}$$

When $n_1 = (n_0, 0) \neq 0$, $\pi(n) = I$.

Define

$$\lambda_1 := +1\,, \qquad \lambda_2 := 0\,, \qquad \lambda_3 := -1\,.$$

The characteristic variety of L is the union of the three lines

$$\ell_j := \Big\{\eta = (\eta_0, \eta_1) \,:\, \eta_0 + \lambda_j \eta_1 = 0\Big\}\,, \qquad j = 1, 2, 3\,.$$

In Figure 10.1.1, the characteristic points of $\mathbb{Z}^2$ are indicated by dots. The circled dots yield two resonance relations, one indicated by a solid triangle and the other dotted,

$$(1,1) + (1,-1) + (-2,0) = 0 \qquad \text{and} \qquad (-1,1) + (-1,-1) + (2,0) = 0\,.$$

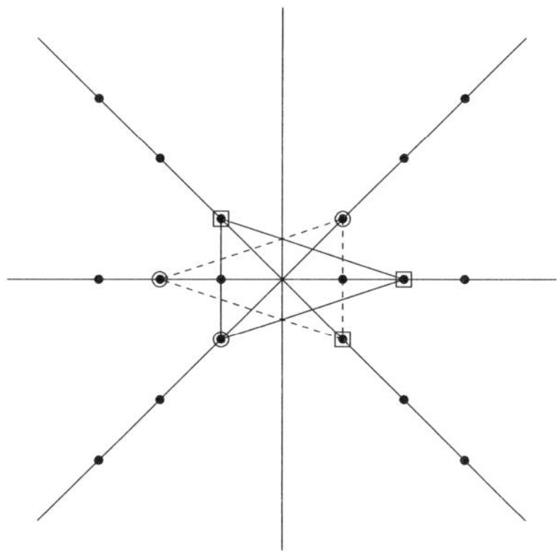

FIGURE 10.1.1. Char L and two resonant triads

Equation (9.4.13) shows that the profile satisfies $\mathbf{E}U_0 = U_0$. In particular, the Fourier coefficients $\widehat{U_0}(y, n_0, n_1)$ vanish unless $n \in \bigcup_j \ell_j$. The coefficients are polarized,

$$(10.1.5) \qquad n \in \ell_j \setminus 0 \quad \Longrightarrow \quad \pi(n) = \pi_j \quad \text{and} \quad \pi_j\, \widehat{U_0}(y,n) = \widehat{U_0}(y,n)\,.$$

Since the π_j sum to I, one has

$$\mathbf{E} \;=\; \sum_1^3 \mathbf{E}_j\,, \qquad \text{where} \qquad \mathbf{E}_j := \pi_j\,\mathbf{E}\,.$$

The definition of $\mathbf{E}$ yields

$$\mathbf{E}_j \sum_{\alpha\in\mathbb{Z}^2} a_\alpha(y)\, e^{i\alpha.Y} \;=\; \sum_{\alpha\in\ell_j\cap\mathbb{Z}^2} \pi_j\, a_\alpha(y)\; e^{i\alpha.Y}\,.$$

For $k \in \mathbb{Z}$, define the scalar Fourier coefficients $\widehat{\sigma_j}$ encoding the spectra of $\widehat{U}_0$ from ℓ_j

$$\begin{aligned}
\widehat{\sigma_1}(y,k) \;&:=\; \big\langle \hat{U}_0(y,(k,-k)\,,\, r_1\big\rangle\,,\\
\widehat{\sigma_2}(y,k) \;&:=\; \big\langle \hat{U}_0(y,(0,k)\,,\, r_2\big\rangle\,,\\
\widehat{\sigma_3}(y,k) \;&:=\; \big\langle \hat{U}_0(y,(k,k)\,,\, r_3\big\rangle\,.
\end{aligned}$$

The corresponding 2π periodic functions are

$$(10.1.6) \qquad \sigma_j(y,\phi) \;:=\; \sum_{k\in\mathbb{Z}} \widehat{\sigma_j}(y,k)\, e^{ik\phi}\,, \qquad j = 1,2,3\,.$$

Then,

$$(10.1.7) \qquad U_0(y,Y_0,Y_1) \;=\; \Big(\sigma_1(y,Y_0-Y_1)\,,\, \sigma_2(y,Y_1)\,,\, \sigma_3(y,Y_0+Y_1)\Big)$$

and

$$\begin{aligned}
\mathbf{E}_1\, U_0 &= r_1 \sum_{k\in\mathbb{Z}} \widehat{\sigma_1}(y,k)\, e^{ik(Y_0-Y_1)}\,, \\
\mathbf{E}_2\, U_0 &= r_2 \sum_{k\in\mathbb{Z}} \widehat{\sigma_2}(y,k)\, e^{ikY_1}\,, \\
\mathbf{E}_3\, U_0 &= r_3 \sum_{k\in\mathbb{Z}} \widehat{\sigma_1}(y,k)\, e^{ik(Y_0+Y_1)}\,.
\end{aligned}$$

The next proposition shows that the projection operators $\mathbf{E}$ are integral operators. This is a special case of a general phenomenon.

Proposition 10.1.2. *For $g(Y) \in \bigcap_s H^s(\mathbb{T}^2)$, the operators $\mathbf{E}_j$ are given by the formulas*

$$\begin{aligned}
(\mathbf{E}_1 g)(Y) &= \int_0^{2\pi} \pi_1\, g(\psi + (Y_0 - Y_1)\,,\, \psi)\, \frac{d\psi}{2\pi}\,, \\
(\mathbf{E}_2 g)(Y) &= \int \pi_2\, g(Y_0, Y_1)\, \frac{dY_0}{2\pi}\,, \qquad (10.1.8)\\
(\mathbf{E}_3 g)(Y) &= \int_0^{2\pi} \pi_3\, g(-\psi + (Y_0 - Y_1)\,,\, \psi)\, \frac{d\psi}{2\pi}\,.
\end{aligned}$$

The expressions depend only on $Y_0 - Y_1$, Y_1, and $Y_0 + Y_1$, respectively.

Proof. One has

$$\mathbf{E}_2(a\, e^{inY}) = \begin{cases} \pi_2\, a\, e^{inY} & \text{when} \quad n_0 = 0, \\ 0 & \text{when} \quad n_0 \neq 0. \end{cases}$$

On such monomials, $\mathbf{E_2}$ agrees with $\pi_2 \int \cdots dY_0/2\pi$. By linearity and density

$$\mathbf{E}_2 g(Y_0, Y_1) = \int \pi_2\, g(Y_0, Y_1)\, \frac{dY_0}{2\pi}\,.$$

Consider next $\mathbf{E}_1$ for which the preserved exponentials are $e^{ik(Y_0-Y_1)}$ with $k \in \mathbb{Z}$. Their phases are constant on the lines $Y_0 - Y_1 = c$. The general monomial is $e^{imY_0} e^{ik(Y_0-Y_1)}$. To kill those with $m \neq 0$, it is sufficient to integrate over $Y_0 - Y_1 = c$. Parametrize $\{Y_0 - Y_1 = c\}$ by Y_1 to obtain

$$\mathbf{E}_1 g = \int_0^{2\pi} \pi_1\, g(Y_1 + c\,,\, Y_1)\, \frac{dY_1}{2\pi}\,.$$

On the domain of integration, $c = Y_0 - Y_1$, and Y_1 is a dummy variable yielding the desired formula

$$\mathbf{E}_1 g = \int_0^{2\pi} \pi_1\, g(\psi + (Y_0 - Y_1)\,,\, \psi)\, \frac{d\psi}{2\pi}\,.$$

For $\mathbf{E}_3$ the exponentials $e^{imY_0}e^{ik(Y_0+Y_1)}$ with $m = 0$ are the ones preserved. We single them out by integrating over $Y_0 + Y_1 = c$ parametrized by Y_1 to find

$$\mathbf{E}_3 g \;=\; \int_0^{2\pi} \pi_3\, g(-Y_1 + c\,,\, Y_1)\, \frac{dY_1}{2\pi}\,.$$

This is equivalent to the formula of the proposition. □

10.2. Semilinear examples

For initial data $U_0(0,x,Y) = \mathbf{E}\,U_0 \in \bigcap_s H^s(\mathbb{R}^d \times \mathbb{T}^{1+1})$, Theorem 9.5.1 proves that the leading profile equation has a unique smooth solution locally in time. Since the small divisor hypothesis is satisfied, Theorem 9.6.2 proves that the corrector profiles U_j exist and are uniquely determined from the initial values of $\mathbf{E}U_j\big|_{t=0}$. As described in §9.7, the semilinear analogues of Theorems 9.6.3 and 9.7.1 imply that they yield infinitely accurate approximate solutions.

The profile equation (9.4.14) is a vector equation with three components. The jth component asserts that

$$\pi_j\,\mathbf{E}\,\Big(L(\partial_y)U_0 + f(U_0)\Big) \;=\; 0\,.$$

The next computation shows that this is an evolution equation for σ_j coupled by lower order terms to the other components.

Use the diagonal structure of L and $[\mathbf{E}, \partial_y] = [\mathbf{E}, \pi_j] = 0$ to find

$$\begin{aligned}\pi_1\,\mathbf{E}\,L(\partial_y)\,U_0 \;&=\; \mathbf{E}_1\,\pi_1 L(\partial_y)\,U_0 \;=\; \mathbf{E}_1\,(\partial_t + \partial_x)\,\pi_1\,U_0\\ &=\; (\partial_t + \partial_x)\,\mathbf{E}_1\,\big(\sigma_1(y, Y_0 - Y_1)\,r_1\big)\\ &=\; (\partial_t + \partial_x)\,\sigma_j(y, Y_0 - Y_1)\,r_1\,.\end{aligned}$$

From the profile equation for U_0 one deduces

$$(\partial_t + \partial_x)\,\sigma_1(y, Y_0 - Y_1) \;+\; \Big\langle \mathbf{E}_1(f(U_0))\,,\, r_1\Big\rangle \;=\; 0\,.$$

Equivalently,

(10.2.1)

$$\begin{aligned}&(\partial_t + \partial_x)\,\sigma_1(y, Y_0 - Y_1)\\ &\qquad + \Big\langle \mathbf{E}_1\Big(f_1(\sigma_1(y, Y_0 - Y_1), \sigma_2(y, Y_1), \sigma_3(y, Y_0 + Y_1))r_1\Big), r_1\Big\rangle = 0.\end{aligned}$$

Similarly, the second and third equations are equivalent to

(10.2.2)

$$\partial_t\sigma_2(y, Y_1) + \Big\langle \mathbf{E}_2\Big(f_2(\sigma_1(y, Y_0 - Y_1)\,,\, \sigma_2(y, Y_1)\,,\, \sigma_3(y, Y_0 + Y_1))r_2\Big)\,,\, r_2\Big\rangle = 0$$

and
(10.2.3)

$$\big(\partial_t - \partial_x\big)\, \sigma_3(y, Y_0 + Y_1)$$
$$+ \Big\langle \mathbf{E}_3\Big(f_3\big(\sigma_1(y, Y_0 - Y_1), \sigma_2(y, Y_1), \sigma_3(y, Y_0 + Y_1)\big)r_3\Big), r_3\Big\rangle = 0.$$

Equations (10.2.1)–(10.2.3) form a coupled system of three integro-differential equations. They are differential in the variables t, x and integral in the variables $(Y_0, Y_1) \in \mathbb{T}^2$. The system is easy to approximate numerically, for example, using the trapezoid rule for the integrations and standard one dimensional hyperbolic methods for t, x discretization.

Example 10.2.1. Continue the computation for the nonlinear term from the three-wave interaction system (9.2.2). In that case, the transport equation for σ_2 is

$$(10.2.4)\qquad \partial_t\, \sigma_2(y, \theta_1) \;=\; c_2 \Big\langle \mathbf{E}_2 \Big(\sigma_1(y, Y_0 - Y_1)\, \sigma_3(y, Y_0 + Y_1)\, r_2\Big)\,,\, r_2 \Big\rangle\,.$$

The right-hand side is best understood terms of Fourier series. Expand to find

$$\sigma_1(y, Y_0 - Y_1)\, \sigma_3(y, Y_0 + Y_1) \;=\; \sum_{\mu,\nu} \widehat{\sigma_1}(y, \nu)\, e^{i\nu(Y_0 - Y_1)}\, \widehat{\sigma_3}(y, \mu)\, e^{i\mu(Y_0+Y_1)}\,.$$

The operator $\mathbf{E}_2$ selects the phases nY with $n_0 = 0$. As the phase is equal to $\nu(Y_0 - Y_1) + \mu(Y_0 + Y_1)$, this yields $\mu = -\nu$, so

$$\mathbf{E}_2\Big(\big(\sigma_1(y, Y_0 - Y_1)\, \sigma_3(y, Y_0 + Y_1)\, r_2\Big) \;=\; \sum_{\nu} \widehat{\sigma_1}(y, \nu)\, \widehat{\sigma_3}(y, -\nu)\, e^{-2i\nu Y_1}\, r_2\,.$$

Therefore, the profile equation (10.2.4) for $\widehat{\sigma_2}(y, n)$ splits according to the parity of $n \in \mathbb{Z}$. For all $\nu \in \mathbb{Z}$,

$$(10.2.5)\quad \partial_t\, \widehat{\sigma_2}(y, -2\nu) = c_2\, \widehat{\sigma_1}(y, \nu)\, \widehat{\sigma_3}(y, -\nu)\,, \qquad \partial_t\, \widehat{\sigma_2}(y, -2\nu + 1) = 0\,.$$

The dynamics for σ_1 is given by

$$(10.2.6)\quad \big(\partial_t + \partial_x\big)\, \sigma_1(y, Y_0 - Y_1) \;=\; c_1 \Big\langle \mathbf{E}_1 \Big(\sigma_2(y, Y_1)\, \overline{\sigma_3}(y, Y_0 + Y_1)\, r_1\Big)\,,\, r_1 \Big\rangle\,.$$

The third profile equation is

$$(10.2.7)\quad \big(\partial_t + \partial_x\big)\, \sigma_3(y, Y_0 + Y_1) \;=\; c_3 \Big\langle \mathbf{E}_3 \Big(\sigma_2(y, Y_1)\, \overline{\sigma_1}(y, Y_0 - Y_1) r_3\Big)\,,\, r_3 \Big\rangle\,.$$

For (10.2.6), use

$$\overline{\sigma_3}(y, \phi) \;=\; \Big(\sum_{\nu} \widehat{\sigma_3}(y, \nu)\, e^{i\nu\phi}\Big)^* \;=\; \sum_{\nu} \widehat{\sigma_3}(y, \nu)^*\, e^{-i\nu\phi} \quad \text{so}$$

$$\sigma_2(y, Y_1)\, \overline{\sigma_3}(y, Y_0 + Y_1) \;=\; \sum_{\mu,\nu} \widehat{\sigma_2}(y, \mu)\, e^{i\mu Y_1}\, \widehat{\sigma_3}(y, \nu)^*\, e^{-i\nu(Y_0+Y_1)}\,.$$

The phase of the product of the exponentals is $-\nu Y_0 + (\mu - \nu)Y_1$. The operator $\mathbf{E}_1$ selects only those phases nY with $n_0 + n_1 = 0$. In this case,

$$-\nu \ + \ (\mu - \nu) \ = \ 0 \qquad \Longleftrightarrow \qquad \mu \ = \ 2\nu.$$

Therefore,

$$\big(\partial_t + \partial_x\big)\, \sigma_3(y, Y_0 + Y_1) \ = \ c_3 \sum_\nu \widehat{\sigma_2}(y, 2\nu)\ \widehat{\sigma_3}(y,\nu)\ e^{-i\nu(Y_0 - Y_1)}\,.$$

In terms of the Fourier coefficients, this is equivalent to

$$(\partial_t + \partial_x)\widehat{\sigma_1}(y,\nu) \ = \ c_3\, \widehat{\sigma_2}(y,-2\nu)\ \widehat{\sigma_3}(y,-\nu)^*\,. \tag{10.2.8}$$

An analogous computation shows that the third profile equation is equivalent to

$$(\partial_t - \partial_x)\widehat{\sigma_3}(y,-\nu) \ = \ c_3\, \widehat{\sigma_1}(y,\nu)^*\ \widehat{\sigma_2}(y,-2\nu)\,. \tag{10.2.9}$$

Exercise 10.2.1. *Verify (10.2.9).*

The equations (10.2.5), (10.2.8), (10.2.9) show that the nonlinear interactions are localized in the triads

$$\Big\{\widehat{\sigma_1}(y,k)\,,\, \widehat{\sigma_2}(y,-2k)\,,\, \widehat{\sigma_3}(y,-k)\Big\}\,. \tag{10.2.10}$$

The corresponding Fourier coefficients of U_0 are

$$\widehat{U_0}(y,k,-k), \qquad \widehat{U_0}(y,0,-2k) \qquad \widehat{U_0}(y,-k,-k)\,.$$

Two such triads are indicated in Figure 10.1.1. The interaction comes about through the resonance relation

$$-2x \ = \ (t-x) \ - \ (t+x)\,, \qquad (0,-2k) = (k,-k) + (-k,-k)\,.$$

For each k, the triple (10.2.10) *satisfies the three-wave interaction partial differential equation decoupled from the other Fourier coefficients.* For k large the interactions involve distant pieces of the Fourier transform of the solution. The fact that the triads are isolated shows that, for this equation, one cannot compound these resonances to generate interactions between distant Fourier coefficients.

Consider three special cases. First is the initial value problem (9.1.1) with $c_1 = c_3 = 0$ and initial data

$$\sigma_1(0,x,\phi) = a_1(x)\, e^{i\phi}\,, \qquad \sigma_2(0,x,\phi) = 0\,, \qquad \sigma_3(0,x,\phi) = a_3(x)\, e^{-i\phi}\,.$$

The initial data ignite a single resonant triad. The function $\sigma(y,\phi)$ is given by

$$\sigma_1 = a_1(t-x)e^{i\phi}\,, \quad \sigma_2 = e^{-2i\phi}\int_0^t a_1(t-x)\, a_3(t+x)\, dt\,, \quad \sigma_3 = a_1(t+x)e^{-i\phi}\,.$$

In this particular case, the approximation of nonlinear geometric optics gives the exact solution.

For the second example modify the third initial datum to

$$u_3(0,x) = a_3(x)e^{inx/\epsilon}\,, \qquad n \in \mathbb{Z}\setminus\{-1\}\,, \tag{10.2.11}$$

to find $\sigma_3(t,x,\phi) = a_3(t+x)\,e^{in\phi}$, and

$$\mathbf{E}_2(\sigma_1(y, Y_0 - Y_1)\,\sigma_3(y, Y_0 + Y_0)r_2) \\ = \mathbf{E}_2\Big(a_1(t-x)\,a_3(t+x)\,e^{i\{(t+x)+n(t-x)\}}r_2\Big) = 0\,.$$

The product inside $\mathbf{E}_2$ always oscillates in time so is annihilated by $\mathbf{E}_2$ to give $\partial_t\sigma_2 = 0$. The oscillations in the second component of U_0 do not change in time and there is no interaction with the oscillations in the other components. This agrees with the nonstationary phase analysis in §9.1.

The third example has real initial data

$$\sigma_1(0,x,\phi) = a_1(x)\,\sin\phi\,, \quad \sigma_2(0,x,\phi) = 0\,, \quad \sigma_3(0,x,\phi) = a_3(x)\,\sin(-\phi)\,.$$

In this case the initial data ignite two resonant triads,

$$\Big\{(0,-2n),(n,n),(-n,-n)\Big\} \qquad \text{and} \qquad \Big\{(0,2n),(-n,-n),(n,n)\Big\}.$$

Each triad of coefficients,

$$\widehat{\sigma_1}(t,x,1)\,,\ \widehat{\sigma_1}(t,x,-2)\,,\ \widehat{\sigma_1}(t,x,-1) \qquad \text{and} \\ \widehat{\sigma_1}(t,x,-1)\,,\ \widehat{\sigma_1}(t,x,2)\,,\ \widehat{\sigma_1}(t,x,1)\,,$$

solves the three-wave interaction partial differential equation. All other coefficients vanish identically.

In the last two cases, the approximate solution is not an exact solution.

Proposition 10.2.1. *Consider the system of profile equations for the three-wave interaction system with $c_j \in \mathbb{R}\setminus 0$. The following are equivalent.*

i. *For arbitrary initial data $\sigma(0,x,\phi) \in \bigcap_s H^s(\mathbb{R}\times\mathbb{T})$, there is a unique global solution $\sigma(t,x,\phi) \in \bigcap_s C^s\big(\mathbb{R}\,;\,H^s(\mathbb{R}\times\mathbb{T})\big)$.*

ii. *The coefficients c_j do not have the same sign.*

Proof. The explosive behavior is proved by considering a single resonant triad that blows up in finite time T_*.

For existence it suffices to observe that the $L^\infty([0,T]\times\mathbb{R})$ bound, for solutions of the three-wave system with c_j not all of the same sign, proves an estimate

$$\|\widehat{\sigma}(t,x,n)\|_{L^\infty([0,T]\times\mathbb{R})} \ \le\ C\Big(\|\widehat{\sigma}(0,x,n)\|_{L^2(\mathbb{R})}\,,\,T\Big)\ \big\|\widehat{\sigma}(0,x,n)\big\|_{L^\infty(\mathbb{R})}\,,$$

with the function $C(\cdot,\cdot)$ independent of n. Use this together with

$$\|\sigma\|_{L^\infty([0,T]\times\mathbb{R}\times\mathbb{T})} \ \le\ \sum_n \|\hat{\sigma}(0,x,n)\|_{L^\infty([0,T]\times\mathbb{R})}$$

and

$$\sum_n \|\hat\sigma(0,x,n)\|^2_{H^s(\mathbb{R})}\,\langle n\rangle^{2s} \ \le\ C(s)\,\|\sigma(0)\|^2_{H^{2s}(\mathbb{R}\times\mathbb{T})}\,.$$

The last inequality for $s=0$ and $s>1/2$, respectively, yields

$$\|\widehat\sigma(0,x,n)\|_{L^2(\mathbb{R})} \ \le\ \|\sigma(0)\|_{L^2(\mathbb{R}\times\mathbb{T})} \quad \text{and}$$
$$\big\|\widehat\sigma(0,x,n)\big\|_{L^\infty(\mathbb{R})} \ \le\ C(s)\,\langle n\rangle^{-s}\big\|\sigma(0)\big\|_{H^{2s}(\mathbb{R}\times\mathbb{T})}\,.$$

Putting these together for $s>1$ shows that

$$\begin{aligned}
&\|\sigma(t,x,\phi)\|_{L^\infty([0,T]\times\mathbb{R}\times\mathbb{T})}\\
&\qquad\le\ \sum_n \|\widehat\sigma(t,x,n)\|_{L^\infty([0,T]\times\mathbb{R})}\\
&\qquad\le\ C\big(s,T,\|\sigma(0)\|_{H^{2s}(\mathbb{R}\times\mathbb{T})}\big)\,\sum_n \langle n\rangle^{-s}\|\sigma(0,x,\phi)\|_{H^{2s}(\mathbb{R}\times\mathbb{T})}\\
&\qquad\le\ C'\big(s,T,\|\sigma(0)\|_{H^{2s}(\mathbb{R}\times\mathbb{T})}\big)\,.
\end{aligned}$$

This a priori estimate implies global solvability using Moser's inequality as in §6.4. □

When the profiles exist globally in time, the semilinear version of Theorem 9.7.1 shows that the approximation of resonant nonlinear geometric optics is accurate on arbitrary long time intervals $0\le t\le T$. In particular the interval of existence of the exact solution grows infinitely long in the limit $\epsilon\to 0$. Theorem 9.2.3.i contains the stronger assertion that for this three-wave interaction system the solutions exist globally. Note that the approximation is not justified on the infinite time interval $0<t<\infty$. One must exercise care in drawing conclusions about the large time behavior of exact solutions from the large time behavior of the profiles.

Similar caution is required for the case of explosive profiles. It is tempting to conclude from profile blowup that there is a parallel blowup of exact solutions. This is not justified. Theorem 9.7.1 justifies the approximation on arbitrary intervals of smoothness, $0<T<T_*$ for the approximate solution. This implies some conclusions that have the flavor of explosion. Denote by v^ϵ the exact solution with the same initial data as the approximate solution u^ϵ. Choosing T very close to T_* implies that

$$\lim_{T\to T_*}\ \liminf_{\epsilon\to 0}\ \big\|v^\epsilon(T,x)\big\|_{L^\infty(\mathbb{R}_x)} \ =\ \infty.$$

This asserts that the family of exact solutions v^ϵ is unbounded, but it does not assert that any given member of the family explodes. Recall the unboundedness of the family of approximate solutions of linear geometric optics near a caustic, for example the incoming spherical waves of §5.4. Those are examples where each member of the family is globally smooth.

Example 10.2.2. Modify equation (9.1.1) so that the right-hand side of the equation for u_2 is changed to a general real quadratic interaction,

$$\partial_t u_2 = \sum_{1\le i\le j\le 3} A_{i,j}\, u_i\, u_j . \tag{10.2.12}$$

The profile equation for σ_2 becomes

$$\partial_t \sigma_2(y, Y_1) = \Big\langle \mathbf{E}_2\Big(\sum_{1\le i\le j\le 3} A_{i,j}\ \sigma_i(y, h_i(Y))\ \sigma_j(y, h_j(Y)) r_2\Big)\,,\, r_2\Big\rangle\,, \tag{10.2.13}$$

with

$$h_1(Y) := Y_0 - Y_1, \qquad h_2(Y) := Y_1, \qquad h_3(Y) := Y_0 + Y_1 .$$

Write U_0 as in (10.1.7). The contribution of the term $A_{1,3}\,\sigma_1\,\sigma_3$ to the profile equation is computed exactly as before and yields

$$\begin{aligned}&\mathbf{E}_2\big(A_{1,3}\,\sigma_1(y, Y_0 - Y_1)\,\sigma_3(y, Y_0 + Y_1)\, r_2\big)\\ &\quad = A_{1,3} \sum_n \widehat{\sigma_1}(y, n)\ \widehat{\sigma_3}(y, -n)\ e^{-2inY_1}\ r_2\\ &\quad = A_{1,3}\big(\sigma_1 * \check{\sigma}_3\big)(y, -2Y_1)\, r_2 .\end{aligned}$$

Denoting with an underline the mean value of a 2π periodic function, one then computes the formulas

$$\begin{aligned}\mathbf{E}_2\big(A_{1,2}\,\sigma_1(y, Y_0 - Y_1)\,\sigma_2(y, Y_1)\, r_2\big) &= A_{1,2}\,\underline{\sigma_1}\,\sigma_2(y, Y_1)\, r_2\,,\\ \mathbf{E}_2\big(A_{2,3}\,\sigma_2(y, Y_1)\,\sigma_3(y, Y_0 + Y_1)\, r_2\big) &= A_{2,3}\,\sigma_2(y, Y_1)\,\underline{\sigma_3}\, r_2\,,\\ \mathbf{E}_2\big(A_{2,2}\,\sigma_2(y, Y_1)\,\sigma_2(y, Y_1)\, r_2\big) &= A_{2,2}\,\sigma_2(y, Y_1)^2\, r_2 .\end{aligned}$$

Combining yields the profile equation

$$\partial_t \sigma_2 = A_{2,2}\sigma_2^2(y,\phi) + A_{1,2}\sigma_2(y,\phi)\underline{\sigma_1} + A_{2,3}\sigma_2(y,\phi)\underline{\sigma_3} + A_{1,3}(\sigma_1 * \check{\sigma}_3)(y, -2\phi) . \tag{10.2.14}$$

Notice that the first three terms are local in y, ϕ while the quadratic convolution interaction term that comes from the resonance is local in Fourier and not in ϕ. For the initial data from (9.1.1), $\underline{\sigma_1} = \underline{\sigma_3} = 0$ and the profile equation simplifies to

$$\partial_t \sigma_2(y, \phi) = A_{2,2}\,\sigma_2^2 + A_{1,3}\, a_1(t - x)\, a_3(t + x)\, e^{-2i\phi} . \tag{10.2.15}$$

Only one Fourier component of σ_2 is affected by the resonant term. The $A_{2,2}\,\sigma_2^2$ broadens the spectrum of σ_2, one of whose Fourier components is influenced by the waves from modes one and three.

With general quadratic interactions in all the equations, one finds coupled integro-differential equations with quadratic self-interaction terms for all j. The resulting 3×3 systems are analogous to

$$\partial_t \sigma = a\,\sigma^2 + b\,\sigma * \sigma\,, \qquad \sigma = \sigma(t,\phi)\,. \tag{10.2.16}$$

Open problem. It would be interesting to understand well the competition between the two quadratic terms on the right of (10.2.16). The term that is local in ϕ is a convolution in n while the convolution in ϕ is local in n.

10.3. Quasilinear examples

The next examples resemble the semilinear examples. An important difference is that the amplitudes of the approximate solutions are smaller. The approximate solution defined by the leading profile is given by

$$u^\epsilon(t,x) = \epsilon\, U_0(t,x,t/\epsilon,x/\epsilon)\,.$$

The prefactor ϵ was absent in the semilinear case. The profiles are periodic in Y, so equation (9.5.1) simplifies to

$$\mathbf{E}\, U_0 = U_0\,, \qquad \mathbf{E}\Big(L(0,\partial_y)\, U_0 + \sum_{\mu=0}^{1} A'_\mu(0)\, U_0\, \frac{\partial U_0}{\partial Y_\mu} \Big) = 0\,. \tag{10.3.1}$$

Example 10.3.1. A quasilinear analogue of the system (9.1.1) is the system of conservation laws

$$\begin{aligned} (\partial_t + \partial_x)u_1 &= 0, \\ \partial_t u_2 + \partial_x(u_1 u_3) &= 0, \\ (\partial_t - \partial_x)u_3 &= 0. \end{aligned} \tag{10.3.2}$$

Proposition 10.1.1 shows that, with $\phi_\mu := y_\mu$, the small divisor hypothesis is satisfied and equations (10.1.1) through (10.1.7) are unchanged. And $A_0(u) = I$. The second component of the profile equation is

$$\partial_t\, \sigma_2 + \Big\langle \mathbf{E}_2\Big(A'_1(0) U_0\, \frac{\partial U_0}{\partial Y_1}\Big)\,,\, r_2 \Big\rangle = 0\,. \tag{10.3.3}$$

Equation (10.1.7) yields

$$\frac{\partial U_0}{\partial Y_1} = \Big(-\sigma'_1(y, Y_0 - Y_1)\,,\, \sigma'_2(y,Y_1)\,,\, \sigma'_3(y, Y_0 + Y_1)\Big)\,. \tag{10.3.4}$$

For (10.3.2)

$$A_1(U) = \begin{pmatrix} 1 & 0 & 0 \\ 0 & 0 & 0 \\ 0 & 0 & -1 \end{pmatrix} + \begin{pmatrix} 0 & 0 & 0 \\ U_3 & 0 & U_1 \\ 0 & 0 & 0 \end{pmatrix} = A_1(0) + A'_1(0)U,$$

so

$$A'_1(0)U_0 = \begin{pmatrix} 0 & 0 & 0 \\ \sigma_3(y, Y_0 + Y_1) & 0 & \sigma_1(y, Y_0 - Y_1) \\ 0 & 0 & 0 \end{pmatrix}\,.$$

Suppressing the y dependence of σ for ease of reading,

$$A_1'(0)\,U_0\,\frac{\partial U_0}{\partial Y_1} = \Big(-\sigma_3(Y_0+Y_1)\,\sigma_1'(Y_0-Y_1) + \sigma_1(Y_0-Y_1)\sigma_3'(Y_0+Y_1)\Big)\,r_2$$
$$= r_2\,\frac{\partial}{\partial Y_1}\Big(\sigma_1(Y_0-Y_1)\sigma_3(Y_0+Y_1)\Big)\,.$$

$\mathbf{E}_2$ commutes with $\partial/\partial_{Y_\mu}$, and $\mathbf{E}_2$ applied to the product is computed as earlier to find

$$\partial_t\sigma_2(t,x,Y_1) \;=\; \frac{\partial}{\partial Y_1}\Big(\sum_\nu e^{-2i\nu Y_1}\,\widehat{\sigma_1}(t,x,\nu)\,\widehat{\sigma_3}(t,x,-\nu)\Big). \tag{10.3.5}$$

The odd Fourier coefficients of σ_2 are stationary and the even ones evolve according to

$$\partial_t\hat\sigma_2(y,-2\nu) = -2i\nu\,\hat\sigma_1(y,\nu)\,\hat\sigma_3(y,-\nu)\,. \tag{10.3.6}$$

The profile equations are

$$\begin{aligned}(\partial_t+\partial_x)\sigma_1 \;&=\; 0\,,\\ \partial_t\sigma_2 \;&=\; \partial_\phi\Big(\big(\sigma_1 * \check\sigma_3\big)(t,x,-2\phi)\Big)\,,\\ (\partial_t-\partial_x)\sigma_3 \;&=\; 0\,.\end{aligned} \tag{10.3.7}$$

The system (10.3.7) is in conservation form. This is a general phenomenon. If the original system is in the conservation form

$$\sum_{\mu=0}^{d}\partial_\mu A_\mu(u) \;=\; 0\,, \tag{10.3.8}$$

then the terms of the equation are $A_\mu'(u)\partial_\mu u$. The coefficients, $A_\mu'(u)$, have the special structure of being derivatives.

Exercise 10.3.1. *If the original system is real and in conservation form (10.3.7) and $A_0(0)=I$, then the profile equation (9.4.14) for periodic profiles can be written in the conservation form*
(10.3.9)

$$\partial_t U_0 + \sum_{j=1}^{d}\frac{\partial}{\partial x_j}\Big(\mathbf{E}\big(A_j(0)U_0\big)\Big) + \sum_{\mu=0}^{d}\sum_{j,k=1}^{N}\frac{\partial}{\partial\theta_k}\Big(\mathbf{E}\,\frac{\partial^2 A_\mu(0)}{\partial u_j\partial u_k}\,U_jU_k\Big) = 0\,.$$

Discussion. For complex equations and quasiperiodic profiles the conservation form also persists.

Equation (10.3.9) implies that in the case of conservation laws a profile as in Theorem 9.5.1 that has mean zero with respect to θ at $\{t=0\}$ remains mean zero throughout its maximal interval of existence. As in Example 10.2.2, the profile equations in the mean zero case are simpler.

Proposition 10.3.1. *Consider a real 3×3 system of conservation laws with $d=1$,*

$$\frac{\partial u}{\partial t} + \frac{\partial}{\partial x} A(u) = 0\,, \qquad A(u) = \big(A_1(u), A_2(u), A_3(u)\big)\,, \tag{10.3.10}$$

satisfying $A'(0) = \operatorname{diag}\{1, 0, -1\}$. Introduce σ_j as in (10.1.7) and six interaction constants

$$b_j := \frac{\partial^2 A_j(0)}{\partial u_j^2}\,, \quad j = 1, 2, 3\,, \tag{10.3.11}$$

$$c_1 := \frac{\partial^2 A_1(0)}{\partial u_2 \partial u_3}\,, \quad c_2 := \frac{\partial^2 A_2(0)}{\partial u_1 \partial u_3}\,, \quad c_3 := \frac{\partial^2 A_3(0)}{\partial u_1 \partial u_2}\,.$$

The profile equation (9.5.1) for periodic profiles (10.1.1) of mean zero is equivalent to the system of equations for the Fourier coefficients

(10.3.12)

$$\begin{aligned}
\big(\partial_t + \partial_x\big)\, \hat\sigma_1(t,x,m) &= b_1\, im\, \widehat{(\sigma_1^2)}(t,x,m) + c_1\, im\, \hat\sigma_2(t,x,2m)\, \hat\sigma_3(t,x,-m)\,,\\
\partial_t\, \sigma_2(t,x,2m) &= b_2\, 2im\, \widehat{(\sigma_2^2)}(t,x,2m) + c_2\, 2im\, \hat\sigma_1(t,x,m)\, \hat\sigma_3(t,x,m)\,,\\
\big(\partial_t - \partial_x\big)\, \hat\sigma_3(t,x,m) &= b_3\, im\, \widehat{(\sigma_3^2)}(t,x,m) + c_3\, im\, \hat\sigma_2(t,x,2m)\, \hat\sigma_1(t,x,-m)\,,\\
\partial_t \hat\sigma_2(t,x,2m+1) &= b_3\, 2im\, \widehat{(\sigma_2^2)}(t,x,2m+1)\,.
\end{aligned}$$

Exercise 10.3.2. *Prove Proposition 10.3.1.*

The next goal is to analyze more closely the resonance terms.

Example 10.3.2. Consider first the case where $b_1 = b_2 = b_3 = 0$. Then for each $m \in \mathbb{Z}$, the three Fourier components $\{\hat\sigma_1(y,m), \hat\sigma_2(y,-2m), \hat\sigma_3(y,-m)\}$ evolve independently of the other Fourier components according to the laws

$$\begin{aligned}
\big(\partial_t + \partial_x\big)\, \hat\sigma_1(t,x,m) &= -c_1\, im\, \hat\sigma_2(t,x,-2m)\, \hat\sigma_3(t,x,\mp m)\,,\\
\partial_t\, \hat\sigma_2(t,x,2m) &= -c_2\, 2im\, \hat\sigma_1(t,x,m)\, \hat\sigma_3(t,x,-m)\,,\\
\big(\partial_t - \partial_x\big)\, \hat\sigma_3(t,x,-m) &= -c_3\, im\, \hat\sigma_2(t,x,-2m)\, \hat\sigma_1(t,x,m)\,.
\end{aligned} \tag{10.3.13}$$

The odd components of σ_2 belong to no such triad and are stationary,

$$\partial_t \widehat{\sigma_2}(t,x,2m+1) = 0\,. \tag{10.3.14}$$

For fixed $m \neq 0$, the triple $(i\sigma_1, i\sigma_2, i\sigma_3)$ satisfies the three-wave interaction partial differential equation that we understand well. In addition to the information already gleaned, one has the following invariance properties.

Proposition 10.3.2. *The profile equations (10.3.13) have the following properties.*

1. *The set of σ, so that for a fixed $m \in \mathbb{Z}$ and all x, ϕ the following three equations hold:*

$$\begin{aligned}\sigma_1(t,x,m) &= -\sigma_1(t,x,-m)\,,\\ \sigma_2(t,x,2m) &= -\sigma_2(t,x,-2m)\,,\\ \sigma_3(t,x,m) &= -\sigma_3(t,x,-m)\,,\end{aligned}$$

is invariant. Imposing this condition for all m shows that the set of σ, so that $\hat\sigma(y,m)$ is odd in m, is invariant under the dynamics. These are exactly the functions $\sigma(y,\phi)$ that are odd in ϕ.

2. *The set of σ, so that for a fixed $m \in \mathbb{Z}$, $\{\hat\sigma_1(y,m), \hat\sigma_2(y,-2m), \hat\sigma_3(y,-m)\}$ are purely imaginary for all x, ϕ, is invariant. Therefore the set of σ, so that $\hat\sigma(y,m)$ is purely imaginary for all $m, x \in \mathbb{Z}\times\mathbb{R}$, is invariant.*

3. *The set of σ, so that for a fixed $m \in \mathbb{Z}$, $\{\hat\sigma_1(t,x,m), \hat\sigma_2(t,x,-2m), \hat\sigma_3(t,x,-m)\}$ do not depend on x, is invariant. Imposing this for all m shows that the set of σ that does not depend on x is invariant.*

Global solvability of the profile equations when $b = 0$ is completely resolved by our analysis of the three-wave interaction partial differential equation.

Proposition 10.3.3. *If $b_1 = b_2 = b_3 = 0$ and the three constants c_1, c_2, and c_3 do not have the same sign, then the profile equations (10.3.13) are globally solvable in the sense that for arbitrary initial data*

$$\sigma(0,x,\phi) \in \bigcap_s \mathrm{Re}\, H^s(\mathbb{R}\times\mathbb{T})$$

there is a unique global solution

$$\sigma(t,x,\phi) \in \bigcap_s C^s\big(\mathbb{R}\,;\, H^s(\mathbb{R}\times\mathbb{T})\big).$$

In contrast, if the b_j vanish, and c_1, c_2, and c_3 have the same sign, then the profile equations (10.3.13) have solutions with finite blowup time $0 < T_* < \infty$.

The blowup is quite striking. Consider for example the case of a profile that is independent of x and whose Fourier series is supported on a single pair of resonant triads as in Figure 10.1.1,

$$\text{(10.3.15)}\quad U_0(t,x,Y) = -\Big(\zeta_1(t)\,\sin m(Y_0 - Y_1)\,,\ \zeta_2(t)\,\sin(-2mY_1)\,,\ \zeta_3(t)\,\sin(-m(Y_0+Y_1))\Big)\,.$$

The exact solution is described by
(10.3.16)

$$u^\epsilon(t,x) \sim -\epsilon\Big(\zeta_1(t)\,\sin\frac{m(t-x)}{\epsilon}\,,\ \zeta_2(t)\,\sin\frac{-2mt}{\epsilon}\,,\ \zeta_3(t)\,\sin\frac{-m(t+x)}{\epsilon}\Big)\,.$$

Suppose that $\zeta(t)$ is a solution of the three-wave interaction ordinary differential equation whose components have the same sign and blow up at time $T_* < \infty$, so that

$$\lim_{t\to T*_-} \big|\zeta(t)\big| = \infty\,. \qquad (10.3.17)$$

The initial data and solutions are periodic in x. The data are bounded in $BV(I)$ for any bounded interval I, and are $O(\epsilon)$ in $L^\infty(\mathbb{R})$. For the exact solutions, Theorem 9.7.1 together with finite speed of propagation yields the following result of unbounded amplification.

Theorem 10.3.4. *Suppose that the system (10.3.12) satisfies $b = 0$ and that c_1, c_2, and c_3 have the same sign. Choose $\zeta(t)$ a real solution of the three wave interaction ordinary differential equation that blows up at time $0 < T_* < \infty$, and define the profile U_0 by (10.3.15). Let u^ϵ be the exact solution with the initial data $U_0(0,0,x/\epsilon) = U_0(t,t/\epsilon,x/\epsilon)|_{\{t=0\}}$. Then for any $T \in\,]0,T_*[$, u^ϵ is smooth on $[0,T]\times\mathbb{R}$ for ϵ small. The data is bounded in the sense that*

$$\|u^\epsilon(0)\|_{L^\infty(\{|x|\le T^*+1\})} \le C\,\epsilon\,, \qquad \|u^\epsilon(0)\|_{BV(\{|x|\le T^*+1\})} \le C\,. \qquad (10.3.18)$$

The family of solutions explodes in BV in the sense that

$$\lim_{T\to T^*_-}\ \lim_{\epsilon\to 0+} \Big|\int_{\{|x|\le T^*+1-T\}} u^\epsilon(T,x)\,\sin\frac{m(x+t)}{\epsilon}\,dx\Big| = \infty\,. \qquad (10.3.19)$$

The BV norm is amplified by as large a constant as one likes in the following sense. For any large $M > 0$ and small $\delta > 0$, one can choose $T \in [0,T^*[$ and $\epsilon_0 > 0$ so that for $0 < \epsilon < \epsilon_0$, u^ϵ is smooth on $[0,T]\times\mathbb{R}$,

$$\big\|u^\epsilon\big\|_{L^\infty([0,T]\times\mathbb{R})} < \delta\,, \qquad (10.3.20)$$

and

$$\big\|u^\epsilon(T)\big\|_{BV\{|x|\le T^*+1-T\}} \ge M\,\big\|u^\epsilon(0)\big\|_{BV\{|x|\le T^*+1\}}\big\|\,. \qquad (10.3.21)$$

Proof. Theorem 9.6.2 together with Theorem 9.7.1 imply that

$$\lim_{\epsilon\to 0+}\int_{\{|x|\le T^*+1\}} u_3^\epsilon(T,x)\,\sin\frac{m(x+t)}{\epsilon}\,dx = \frac{T^*+1}{\pi}\,\zeta_3(T,m)\,.$$

Exercise 9.3.1 shows that each component of ζ must explode as $T\to T^*$, and (10.3.19) follows.

To prove the last assertion of the theorem, choose $T < T^*$ and then ϵ_0 so that

$$|\zeta(T)| > M \qquad \text{and} \qquad \sup_{t\in[0,T]} \epsilon_0 \big|\zeta(t)\big| < \delta\,.$$

Theorems 9.6.2 and 9.7.1 complete the proof. □

A weakness of this result demonstrating unbounded amplification of the BV norm of a family of solutions with sup norm tending to zero and initial BV norms bounded is that the hypothesis $b = 0$ implies that the system is *not* genuinely nonlinear. In [Joly, Métiver, and Rauch, Comm. Math. Phys. 1994], it is proved that for b sufficiently small, the profile equations have explosive solutions near those just constructed. In this way one has examples of families of solutions of a fixed genuinely nonlinear system that are uniformly small in L^∞, uniformly bounded in BV, and for which the BV norm at time $t = 1$ is as large a multiple of the BV norm at $t = 0$ as one likes. For genuinely nonlinear 3×3 systems, a desirable estimate is

$$\big\|u(1)\big\|_{BV} \leq C\,\big\|u(0)\big\|_{BV}\,. \tag{10.3.22}$$

The construction shows that this estimate is not valid for some genuinely nonlinear equations and L^∞ small solutions. Estimate (10.3.22) was proved for scalar equations even for $d > 1$ by Conway and Smoller (1966) and Kruzkov (1970). Glimm and Lax (1970) proved such estimates for 2×2 genuinely nonlinear systems when $d = 1$. The above examples show that the Glimm–Lax result does not extend to general genuinely nonlinear 3×3 systems. After its discovery using nonlinear geometric optics, alternate constructions of such amplification were found by Young (1999, 2003) and by Jenssen (2000).

Chapter 11

Dense Oscillations for the Compressible Euler Equations

In this chapter it is proved that the compressible inviscid Euler equations have a cascade of resonant nonlinear interactions that can create waves moving in a dense set of directions from three incoming waves.

11.1. The $2-d$ isentropic Euler equations

This system describes compressible inviscid fluid flow with negligible heat conduction. The velocity and density are denoted $v = (v(t,x), v(t,x))$ and $\rho(t,x)$, respectively. The pressure is assumed to be a function, $p(\rho)$, of the density. The governing equations for flows without shocks are

$$
\begin{aligned}
&(\partial_t + v_1\partial_1 + v_2\partial_2)v_1 + (p'(\rho)/\rho)\partial_1\rho = 0\,,\\
&(\partial_t + v_1\partial_1 + v_2\partial_2)v_2 + (p'(\rho)/\rho)\partial_2\rho = 0\,,\\
&(\partial_t + v_1\partial_1 + v_2\partial_2)\rho + \rho(\partial_1 v + \partial_2 v_2) = 0\,,
\end{aligned}
\tag{11.1.1}
$$

where $x = (x_1, x_2)$ and $\partial_j := \partial/\partial x_j$. Denote by

$$
u \;:=\; (v_1, v_2, \rho) \qquad \text{and} \qquad f(\rho) \;:=\; \frac{p'(\rho)}{\rho}\,.
$$

The system (11.1.1) is of the form $L(u,\partial)u = 0$ with coefficient matrices
(11.1.2)

$$A_0 = I, \quad A_1(u) = \begin{pmatrix} v_1 & 0 & f(\rho) \\ 0 & v_1 & 0 \\ \rho & 0 & v_1 \end{pmatrix}, \quad A_2(u) = \begin{pmatrix} v_2 & 0 & 0 \\ 0 & v_2 & f(\rho) \\ 0 & \rho & v_2 \end{pmatrix}.$$

The system is symmetrized by multiplying by

$$D(\rho) \ := \ \operatorname{diag}\big(\rho, \rho, f(\rho)\big) \ := \ \begin{pmatrix} \rho & 0 & 0 \\ 0 & \rho & 0 \\ 0 & 0 & f(\rho) \end{pmatrix}.$$

At a constant background state $\underline{u} := (\underline{v}, \underline{\rho})$, the linearized operator has symbol

$$L(\underline{u}, \tau, \xi) \ = \ \begin{pmatrix} \tau + \underline{v}\xi & 0 & \underline{f}\,\xi_1 \\ 0 & \tau + \underline{v}\xi & \underline{f}\,\xi_2 \\ \underline{\rho}\,\xi_1 & \underline{\rho}\,\xi_2 & \tau + \underline{v}\xi \end{pmatrix}, \qquad \text{where} \qquad \underline{f} := f(\underline{\rho}).$$

The linearized operator is symmetrized by multiplying by the constant matrix $D(\underline{\rho})$. Its characteristic polynomial is

(11.1.3) $$\det L(\underline{u}, \tau, \xi) = (\tau + \underline{v}\xi)\big[(\tau + \underline{v}\xi)^2 - c^2|\xi|^2\big],$$

where the sound speed c is defined by

(11.1.4) $$c^2 \ := \ p'(\rho), \qquad c > 0.$$

Convention. *By a linear change of time variable $t' = ct$, we may assume without loss of generality that $c = 1$.*

The asymptotic relations

$$L(u^\epsilon, \partial)u^\epsilon \ \sim \ 0 \qquad \text{and} \qquad D(\rho)\, L(u^\epsilon, \partial)u^\epsilon \ \sim \ 0$$

are equivalent. The latter is symmetric. Solutions are constructed in §9.4 and §9.6. The construction was carried out after the change of variable $\tilde{u} = D^{1/2}u$ with new coefficients $D^{-1/2}\, D\, A_\mu\, D^{-1/2}$. As the formulae are somewhat simpler, we compute with the equations (11.1.1). The background state $\underline{u}$ is constant.

The method of §9.4 is to expand the symmetric hyperbolic expression

$$(D(\rho)\, L)\Big(\underline{u} + \epsilon U^\epsilon(y, Y), \ \frac{\partial}{\partial y} + \frac{1}{\epsilon}\frac{\partial}{\partial Y}\Big)\ \Big(\underline{u} + \epsilon\, U^\epsilon(y, Y)\Big)$$

in powers of ϵ and to determine the profiles in the expansion of U^ϵ so that coefficients of each power of ϵ vanishes. This holds if and only if

$$L\Big(\underline{u} + \epsilon U^\epsilon(y, Y), \ \frac{\partial}{\partial y} + \frac{1}{\epsilon}\frac{\partial}{\partial Y}\Big)\ \Big(\underline{u} + \epsilon\, U^\epsilon(y, Y)\Big) \ \sim \ 0.$$

Expanding as in §9.4, the leading two terms yield

$$L(\underline{u}, \partial_Y)\, U_0(y, Y) \;=\; 0 \tag{11.1.5}$$

and

$$L(\underline{u}, \partial_y)\, U_0 + \sum_\mu A'_\mu(\underline{u}) U_0\, \partial_{Y_\mu} U_0 + L(\underline{u}, \partial_Y)\, U_1 \;=\; 0\,. \tag{11.1.6}$$

On Fourier series,

$$L(\underline{u}, \partial_Y) \sum_\alpha a_\alpha(y)\, e^{i\alpha Y} \;=\; i \sum_\alpha L(\underline{u}, \alpha)\, a_\alpha(y)\, e^{i\alpha Y}\,.$$

The result of the next lemma is trivial in the symmetric case.

Lemma 11.1.1. *For $(\tau, \xi) \in \mathbb{R}^{1+d}$, $\mathbb{C}^N$ is the direct sum of the image and kernel of $L(\underline{u}, \tau, \xi)$. The norms of the spectral projections $\pi(\tau, \xi)$ along the image onto the kernel are bounded independent of τ, ξ.*

Proof. Write $L(\tau, \xi) = D^{-1}\big(DL(\tau, \xi)\big)$. Since DL is hermitian and $D = D^* > 0$, it follows that the matrix L is hermitian in the scalar product

$$(u, v)_D \;:=\; (Du, v)\,.$$

In fact,

$$(Lu, v)_D := (DLu, v) = (u, DLv) = (Du, Lv) := (u, Lv)_D\,.$$

The image and kernel are orthogonal in this scalar product and therefore complementary.

The spectral projections are orthogonal with respect to the scalar product $(\ ,\ )_D$. They have norm equal to 1 in the corresponding matrix norm.

Since this norm is equivalent to the euclidean matrix norm, the $\pi(\tau, \xi)$ are uniformly bounded. □

The profile U is constructed as a periodic function of $Y = (T, X_1, X_2)$. In the language of §9.4, that means choosing

$$\phi_1(T) \;:=\; T, \qquad \phi_2(Y) \;:=\; X_1\,, \quad \text{and} \qquad \phi_3(Y) \;:=\; X_2\,.$$

With this choice $\mathbf{E} = \mathcal{E}$,

$$\mathbf{E}\Big(\sum a_\alpha(y)\, e^{i\alpha Y}\Big) \;:=\; \sum \pi(\alpha)\, a_\alpha(y)\, e^{i\alpha Y}\,.$$

The lemma shows that $\mathbf{E}$ is bounded on all $H^s([0, T]_t \times \mathbb{R}^d_x \times \mathbb{T}^{1+d}_Y)$. It projects onto the kernel of $L(\underline{u}, \partial_Y)$, and

$$\mathbf{E}\, L(\underline{u}, \partial_Y) \;=\; 0\,.$$

Thus (11.1.5) is equivalent to

$$\mathbf{E}\, U_0 \;=\; U_0\,. \tag{11.1.7}$$

Multiplying (11.1.6) by $\mathbf{E}$ yields

$$\mathbf{E}\Big(L(\underline{u},\partial_y) \;+\; \sum_\mu A'_\mu(\underline{u})U_0\partial_{Y_\mu}\Big)U_0 \;=\; 0\,. \tag{11.1.8}$$

We use equations (11.1.7)–(11.1.8) for the principal profile. The advantage is that the coefficients of the nonsymmetric system L are simple.

The equations (11.1.7)–(11.1.8) have the exact same form as the equations derived for the symmetric operator DL, only the operator $\mathbf{E}$ has changed.

11.2. Homogeneous oscillations and many wave interaction systems

In this section we prove that for homogeneous oscillations, the equations for the leading profile are equivalent to an infinite system of coupled ordinary differential equations for its Fourier coefficients. This is a far-reaching generalization of of the three-wave interaction ordinary differential equation.

Homogeneous oscillations refers to approximate solutions of geometric optics type whose leading profile is independent of x. The profile has the form $U_0(t,Y)$. For simplicity restrict attention to profiles $2\pi\times 2\pi\times 2\pi$ periodic in (Y_0,Y_1,Y_2). The approximate solution then has the form

$$u^\epsilon(t,x) \;=\; u^\epsilon(t\,,\,x_1\,,\,x_2) \;\sim\; \epsilon\, U_0(t\,,\,t/\epsilon\,,\,x/\epsilon) \;=\; \epsilon\, U_0\big(t\,,\,t/\epsilon\,,\,x_1/\epsilon\,,\,x_2/\epsilon\big)\,.$$

The profile equations are

$$\mathbf{E}\,U_0 \;=\; U_0\,, \qquad \mathbf{E}\Big(\partial_t U_0 + \sum_{\mu=0}^{2}\big(A'_\mu(\underline{u})\,U_0\big)\,\frac{\partial U_0}{\partial Y_\mu}\Big)\,. \tag{11.2.1}$$

Define

$$B_\mu(V) \;:=\; \big(A'_\mu(\underline{u})\big)(V) \tag{11.2.2}$$

so that B_μ is a linear matrix valued function of the vector V. Since $A_0(u)=I$, one has $B_0=0$.

For convenience in denoting Fourier coefficients, suppress the subscript 0 and expand the leading profile,

$$U(t,Y) \;=\; U(t,Y_0,Y_1,Y_2) \;=\; \sum_{\alpha\in\mathbb{Z}\times\mathbb{Z}^2} U_\alpha(t)\,e^{i\alpha Y}\,. \tag{11.2.3}$$

Inserting this in (11.2.1), the nonlinear term is equal to

$$\sum_\mu\Big(\sum_\alpha B_\mu(U_\alpha)\;e^{i\alpha Y}\Big)\Big(\sum_\beta i\,\beta_\mu\,U_\beta\;e^{i\beta Y}\Big)\,.$$

Setting the coefficient of $e^{i\gamma Y}$ equal to zero yields

$$\frac{dU_\gamma}{dt} + \pi(\gamma) \sum_{\alpha+\beta=\gamma} \Big[\sum_{\mu=1}^{2} B_\mu(U_\alpha)(i\beta_\mu)U_\beta\Big] = 0 . \tag{11.2.4}$$

The factor $\pi(\gamma)$ is from the operator $\mathbf{E}$.

When $\alpha+\beta=\gamma$, the mapping

$$U_\alpha, U_\beta \quad \longmapsto \quad -\pi(\gamma)\Big[\sum_{\mu=1}^{2} B_\mu(U_\alpha)\,\beta_\mu\,U_\beta\Big] \tag{11.2.5}$$

defines a bilinear map

$$\ker\big(\pi(\alpha)\big) \times \ker\big(\pi(\beta)\big) \quad \longmapsto \quad \ker\big(\pi(\gamma)\big) . \tag{11.2.6}$$

In the important special case when the kernels are one dimensional, choose bases r_α homogeneous of degree zero in α. Define scalar valued $a_\alpha(t)$ and $\Gamma(\alpha,\beta)$ by

$$U_\alpha(t) = a_\alpha(t)\, r_\alpha , \qquad \Gamma(\alpha,\beta)\, r_\gamma := -\pi(\gamma)\Big[\sum_{\mu=1}^{2} B_\mu(r_\alpha)\,\beta_\mu\, r_\beta\Big] . \tag{11.2.7}$$

The coefficient $\Gamma(\alpha,\beta)$ measures one of the two quadratic terms by which the α and β components can influence the $\gamma=\alpha+\beta$ component. The second such term is $\Gamma(\beta,\alpha)$.

In (11.2.4) the contribution of these two terms yields

$$\frac{da_\gamma(t)}{dt} = \frac{i}{2} \sum_{\alpha+\beta=\gamma} \Big(\Gamma(\alpha,\beta)+\Gamma(\beta,\alpha)\Big)\, a_\alpha(t)\, a_\beta(t) , \tag{11.2.8}$$

where the *interaction coefficient*, $\Gamma(\alpha,\beta)+\Gamma(\beta,\alpha)$, is scalar valued, homogeneous of degree 1 in α,β,γ, and symmetric in α,β. Equation (11.2.8) is a three-wave equation with quadratic interactions among a possibly infinite family of oscillations.

Denote by $\mathbb{S}$ the smallest $\mathbb{Z}$-module containing the spectrum of $U(0,\cdot)$. Theorem 9.5.2 implies that

$$\forall\, y \in [0,T_*[\, \times \mathbb{R}^d , \qquad \operatorname{spec}\big(U_0(y,\,\cdot\,)\big) \;\subset\; \operatorname{Char} L(\underline{u},\partial) \;\cap\; \mathbb{S}. \tag{11.2.9}$$

Convention. *For the rest of the chapter the background velocity $\underline{v}$ is assumed to vanish.*

Thanks to gallilean invariance, this not an essential restriction. The main result of this chapter asserts that there are three initial waves at $t=0$ for Euler's equation so that nonlinear interaction generates waves in the leading profile that have velocities dense in the sphere. There are two articles by Joly, Métivier, and Rauch ([World Scientific, 1996] and [Longman, 1998]) dedicated to this problem. The present proof is shorter, yields a sharper result, and clarifies some rough spots in the 1998 article.[1] Recall that the group velocity associated to $(\alpha_0, \alpha_1, \alpha_2)$ with $\alpha_0 > 0$ is $-(\alpha_1, \alpha_2)/\|(\alpha_1, \alpha_2)\|$.

Theorem 11.2.1. *There is a real smooth local in time solution $U_0(t, Y)$ of the leading profile equation that is $2\pi \times 2\pi \times 2\pi$ periodic in (Y_0, Y_1, Y_2) and that satisfies*

$$U_\alpha(0) = 0 \quad \text{except for} \quad \alpha \in \mathbb{Z}(1,1,0) \cup \mathbb{Z}(0,1,0) \cup \mathbb{Z}(0,3,4)\,. \tag{11.2.10}$$

Define a set of illuminated group velocities $\mathcal{I} \subset \{|\xi| = 1\}$ by

$$\mathcal{I} := \left\{ \frac{(\alpha_1, \alpha_2)}{\|(\alpha_2, \alpha_3)\|} \;:\; \exists (\alpha_0, \alpha_1, \alpha_2) \in \mathbb{Z}^3, \quad \alpha_0 > 0, \quad \frac{d^2 U_\alpha(0)}{dt^2} \neq 0 \right\}. \tag{11.2.11}$$

If $|\underline{\xi}| = 1$ has rational slope not in $\{\pm 1/5, -1/3, 0, -24/7\}$, then $\underline{\xi} \in \mathcal{I}$.

Remarks. 1. Condition (11.2.10) shows that at $\{t = 0\}$ there are only three oscillating wave trains.

2. Since the spectrum of U_0 belongs to $\mathbb{Z}^3$, it follows that the group velocities present all have rational slope. The theorem asserts that virtually all rational slopes are attained. In the future there are waves traveling in directions dense in the unit circle.

3. The condition on the rational slopes is equivalent to the five inequality constraints

$$\underline{\xi}_1 + 5\underline{\xi}_2 \neq 0, \quad \underline{\xi}_1 - 5\underline{\xi}_2 \neq 0, \quad \underline{\xi}_1 + 3\underline{\xi}_2 \neq 0, \quad \underline{\xi}_2 \neq 0, \quad 24\underline{\xi}_1 + 7\underline{\xi}_2 \neq 0. \tag{11.2.12}$$

4. In [Joly, Métivier, and Rauch, World Scientific, 1996] it is shown that if $g(\theta)$ chosen below is real analytic, then the $U_\alpha(t)$ are real analytic in time. The U_α with $d^2 U_\alpha / dt^2 \neq 0$ therefore vanish at most at a discrete set of points. Thus the dense set of wave trains are simultaneously illuminated with at most a countable set of exceptional times.

Open problem. We conjecture that there is a $\delta > 0$ so that $U_\alpha(t) \neq 0$ for $0 < t < \delta$ and α in a set large enough so that the group velocities are dense in the unit circle.

11.3. Linear oscillations for the Euler equations

For the background state $\underline{v} = 0$, equation (11.1.3) simplifies to $\tau(\tau - |\xi|^2)$, so

$$\text{Char}\, L(0, \partial) \;=\; \{\tau = 0\} \cup \{\tau^2 = |\xi|^2\}$$

is the union of a horizontal plane and a light cone as for Maxwell's equations.

[1]The first article was written in 1992 but appeared in 1998. The second article was written nearly two years later but has publication date 1996 so looks like it is earlier than the first.

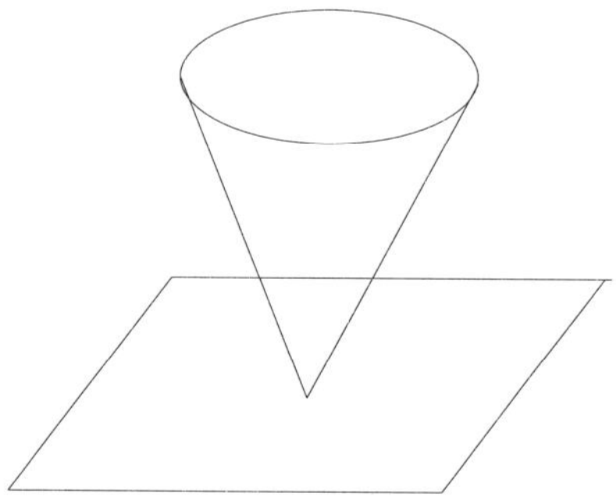

FIGURE 11.3.1. The characteristic variety of linearized Euler

In contrast to the case of electrodynamics, all the sheets of the characteristic variety are relevant. For Maxwell's equations, the divergence free constraints eliminate the plane.

Conventions. *For the rest of the chapter the background state is* $\underline{v} = 0$ *and* $L(0, \partial)$ *is abbreviated to* $L(\partial)$.

Proposition 11.3.1. *The small divisor hypothesis of §9.6 is satisfied.*

Proof. Compute

$$\det\Big(L(\tau,\xi) - \sigma I\Big) \;=\; (\tau - \sigma)\,\big((\tau-\sigma)^2 - |\xi|^2\big)\,.$$

For

$$(\tau, \xi) \;=\; (n_0, n_1, n_2) \;\in\; \mathbb{Z}^3,$$

one has

$$\det\Big(L(\tau,\xi) - \sigma I\Big) \;=\; (n_0 - \sigma)\,\big((n_0-\sigma)^2 - n_1^2 - n_2^2\big)\,.$$

Therefore σ is a nonzero eigenvalue if and only if

$$\sigma = n_0 \in \mathbb{Z}\setminus 0 \qquad \text{or} \qquad (n_0-\sigma)^2 = n_1^2 + n_2^2 \quad \text{with} \quad n_0 - \sigma \neq 0\,.$$

In the first case one has the lower bound $|\sigma| \geq 1$.

In the second case,

$$2n_0\sigma - \sigma^2 \;=\; n_0^2 - n_1^2 - n_2^2\,.$$

If the left-hand side vanishes, then $\sigma = 2n_0$ is a nonzero integer hence of modulus ≥ 1. When it is nonzero, one has

$$|2n_0\sigma - \sigma^2| \;\geq\; 1.$$

If $|\sigma| < 1/4|n_0|$, then each summand on the left is smaller than $1/2$, so the sum cannot be larger than 1. Therefore the eigenvalue satisfies

$$|\sigma| \;\geq\; \frac{1}{4|n_0|}\,,$$

proving the small divisor hypothesis. □

Theorems 9.5.1 and 9.6.2 then yield profiles, and Theorem 9.6.3 shows that the residual is infinitely small. Theorem 9.7.1 shows that these solutions are infinitely close to the exact solutions with the same initial data. We analyze the resonance relations and the profile equations in detail in order to prove Theorem 11.2.1.

It is sometimes confusing that (t,x), (τ,ξ), (v,ρ) and dual vectors to the (v,ρ) space are all objects with three components. To maintain some distinction, we use round brackets (t,x), (τ,ξ), for the first two and Dirac brackets $|v,\rho\rangle$ for the third. Dual vectors to the $|v,\rho\rangle$ are denoted, with reversed brackets $\langle\cdot\,,\cdot\,|$. The pairings of $\mathbb{R}^2_x$ and $\mathbb{R}^2_\xi$ and of $\mathbb{R}^3_{t,x}$ and $\mathbb{R}^3_{\tau,\xi}$ are indicated with a dot, e.g., $(t,x)\cdot(\tau,\xi)$. Dirac's notation $\langle\cdot|\cdot\rangle$ is used for the duality of the brackets.

Begin by computing $\ker L(\tau,\xi)$, $\operatorname{range} L(\tau,\xi)$, and the spectral projection $\pi(\tau,\xi)$ for each $(\tau,\xi)\in \operatorname{Char} L$. Since $L(\tau,\xi)$ is homogeneous of degree one with $\det L(1,0,0)\neq 0$, it suffices to consider $|\xi|=1$. For ξ fixed there are three points in the characteristic variety, $\tau=0$ and $\tau=\pm|\xi|$.

For $\tau=0$ one has

$$L(0,\xi) = \begin{pmatrix} 0 & 0 & \underline{f}\,\xi_1 \\ 0 & 0 & \underline{f}\,\xi_2 \\ \underline{\rho}\,\xi_1 & \underline{\rho}\,\xi_2 & 0 \end{pmatrix},$$

$$\ker L(0,\xi) = \mathbb{R}\,|\xi_2\,,\,-\xi_1\,,\,0\rangle\,, \tag{11.3.1}$$

$$\operatorname{range} L(0,\xi) = \operatorname{Span}\big\{\,|\xi_1,\xi_2,0\rangle\,,\,|0,0,1\rangle\,\big\}. \tag{11.3.2}$$

Since the range is orthogonal to the kernel, the spectral projection is the orthogonal projection

$$\pi(0,\xi) := |\xi_2\,,\,-\xi_1\,,\,0\rangle\langle\xi_2\,,\,-\xi_1\,,\,0|\,. \tag{11.3.3}$$

Similarly,

$$L(\pm1,\xi) = \begin{pmatrix} \pm1 & 0 & \underline{f}\xi_1 \\ 0 & \pm1 & \underline{f}\xi_2 \\ \underline{\rho}\xi_1 & \underline{\rho}\xi_2 & \pm1 \end{pmatrix},$$

$$\ker\big(L(\pm1,\xi)\big) = \mathbb{R}\,|\xi_1,\xi_2,\mp1/\underline{f}\rangle,$$

$$\operatorname{range}\big(L(\pm1,\xi)\big) = \operatorname{Span}\big\{|\pm1,0,\underline{\rho}\xi_1\rangle, |0,\pm1,\underline{\rho}\xi_2\rangle\big\},$$

$$\pi(\pm1,\xi) = \frac{1}{2}\Big|\xi_1,\xi_2,\mp1/\underline{f}\Big\rangle\Big\langle\xi_1,\xi_2,\mp1/\underline{\rho}\Big|\,.$$

These computations yield the plane wave solutions of the linearized equation

$$|\xi_2,-\xi_1,0\rangle\,F(\xi x), \qquad |\xi_1,\xi_2,\mp1/\underline{f}\rangle\,F(\pm|\xi|t+\xi x),$$

where F is an arbitrary real valued function of one variable.

The first family of waves satisfies div $v = 0$, $|\mathrm{curl}\, v| \sim |\xi|^2$, and has no variation in density. They are standing waves. Given the background state with velocity zero, this means that they are convected with the background fluid velocity. They are called *vorticity waves.*

Waves of the second family have $\mathrm{curl}\, v = 0$ and $|\mathrm{div}\ v| \sim |\xi|^2$. The group velocity associated to $\tau = \pm|\xi|$ is equal to $-\nabla_\xi(\pm|\xi|) = \mp\xi/|\xi|$. These solutions are called *acoustic waves* or *compression waves.* They can move in any direction with speed one.

This prediction of the speed of sound, $c = p'(\rho)^{1/2}$, from the static measurement of $p(\rho)$ is an early success of continuum mechanics. It is also a model of what is found in science text treatments of a nonlinear hyperbolic model, that is, linearization at constant states, and computation of plane wave solutions and group velocities for the resulting constant coefficient equation governing small perturbations.

The solution of the linear oscillatory initial value problem

$$L(\partial_t, \partial_x)w = 0\,, \qquad w(0,x) = g\, e^{i\zeta x}\,, \qquad g \in \mathbb{C}^3\,, \quad \zeta \in \mathbb{R}^2\,,$$

is given by the exact formula

$$w \;=\; \big(\pi(0, \zeta/|\zeta|)g\big)\ e^{i\zeta x} \;+\; \sum_{\pm} \big(\pi(\pm c, \zeta/|\zeta|)g\big)\ e^{\pm|\zeta|t + \zeta x}\,.$$

11.4. Resonance relations

Quadratic nonlinear interaction of oscillations $r_\alpha e^{i\alpha\cdot(t,x)}$ and $r_\beta e^{i\beta\cdot(t,x)}$ with α and β belonging to $\mathrm{Char}\, L$ produce terms in $e^{i(\alpha+\beta)\cdot(t,x)}$ that will propagate whenever $\alpha + \beta$ is characteristic.

Definition. A *(quadratic) resonance* is a linear relation $\alpha + \beta + \eta = 0$ between three nonzero elements of $\mathrm{Char}\, L$.

These are sometimes called resonances of order 3, as they involve three points of the characteristic variety. The corresponding interactions are called *three-wave interactions.* The simplest are collinear resonances, always present for homogeneous systems L, when α, β, and η are multiples of a fixed vector.

For semilinear problems one must consider linear relations among any number of characteristic covectors. Treating small amplitude oscillations for quasilinear problems yields only quadratic nonlinearities in the equations for the leading profile. This permits us to consider only triples.

Theorem 11.4.1. *Quadratic resonances for the Euler equations belong to three families:*

i. Collinear vectors satisfying $\tau^2 = |\xi|^2$,

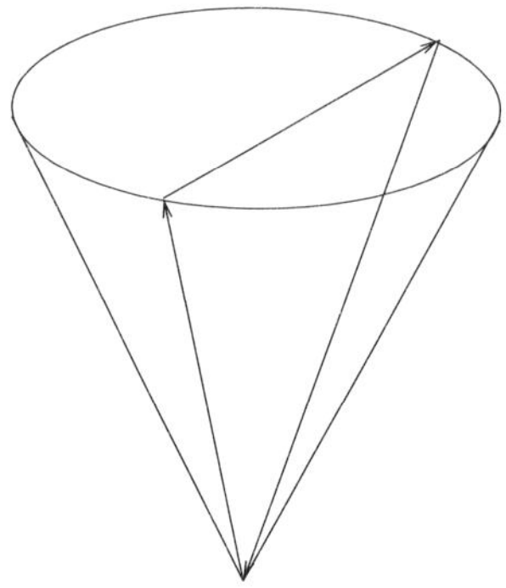

FIGURE 11.4.1. Resonance of type iii

ii. Triples α, β, η that belong to $\{\tau = 0\}$,

iii. Relations equivalent by $\mathbb{R}$-dilation, x-rotation, x-reflection, and permutation of the three covectors to

$$(\pm 1, \alpha_1, \alpha_2) + (0, 0, -2\alpha_2) + (\mp 1, -\alpha_1, \alpha_2) \;=\; 0, \qquad \alpha_1^2 + \alpha_2^2 = 1, \quad \alpha_1 \geq 0\,.$$

Proof. Seek α, β, η in $\{\tau(\tau^2 - |\xi|^2) = 0\}$ whose sum is zero. The classification above depends on counting how many of the α, β, η belong to $\{\tau = 0\}$.

If all three belong, it is case ii.

If exactly two belong, the relation $\alpha + \beta + \eta = 0$ is impossible since the τ component of the sum will equal the τ component of the covector which does not lie in $\tau = 0$.

If exactly one belongs, rotate axes so that the covector in $\tau = 0$ is parallel to $(0, 0, 1)$. The relation then is a multiple of case iii, except possibly for the sign of α_1, which can be adjusted by a reflection in x_1.

If none of the covectors belong to $\tau = 0$, we must show that the only possibility is collinear resonance.

A rotation followed by multiplication by a nonzero real reduces to the case $\alpha = (1, 1, 0)$.

Seek $\beta \in \mathrm{Char}\,(L)$ such that $\alpha + \beta \in \mathrm{Char}\,(L)$. If the τ component of β is positive then $\alpha + \beta$ belongs to the interior of the positive light cone unless α and β are collinear.

Thus it suffices to look for $\beta = (-|\xi|, \xi)$ with

$$(1, 1, 0) \;+\; (-|\xi|, \xi) \;\in\; \{\tau^2 = |\xi|^2\}\,.$$

This holds if and only if

$$(1 - |\xi|)^2 \;=\; (1 + \xi_1)^2 + \xi_2^2\,.$$

Cancelling common terms shows that this holds if and only if $-2|\xi| = 2\xi_1$, so $\xi_2 = 0$ and $\xi_1 < 0$. Thus $(1, 1, 0)$ and $(-|\xi|, \xi)$ are collinear. □

11.5. Interaction coefficients for Euler's equations

To define interaction coefficients, basis vectors $r_{(\tau,\xi)}$ for $\ker L(\tau,\xi)$ with $(\tau,\xi) \in \operatorname{Char} L$ are needed. Choose vectors that are the extensions of formulas (11.3.1), (11.3.2) homogeneous of degree zero,

$$r_{(0,\xi)} := |\xi|^{-1} \Big|\xi_2, -\xi_1, 0\Big\rangle, \tag{11.5.1}$$

$$r_{(\pm|\xi|,\xi)} := |\xi|^{-1} \Big|\xi_1, \xi_2, \mp|\xi|\underline{\rho}\Big\rangle . \tag{11.5.2}$$

Theorem 11.5.1. *Suppose that α, β, and $\alpha+\beta := \gamma$ are nonzero elements of* $\operatorname{Char} L$.

i. *If α and β are collinear elements of $\{\tau^2 = |\xi|^2\}$ with $\beta = a\alpha$, $a \neq 0, -1$, then the interaction coefficient $\Gamma(\alpha,\beta)+\Gamma(\beta,\alpha)$ is given by*

$$\Gamma(\alpha,\beta)+\Gamma(\beta,\alpha) = (\gamma_1^2+\gamma_2^2)^{1/2}\,(3+h\underline{\rho}^2)\,\operatorname{sgn}(a)/2\,, \tag{11.5.3}$$

where h is defined in (11.5.6).

ii. *If α and β belong to $\{\tau = 0\}$, the interaction coefficient is given by*

$$\Gamma(\alpha,\beta)+\Gamma(\beta,\alpha) = -(1/2)\,|\beta-\alpha|\,\sin\big(\angle(\alpha,\beta)\big)\,\cos\big(\angle(\alpha+\beta,\alpha-\beta)\big), \tag{11.5.4}$$

where $\angle(\alpha,\beta) \in \mathbb{R}/2\pi\mathbb{Z}$ denotes the angle between α and β measured in the positive sense.

iii. *If $\alpha = (\pm 1, \alpha_1, \alpha_2)$, $\alpha_1 > 0$, and $\beta = (0,0,-2\alpha_2)$ as in Theorem 1.4.1.iii, then with $\phi := \angle\big((\gamma_1,\gamma_2),(\alpha_1,\alpha_2)\big)$ the interaction coefficient is given by*

$$\Gamma(\alpha,\beta)+\Gamma(\beta,\alpha) = \cos(\phi/2)\,\cos(\phi)\,(\gamma_1^2+\gamma_2^2)^{1/2}\,\operatorname{sgn}(-\alpha_2)\,. \tag{11.5.5}$$

iiibis. *If $\alpha = (\pm 1, \alpha_1, \alpha_2)$, $\beta = (\mp 1, -\alpha_1, \alpha_2)$, $\gamma = (0,0,2\alpha_2)$, then the interaction coefficient $\Gamma(\alpha,\beta)+\Gamma(\beta,\alpha)$ vanishes.*

The exceptional case, **iiibis**, asserts that *for the creation of a vorticity wave by the interaction of two acoustic waves, the interaction coefficient vanishes.*

Proof. First compute the matrices $B_j = D_u A_j(\underline{u})$. Define the constant h by

$$h := \frac{d}{d\rho}\left(\frac{p'(\rho)}{\rho}\right)\Big|_{\rho=\underline{\rho}} . \tag{11.5.6}$$

From (11.1.4) one finds

$$(11.5.7)\qquad B_1\big(|\delta v_1,\delta v_2,\delta\rho\rangle\big) = \begin{pmatrix}\delta v_1 & 0 & h\,\delta\rho\\ 0 & \delta v_1 & 0\\ \delta\rho & 0 & \delta v_1\end{pmatrix},$$

$$(11.5.8)\qquad B_2\big(|\delta v_1,\delta v_2,\delta\rho\rangle\big) = \begin{pmatrix}\delta v_2 & 0 & 0\\ 0 & \delta v_2 & h\,\delta\rho\\ 0 & \delta\rho & \delta v_2\end{pmatrix}.$$

Case ii. When $\alpha = (0,\alpha_1,\alpha_2)$ and $\beta = (0,\beta_1,\beta_2)$ belong to $\{\tau = 0\}$, (11.5.1) shows that $\delta\rho = 0$ for r_α and r_β, so

$$B_1(r_\alpha) = |\alpha|^{-1}\alpha_2\, I,\qquad B_2(r_\alpha) = -|\alpha|^{-1}\alpha_1 I,$$

and (11.3.3) yields $\pi(\gamma) = |r_\gamma\rangle\langle r_\gamma|$. Thus

$$\begin{aligned}\Gamma(\alpha,\beta) &= \langle r_\gamma|[\beta_1 B_1(r_\alpha) + \beta_2 B_2(r_\alpha)]r_\beta\rangle = |\alpha|^{-1}(\beta_1\alpha_2 - \beta_2\alpha_1)\langle r_{\alpha+\beta}|r_\beta\rangle\\ &= |\alpha|^{-1}(\beta_1\alpha_2-\beta_2\alpha_1)|\alpha+\beta|^{-1}|\beta|^{-1}\\ &\qquad\qquad \langle \alpha_2+\beta_2, -\alpha_1-\beta_1, 0\,\big|\,\beta_2,-\beta_1,0\rangle\,.\end{aligned}$$

The last duality is equal to the scalar product $\langle\alpha+\beta|\beta\rangle$, so

$$\Gamma(\alpha,\beta) \;=\; |\alpha|^{-1}\,(\beta_1\alpha_2-\beta_2\alpha_1)\,|\alpha+\beta|^{-1}\,|\beta|^{-1}\,\langle\alpha+\beta|\beta\rangle\,.$$

Interchanging the role of α and β and summing yields

$$(11.5.9)\qquad \Gamma(\alpha,\beta)+\Gamma(\beta,\alpha) \;=\; |\alpha|^{-1}\,(\beta_1\alpha_2-\beta_2\alpha_1)\,|\alpha+\beta|^{-1}\,|\beta|^{-1}\,\langle\alpha+\beta|\beta-\alpha\rangle\,.$$

Formula (11.5.4) follows since

$$\begin{aligned}|\alpha|^{-1}\,(\beta_1\alpha_2-\beta_2\alpha_1)\,|\beta|^{-1} &= -\sin\big(\angle(\alpha,\beta)\big),\\ |\alpha+\beta|^{-1}\,|\beta-\alpha|^{-1}\,\langle\alpha+\beta|\beta-\alpha\rangle &= \cos\big(\angle(\alpha+\beta,\beta-\alpha)\big)\,.\end{aligned}$$

Case i. By homogeneity and euclidean invariance, it suffices to compute the case

$$\alpha = (\pm 1, 1, 0),\qquad \beta = a\alpha,\qquad a\in\mathbb{R}\setminus 0\,.$$

Then r_α, r_β, and $\pi(\gamma)$ are given by

$$r_\alpha \;=\; |1,0,\mp\underline{\rho}\rangle\,,\qquad r_\beta \;=\; \operatorname{sgn}(a)\; r_\alpha\,,\qquad \pi(\gamma) \;=\; \frac{1}{2}\; r_\alpha\;\langle 1,0,\mp 1/\underline{\rho}|\,.$$

Since $\beta_2 = 0$, one has

$$\begin{aligned}\Gamma(\alpha,\beta)r_\gamma = \;&\pi(\gamma)\Big[\beta_1 B_1(r_\alpha)r_\beta\Big] = a\,\mathrm{sgn}\,(a)\,\pi(\gamma)\begin{pmatrix}1 & 0 & \mp h\,\underline{\rho}\\ 0 & 1 & 0\\ \mp\underline{\rho} & 0 & 1\end{pmatrix}\\ &= |a|\,\frac{r_\alpha}{2}\Big\langle 1\,,\,0\,,\,\mp 1/\underline{\rho}\;\Big|\;1+h\,\underline{\rho}^2\,,\,0\,,\,\mp 2\underline{\rho}\Big\rangle \;=\; |a|\,\big(3+h\,\underline{\rho}^2\big)\,\frac{r_\alpha}{2}\,.\end{aligned}$$

Now $r_\alpha = \pm r_\gamma$ the sign depending on whether $\gamma = (1+a)\alpha$ is a positive or negative multiple of α, that is by sgn $(1+a)$. Thus

$$\Gamma(\alpha,\beta) \;=\; \mathrm{sgn}(1+a)\;|a|\;(3+h\,\underline{\rho}^2)/2.$$

By homogeneity the case of general α is given by

$$\begin{aligned}\Gamma(\alpha,\beta) &= (\alpha_1^2+\alpha_2^2)^{1/2}\,\mathrm{sgn}(1+a)\;|a|\;(3+h\,\underline{\rho}^2)/2\\ &= (\beta_1^2+\beta_2^2)^{1/2}\,\mathrm{sgn}(1+a)\;(3+h\,\underline{\rho}^2)/2\,,\end{aligned}$$

since $|a|\|\alpha_1,\alpha_2\| = \|\beta_1,\beta_2\|$. Noting that $\alpha = a^{-1}\beta$, the reversed coefficient is given by

$$\Gamma(\beta,\alpha) \;=\; (\alpha_1^2+\alpha_2^2)^{1/2}\,\mathrm{sgn}(1+a^{-1})\;(3+h\,\underline{\rho}^2)/2\,.$$

Adding yields

$$\begin{aligned}&\Gamma(\alpha,\beta)+\Gamma(\beta,\alpha)\\ &= \Big[(\beta_1^2+\beta_2^2)^{1/2}\,\mathrm{sgn}(1+a)+(\alpha_1^2+\alpha_2^2)^{1/2}\,\mathrm{sgn}(1+a^{-1})\Big](3+h\,\underline{\rho}^2)/2\,.\end{aligned}$$

In the three cases $a>0$, $-1<a<0$, and $a<-1$, the factor in square brackets is given by

$$\|\beta\|+\|\alpha\|\,,\qquad \|\beta\|-\|\alpha\|\,,\qquad \text{and}\qquad -\|\beta\|+\|\alpha\|\,,$$

respectively. In all cases this is equal to sgn $(a)\|\gamma\|$, proving (11.5.3).

Case iii. It is sufficient to consider α with $\alpha_1^2+\alpha_2^2 = 1$ and $a>0$. Then

$$\alpha = (\pm 1,\alpha_1,\alpha_2)\,,\quad \beta = (0,0,-2\alpha_2)\,,\quad \gamma = (\pm 1,\alpha_1,-\alpha_2)\,,$$
$$r_\alpha = |\alpha_1,\alpha_2,\mp\underline{\rho}\rangle,\quad r_\beta = |\mathrm{sgn}\,(-\alpha_2),0,0\rangle\,,\quad r_\gamma = |\alpha_1,-\alpha_2,\mp\underline{\rho}\rangle\,,$$
$$\pi(\gamma) = \frac{r_\gamma}{2}\,\langle\alpha_1,-\alpha_2,\pm 1/\underline{\rho}|\,.$$

Since $\beta_1 = 0$,

$$\begin{aligned}\Gamma(\alpha,\beta)r_\gamma &= \pi(\gamma)\,\Big[\beta_2 B_2(r_\alpha)r_\beta\Big]\\ &= \pi(\gamma)\left[-2\alpha_2\begin{pmatrix}\alpha_2 & 0 & 0\\ 0 & \alpha_2 & \mp h\underline{\rho}\\ 0 & \mp\underline{\rho} & \alpha_2\end{pmatrix}\begin{pmatrix}\mathrm{sgn}\,(-\alpha_2)\\ 0\\ 0\end{pmatrix}\right]\\ &= \frac{r_\gamma}{2}\Big\langle\alpha_1\,,-\alpha_2\,,\mp 1/\underline{\rho}\;\Big|\;-2\alpha_2^2\,\mathrm{sgn}(-\alpha_2)\,,0\,,0\Big\rangle\,.\end{aligned}$$

Therefore $\Gamma(\alpha\,,\beta) = -\alpha_1\alpha_2^2\,\mathrm{sgn}(-a)$.

To compute the coefficient $\Gamma(\beta,\alpha)$, note first that $B_2(r_\beta) = 0$ and $B_1(r_\beta) = \mathrm{sgn}\,(-\alpha_2)I$, since for r_β, $\delta v_2 = \delta\rho = 0$ and $\delta v_1 = \mathrm{sgn}\,(-\alpha_2)$.

Therefore

$$\Gamma(\beta,\alpha)r_\gamma = \pi(\gamma)\Big[\alpha_1 B_1(r_\beta)r_\alpha\Big] = \pi(\gamma)\Big[\alpha_1 \mathrm{sgn}\,(-\alpha_2)r_\alpha\Big]$$
$$= \frac{\alpha_1}{2}\,\mathrm{sgn}\,(-\alpha_2)\,r_\gamma\,\Big\langle \alpha_1\,,\,-\alpha_2\,,\,\mp 1/\underline{\rho}\;\Big|\;\alpha_1\,,\,\alpha_2\,,\,\mp\underline{\rho}\Big\rangle.$$

Therefore,

$$\Gamma(\beta,\alpha) \;=\; \frac{\alpha_1}{2}\mathrm{sgn}(-\alpha_2)\,(\alpha_1^2-\alpha_2^2+1) = \alpha_1\,\mathrm{sgn}\,(-\alpha_2)\;\alpha_1^2\,.$$

Adding the previous results yields

$$\Gamma(\alpha,\beta)+\Gamma(\beta,\alpha) = \alpha_1\,(\alpha_1^2-\alpha_2^2)\,\mathrm{sgn}(-\alpha_2)$$
$$= \cos(\phi/2)\,\big(\cos^2(\phi/2)-\sin^2(\phi/2)\big)\,\mathrm{sgn}(-\alpha_2),$$

since $(\alpha_1,\alpha_2)=(\cos(\phi/2),\sin(\phi/2))$. Formula (11.5.5) for the case $\alpha_1^2+\alpha_2^2 = 1$ follows. The general case follows by homogeneity.

Case iiibis. When $|\alpha_1,\alpha_2| = 1$, one has

$$r_\alpha = |\alpha_1,\alpha_2,\mp\underline{\rho}\rangle\,, \qquad r_\beta = |-\alpha_1,\alpha_2,\mp\underline{\rho}\rangle\,,$$
$$r_\gamma = \mathrm{sgn}(\alpha_2)\,|1,0,0\rangle\,, \qquad \pi(\gamma) = |1,0,0\rangle\langle 1,0,0|\,.$$

By definition, $\mathrm{sgn}(\alpha_2)\,\Gamma(\alpha,\beta)$ is equal to the first component of the vector

$$\left[-\alpha_1\begin{pmatrix}\alpha_1 & 0 & \mp h\underline{\rho}\\ 0 & \alpha_1 & 0\\ \mp\underline{\rho} & 0 & \alpha_1\end{pmatrix} + \alpha_2\begin{pmatrix}\alpha_2 & 0 & 0\\ 0 & \alpha_2 & \mp h\,\underline{\rho}\\ 0 & \mp\underline{\rho} & \alpha_2\end{pmatrix}\right]\begin{pmatrix}-\alpha_1\\ \alpha_2\\ \mp\underline{\rho}\end{pmatrix}.$$

Therefore,

$$\Gamma(\alpha,\beta) = -\alpha_1\,\mathrm{sgn}(\alpha_2)\,\big(-\alpha_1^2+\alpha_2^2+h\,\underline{\rho}\big)\,.$$

To compute the coefficient $\Gamma(\beta,\alpha)$, it suffices to change the sign of α_1 in the above computation which simply changes the sign of the result. Thus $\Gamma(\beta,\alpha) = -\Gamma(\alpha,\beta)$ proving **iiibis**. □

11.6. Dense oscillations for the Euler equations

11.6.1. The algebraic/geometric part. The characteristic variety is given by the equation

$$\tau\big(\tau^2-(\xi_1^2+\xi_2^2)\big) = 0.$$

Definition. *In the* (τ,ξ) *space denote by* Λ *the lattice of integer linear combinations of the characteristic covectors*

$$(1,1,0)\,,\quad (0,1,0)\,,\quad \textit{and}\quad (0,3,4)\,. \tag{11.6.1}$$

The next result is stronger than the corresponding result in [Joly, Métivier, and Rauch, Longman, 1998] and the proof is much simpler.

Theorem 11.6.1. *For every rational r there is a point*

$$(\tau, \xi) \in \Lambda \cap \{\tau = |\xi| > 0\},$$

with $\xi_1/\xi_2 = r$.

The group velocity associated to (τ, ξ) is equal to $-\xi/|\xi|$, so this asserts that there are wave numbers in $\Lambda \cap \{\tau = |\xi| > 0\}$ with group velocities with arbitrary rational slope.

Proof. The points of Λ are the integer linear combinations

(11.6.2)

$$(\tau, \xi) = n_1(1,1,0) + n_2(0,1,0) + n_3(0,3,4) = \Big(n_1\,,\, n_1 + n_2 + 3n_3\,,\, 4n_3\Big)\,.$$

This (τ, ξ) belongs to $\{\tau = |\xi| > 0\}$ if and only if

$$n_1^2 = (n_1 + n_2 + 3n_3)^2 + (4n_3)^2 \qquad \text{and} \qquad n_1 > 0\,. \tag{11.6.3}$$

Dividing by n_1^2 and setting

$$q_2 := \frac{n_2}{n_1}, \qquad q_3 := \frac{n_3}{n_1},$$

shows that $q_j \in \mathbb{Q}$ satisfy

$$1 = (1 + q_2 + 3q_3)^2 + (4q_3)^2\,. \tag{11.6.4}$$

For any rational s, the line $q_2 = sq_3$ intersects the ellipse (11.6.4) when

$$1 = (1 + (s+3)q_3)^2 + (4q_3)^2, \quad \text{equivalently}$$

$$q_3\Big((4^2 + (s+3)^3)q_3 + 2(s+3)\Big) = 0\,.$$

Therefore, $q_3 = -2(s+3)/(4^2 + (s+3)^2)$ and $q_2 = sq_3$ is a solution.

Multiplying by the greatest common multiple of the denominators of the q_j gives an integer solution of (11.6.3). Multiplying by ± 1 gives the desired solution with $n_1 > 0$. □

11.6.2. Construction of the profiles. We construct a solution of the profile equation (11.2.1) satisfying the conditions of Theorem 11.2.1. Introduce $g \in C^\infty(S^1)$ all of whose Fourier coefficients are nonzero,

$$g(\theta) := \sum_{n\in\mathbb{Z}} g_n\, e^{in\theta}\,, \qquad g_n \neq 0\,, \tag{11.6.5}$$

with g_n rapidly decreasing.

Let $U^0(T, X_1, X_2)$ be the solution of the linearized Euler equation consisting of three plane waves and given by

$$\begin{aligned} U^0(T,X) \;&:=\; |1,0,-1/\underline{f}\rangle\, g\big((1,1,0)\cdot(T,X_1,X_2)\big) \\ &\quad + |0,1,0\rangle\, g\big((0,1,0)\cdot(T,X_1,X_2)\big) \\ &\quad + |4,-3,0\rangle\, g\big((0,3,4)\cdot(T,X_1,X_2)\big)\,. \end{aligned}$$

The spectrum of $U^0(T, X)$ is exactly equal to the integer multiples of the covectors in (11.6.1).

Theorem 9.5.1 implies that there is a unique local solution

$$U = \sum U_\alpha(t) e^{i\alpha\cdot(T,X)}$$

of the profile equations (11.2.1) (equivalently (11.2.8)) with initial value $U(0, T, X) = U^0(T, X)$. Theorem 9.5.2 shows that

$$\operatorname{spec} U(t) \subset \Lambda \cap \operatorname{Char} L(\partial).$$

As in §11.1, define $a_\alpha(t)$ by $U_\alpha(t) = a_\alpha(t) r_\alpha$.

The next result also improves on [Joly, Métivier, and Rauch, Longman, 1998]. The proof of Theorem 11.2.1 is finished afterward.

Theorem 11.6.2. *If*

$$(\tau,\xi) \;=\; n_1(1,1,0) + n_2(0,1,0) + n_3(0,3,4) \;\in\; \Lambda \cap \{\tau = |\xi| > 0\}$$

with

$$n_1 n_2 n_3 \neq 0, \qquad \xi_1 + 5\,\xi_2 \neq 0, \qquad \xi_1 - 5\,\xi_2 \neq 0, \qquad \textit{and} \qquad \xi_1 + 3\,\xi_2 \neq 0\,,$$

then

$$U_{(\tau,\xi)}(0) = \frac{dU_{(\tau,\xi)}(0)}{dt} = 0 \qquad \textit{and} \qquad \frac{d^2 U_{(\tau,\xi)}(0)}{dt^2} \neq 0\,.$$

Proof. With $U_\alpha(t) = a_\alpha(t)\, r_\alpha$ as above, the dynamics is given by

$$\frac{da_\delta}{dt} \;=\; \sum_{\alpha+\beta=\delta} K(\alpha,\beta)\, a_\alpha\, a_\beta\,, \qquad K(\alpha,\beta) \;=\; \frac{i}{2}\big(\Gamma(\alpha,\beta) + \Gamma(\beta,\alpha)\big)\,.$$

The function K is symmetric in α, β and $K(\alpha, \beta) = 0$ if at least one of the three vectors α, β, and $\alpha + \beta$ is not characteristic for $L(\partial)$.

Consider

$$\delta \;=\; (\tau,\xi) \;=\; n_1(1,1,0) + n_2((0,1,0) + n_3(0,3,4) := n_1\alpha + n_2\beta + n_3\gamma, \qquad \text{with} \quad n_1 n_2 n_3 \neq 0\,.$$

This is the unique representation of δ as a linear combination of α, β, γ. Thus δ is never equal to a linear combination of one (resp., two) of the vectors. This implies that $a_\delta(0) = 0$ (resp., $da_\delta(0)/dt = 0$).

Compute

$$\begin{aligned}
\frac{d^2 a_\delta}{dt^2} &= \sum_{\alpha+\beta=\delta} K(\alpha,\beta)\Big((\partial_t a_\alpha)a_\beta + a_\alpha \partial_t a_\beta\Big) \\
&= \sum_{\alpha+\beta=\delta} K(\alpha,\beta)\Big(a_\beta \sum_{\mu+\nu=\alpha} K(\mu,\nu)a_\mu\, a_\nu + a_\alpha \sum_{\kappa+\lambda=\beta} K(\kappa,\lambda)a_\kappa\, a_\lambda\Big) \\
&= \sum_{\substack{\alpha+\beta=\delta \\ \mu+\nu=\alpha}} K(\alpha,\beta)\, K(\mu,\nu)\ a_\beta a_\mu\, a_\nu \ + \sum_{\substack{\alpha+\beta=\delta \\ \kappa+\lambda=\beta}} K(\alpha,\beta)\, K(\kappa,\lambda)\ a_\alpha\, a_\kappa\, a_\lambda \\
&= \sum_{\beta+\mu+\nu=\delta} K(\mu+\nu,\beta)\, K(\mu,\nu)\, a_\beta\, a_\mu\, a_\nu \\
&\qquad + \sum_{\alpha+\kappa+\lambda=\delta} K(\alpha,\kappa+\lambda)\, K(\kappa,\lambda)\, a_\alpha\, a_\kappa\, a_\lambda\,.
\end{aligned}$$

Since K is symmetric, the two sums are equal. Therefore

$$\frac{d^2 a_\delta}{dt^2} = \sum_{\alpha+\mu+\nu=\delta} J(\alpha,\mu,\nu)\, a_\alpha\, a_\mu\, a_\nu\,,$$

with

$$J(\alpha,\mu,\nu) := 2\, K(\alpha\,,\,\mu+\nu)\, K(\mu\,,\,\nu)\,.$$

The interaction coefficient, $J(\alpha,\mu,\nu)$, is homogeneous of degree two in α,μ,ν and symmetric in μ,ν.

The derivative $d^2a_\delta(0)/dt^2$ is the sum of six terms corresponding to the permutations of the triple α,β,γ. At $t=0$ each of the six terms is the product of

$$a_{n_1(1,1,0)}\, a_{n_2(0,1,0)}\, a_{n_3(0,3,4)}\big|_{t=0} = g_{n_1}\, g_{n_2}\, g_{n_3}$$

and an interaction coefficient J. A typical interaction coefficient is

$$\begin{aligned}
(11.6.6)\quad & J\Big(n_1(1,1,0)\,,\, n_2(0,1,0)\,,\, n_3(0,3,4)\Big) \\
&\qquad = n_1^2\, J\Big((1,1,0)\,,\, q_2(0,1,0)\,,\, q_3(0,3,4)\Big)\,,
\end{aligned}$$

with q_j from (11.6.3). The other five terms come from permuting the arguments of J. We need to compute the sum of the six terms from the permutations of the arguments.

Fortunately, four of the six summands vanish, and the other two are equal. Since $K(\kappa,\lambda)$ vanishes if λ is not characteristic, J vanishes whenever the sum of its last two arguments is noncharacteristic. Of the six summands that leaves only

$$(11.6.7)\quad J(\alpha,\beta,\gamma)+J(\alpha,\gamma,\beta) = 2\, J(\alpha,\beta,\gamma) = 2\, K(\alpha,\beta+\gamma)\, K(\beta,\gamma).$$

Formula (11.5.4) implies that

$$(11.6.8)\qquad K(\beta,\gamma)=0 \quad\Longleftrightarrow\quad \|\beta\|^2=\|\alpha\|^2 \quad\Longleftrightarrow\quad q_2 \;=\; \pm\, 5q_3\,.$$

Formula (11.5.5) implies that

$$(11.6.9)\ K(\alpha,\beta+\gamma)=0 \quad\Longleftrightarrow\quad (1,0)\perp(\beta+\gamma) \quad\Longleftrightarrow\quad q_2+3\,q_3=0\,.$$

Conditions (11.6.8)–(11.6.9) show that the three inequality hypotheses on ξ imply that the interaction coefficient is not equal to zero. This completes the proof of the theorem. □

Proof of Theorem 11.2.1. Suppose that $(\underline{\xi}_1,\underline{\xi}_2)$ is a point on the unit circle with rational slope. Theorem 11.6.1 implies that there is a point $\alpha=\sigma(1,\underline{\xi})\in\Lambda\cap\{\tau=|\xi|>0\}$. Expand

$$(11.6.10)\qquad \alpha \;=\; n_1(1,1,0)+n_2(0,1,0)+n_3(0,3,4)\,,\qquad n_1=\sigma>0\,.$$

The first three inequalities in (11.2.12) are identical to the three inequality constraints in Theorem 11.6.2. To apply Theorem 11.6.2 it remains to verify that $n_1n_2n_3\neq 0$.

That $n_1>0$ is noted in (11.6.10). If $n_3=0$, then $\underline{\xi}_2=0$. This is excluded by the fourth inequality in (11.2.12). Finally, if $n_2=0$, then (11.6.10) shows that

$$(1,1,0)+c(0,3,4),\qquad c:=n_3/n_1$$

is characteristic for L. Equivalently

$$(11.6.11)\qquad \|(1,0)+c(3,4)\|^2=1\,.$$

Compute $\|(1,0)+c(3,4)\|^2-1=c(25c+6)$. Equation (11.6.11) has two roots, $c=0$ and $c=-6/25$. The root $c=0$ is excluded by the condition $\underline{\xi}_2\neq 0$. The root $c=-6/25$ yields

$$(1,0)-\frac{6}{25}\,(3,4)\;=\;\frac{1}{25}\,(7,-24)\,.$$

That is ruled out by the fifth linear inequality in (11.2.12). Thus, $n_2\neq 0$.

Since the hypotheses of Theorem 11.2.1 imply that $n_1n_2n_3\neq 0$, Theorem 11.6.2 implies that $U_\alpha(0)=dU_\alpha(0)/dt=0$ and $d^2U_\alpha(0)/dt^2\neq 0$. This completes the proof. □

To readers reaching this point, I thank you for your interest and energy. I hope that you derive great satisfaction from your further studies of mathematics, hopefully including hyperbolic partial differential equations and/or multiscale asymptotics.

Bibliography

M. Alber, G. Luther, J. Marsden and J. Robbins, *Geometric phases, reduction and Lie–Poisson structure for the resonant three wave interaction*, Physica D **123** (1998), 271–290.

S. Alinhac, *Équations differentielles: étude asymptotique; applications aux équations aux dérivées partielles* (French). [Differential equations: asymptotic study; applications to partial differential equations.] Publications Mathématiques d'Orsay, 80, 4. Université de Paris–Sud, Département de Mathématique, Orsay, 1980.

S. Alinhac, *Blowup for Nonlinear Hyperbolic Equations*, Birkhäuser, 1995.

S. Alinhac and P. Gérard, *Pseudo-differential Operators and the Nash–Moser Theorem.* (Translated from the 1991 French original by Stephen S. Wilson.) Graduate Studies in Mathematics, 82. American Mathematical Society, Providence, RI, 2007.

S. Banach, *Théorie des Operations Linéaires*, reprint, Chelsea Publishing Co., New York, 1955.

S. Basu, R. Pollack and M.-F. Roy, *Algorithms in Real Algebraic Geometry, Second ed.* Algorithms and Computation in Mathematics, **10**, Springer-Verlag, Berlin, 2006.

M. Beals, *Propagation and Interaction of Singularities in Nonlinear Hyperbolic Problems*, Birkhäuser, Boston, 1989.

R. Benedetti and J.-L. Risler, *Real Algebraic and Semialgebraic Sets*, Actualités Mathématiques, Hermann, 1990.

S. Benzoni-Gavage and D. Serre, *Multidimensional Hyperbolic Partial Differential Equations. First-order Systems and Applications.* Oxford Mathematical Monographcs. The Clarendon Press, Oxford University Press, Oxford, 2007.

N. Bloembergen, *Nonlinear Optics*, Benjamin, New York, 1964.

J.-M. Bony, *Calcul symbolique et propagation des singularités pour les équations aux dérivées nonlinéaires*, Ann. Sci. Ecole Norm. Sup. **14** (1981), 209–246.

P. Brenner, *The Cauchy problem for symmetric hyperbolic systems in L_p*, Math. Scand. **19** (1966), 27–37.

P. Brenner and W. von Wahl, *Global classical solutions of nonlinear wave equations*, Math. Z. **176** (1981), 87–121.

A. Bressan, *Hyperbolic Systems of Conservation Laws. The One-dimensional Cauchy Problem.* Oxford Lecture Series in Mathematics and Its Applications, 20. Oxford University Press, Oxford, 2000.

M. D. Bronšteĭn, *The Cauchy problem for hyperbolic operators with characteristics of variable multiplicity.* (Russian) Trudy Moskov. Mat. Obshch. **41** (1980), 83–99.

A. Cauchy, *Mémoire sur l'intégration des équations differentielles.* Exercises d'analyse et de physique mathématiques, vol. 1, Paris, 1840.

J. Chazarain and A. Piriou, *Introduction to the Theory of Linear Partial Differential Equations.* (Translated from the French.) Studies in Mathematics and Its Applications, 14. North-Holland Publishing Co., Amsterdam, 1982.

J.-Y. Chemin, B. Desjardins, I. Gallagher and E. Grenier, *Mathematical Geophysics. An Introduction to Rotating Fluids and the Navier–Stokes Equations.* Oxford Lecture Series in Mathematics and Its Applications, 32. The Clarendon Press, Oxford University Press, Oxford, 2006.

Y. Choquet-Bruhat, *Ondes asymptotiques et approchées pour les systèmes d'équations aux dérivées partielles nonlinéaires*, J. Math. Pures. Appl. **48** (1969), 117–158.

E. Conway and J. Smoller, *Global solutions of the Cauchy problem for quasi-linear first-order equations in several space variables.* Comm. Pure Appl. Math. **19** (1966), 95–105.

R. Courant and D. Hilbert, *Methods of Mathematical Physics. Vol. I.* Interscience Publishers, Inc., New York, 1953.

R. Courant and D. Hilbert, *Methods of Mathematical Physics. Vol. II.* (Vol. II by R. Courant.) Interscience Publishers (a division of John Wiley & Sons), New York, 1962.

R. Courant, K. O. Friedrichs, and H. Lewy, *Über dei partiellen differenzengleichungen der physik*, Math. Ann. **100** (1928–1929), 32–74.

C. M. Dafermos, *Hyperbolic Conservation Laws in Continuum Physics*, Third edition. Grundlehren der Mathematischen Wissenschaften [Fundamental Principles of Mathematical Sciences], 325. Springer-Verlag, Berlin, 2010.

P. Donnat, "Quelque contributions mathématiques in optique nonlinéaire", Ph.D. Thesis, Ecole Polytéchnique, Paris, 1994.

P. Donnat, J. L. Joly, G. Métivier and J. Rauch, *Diffractive nonlinear geometric optics.* Séminaire Equations aux Dérivées Partielles, Ecole Polytéchnique, Palaiseau, 1995–1996.

P. Donnat and J. Rauch, *Modelling the dispersion of light*, Singularities and Oscillations, (J. Rauch and M. Taylor, eds.), 17–35, IMA Volumes in Mathematics and Its Applications, 91, Springer-Verlag, New York, 1997.

P. Donnat and J. Rauch, *Dispersive nonlinear geometric optics*, J. Math. Phys. **38(3)** (1997), 1484–1523.

J. J. Duistermaat, *Fourier Integral Operators.* Courant Institute of Mathematical Sciences, New York University, New York, 1973.

L. C. Evans, *Partial Differential Equations*, Graduate Studies in Mathematics, 19. American Mathematical Society, Providence, RI, 1998

L. C. Evans and M. Zworski, *Semiclassical Aanlysis*, book manuscript available at Zworski's home page `http://math.berkeley.edu/~zworski/`, 2010.

R. Feynman, R. Leighton and M. Sands, *The Feynman Lectures on Physics*, Addison-Wesley, Redwood City, CA, 1964.

K. O. Friedrichs, *Symmetric hyperbolic linear differential equations*, Comm. Pure Appl. Math. **7** (1954), 345–392.

A. Gabor, *Remarks on the wave front set of distribution*, Trans. AMS **170** (1972), 239–244.

P. Garabedian, *Partial Differential Equations*, John Wiley & Sons, Inc., New York-London-Sydney, 1964.

L. Gårding, *Linear hyperbolic partial differential equations with constant coefficients.* Acta Math. **85** (1951), 1–62.

V. Georgiev, S. Lucente and G. Zillotti, *Decay estimates for hyperbolic systems*, Hokkaido Math. J. **33** (2004), 83–113.

D. Gérard-Varet, *A geometric optics type approach to fluid boundary layers*, Comm. Part. Diff. Eq., **28** (2003), 1605–1626.

J. Ginibre and G. Velo, *Generalized Strichartz inequalities for the wave equation.* J. Funct. Anal. **133** (1995), 50–68.

J. Ginibre and G. Velo, *The global Cauchy problem for the nonlinear Klein-Gordon equation.* Math. Z. **189** (1985), 487–505.

J. Glimm and P. Lax, *Decay of solutions of systems of nonlinear hyperbolic conservation laws*, Memoirs of the American Mathematical Society, No. 101, American Mathematical Society, Providence, RI, 1970.

E. Grenier and O. Guès, *Boundary layers for viscous perturbations of noncharacteristic quasilinear hyperbolic problems*, J. Diff. Eq. **143** (1998) 110–146.

O. Guès, *Ondes multidimensionelles epsilon stratifiées et oscillations*, Duke Math. J. **68** (1992), 401–446.

O. Guès, *Dévelopment asymptotique de solutions exactes de systèmes hyperboliques quasilinéaires*, Asympt. Anal. **6** (1993), 241–269.

O. Guès and J. Rauch, *Hyperbolic L^p multipliers are translations*, Comm. PDE **31** (2006), 431–443.

V. Guillemin and A. Pollack, *Differential Topology*, Prentice Hall, Inc., Englewood Cliffs, NJ, 1974.

A. Haar, *Über eindeutikeit un analytizität der lösungen parieler differenzialgleichen*, Atti del Congr. Intern. dei Mat., Bologna, vol. **3** (1928), 5–10.

J. Hadamard, *Lectures on the Cauchy Problem in Linear Partial Differential Equations*, reissue of 1923 original, Dover Publ. New York, 2003.

L. Hörmander, *Fourier integral operators I*, Acta Math. **127** (1971), 79–183.

L. Hörmander, *The Analysis of Linear Partial Differential Operators vols. I, II*, Springer-Verlag, Berlin, 1983.

L. Hörmander, *Lectures on Nonlinear Hyperbolic Differential Equations*, Springer-Verlag, Berlin-New York-Heidelberg, 1997.

J. Hunter and J. Keller, *Weakly nonlinear high frequency waves*, Comm. Pure Appl. Math. **36** (1983) no. 5, 547–569.

J. Hunter and J. Keller, *Caustics of nonlinear waves.* Wave Motion **9** (1987) no. 5, 429–443.

J. Hunter, A. Majda and R. Rosales, *Resonantly interacting weakly nonlinear hyperbolic waves II*, Stud. Appl. Math. **75** (1986), 187–226.

N. Iwasaki, *Local decay of solutions for symmetric hyperbolic systems with dissipative and coercive boundary conditions in exterior domains*, Publ. RIMS Kyoto Univ. **5** (1969), 193–218.

H. K. Jenssen, *Blowup for systems of conservation laws*, SIAM J. Math. Anal. **31** (2000), no. 4, 894–908.

F. John, *Partial Differential Equations, 4th ed.*, Springer-Verlag, New York, 1982.

J. L. Joly, *Sur la propagation des oscillations par un système hyperbolique semi-linéaire en dimension* 1 *d'espace.* (French) [On the propagation of oscillations by a hyperbolic semilinear system in one space dimension.] C. R. Acad. Sci. Paris Sér. I Math. **296** (1983), no. 15, 669–672.

J. L. Joly, G. Métivier and J. Rauch, *Formal and rigorous nonlinear high frequency hyperbolic waves*, in Nonlinear Hyperbolic Waves and Field Theory, (M. K. Murthy and S. Spagnolo eds.), Pitman Research Notes in Math, **253** (1992), 121–143.

J. L. Joly, G. Métivier and J. Rauch, *Resonant one dimensional nonlinear geometric optics*, J. Funct. Anal. **114** (1993), 106–231.

J. L. Joly, G. Métivier and J. Rauch, *Generic rigorous asymptotic expansions for weakly nonlinear multidimensional oscillatory waves*, Duke Math. J. **70** (1993), 373–404.

J. L. Joly, G. Métivier and J. Rauch, *Coherent nonlinear waves and the Wiener algebra*, Ann. Inst. Fourier, **44** (1994), 167–196.

J. L. Joly, G. Métivier and J. Rauch, *A nonlinear instability for* 3×3 *systems of conservation laws*, Comm. Math. Phys. **162** (1994), 47–59.

J. L. Joly, G. Métivier and J. Rauch, Dense oscillations for the compressible 2-d Euler equations. *Nonlinear partial differential equations and their applications. Collége de France Seminar, Vol. XIII (Paris, 1994/1996)*, 134–166, Pitman Res. Notes Math. Ser., 391, Langman, Harlow, 1998.

J. L. Joly, G. Métivier and J. Rauch, Dense oscillations for the Euler equations II, *Hyperbolic problems: theory, numerics, applications*, (Stony Brook, NY, 1994), 425–430, World Sci. Publ., River Edge, NJ, 1996.

J. L. Joly, G. Métivier and J. Rauch, *Diffractive nonlinear geometric optics with rectification*, Indiana U. Math. J. **47** (1998), 1167–1242.

J. L. Joly, G. Métivier and J. Rauch, *Dense oscillations for the compressible two dimensional Euler equations.* Nonlinear Partial Differential Equations and their Applications, College de France Seminar 1992–1993, (D. Cioranescu and J. L. Lions, eds.), Pitman Research Notes in Math. **391**, Longman Publs. 1998.

J. L. Joly, G. Métivier and J. Rauch, *Hyperbolic domains of dependence and Hamilton–Jacobi equations*, J. Hyp. Diff. Eq., **2** (2005), 713–744.

J. L. Joly and J. Rauch, *Justification of multidimensional single phase semilinear geometric optics*, Trans. AMS **330** (1992), 599–623.

K. Kajitani and A. Satoh, *Time decay estimates for linear symmetric hyperbolic systems with variable coefficients and its applications.* Phase Space Analysis of Partial Differential Equations, (F. Colombini and L. Pernazza, eds.), Centro De Giorgi, Scuola Normale Superiore, Pisa, 2004.

M. Keel and T. Tao, *Endpoint Strichartz estimates*, Amer. J. Math. **120** (1998), 955–980.

J. Keller, *On solutions of nonlinear wave equations*, Comm. Pure Appl. Math. **10** (1957), 523–530.

S. Klainerman, *On the work and legacy of Fritz John, 1934–1991. Dedicated to the memory of Fritz John.* Comm. Pure Appl. Math. **51** (1998), no. 9-10, 991–1017.

H. O. Kreiss, *Über sachgemässe Cauchyprobleme*, Math. Scand. **13** (1963), 109–128.

S. Kruzkhov, *First order quasilinear equations with several space variables*, Math. USSR, Sbornik **10** (1970), 217–243.

P. D. Lax, *On Cauchy's problem for hyperbolic equations and the differentiability of solutions of elliptic equations*, Comm. Pure Appl. Math. **8** (1955), 615–633.

P. D. Lax, *Hyperbolic systems of conservation laws II*, Comm. Pure. Appl. Math. **10** (1957), 537–566.

P. D. Lax, *Asymptotic solutions of oscillatory initial value problems*, Duke Math. J. **24** (1957), 627–646.

P. D. Lax, *Lectures on Hyperbolic Partial Differential Equations*, Stanford University Lecture Notes, 1963.

P. D. Lax, *Shock waves and entropy*, Contributions to Nonlinear Functional Analysis, (E. Zarantonello, ed.), Academic Press, New York, 1971.

P. D. Lax, *Hyperbolic Partial Differential Equations*, Courant Lecture Notes, 14, American Mathematical Society, Providence, RI, 2006.

G. Lebeau, *Nonlinear optic and supercritical wave equation, Hommage à Pascal Laubin.* Bull. Soc. Roy. Sci. Liège **70** (2001), no. 4-6, 267–306.

G. Lebeau, *Perte de régularité pour les équations d'ondes sur-critiques*, Bull. Soc. Math. France **133** (2005) no. 1, 145–157.

J. Leray, *Lectures on Hyperbolic Partial Differential Equations*, Institute for Advanced Study, Princeton, 1953.

H. Levy, *An example of a smooth linear partial differential equation without solution*, Ann. of Math. **66** (1957), 155–158.

S. Lucente and G. Zillotti, *A decay estimate for a class of hyperbolic pseudo-differential equations*, Math. App. **10** (1999), 173–190, Atti, Acc. Naz. Linccei Cl. Sci. Fis. Mat. Natur. Rend. Lincei (9).

D. Ludwig, *Exact and asymptotic solutions of the Cauchy problem*, Comm. Pure Appl. Math. **13** (1960), 473–508.

D. Ludwig, *Conical refraction in crystal optics and hydromagnetics*, Comm. Pure Appl. Math. **14** (1961), 113–124.

R. K. Luneburg, *The Mathematical Theory of Optics*, Brown Univ. Press, Providence, RI, 1944.

A. Majda, *Compressible Fluid Flow and Systems of Conservation Laws in Several Space Variables.* Applied Mathematical Sciences, 53. Springer-Verlag, New York, 1984.

A. Majda, *Nonlinear geometric optics for hyperbolic systems of conservation laws*, Oscillations Theory, Computation, and Methods of Compensated Compactness (C. Dafermos, J. Ericksen, D. Kinderlehrer, and I. Muller, eds.), IMA Volumes on Mathematics and its Applications vol. 2 pp. 115–165, Springer-Verlag, New York, 1986.

A. Majda and R. Rosales, *Resonantly interacting weakly nonlinear hyperbolic waves*, Stud. Appl. Math. **71** (1986), 149–179.

A. Majda, R. Rosales, and M. Schonbek, *A canonical system of integro-differential equations in nonlinear acoustics*, Stud. Appl. Math. **79** (1988), 205–262.

D. McLaughlin, G. Papanicolaou and L. Tartar, *Weak limits of semilinear hyperbolic systems with oscillating data*, Macroscopic Modeling of Turbulent Flows, Lecture Notes in Physics, 230, Springer-Verlag, 1985, pp. 277–298.

G. Métivier, *The Mathematics of Nonlinear Optics*, Handbook of differential equations: evolutionary equations. vol. **V**, Elsevier/North-Holland, Amsterdam, 2009, pp. 169–313.

G. Métivier, *Problèmes de Cauchy et ondes non linéaires.* (French) [Cauchy problems and nonlinear waves] Journées "Équations aux dérivées partielles" (Saint Jean de Monts, 1986), No. I, École Polytéchnique, Palaiseau, 1986.

G. Métivier, *Para-differential Calculus and Applications to the Cauchy problem for Nonlinear Systems.* Centro di Ricerca Matematica Ennio De Giorgi (CRM) Series, 5. Edizioni della Normale, Pisa, 2008.

G. Métivier and J. Rauch, *Real and complex regularity are equivalent for hyperbolic characteristic varieties*, Differential Integral Equations, **16** (2003), 993–999.

G. Métiver and J. Rauch, *Invariance and stability of the profile equations of geometric optics*, Acta Math. Sci. **31** B6 (2011), 2141–2158.

Y. Meyer, *Remarque sur un théoreme de J.-M. Bony*, Rend. Circ. Mat. Palermo, II. Ser. 1981, Suppl. **1** (1981) 1–20.

S. Mizohata, *Lectures on the Cauchy Problem*, Notes by M. K. V. Murthy and B. V. Singbal. Tata Institute of Fundamental Research Lectures on Mathematics and Physics, No. 35, Tata Institute of Fundamental Research, Bombay, 1965.

T. S. Motzkin and O. Tausky, *Pairs of matrices with property L*, Trans. AMS **73** (1952), 108–114.

G. Peano, *Sull'inegrabilità della equazioni differenziali di primo ordine*, Atti. Acad. Torino, **21A** (1886), 677–685.

J. Ralston, *Solutions of the wave equation with localized energy*, Comm. Pure Appl. Math. **22** (1969) 807–823.

J. Rauch, *An L^2 proof that H^s is invariant under nonlinear maps for $s > n/2$.* Global Analysis - Analysis on Manifolds (T, Rassias, ed.), Teubner Texte zur Math., band 57, Leipzig, 1983.

J. Rauch, *Partial Differential Equations*, Graduate Texts in Math., 128, Springer-Verlag, New York, 1991.

J. Rauch, *Lectures on geometric optics.* Nonlinear Wave Phenomena (L. Caffarelli and W. E, eds.), IAS/Park City Math. Ser., 5, American Mathematical Society, Providence, RI, 1998.

J. Rauch, *Precise finite speed with bare hands.* Methods Appl. Anal. **12** (2005), no. 3, 267–277.

J. Rauch, *Á Travers un Prism.* Lecons de Mathématiques d'Aujourd'hui, vol. III (E. Charpentier and N. Nikolski, eds.), Cassini Pubs., Paris, 2007, pp. 35–68.

J. Rauch, *Precise finite speed and uniqueness in the Cauchy problem for symmetrizable hyerbolic systems*, Trans. AMS, **363** no. 3 (2011), 1161–1182.

J. Rauch and F. Massey, *Differentiability of solutions to hyperbolic initial-boundary value problems*, Trans. AMS, **189** (1974), 303–318.

J. Rauch and M. Reed, *Nonlinear microlocal analysis of semilinear hyperbolic systems in one space dimension*, Duke Math. J. **49** (1982), 379–475.

J. Rauch and M. Reed, *Bounded, stratified, and striated solutions of hyperbolic systems.* Nonlinear Partial Differential Equations and Their Applications Vol. IX (H. Brezis and J. L. Lions, eds.), Pitman Research Notes in Math., **181** (1989), 334–351.

J. Rauch and M. Taylor, *Exponential decay of solutions to hyperbolic equations in bounded domains*, Indiana U. Math. J. **24** (1974), 79–86.

M. Reed and B. Simon, *Scattering Theory, Methods of Mathematical Physics, Vol. III*, Academic Press, 1983.

R. Sakamoto, *Hyperbolic Boundary Value Problems*, Translated from the Japanese by Katsumi Miyahara, Cambridge University Press, Cambridge-New York, 1982.

A. Satoh, *Scattering for nonlinear symmetric hyperbolic systems*, Tsukuba J. Math. **29** (2005), no. 2, 285–307

J. Schauder, *Das anfangswertprobleme einer quasilinearen hyperbolishen differentialgleichung zweiter ordnung*, Fund. Math. **24** (1935), 213–246.

S. Schochet, *Fast singular limits of hyperbolic equations*, J. Diff. Eq. **114** (1994), 476–512.

D. Serre, *Systems of Conservation Laws. 1. Hyperbolicity, Entropies, Shock Waves.* (Translated from the 1996 French original by I. N. Sneddon.) Cambridge University Press, Cambridge, 1999.

D. Serre, *Systems of Conservation Laws. 2. Geometric Structures, Oscillations, and Initial-boundary Value Problems.* (Translated from the 1996 French original by I. N. Sneddon.) Cambridge University Press, Cambridge, 2000.

J. Shatah and M. Struwe, *Geometric Wave Equations*, Courant Lecture Notes in Mathematics, New York University, Courant Institute of Mathematical Sciences, New York; American Mathematical Society, Providence, RI, 1998.

J. Smoller, *Shock Waves and Reaction-Diffusion Equations.* Grundlehren der Mathematischen Wissenschaften [Fundamental Principles of Mathematical Sciences], 258. Springer-Verlag, Berllin, 1983.

C. Sogge, *Lectures on nonlinear wave equations*, Second Edition. International Press, Boston, MA, 2008.

E. Stein, *Singular Integrals and Differentiability Properties of Functions*, Princeton University Press, Princeton, 1970.

E. Stein, *Harmonic Analysis, Real Variable Methods, Orthogonality, and Oscillatory integrals*, with the assistance of T. Murphy, Princeton Mathematical Series, 43, Princeton University Press, 1993.

W. Strauss, *Nonlinear Wave Equations*, CBMS Regional Conf. Series, 73, American Mathematical Society, Providence, RI, 1989.

T. Tao, *Nonlinear Dispersive Equations. Local and Global Analysis.* CBMS Regional Conference Series in Mathematics, 106, American Mathematical Society, Providence, RI, 2006.

M. Taylor, *Pseudodifferential Operators*, Princeton Math. Series, 34, Princeton University Press, Princeton, 1981.

M. E. Taylor, *Pseudodifferential operators and nonlinear PDE.* Progress in Mathematics, 100. Birkhäuser Boston Inc., Boston, 1991.

M. Taylor, *Partial Differential Equations, Basic Theory*, Texts in Applied Math., 23, Springer-Verlag, New York, 1996.

M. Taylor, *Partial Differential Equations. II. Qualitative Studies of Linear Equations*, Applied Mathematical Sciences, 116. Springer-Verlag, New York, 1996.

M. Taylor, *Partial Differential equations. III. Nonlinear equations*, Applied Mathematical Sciences, 117. Springer-Verlag, New York, 1997.

T. Wagenmaker, "Analytic solutions and resonant solutions of hyperbolic partial differential equations", Ph.D. Thesis, University of Michigan, 1993.

G. Whitham, *Linear and Nonlinear Waves*, Wiley-Interscience, New York, 1974.

R. Young, *Exact solutions to degenerate conservation laws*, SIAM J. Math. Anal. **30** (1999), no. 3, 537–558.

R. Young, *Blowup of solutions and boundary instabilities in nonlinear hyperbolic equations*, Commun. Math. Sci. **1** (2003), no. 2, 269–292.

V. Yudovich, *Nonstationary flows of an ideal incompressible fluid*, Zh. Vycg. Math. **3** (1963), 1032–1066.

Index